Reluctance Electric Machines

Design and Control

Reluctance Electric Machines
Design and Control

Ion Boldea
Lucian Tutelea

CRC Press
Taylor & Francis Group
Boca Raton London New York

CRC Press is an imprint of the
Taylor & Francis Group, an **Informa** business

CRC Press
Taylor & Francis Group
6000 Broken Sound Parkway NW, Suite 300
Boca Raton, FL 33487-2742

Printed on acid-free paper

International Standard Book Number-13: 978-1-4987-8233-3 (Hardback)

Library of Congress Cataloging-in-Publication Data

Names: Boldea, I., author. | Tutelea, Lucian, author.
Title: Reluctance electric machines : design and control / Ion Boldea and Lucian Tutelea.
Description: Boca Raton : Taylor & Francis, a CRC title, part of the Taylor & Francis imprint, a member of the Taylor & Francis Group, the academic division of T&F Informa, plc, 2018. | Includes bibliographical references.
Identifiers: LCCN 2018010557| ISBN 9781498782333 (hardback : acid-free paper) | ISBN 9781498782340 (ebook)
Subjects: LCSH: Reluctance motors--Design and construction. | Electric motors--Electronic control.
Classification: LCC TK2781 .B65 2018 | DDC 621.46--dc23
LC record available at https://lccn.loc.gov/2018010557

Visit the Taylor & Francis Web site at
http://www.taylorandfrancis.com

and the CRC Press Web site at
http://www.crcpress.com

Contents

Preface

Electric energy is arguably a key agent for our material prosperity. With the notable exception of photovoltaic generators, electric generators are exclusively used to produce electric energy from mechanical energy. Also, more than 60% of all electric energy is used in electric motors for useful mechanical work in various industries.

Renewable energy conversion is paramount in reducing the CO_2 quantity per kWh of electric energy and in reducing the extra heat on Earth.

Electrical permanent magnet machines developed in the last two decades with torques up to the MNm range are showing higher efficiency for smaller weights. However, the temptation to use them in all industries—from wind and hydro generators; to ships, railroad, automotive, and aircraft propulsion (integral or assisting); to small electric drives in various industries with robotics, home appliances, and info-gadgets has led recently to a strong imbalance between high–specific energy magnet demand and supply, which has been "solved" so far mainly by stark increases in the price of high-quality permanent magnets (PMs).

In an effort to produce high-performance electric motors and generators, as well as drives for basically all industries, but mainly for renewable energy conversion, electric mobility, robotics and so on, the variable reluctance concept in producing torque in electric machines, eventually assisted by lower-total-cost PMs, has shown a spectacular surge in research and development (R&D) worldwide in the last two decades.

The extension of electric machines to lower-speed applications—with less or no mechanical transmission—in an effort to keep performance high but reduce initial and maintenance costs has found a strong tool in the *variable reluctance* concept; it aims to produce electromagnetic torque (and power) by creating strong magnetic anisotropies in electric machines (rather than by PM- or direct current (DC)-excited or induced-current rotors).

Though the principle of the variable reluctance motor was patented in the late nineteenth century, it was not until power electronics developed into a mature industry in the 1970s that reluctance electric machines became a strong focus point in R&D and industry. Delayed by the spectacular advent of PM electric machines for a few decades, only in the last 10 years have reluctance electric machines enjoyed increased attention.

Very recently, such *reluctance synchronous motor drives* for variable speeds have reached mass production from 10 kW to 500 kW.

Given the R&D results so far and the current trends in the industry, reluctance electric machines and drives are expected to penetrate most industries.

Because of this, we believe an overview of recent progress with classifications, topologies, principles, modeling for design, and control is timely, and this is what the present monograph intends to do.

After an introductory chapter (Chapter 1), the book is divided into two parts:

Part 1. One- and three-phase reluctance synchronous motors in line-start (constant speed) and then in variable-speed applications, with PM assistance to increase efficiency at moderate extra initial cost and in variable-speed drives (Chapters 2 through 5).

Part 2. Reluctance motors and generators in pulse width modulation (PWM) converter-fed variable speed drives, where high efficiency at a moderate power factor and moderate initial system and ownership costs are paramount (Chapters 6 through 14).

Part 2 includes a myriad of topologies under the unique concept of *flux modulation* and includes:

- Claw pole rotor synchronous motors, Chapter 6
- Brushless DC–multiple phase reluctance machines (BLDC-MRMs), Chapter 7

- Brushless doubly fed reluctance machines (BDFRMs), Chapter 8
- Switched flux PM synchronous machines (SF-PMSMs), Chapter 9
- Flux reversal PMSMs (FR-PMSMs), Chapter 10
- Vernier PM machines, Chapter 11
- Transverse-flux PMSMs (TF-PMSMs), Chapter 12
- Magnetic-gear dual-rotor reluctance electric machines (MG-REMs), Chapter 13
- DC+alternating current (AC) doubly salient electric machines, Chapter 14

The structure of all chapters is unitary to treat:

- Topologies of practical interest.
- Principles, basic modeling, performance, and preliminary design with numerical examples.
- Advanced modeling by magnetic equivalent circuit (MEC) with analytical optimal design (AOD).
- Key finite elements method (FEM) validation of AOD or direct FEM-based geometrical optimization design.
- Basic and up-to-date control of electric motors and generators with and without encoders.
- Sample representative results from recent literature and from authors' publications are included in all chapters.

Though it is a monograph, the book is conceived with self-sufficient chapters and thus is suitable to use as a graduate textbook and a design assistant for electrical, electronics, and mechanical engineers in various industries that investigate, design, fabricate, test, commission, and maintain electric motor-generator drives in most industries that require digital motion (energy) control to reduce energy consumption and increase productivity.

Ion Boldea and Lucian Tutelea
Timisoara, Romania

1 Reluctance Electric Machines
An Introduction

1.1 ELECTRIC MACHINES: WHY AND WHERE?

Electric machines provide the conversion of electric energy (power) into mechanical energy (power) in motor operation, and vice versa in generating operation. This dual energy conversion presupposes energy storage in the form of magnetic energy located mainly in the airgap between the stator (fixed) and rotor (rotating) parts and in the permanent magnets (if any).

Energy is the capacity of a conservative system to produce mechanical (useful) work. As all we do in industry is, in fact, temperature and motion control, we may reduce it to energy control for better productivity and energy savings for a given variable useful output.

The main forms of energy are mechanical (kinetic and potential), thermal, electromagnetic (and electrostatic), and electrochemical.

According to the law of energy conservation, energy may not be created or disappear but only converted from one form to another. In addition, each form of energy has its own fundamental laws that govern its conversion and control, all discovered experimentally over the last three centuries.

Energy conversion starts with a primary source:

- Fossil fuels (coal, gas, petroleum, nuclear, waste)
- Solar thermal irradiation (1 kW/m^2, average)
- Solar photovoltaic source
- Wind (solar, ultimately) source
- Geothermal source
- Hydro (potential or kinetic) sources: potential (from water dams) and kinetic (from rivers, marine currents, and waves)

In the process of energy conversion, electric machines are used to produce electric power, which is very flexible in transport and rather clean, and allows for fast digital control. Only in solar photovoltaic energy conversion (less than 1% of total) are electric machines not involved.

Electric machines are ubiquitous: there is one in every digital watch; three (the loudspeaker, microphone, and ring actuator) in a cellular phone; tens in any house (for various appliances) and in automobiles; and hundreds on trains, streetcars, subway systems, ships, aircraft, and space missions, not to mention most industrial processes that imply motion control to manufacture all the objects that make for our material prosperity.

So far, the more electricity (installed power) used per capita (Norway leads the way), the higher the material living standard. Environmental concerns are shifting this trend toward an optimal (limited) installed electric power (kW/person) in power plants per citizen, which, through distributed systems that make use of lower-pollution energy conversion processes (known as renewable), should provide material comfort but keep the environment clean enough to sustain a good life.

This new, environmentally responsible trend of reducing chemical and thermal pollution will emphasize electric energy conversion and control even more because of its natural merits in cleanness, ease of transport, distribution, and digital control.

But this means ever better electric machines as generators (for almost all produced electricity) and motors (60% of all electric power is used in motion control), all with better power electronics digital control systems.

Here, "better" means:

- Lower initial (materials plus fabrication) costs in USD/Nm for motors and USD/kVA in generators and static power converters.
- Lower global costs (initial motor [plus power electronics converter] losses, plus maintenance costs) for the entire life of the electric machine/drive.
- Sometimes (for low power), the lower global costs are replaced by a simplified energy conversion ratio called efficiency (power or energy efficiency) to output power (energy)/input power (energy). There are even more demanding standards of energy conversion, mainly for line-start (grid-connected) constant-speed electric motors—standards expressed in high efficiency classes.

1.2 ELECTRIC MACHINE (AND DRIVE) PRINCIPLES AND TOPOLOGIES

Electric machines are built from milliwatt to 2 GW power per unit. With more than 12,000 GW in installed electric power plants in the world (400 GW in wind generators already), the total electrical energy usage should still increase at a rate of a single-digit percentage in the next decades, mainly due to strong anticipated developments in high-population areas (China, India, Africa, South America) as they strive for higher prosperity.

Existing electric machines, generators, and motors alike, as they are reversible, may be classified by principle as mature technologies that started around 1830, in three main (standard) categories:

- Brush-commutator (fixed magnetic field) machines (Figure 1.1): brush-DC machines
- Induction (traveling magnetic field) machines: AC machines (Figures 1.2 and 1.3)
- DC plus AC (synchronous)-traveling magnetic field machines (Figure 1.4)

Each type of standard electric machine in Figures 1.1 through 1.4 has merits and demerits.

For constant-speed (line-start) applications (still above 50% of all electric motors and most generators), the AC induction and synchronous machines are dominant in three-phase topologies at

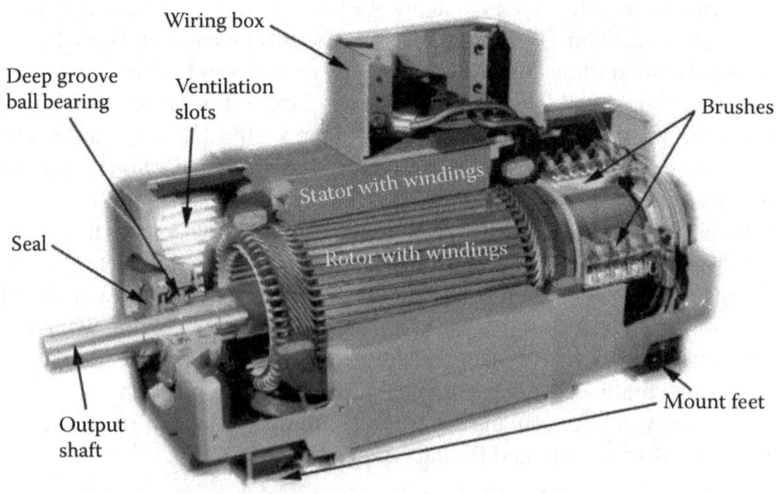

FIGURE 1.1 DC brush commutator machine.

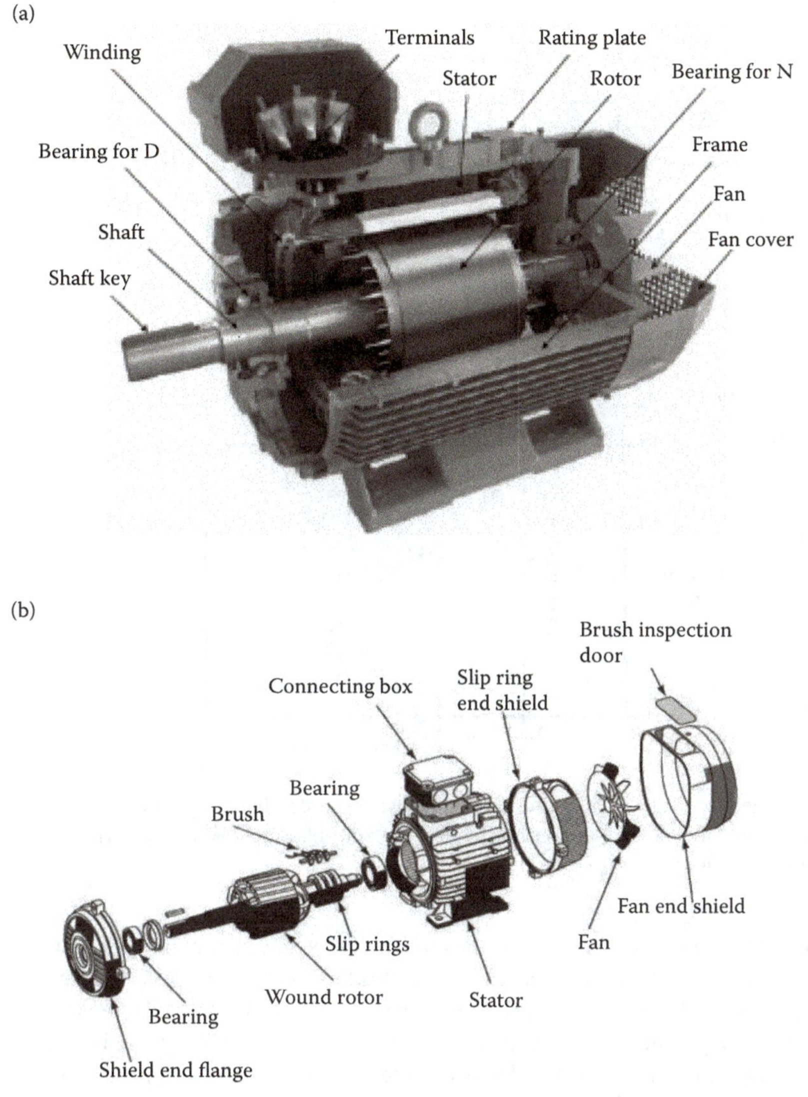

FIGURE 1.2 Three (or more)-phase variable-speed induction machines (plus static converter control): (a) with cage rotor (brushless); (b) with wound rotor (with slip rings and brushes, in standard topology).

medium and large power/units and in split-phase capacitor topologies for one-phase AC sources at low powers (motors less than 1 Nm).

While PM DC-brush motors are still used as low-power actuators, mainly in vehicles and robotics, AC variable-speed motors with static power converter (variable voltage and frequency) control are dominant in the markets for all power levels.

The development of high-energy PMs in the last decades—based on sintered Sm_xCo_y (up to 300°C) and NdFeB (up to 120°C), with permanent flux densities above 1.1 T and coercive force of around (and above) 900 kA/m (recoil permeability $\mu_{rec} = 1.05$–$1.07\ \mu_0$; μ_0—air magnetic permeability: $\mu_0 = 1.256 \times 10^{-6}$ H/m), which are capable of large magnetic energy storage of a maximum of 210/250 kJ/m^3, have led to formidable progress in high-efficiency electric generators and motors both in line-start and variable-speed (with static power converter control) electric drives.

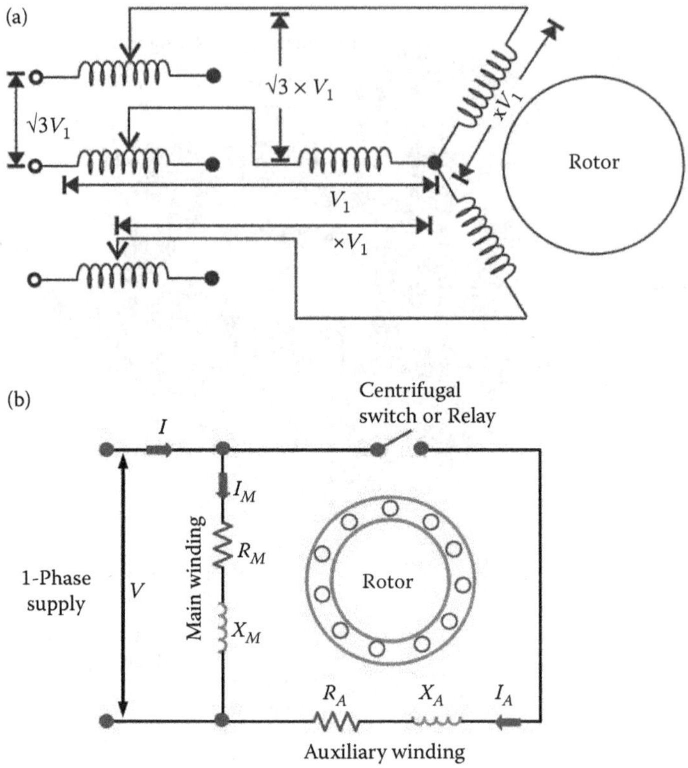

FIGURE 1.3 Line start cage-rotor induction machines: (a) three-phase type (http://www.electrical4u.com/starting-methods-for-polyphase-induction-machine/); (b) split-phase capacitor type (http://circuitglobe.com/split-phase-induction-motor.html).

The recent trends of using PM synchronous machines as wind generators or standalone generators in vehicles or electric propulsion in electric and hybrid electric vehicles (EVs and HEVs) has debalanced the equilibrium between demand and offer in rare-earth high-energy PMs.

Additionally, even betting on the hope that new high-energy PMs will be available abundantly at less than 25 USD/kg (now 75–100 USD/kg), it seems wise to follow approaches that:

- Use high-energy, rare-earth PMs mostly in low-power machines and in medium- and high-power machines only at high speeds such that the PM kgs/kW remains less than about 25% of the initial active materials cost of the electric machine.
- Reduce or eliminate high-energy, rare-earth PM usage in high-demand machines by replacing them with induction machines, DC-excited synchronous machines, or reluctance electric machines.
- And the middle way: use reluctance electric machines with low-cost (Ferrite [$B_r = 0.45$ T] or bonded NdFeB [$B_r = 0.6$ T]) PM assistance, not so much for additional torque but for a wider constant power–speed range (better power factor, also, with lower static power converter kVA ratings).

1.3 RELUCTANCE ELECTRIC MACHINE PRINCIPLES

Electric machines produce electromagnetic torque either to oppose motion (in generator mode) or promote motion (in motor mode). In general, the torque formula, based on the definitions of forces

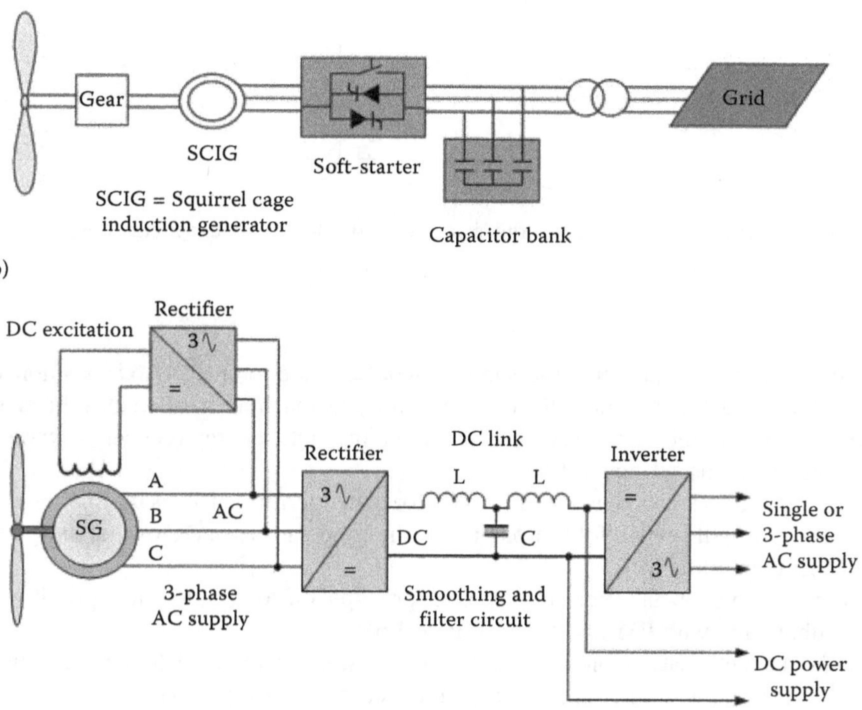

FIGURE 1.4 Electric generators: (a) cage rotor induction generator—constant speed/grid connected (http://powerelectronics.com/alternative-energy/energy-converters-low-wind-speed); (b) three-phase DC excited synchronous motor—variable speed, by static converter control (http://www.alternative-energy-tutorials.com/wind-energy/synchronous-generator.html).

in electromagnetic fields, has the general expressions per phase (Figure 1.5):

$$T_e = +\left(\frac{\partial W_{\text{co mag}}(i, \theta_r)}{\partial \theta_r}\right)_{i=ct}; \; W_{\text{mag}} = \int_0^{\Psi} i d\Psi = \Psi i - W_{\text{co mag}} \tag{1.1}$$

or

$$T_e = -\left(\frac{\partial W_{\text{mag}}(\Psi, \theta_r)}{\partial \theta_r}\right)_{\Psi=ct}; \; W_{\text{co mag}} = \int_0^{i} \Psi di \tag{1.2}$$

T_e is the electromagnetic torque (in Nm), θ_r is the rotor axis position (in radians), W_{mag} and $W_{\text{co mag}}$ are machine-stored magnetic energy and coenergy, Ψ is the flux linkage (per machine phase), and i is the machine phase current.

In general, the flux linkage per phase Ψ is dependent on machine phase inductances and currents, and, if the AC machine has a DC excitation system or PMs on the rotor, the latter contribution to the flux linkage has to be observed. In matrix form for a three-phase AC machine:

$$\left|\Psi_{a,b,c}\right| = |L_{abc}(\theta_r, i_{abc})| \cdot |i_{abc}| + |\Psi_{\text{PM}\,abc}(\theta_r)| \tag{1.3}$$

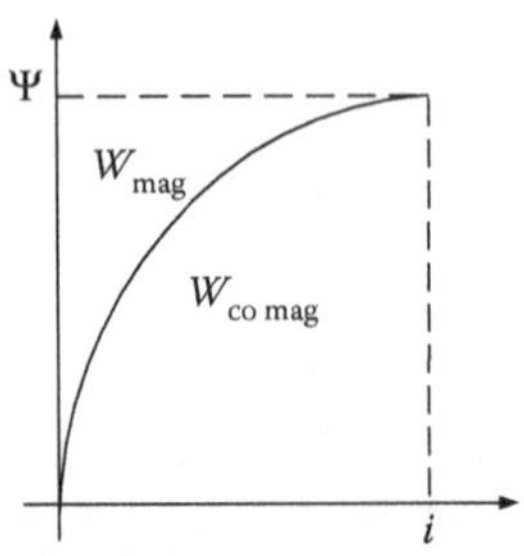

FIGURE 1.5 Magnetic energy and coenergy from magnetic flux linkage/current (mmf) curve of electric machines.

When the self (and mutual if any) inductances depend on the rotor (PM) axis position, even in the absence of a PM (or DC excitation on the rotor or stator), the machine is called the reluctance type and is capable of producing torque because, implicitly, the magnetic energy (coenergy) varies with rotor position, as in Equations 1.1 and 1.2.

A reluctance machine may be assisted by PMs for better performance, as the energy input to magnetize once the PMs will be manifold surplanted by the copper loss in a DC excitation system over the machine's operational life.

Let us illustrate the reluctance electric machine principle on AC single-phase primitive reluctance machines without and with PM assistance (Figure 1.6).

Neglecting magnetic saturation in the flux Ψ expressions in Figure 1.6, the magnetic co-energy Equation 1.1 may be calculated simply for the three cases (with Equation 1.3): pure variable reluctance rotor, PM rotor, and PM-stator assistance:

$$(W_{\text{co mag}})_{\text{Re}\,l} = \int_0^i \Psi di = (L_0 + L_2 \cos 2\theta_r)\frac{i^2}{2} \tag{1.4}$$

$$(W_{\text{co mag}})_{\text{PM Re}\,l} = (L_0 + L_2 \cos 2\theta_r)\frac{i^2}{2} - \Psi_{\text{PM}} i \sin \theta_\gamma \tag{1.5}$$

$$(W_{\text{co mag}})_{\text{PMS Re}\,l} = (L_0' + L_2' \cos 2\theta_r)\frac{i^2}{2} + \Psi_{\text{PM}} \cdot i \cdot \cos \theta_\gamma \tag{1.6}$$

The electromagnetic torque T_e in the three cases (from Equation 1.1) is

$$T_{e(\text{Re}\,l)} = -L_2 \cdot i^2 \cdot \sin 2\theta_r \tag{1.7}$$

$$T_{e(\text{PM Re}\,l)} = -L_2 \cdot i^2 \cdot \sin 2\theta_r - \Psi_{\text{PM}} \cdot i \cdot \cos \theta_r \tag{1.8}$$

$$T_{e(\text{PM S Re}\,l)} = -L_2 \cdot i^2 \cdot \sin 2\theta_r - \Psi_{\text{PM}\,a} \cdot i \cdot \sin \theta_r \tag{1.9}$$

A few remarks are in order:

- The average torque per revolution is zero, so the machine may not start in this primitive topology (constant airgap and symmetric poles in stator and rotor). Something has to be done to place the rotor in a safe self-starting position.
- The torque may be positive or negative (motor or generator).

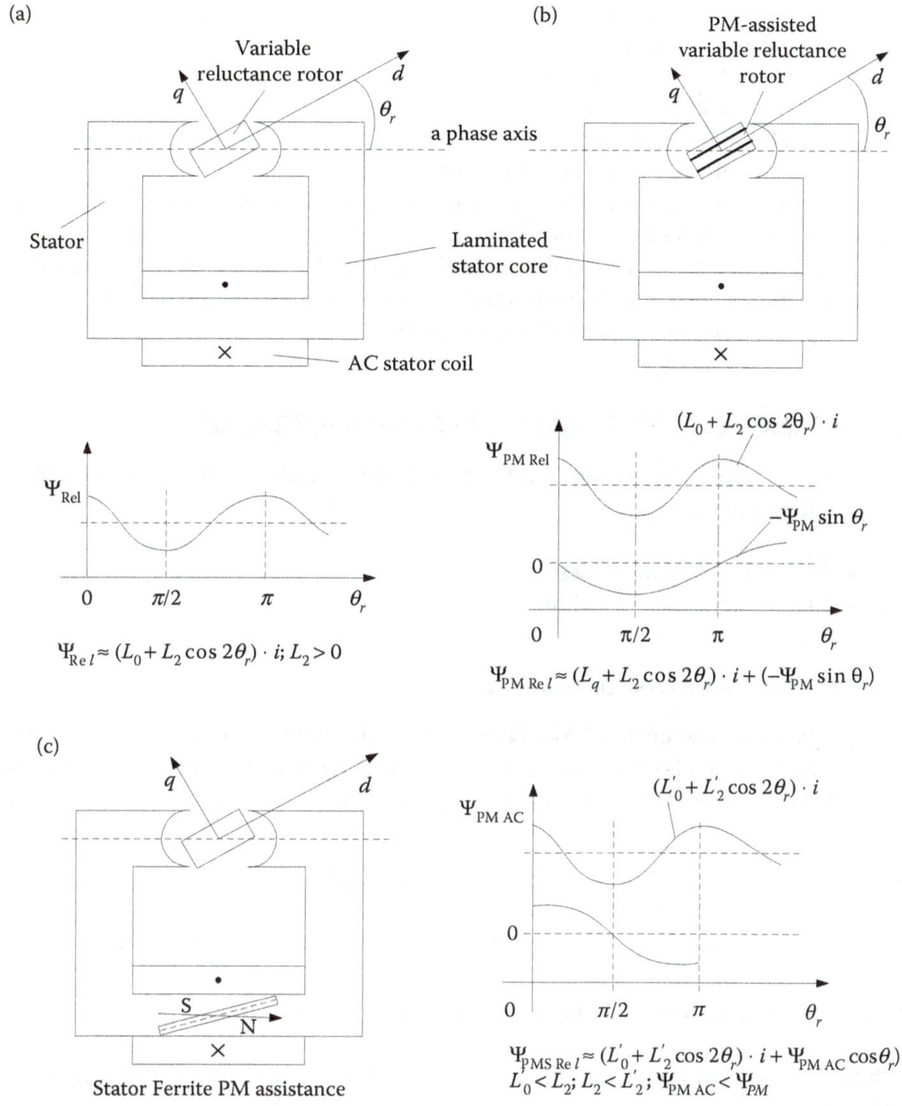

FIGURE 1.6 Single-phase AC reluctance primitive machine: (a) pure variable reluctance rotor; (b) with rotor PM; (c) with stator PM.

- A bipolar current shape "intact" with rotor position should be provided to produce nonzero average torque, but only in multiphase topologies with nonzero mutual variable inductances. Monopolar current may also be used, but then only one of the inductance ramps (the positive one) may be used for torque production (roughly half torque).
- The PM placed in the minimum inductance axis for the PM rotor assists in producing torque while it decreases both L_0 and L_2.
- Only the L_2 term produces reluctance torque.
- The PMs may also be placed on the stator, but then the maximum available AC PM flux linkage is less than half that obtained with rotor PMs for the same magnet volume for the configuration in Figure 1.6c. In alternative configurations, the PM flux linkage may reverse polarity in the AC stator coil, but then only half the latter's span is active: this is how stator

PM–assisted (switched flux or flux reversal) reluctance machines experience, one way or another, a kind of "homopolar" effect (limit). Practical reluctance machines are built in more involved configurations: single phase and multiphase; with and without AC excitation or PM assistance; with distributed or tooth-wound AC coils; with a stator and a rotor (single airgap) or with an additional part, a flux modulator (for low-speed torque magnification: magnetic gearing effect), which may be fixed or rotating.

- This is how reluctance synchronous machines (pure or with PM assistance) and reluctance flux-modulation machines come into play to:
  - Reduce initial costs (by reducing [or eliminating] PM costs) in variable-speed drives
  - Increase efficiency at only 5–6 p.u. starting current in line-start applications when a cage is also present on the variable reluctance rotor

1.4 RELUCTANCE ELECTRIC MACHINE CLASSIFICATIONS

We attempt here to classify the myriad of reluctance electric machines, many introduced in the last two decades, first into two categories:

1. Reluctance synchronous machines
2. Flux-modulation reluctance machines

1.4.1 Reluctance Synchronous Machines

Reluctance synchronous machines (RSMs) have a uniformly slotted stator core with distributed (with almost pole–pitch τ span coils) windings connected in two (split phase), three, or more phases, fed with sinusoidal currents that produce, basically, a travelling magnetomotive force (mmf) (Figure 1.7).

$$F_s = F_{1m} \cdot \cos\left(\omega_1 t - \frac{\pi}{\tau}x\right) \tag{1.10}$$

They exhibit $2p_s$ poles per rotor periphery. The rotor is of the variable reluctance type, showing the same number of poles $2p_r = 2p_s = 2p$.

The variable reluctance rotor may be built in quite a few configurations:

- With regular laminations and concentrated (Figure 1.8a) or distributed magnetic anisotropy (Figure 1.8b)
- With axial–lamination poles with through shaft $2p \geq 4$ (Figure 1.9a) and, respectively, nonthrough shaft for $2p = 2$ poles

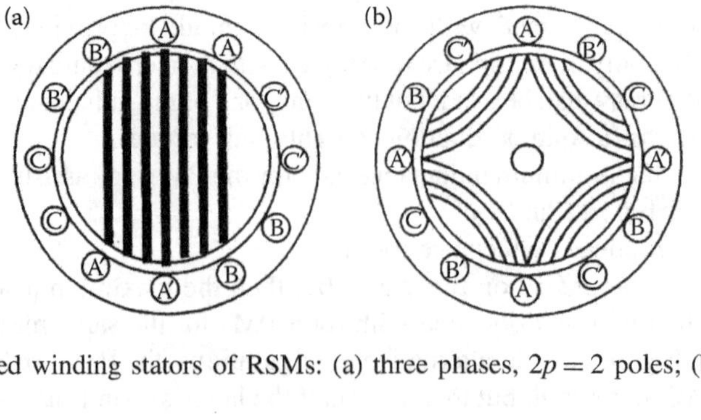

FIGURE 1.7 Distributed winding stators of RSMs: (a) three phases, $2p = 2$ poles; (b) four phases, $2p = 4$, $q = 1$.

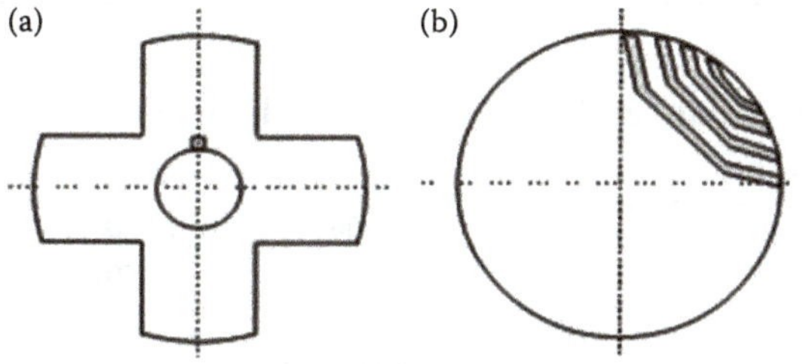

FIGURE 1.8 Regular-lamination variable reluctance rotors: (a) with concentrated anisotropy (with salient standard poles); (b) with distributed anisotropy (with multiple flux barriers).

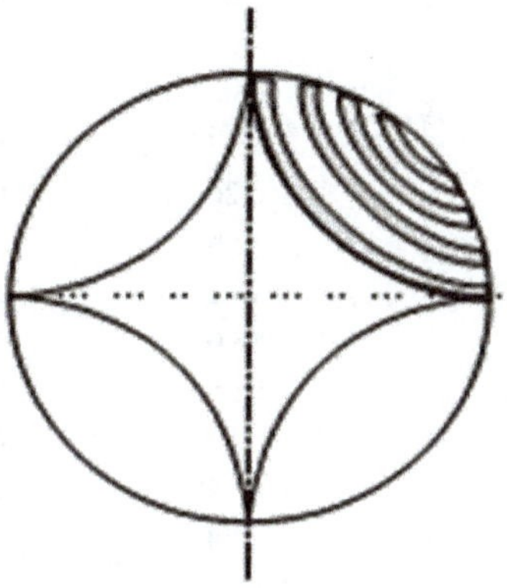

FIGURE 1.9 Axially laminated anisotropic rotors with through shaft ($2p \geq 4$).

As demonstrated later, large magnetic anisotropy (magnetization inductance differential L_{dm}/L_{qm}) in the two orthogonal (d, q) axes is required for high/competitive torque density (Nm/kg or Nm/liter), while a high ratio $(L_{dm}+L_{sl})/(L_{qm}+L_{sl})=L_d/L_q$ is required for an acceptable power factor and thus a wider constant power speed range [1,2].

The first RSMs, using salient (standard) rotor poles, exhibited a low total saliency ratio $L_d/L_q \approx$ 1.8–2.5 [3], but their rotor simplicity and ruggedness proved practical, mainly in low-power (especially high-speed) applications.

Multiple flux-barrier rotors (Figure 1.8) making use of regular laminations, proposed in the late 1950s [4], have gained recent industrial acceptance (ASEA Brown Boveri [ABB] site [5]) to powers up to 500 kW in variable-speed inverter fed drives, with 1.5%–3% more efficiency than induction motors with the same stator but a power factor smaller by 8%–10% (segmented rotor configurations retain only a historical interest).

The axially laminated anisotropic (ALA) rotor (Figure 1.9) [6] is still considered less practical (unless performance is far more important than initial cost and materials plus fabrication) because the stator q axis current magnetic field space harmonics produce additional losses in the ALA rotor. While the second objection has been solved by using thin slits (three to four per entire rotor axial length), the first objection still hampers its commercialization.

Typical data on recent commercial RSMs for variable-speed drives (no cage on the rotor) are shown in Table 1.1 (after [5]).

The RSMs may be used also in line-start (constant-speed) machines; however, a cage has to be added on the rotor for asynchronous starting (Figure 1.10).

Higher efficiency in contemporary line-start RSMs (split-phase and three-phase) has been more vigorously pursued, especially with PM (even Ferrite) assistance than induction motors but keeping the starting current lower than 6.5 p.u. (per unit) (Table 1.2).

TABLE 1.1
Sample Performance of RSMs for Variable Speed

Machine Type	IM	SynSRM	SynSRM
Evaluation Type	**Measur.**	**Measur.**	**Calc.**
Speed [rpm]	1498	1501	1500
f1 [Hz]	51	50	50
Slip [%]	2099	0	0
V1ph [V rms]	215	205	219
I1 [A rma]	29	31	31
Losses [W]	1280	928	1028
Stator Copper Losses [W]	536	613	611
Friction Losses [W]	86	86	78
Rest Losses [W]	658	229	338
T [Nm]	88.3	87.7	87.2
T/I1 [Nm/A]	0.26	0.23	0.23
Pout [kW]	13.9	13.8	13.7
Pin [kW]	15.1	14.7	14.7
S1in [kVA]	18.6	19.2	20.6
Motor η [%]	91.5	93.7	93.0
PF1	0.82	0.77	0.71
$\eta \cdot$ PF1	0.746	0.717	0.664
I/($\eta \cdot$ PF1)	1.34	1.39	1.5
Winding Temp. Rise [K]	66	61	61
Housing Temp. Rise [K]	42	36	NA
Shaft on N-side Temp. Rise [K]	32	19	NA
Shaft on D-side Temp. Rise [K]	NA	21	NA
Inv. Avr. Switch Freq. [kHz]	4	4	NA
Inverter Losses [W]	442	442	NA
Inverter η [%]	97.2	97.1	NA
System η [%]	88.9	91.0	NA

The nonsymmetric cage in Figure 1.10a is admittedly producing additional torque pulsations during asynchronous starting, but it uses the rotor cross-section better while leaving room for the assisting PMs.

RSMs are in general provided with distributed windings in the stator to produce higher magnetic anisotropy. Good results may be obtained with τ/g (pole pitch/airgap) > 300 and $q \geq 3$ slots/pole/phase [1]: $L_{dm}/L_{qm} > 20$.

However, at least for variable speed, RSM tooth-wound coil stator windings have also been proposed with variable reluctance rotors only for it to be found that $L_d/L_q < 2.5$ in general, because the high differential leakage inductance of such windings reduces the L_d/L_q ratio to small values. All RSMs are operated with sinusoidal stator currents. For variable speed, field-oriented or direct torque control is applied. Noticing the lower power factor of RSMs for variable-speed drives, a new family of multiphase reluctance machines with two-level bipolar flat-top current control was introduced [7–10]. This new breed of REM is called here brushless DC multiphase reluctance machines, which operate in fact as DC brush machines, utilizing the inverter ceiling voltage better to produce an efficiently wide constant power speed range. BLDC-MRMs act in a way as synchronous machines in the sense that the number of poles of diametrical (pole pitch span) coil stator winding and magnetically anisotropic rotor are the same: $p_s = p_r = p$.

So the stator current frequency $f_1 = n \cdot p$; n-speed.

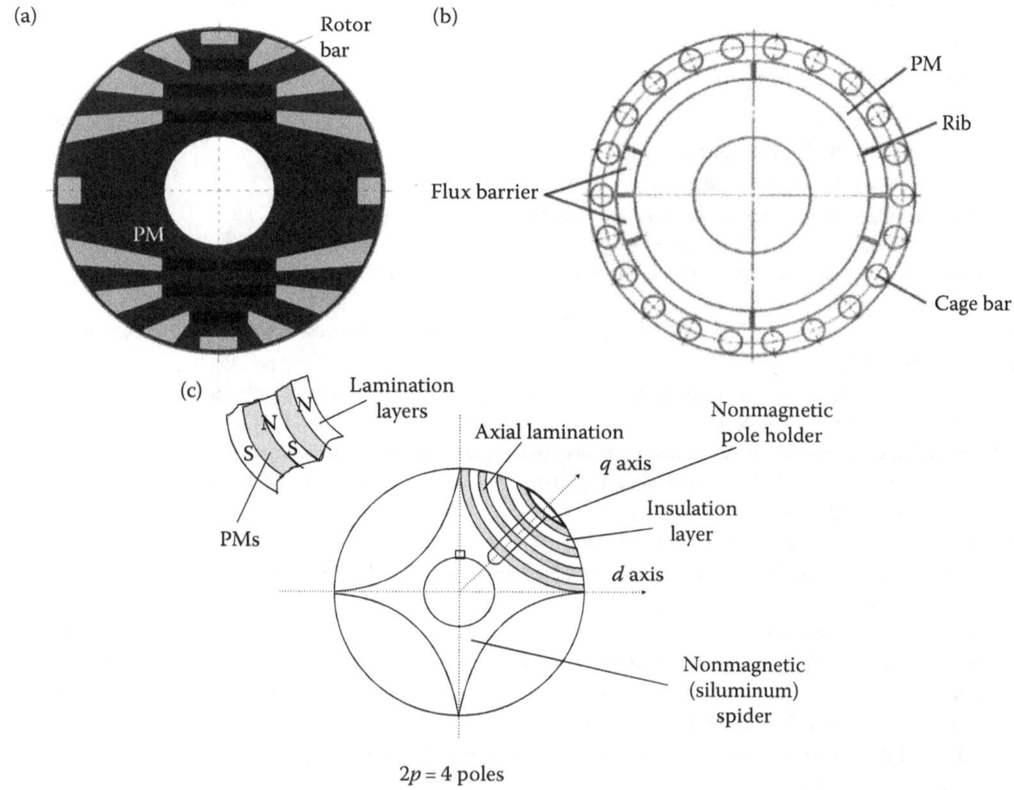

FIGURE 1.10 Cage rotor with magnetic anisotropy: (a) with cage inside flux barriers (and PMs); (b) with cage above PM (and low magnetic saliency $L_{dm}/L_{qm} = 1.3$–1.6); (c) with cage (no PMs) as thin conductor sheets between axial laminations.

1.4.2 Brushless Direct Current-Multiphase Reluctance Machines

BLDC-MRMs were introduced in the late twentieth century [7–9] and revived recently [10] when fault tolerance became an important practical issue and multiphase inverter technologies reached mature stages.

The machine topology springs from an inverting (swapping) stator with a rotor in a DC brush machine stripped of DC excitation and with brushes moved at the pole corners, with the mechanical commutator replaced by inverter control of bipolar two-level flat-top currents (Figure 1.11).

The BLDC-MRM is characterized by the following:

- A $q = 1$ multiphase diametrical coil stator winding ($m = 5, 6, 7, \ldots$).
- The rotor has a large magnetic reluctance in axis q so that $L_{qm}/L_{dm} \leq (4$–$5)$ to reduce the armature reaction of m_T torque phases placed temporarily under rotor poles.
- The $m_F = m$–m_T phases, placed temporarily between rotor poles, produce an electromagnetic force (emf) in the torque phases at an optimum airgap of the machine.
- A too-small airgap will delay the commutation of current, while one that is too large will increase the losses for the AC excitation process in the m_F phases. This not-so-small airgap may be useful for high-speed applications to reduce mechanical losses for good performance, even if a carbon fiber shell is mounted to enforce the rotor mechanically. A small airgap, we should remember, is a must in both RSMs (with sinusoidal current) and switched reluctance motors.

TABLE 1.2
Typical Standard Efficiency Classes

Standard & Year Published	State
IEC 60034-1, Ed. 12, 2010. *Rating and performance.*	
Application: Rotating electrical machines.	Active.
IEC 60034-2-1, Ed. 1, 2007, *Standard methods for determining losses and efficiency from tests (excluding machines for traction vehicles).*	
Establishes methods of determining efficiencies from tests, and also specifies methods of obtaining specific losses.	
Application: DC machines and AC synchronous and induction machines of all sizes within the scope of IEC 60034-1.	Active, but under revision.
IEC 60034-2-2, Ed.1, 2010, *Specific methods for determining separate losses of large machines from tests—supplement to IEC60034-2-1.*	
Establishes additional methods of determining separate losses and to define an efficiency supplementing IEC 60034-2-1. These methods apply when full-load testing is not practical and results in a greater uncertainty.	
Application: Special and large rotating electrical machines	Active.
IEC 60034-2-3, Ed. 1, 2011, *Specific test methods for determining losses and efficiency of converter-fed AC motors.*	
Application: Converter-fed motors	Not active. Draft.
IEC 60034-30, Ed. 1. 2008, *Efficiency classes of single-speed, three-phase, cage induction motors (IE code).*	Active, but under revision.
Application: 0.75–375 kW, 2, 4, and 6 poles, 50 and 60 Hz.	
IEC 60034-31, Ed. 1, 2010, *Selection of energy-efficient motors including variable speed applications—Application guide.*	
Provides a guideline of technical aspects for the application of energy-efficient, three-phase, electric motors. It not only applies to motor manufacturers, original equipment manufacturers, end users, regulators, and legislators, but to all other interested parties.	
Application: Motors covered by IEC 60034-30 and variable frequency/speed drives.	Active.
IEC 60034-17, Ed. 4, 2006, *Cage induction motors when fed from converters—Application guide.*	
Deals with the steady-state operation of cage induction motors within the scope of IEC 60034-12 when fed from converters. Covers the operation over the whole speed setting range, but does not deal with starting or transient phenomena.	
Application: Cage induction motors fed from converters.	Active.

- The BLDC-MRM operates more like a DC brush machine with a flat-top bipolar current, at a power angle that fluctuates around 90%.
- It is to be noted that the excitation of this machine is performed in AC, but the flat-top bipolar currents and the division of phases into torque (m_T) and field phases (m_F) may produce better inverter voltage ceiling utilization. This translates into a wider constant power speed range (CPSR > 4, if $L_{dm}/L_{qm} > 4$) where the excitation current is reduced for flux weakening and not increased as in interior PM synchronous motors (IPMSMs); this should mean better efficiency during flux weakening for larger constant power speed ranges (CCPSRs); much as in DC-excited synchronous motors.

1.4.3 The Claw Pole–Synchronous Motor

The claw pole–synchronous motor (CP-SM) (Figure 1.12) may also be assimilated into a reluctance machine, as the rotor magnetic claws host a single DC coil (and/or a ring-shaped PM) to produce a $2p$ pole airgap flux density distribution, with a stator with a typical distributed three-phase winding

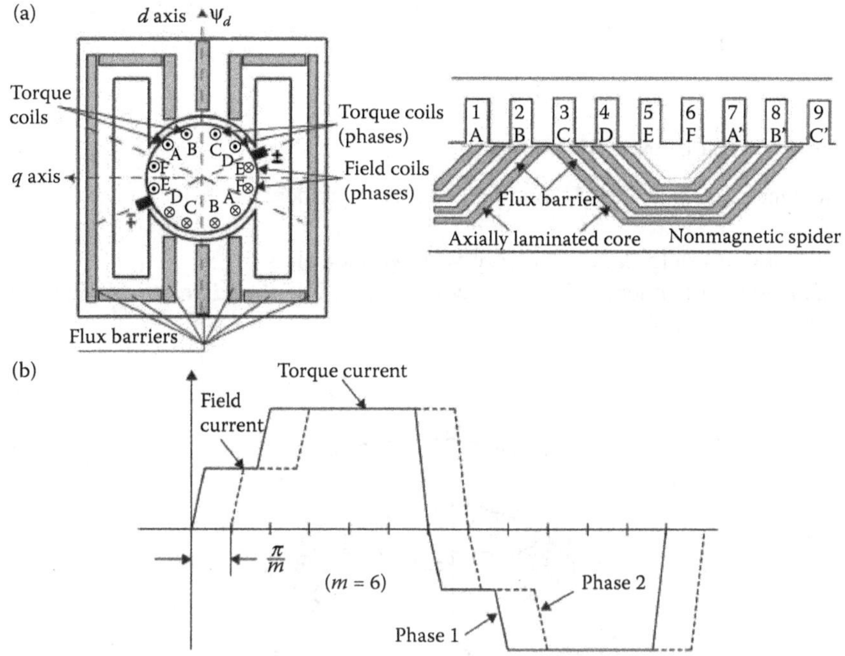

FIGURE 1.11 Six-phase BLDC-MRM: (a) equivalent exciterless DC-brush machine and rotary (or linear) simplified geometry; (b) typical flat-top bipolar phase current wave forms. (After D. Ursu et al., *IEEE Trans*, vol. IA–51, no. 3, 2015, pp. 2105–2115. [10])

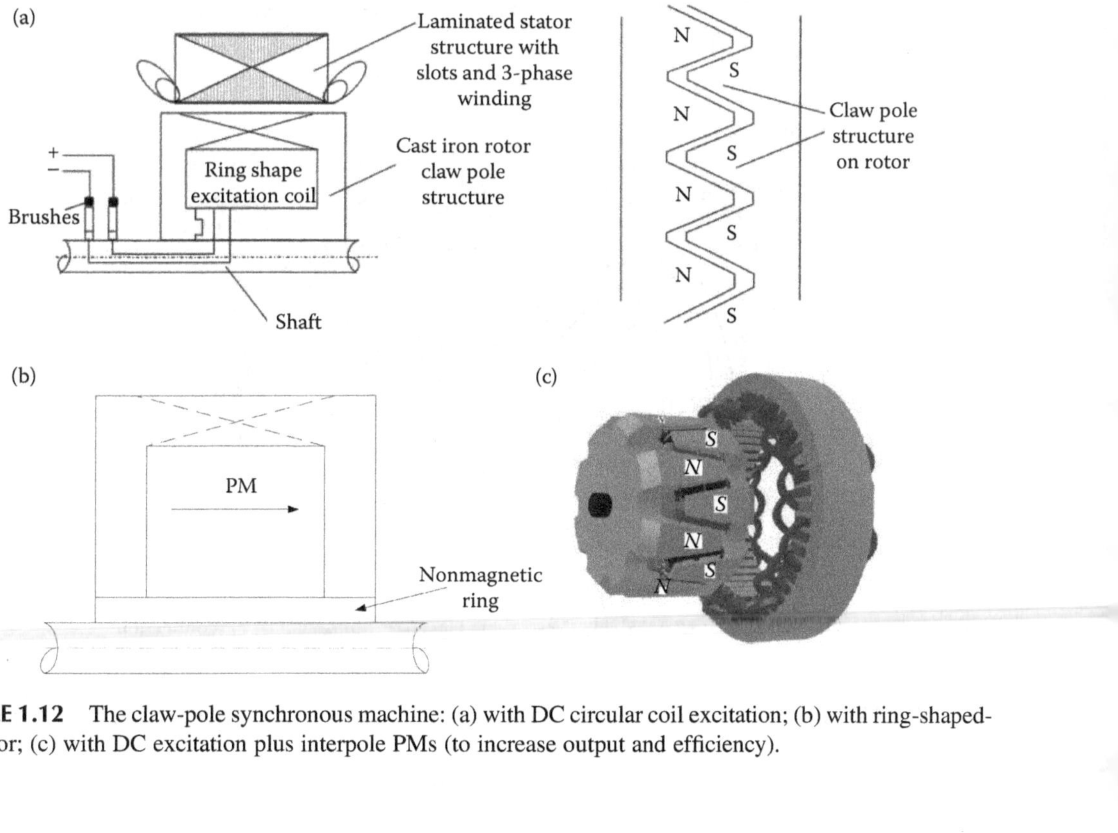

FIGURE 1.12 The claw-pole synchronous machine: (a) with DC circular coil excitation; (b) with ring-shaped-PM rotor; (c) with DC excitation plus interpole PMs (to increase output and efficiency).

[11]. Its main merit is the reduction of DC excitation losses and rotor ruggedness with a large number of poles, $2p > 8$ in general.

1.4.4 Switched Reluctance Machines

Switched reluctance machine (SRM) topologies [12,13] include:

- A single (or dual) stator laminated core with N_s open slots
- The stator slots host either tooth-wound coils (Figure 1.13a) or diametrical (distributed) coil winding (Figure 1.13b)

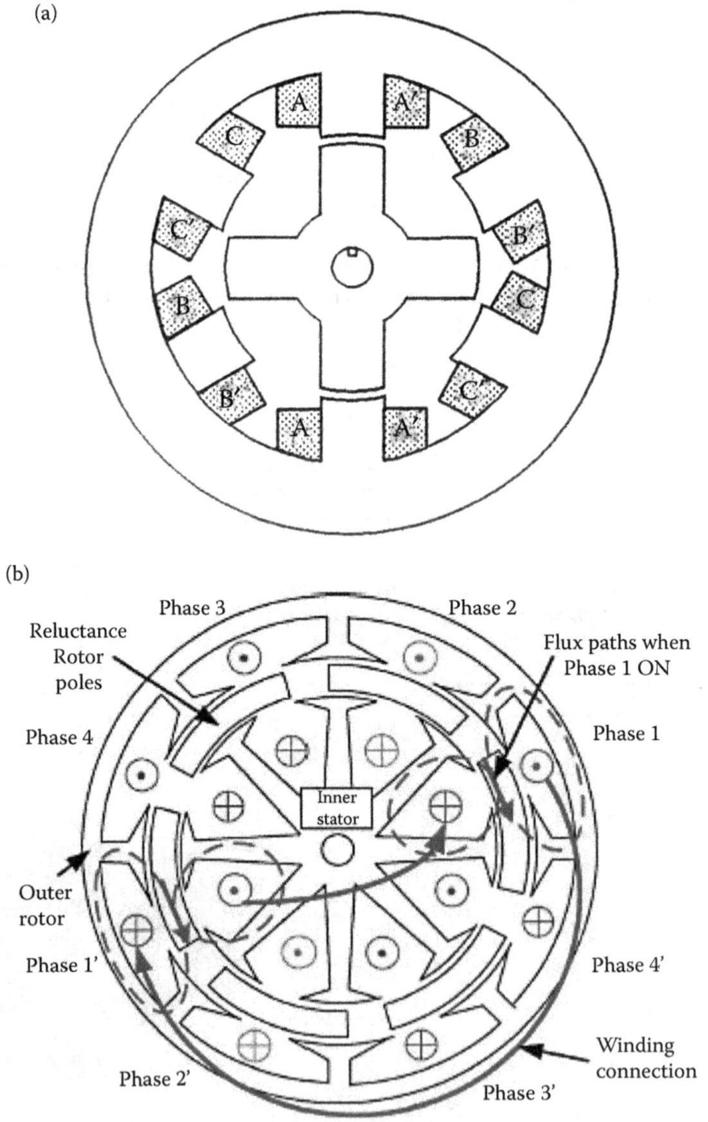

FIGURE 1.13 SRM: stator with multiphase coils: (a) single stator with tooth-wound three-phase coils; (b) dual stator with diametrical four-phase coils and segmented rotor.

- The rotor laminated core shows N_r salient (Figure 1.13a) or distributed anisotropy poles (Figure 1.14b)
- In general,

$$N_s - N_r = \pm 2K; \quad K = 1, 2, \ldots. \tag{1.11}$$

SRMs have a rugged structure and are not costly, being capable of high efficiency (to suit even automotive traction standards [14]). However, being fed with one single-polarity current, with one

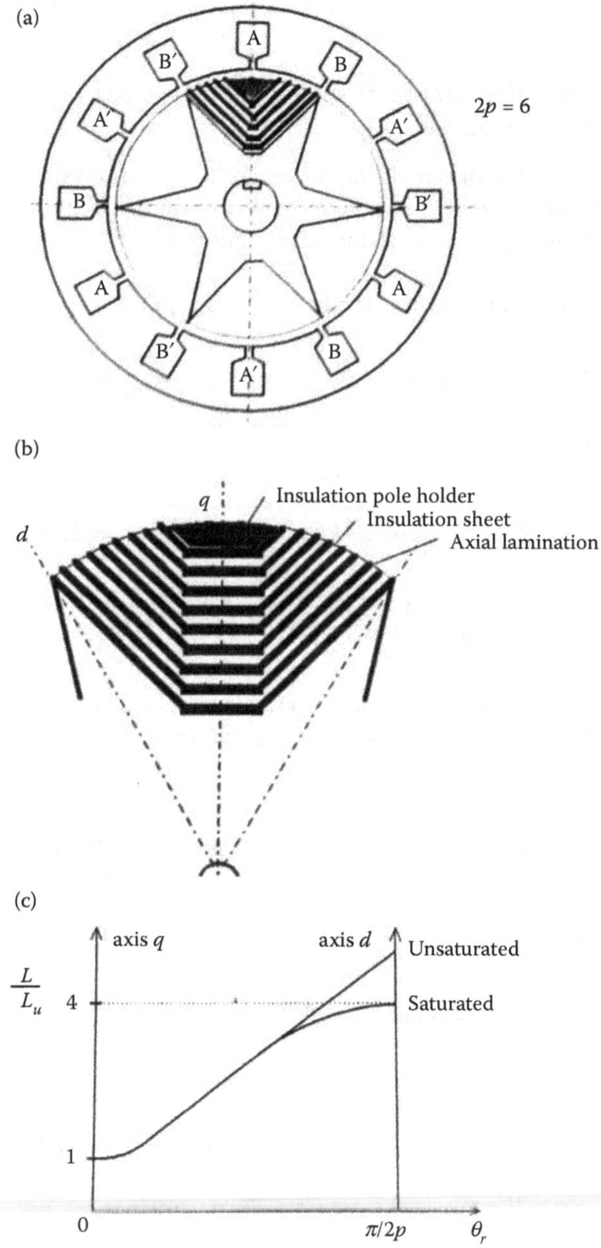

FIGURE 1.14 Six-pole two-phase SRM: (a) cross-section; (b) ALA rotor; (c) phase inductance L versus rotor position θ_r.

(or two) phases active at any time, and exposing a rather large inductance (needed for a high saliency ratio $[L_{dm}/L_{qm} > (4\text{–}6)]$), the static power converters that feed SRMs are made as dedicated objects to handle large kVA (due to low $\text{kW}/\text{kVA} = 0.6 - 0.7$).

The SRM may be considered synchronous because the fundamental frequency of the stator currents is $f_1 = n \cdot N_r$: N_r rotor salient poles (pole pairs) and n speed, though the current pulses are monopolar. On the other hand, if we calculate the stator mmf fundamental number of poles, we may consider it $p_a = 2\,\text{K}$ and thus Equation 1.11 translates into $N_s - N_r = \pm p_a$.

But this is the synchronism condition for the so-called "flux modulation" or Vernier reluctance machines, with N_s the pole pairs of a fixed flux-modulator and N_r the pole pairs of the variable reluctance rotor.

1.4.5 When and Where Reluctance Synchronous Machines Surpass Induction Machines

Any comparison between electric machines should be carefully based on similar performance criteria, the same stator, and similar inverters for variable-speed drives, or based on the efficiency × power factor for line-start machines with same stator, maximum power factor, voltage, flux linkage, and winding losses.

The torque of the IM has, in the dq-model, the standard formula:

$$(T_e)_{\text{IM}} = \frac{3}{2} p \cdot (L_s - L_{sc}) i_d i_q; \quad \cos\varphi_{\max} \approx \frac{1 - L_{sc}/L_s}{1 + L_{sc}/L_s}; \tag{1.12}$$

with:

$$V_s \approx \Psi_s \cdot \omega_1; \ \Psi_s = \sqrt{(L_s i_d)^2 + (L_{sc} i_q)^2}; \ \Psi_r = L_m \cdot I_d$$

$$P_{\text{copper}} = \frac{3}{2} R_s (i_d^2 + i_q^2) + \frac{3}{2} R_r \left(\frac{L_s}{L_m}\right)^2 i_q^2; \ L_s = L_{sc} + L_m; \ L_{sc} = L_{sl} + L_{rl}$$

Core losses are neglected above, and L_s and L_{sc} are the no load and, respectively, the short-circuit inductance of IM; R_s, R_r are stator and rotor phase resistances.

Similarly, for the RSM:

$$(T_e)_{\text{RSM}} = \frac{3}{2} p \cdot (L_d - L_q) i_d i_q; \ \cos\varphi_{\max} \approx \frac{1 - L_q/L_d}{1 + L_q/L_d}; \tag{1.13}$$

$$V_s \approx \Psi_s \cdot \omega_1; \ \Psi_s = \sqrt{(L_d i_d)^2 + (L_q i_q)^2}; \ L_d = L_{sl} + L_{dm}; \ L_q = L_{sl} + L_{qm}$$

$$P_{\text{copper}} = \frac{3}{2} R_s (i_d^2 + i_q^2)$$

From Equations 1.12 and 1.13 we may infer that:

- The IM under field-oriented control ($\Psi_r = \text{const}$) operates like an RSM with $L_s \to L_d$; $L_{sc} \to L_q$.
- For the same stator and $2p = 2$, 4, 6, it is feasible to expect $L_s \approx L_d$ and $L_{sc} \approx L_q$; thus, the same torque is obtained with the same stator currents i_d and i_q, but in RSM, there are no rotor winding losses and thus $(p_{\text{co}})_{\text{RSM}} < (p_{\text{co}})_{\text{IM}}$. Consequently, the efficiency can be 3%–6% better for RSMs.

- Now, concerning the maximum power factor, it is not easy to make $(L_q)_{\text{RSM}} = (L_{sc})_{\text{IM}}$ for the same stator (unless an ALA rotor is used) and thus $L_q > L_{sc}$; consequently, the power factor is 6%–10% lower for RSMs, even with multiple-flux-barrier rotors.
- A smaller power factor with a better efficiency would require about the same inverter for the same stator machine; thus, we remain with the advantage of a few percentage points more efficiency for a cageless rotor in variable-speed inverter-fed drives.
- For line-start applications, only the advantage of a few percentage points more efficiency over the IM of the same stator remains; fortunately, the starting current remains for RSM in the standard range of 5.5–6.5 p.u. This means that the line-start RSM may qualify for a superior-class efficiency motor with reasonable starting current at reasonable costs (about the same stator as the lower efficiency class IM). Moreover, adding PMs on the cage rotor in addition to magnetic saliency may further improve the performance (efficiency × power factor) of line-start RSMs. The standard value of the starting current level of RPM allows use of the existing local lower grid ratings in contrast to high class efficiency recent IMs, which require 7–8 p.u. starting currents.

1.5 FLUX-MODULATION RELUCTANCE ELECTRIC MACHINES

RSMs and CP-SMs are typical multiphase traveling field sinusoidal current synchronous machines. BLDC-MRMs and SRMs are quasisynchronous machines with impure traveling/jumping magnetic-airgap fields and with bipolar (monopolar, respectively) flat current waveforms.

Quite a few other quasisynchronous machines have been proposed, some recently, but some decades ago. In general, they combine PM torque contribution to reluctance torque effects, almost all being most suitable for low/moderate speeds, but where fundamental frequencies as high as 1 kHz can be handled by the PWM inverter that controls such machines.

They basically contain not two but three elements:

- The exciter (eventually with PMs) part: with p_e pole-pairs
- The flux modulator with variable reluctance part: with p_m pole-pairs
- The armature m.m.f. part: with p_a pole-pairs

At least one of these parts should be, in general, stationary. Also, only the constant and first space harmonic components of airgap permeance are considered.

Let us consider a traveling excitation m.m.f. $F_e(\theta_s, t)$:

$$F_e(\theta_s, \Omega_e, t) = F_{e1} \cdot \cos(p_e\theta_s - p_e\Omega_e t) \tag{1.14}$$

And the airgap permeance function Λ_g (produced by the flux modulator side).

$$\Lambda_g(\theta_s, \Omega_m t) = \Lambda_0 + \Lambda_1 \cos(p_m\theta_s - p_m\Omega_m t) \tag{1.15}$$

In general,

$$\Lambda_1 < \Lambda_0/2 \tag{1.16}$$

Now, the airgap flux density $B_{ge}(\theta_s,t)$ is:

$$B_{ge}(\theta_s, t) = F_{e1}(\theta_s, t) \cdot \Lambda_g(\theta_s, \Omega_m t) \tag{1.17}$$

or

$$B_{ge} = F_{e1}\Lambda_0 \cos(p_e\theta_s - p_e\Omega_e t) + \frac{1}{2}F_{e1}\Lambda_1 \cos[(p_e \pm p_m)\theta_s - (p_e\Omega_e \pm p_m\Omega_m)t] \quad (1.18)$$

The armature-side m.m.f. is also considered a traveling wave:

$$F_a(\theta_r, t) = F_{a1} \cdot \cos(p_a\theta_s - p_a\Omega_a t) \quad (1.19)$$

To produce constant torque, $B_{ge}(\theta_r,t)$ and $F_a(\theta_s,t)$ should act in synchronicity; thus, they should have:

- The same pole pairs: $p_a = |p_e \pm p_m|$
- The same angular speed $\Omega_a = (p_e\Omega_e \pm p_m\Omega_m)/(p_e \pm p_m)$
- They should not be aligned with each other (B_{ge} and F_a)

Such a typical flux-modulation REM, shown in Figure 1.15, comprises three parts (not two as do RSMs).

Both the flux modulator and the exciter may rotate at speeds Ω_m and Ω_e, or the flux modulator (and the armature) may be stationary. Only if the flux modulator is stationary ($\Omega_m = 0$) will the armature and excitation m.m.fs exhibit the same frequency ($\Omega_e = \Omega_a$).

If the speed of coordinates is ω_f and equal to the speed of the flux modulator (Ω_m) and Ω_a^r and Ω_e^r are the speeds of the armature and excitation m.m.fs with respect to it, the power balance offers the relationships:

$$\frac{\Omega_e^r}{\Omega_a^r} = \frac{\omega_e - \omega_f}{\omega_a - \omega_f} = -\mathrm{sgn}\frac{p_e}{p_a};\ \beta = \mathrm{sign}\frac{p_e}{p_a} = \begin{cases} 1 \text{ for } p_m = p_e + p_a \\ -1 \text{ for } p_m = p_e - p_a \end{cases} \quad (1.20)$$

$$T_m = -\left(1 + \mathrm{sgn}\frac{p_e}{p_m}\right)T_a;\ T_e = \mathrm{sgn}\frac{p_e}{p_m}T_a,$$

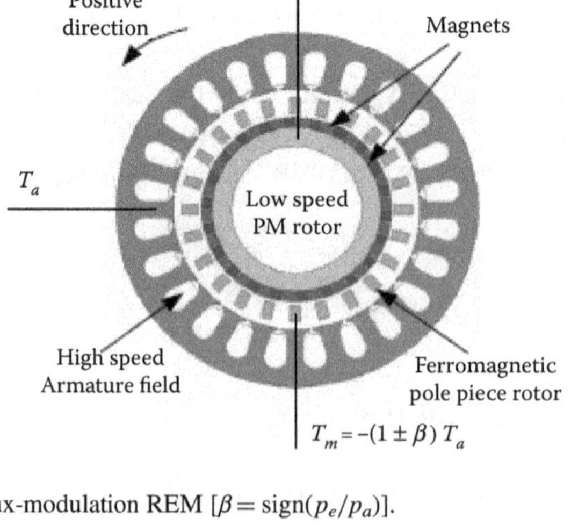

FIGURE 1.15 Generic flux-modulation REM [$\beta = \mathrm{sign}(p_e/p_a)$].

Where T_m, T_a, and T_e are the torques exerted on the flux-modulator, armature, and exciter, respectively.

As visible in Equation 1.20, which exhibits two terms, to maximize the torque: $\beta = +1$; thus:

$$p_a = |p_e - p_m|; \ \Omega_a = \frac{p_e\Omega_e - p_m\Omega_m}{p_e - p_m} \tag{1.21}$$

An apparent torque magnification effect occurs in flux-modulation reluctance electric machines (FM-REMs) via the first harmonic of airgap permeance and here from the expected increased torque density, provided the numbers of pole pairs p_e and p_m are large, with $p_a = 1,2$, as proved later in the book, for a reasonable power factor and efficiency. But, for a given motor/generator speed, this means a rather large fundamental current frequency to be handled thorough the PWM converter. This is how low- (hundreds and tens of rpms) and medium- (a few thousand rpms) speed applications are considered the main targets for FM-REMs for wind and hydro generators and automotive and industrial variable-speed drives.

To go the natural way, we will first present the main categories of FM-REMs as introduced in the literature and then at the end treat a few aspects that are common to them all:

- Vernier machine drives
- Flux-switched (stator PM) machine drives
- Flux reversal machine drives
- Transverse flux machine drives
- Dual stator-winding REM drives
- Magnetically geared REM drives

They will have dedicated separate chapters later in this book.

1.5.1 Vernier Permanent Magnet Machine Drives

Vernier machines are multiphase surface or spoke-shaped PM-rotor REMs (Figure 1.16a and b) [15–17], where the flux modulator role is played by the stator N_s teeth and open slots, while the rotor holds $2p$ PM poles and the stators hold a $2p_a$ pole distributed (standard) AC winding such that:

$$N_s = p + p_a \tag{1.22}$$

It may be built with a radial airgap (Figure 1.16) or an axial airgap.

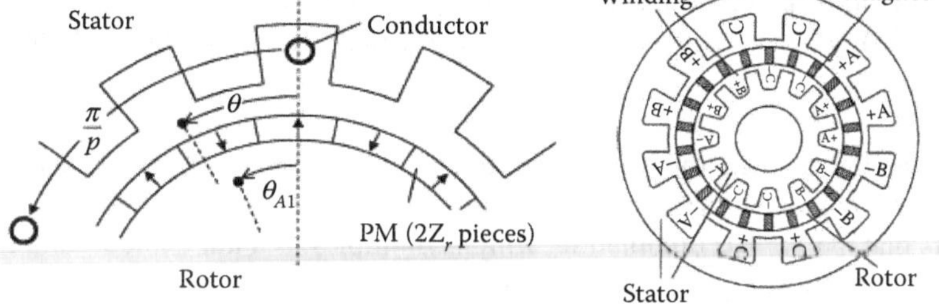

FIGURE 1.16 Vernier PM machines: (a) with surface PM-rotor. (After D. Li et al., *IEEE Trans*, vol. IA–50, no. 6, 2014, pp. 3664–3674. [17]); (b) with spoke–PM–rotor and distributed winding.

The two pole stator windings (Figure 1.16b) lead to an increase of torque density in Nm/liter of the active machine part but imply long end connection coils ($2p_a = 2, 4$) in general to keep the power factor within reasonable limits (0.75–0.85). Also, the frame is long and the between-stators rotor implies a sophisticated fabrication process. From this latter point of view, the axial airgap version with two identical stators [18] seems more manufacturable. For $N_s = 24$ slots, $p_a = 2$ pole pairs ($q = 2$ slots/pole/phase, three phases), $p = 22$ PM pole pairs ($24 = 22 + 2$), an outside stator diameter of 630 mm, stack length $l_{stack} = 50$ mm, 1000 Nm, 30 rpm machine [16], with the spoke–magnet–rotor dual-stator configuration (Figure 1.16b), the torque density has been increased 1.73 times with respect to the surface–PM–rotor single-stator topologies (Figure 1.16a), for a measured efficiency of 0.85 and a power factor of 0.83 at 6.6 Nm/liter of stator stack (active) total volume for 1.2 A/mm^2 current density. The latter indicates the use of a lot of copper for this good performance. There are a few advantages to Vernier PM machines:

- The occurrence of two terms in the emf, torque and power: one related to constant (conventional) airgap permeance and the other related to its first harmonic (addition), produced by the flux modulator (stator open slotting structure), is characteristic of Vernier machines.
- However, the conventional term of torque refers to a machine component with $2\,p_a = 2, 4$ poles, which does not produce a great amount of torque.
- The usage of all magnets all the time in the spoke–PM–rotor with two stators, when less PM flux is also wasted by fringing between rotor poles, is another clear benefit.
- Vernier machines offer the possibility of improving the power factor by carefully designing the machine with spoke magnets and dual stators, with high (1.4 T) stator teeth flux density and radially long magnets to produce a large PM flux concentration, which reduces the armature reaction to reasonable values.
- All these are corroborated with a small number of poles of armature (stator) m.m.f. $2p_a$ and slots /pole/phase($q = 1, 2$).
- However, the fabrication difficulties and the long coil end connections, inevitable for the rather short stack length imposed by the cantilever configuration of the rotor, constitute strong challenges for the radial airgap dual-stator Vernier machine. The situation is slightly better for the axial airgap version [18].

1.5.2 Flux-Switched (Stator Permanent Magnet) Reluctance Electric Machines

The placement of PMs in the stator offers advantages in terms of a rugged, passive magnetically anisotropic rotor where the stator PMs may be protected mechanically and cooled more easily [19,20]; see Figure 1.17a and b. However, the PM eddy currents have to be considered since all AC armature fields may go through them.

The PM flux produced in the stator changes polarity in the stator coils in Figure 1.17 and thus an apparently larger AC emf is produced; however, only a small part of the coil span is traveled by PM flux density, so the iron and coil utilization is not large in this machine. A contemporary three-phase flux-switch (FS) PMSM, where the key factor is also the magnetic anisotropy of both stator and rotor, is shown in Figure 1.18 [21].

- Again, in such machines, the flux-modulation principle operates but in general is hidden in the literature by so many stator teeth to salient pole combinations [21].
- It appears that the 12/10 combination, with its multiples, is a fair choice/compromise when torque density, efficiency, power factor radial forces, noise, and vibration are all considered.
- Even in this case [22], a rather complete comparison with the typical IPMSM with spoke–PM–rotor for the 12 slot/10 pole combination yields conclusions such as:
    - About the same torque per weight, but for more PM weight in the case of FS-PMSM.

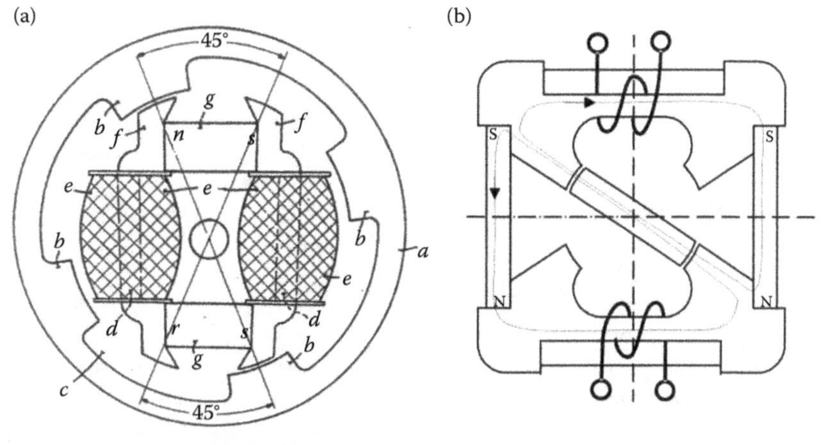

FIGURE 1.17 Original flux-switch single-phase PM machines: (a) with interior stator. (After E. Binder, "Magnetoelectric rotating machine" (Magnetelektrische Schwungradmaschine), German Patent no. DE741163 C, 5 Nov. 1943. [19]); (b) with interior rotor. (After S.E. Rauch, L. J. Johnson, *AIEE Trans*, vol. 74 III, 1955, pp. 1261–1268. [20])

- ◦ FS-PMSM is slightly better in terms of CPSR capability (Figure 1.19), but the core losses are larger.
- ◦ The use of ferrites in FS-PMSM is an option (less torque, less torque ripple) for a given volume, but the risk of PM demagnetization is larger.

- In spite of this tight competition with IPMSM, the FS-PMSM is thoroughly investigated in combined (hybrid) excitation—(DC + PM)-stator—to yield wider CPSR for tractionlike applications.
- Moreover, FS-DC-excited SMs are being proposed for motoring and generating in different stator/rotor (slot/pole) combinations [23,24] (Figure 1.20).
- While acceptable torque density may be produced, DC-excited FS-RMs are characterized by:
  - A large machine reactance in p.u., which is useful for a wide CPSR in motoring but inflicts a low power factor (0.6) for competitive torque density: the same also leads to

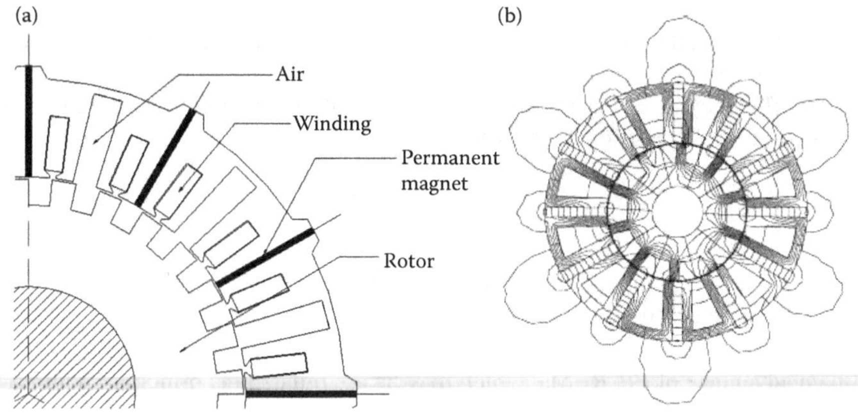

FIGURE 1.18 Typical three-phase flux-switch PMSM: (a) with separated phase coils. (After C. H. T. Lee et al., *IEEE Trans*, vol. EC–30, no.4, 2015, pp. 1565–1573. [27]); (b) with neighboring phase coils (with magnetic flux lines under load). (After A. Fasolo et al., *IEEE Trans*, vol. IA–50, no. 6, 2014, pp. 3708–3716. [22])

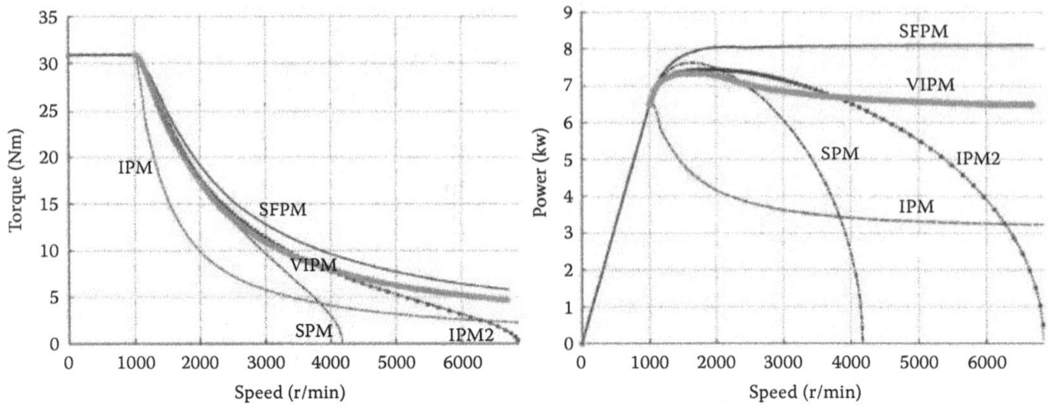

FIGURE 1.19 Flux weakening control results of: SPMSM, V-IPMSM (Vernier - interior permanent magnet synchronous motor), IPM 2 (spoke PMs), IPM (single flux barrier rotor) and SF-PMSM. (A. Fasolo et al., *IEEE Trans*, vol. IA–50, no. 6, 2014, pp. 3708–3716. [22])

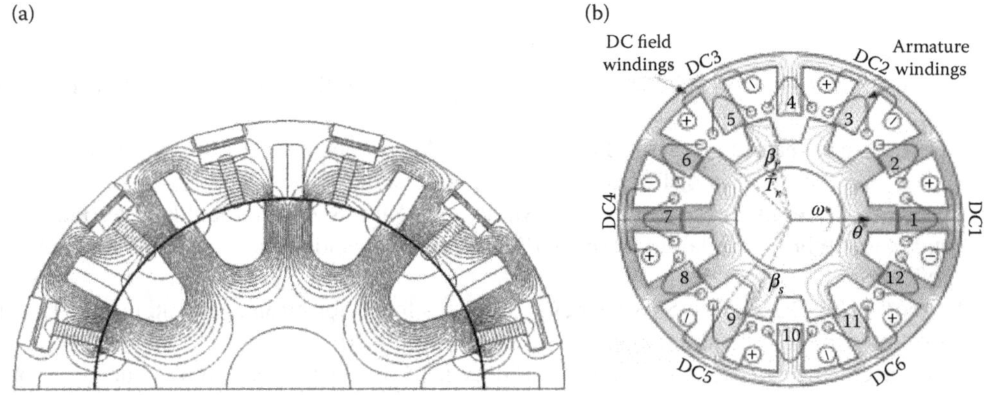

FIGURE 1.20 DC excitation FS-REM: (a) 24/10 combination. (After E. Sulaiman et al., *Record of EPE—2011*, Birmingham, UK. [23]); (b) 12/10 combination. (After X. Liu, Z. Q. Zhu, *Record of IEEE—ECCE*, 2013. [24])

large voltage regulation in autonomous generator mode with PWM converter (or diode rectifier) output.

- Lower efficiency at low speeds (due to DC excitation losses) but acceptable efficiency during flux weakening for wide CPSR.

1.5.3 Flux-Reversal Reluctance Electric Machines

In yet another slight topology modification, flux-reversal (FR) REMs have been introduced, with PMs on the stator [25] or rotor [26]—Figure 1.21a and b:

- The main advantage of FR-REMs, which may be assimilated (as principle) with FS-REMs with stator PMs, is that they make use of conventional laminations in their cores and thus are supposed to be easier to fabricate. The large PM flux fringing in stator–SPM configurations (Figure 1.21a) is partially compensated by the topology with rotor PMs (Figure 1.21b), where all magnets are active all the time. Larger tooth (pole)-wound coils in the stator are

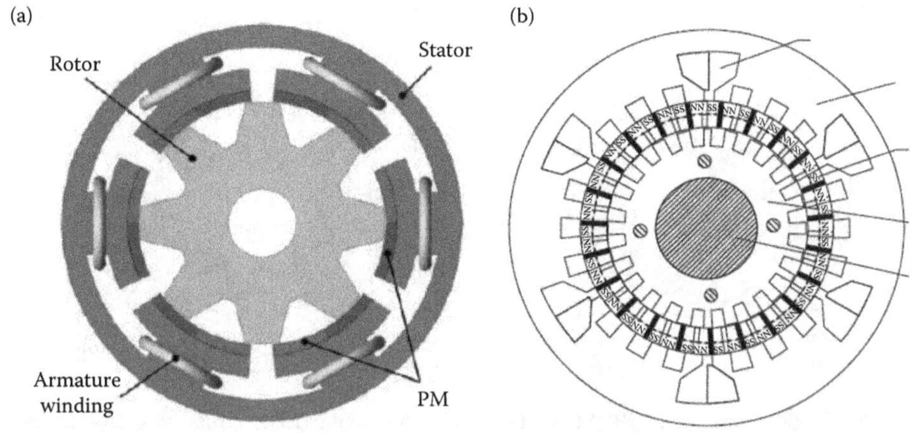

FIGURE 1.21 Flux reversal REMs: (a) with stator PMs; (b) with rotor PMs.

presupposed in all FR-REMs. A DC stator-excited version of an FR-REM was introduced recently [27]—Figure 1.22—in an effort to eliminate PMs in a large machine, typical for wind generator applications. Apparently, trapezoidal (BLDC) current control is more suitable than sinusoidal current, as the emf waveform [27] tends to be trapezoidal.

In yet another effort to improve FR-REMs, the stator was partitioned, with the PMs placed on a dedicated (additional) interior stator while the flux-modulation anisotropic rotor runs between the two stators [28]—Figure 1.23. Even a consequent pole PM interior stator will do to reduce PM weight per Nm notably.

- While the configuration with spoke magnets exhibits the highest torque density, it does so with poor PM usage.
- The configuration with V-shaped magnets is, in terms of efficiency and PM cost, competitive [28]. Still, the fabrication complexity with an in-between rotor is not to be overlooked.
- However, axial airgap topologies of rotor—PM or portioned—stator PM FR-REMs might be easier to fabricate, though they imply spooled axial lamination cores.

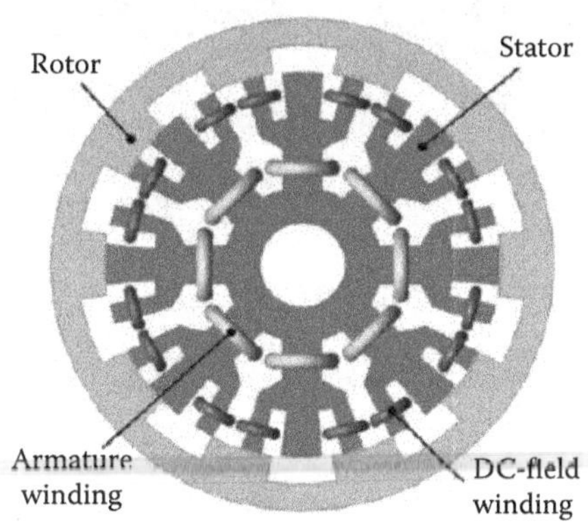

FIGURE 1.22 DC stator excited four-phase FR-REM. (After C. H. T. Lee et al., *IEEE Trans*, vol. EC–30, no. 4, 2015, pp. 1565–1573. [27])

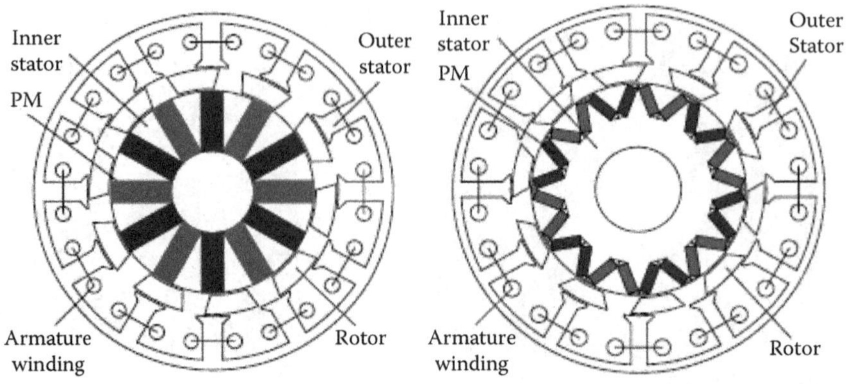

FIGURE 1.23 Portioned stator FR-REM. (After Z. Q. Zhu et al., *IEEE Trans*, vol. IA–52, no. 1, 2016, pp. 199–208. [28])—12/10—combination: (a) with spoke magnets; (b) with V-shaped magnets.

1.5.4 The Transverse Flux-Reluctance Electric Machine

In an effort to decouple the magnetic circuit (core) and electric circuit (winding) design, the transverse flux (TF) machine was introduced [29–31]—Figures 1.24 and 1.25.

Here, the circular phase coil imposes three (m) one-phase machines placed in a row axially on the shaft. But the circumferential coil embraces all stator poles N_s, which are equal to the number of rotor (stator) PM pole pairs p or the number of rotor salient poles N_r (with interpoles such that the machine acts as a synchronous machine). We may also count it as a flux-modulation REM with the number of

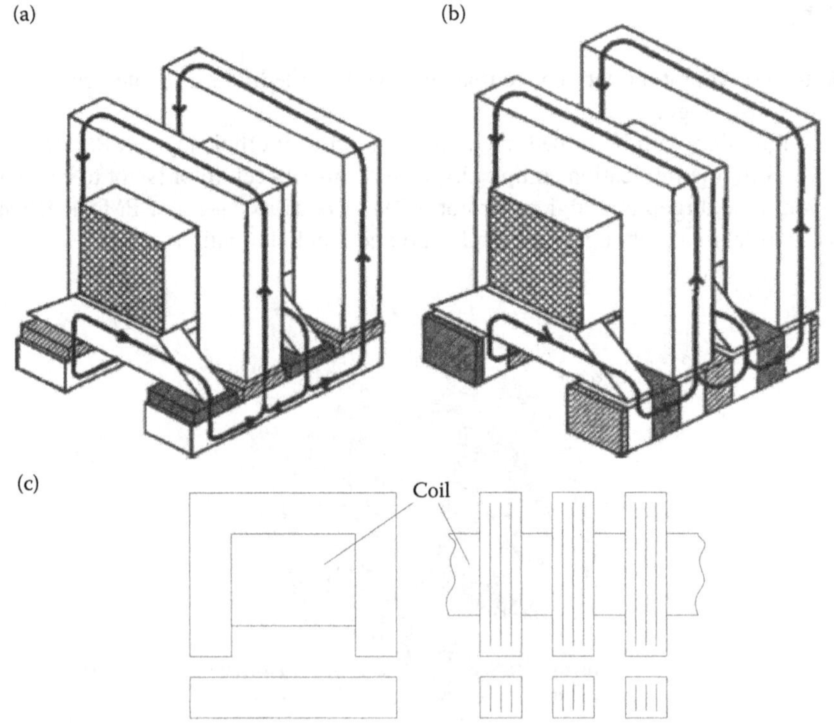

FIGURE 1.24 (a) Transverse flux REM; (b) with PM rotor poles; (c) salient pole rotor. (After H. Weh, H. May, *Record of ICEM—1986*, Munchen, Germany, vol. 3, pp. 1101–1111. [29])

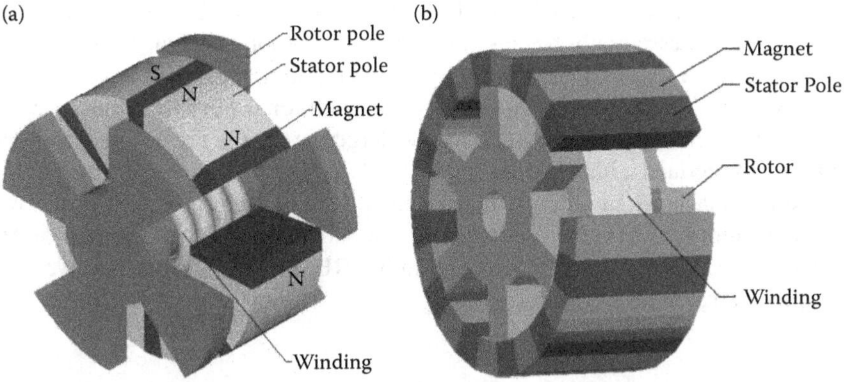

FIGURE 1.25 TF-REM with stator PMs. (After J. Luo et al., *Record of IEEE—IAS—2001 Annual Meeting*, Chicago IL. [30]): (a) with axial airgap; (b) with radial airgap.

flux modulator (stator) salient poles equal to the rotor pole pairs p. Consequently, the pole pairs of stator mmf p_a are expected to be: $p_a = N_s - p = 0$, but this means that the stator mmf is homopolar along the rotor periphery; and so it is!

The TF-REM with PMs is credited with high torque density for small pole pitches $\tau \approx$ 10–12 mm for mechanical airgaps of 1–2 mm. The spoke-type PM rotor configuration (Figure 1.23b) makes better use of PMs but leads to a higher armature reaction (and thus higher machine inductance), which means, again, a lower power factor or higher voltage regulation as an autonomous generator.

- As the number of poles increases for a given geometry circumferential coil (mmf), the torque increases, while the PM flux fringing (leaking between rotor poles) also increases. At some low pole pitch $\tau \approx$ 10–12 mm with mechanical airgap $g = 1$–2 mm, the two effects cancel each other out and thus no further torque "magnification" is noticed.
- And there is the problem of fabricating the TF-REM that involves multiple axial lamination cores mounted together or solid magnetic composite (SMC) cores; in the latter case, the performance (torque density) is reduced by $\mu_{SMC} \approx (500\text{–}700)\mu_0$ at B = 1.5 T, with about 20% for same machine geometry.
- Still, offering practically the lowest copper loss torque (W/Nm) and high torque density–by torque magnification (multiple pole embrace stator coils), the TF-REM with PMs is to be pursued in applications where torque density/copper losses should be high, mainly in high torque–low speed drives, to limit the fundamental frequency $f_1 = n \cdot p$ and thus keep the core loss under control while having also enough switching frequency f_{sw}/f_1 ratio in the PWM converter that controls the machine as a motor or generator via vector or direct torque control.

1.5.5 Note on Power Factor in Flux-Modulation-Reluctance Electric Machines

Basically, the power factor in all flux-modulation machines—which, in spite of being salient pole machines, do not exhibit notably magnetic anisotropy $1 < L_d/L_q < 1.1$—as in synchronous machines is given approximately (losses neglected) by:

$$\cos \varphi_{\text{rated}} \approx \frac{1}{\sqrt{1 + \left(\frac{\omega_1 L_s I_{q\,\text{rated}}}{\omega_1 \Psi_{\text{PM1}}}\right)^2}}; \quad \omega_1 = 2\pi f_1 \tag{1.23}$$

- As shown in detail in Reference 32, all FM-REMs are plagued by a large inductance L_s in p.u. values. This is essentially because the constant component of airgap magnetic permeance Λ_0 produces a smaller portion of torque than the first harmonic Λ_1, which is small in general ($\Lambda_1 \leq \Lambda_0/2$). Moreover, Λ_0 imposes an additional large inductance component; thus, with a large number of poles, L_s becomes large in p.u. and the power factor decreases.
- The attempt to produce a high PM flux linkage Ψ_{PM1} by spoke-magnet rotors and dual stators in Vernier machines in order to increase the power factor is an example of how to alleviate the situation; provided the airgap is increased over 1–1.5 mm, even for 1000-Nm machines, to keep L_s reasonably large, as in TF-REMs, the machine behaves as having two magnetic airgaps.

1.5.6 The Dual Stator Winding (Doubly Fed Brushless) Reluctance Electric Machine

A variable-reluctance rotor with p_r salient poles is integrated in a uniformly slotted stator core that hosts two distributed three-phase AC windings of different pole pairs p_1 and p_2 (Figure 1.25) such as:

$$p_r = p_1 + p_2 \tag{1.24}$$

Equation 1.24 again fulfills the condition of continuous torque (the *frequency theorem*) valid for all flux-modulation machines.

- The interaction between the main winding (designed for 1 p.u. power ratings) connected to the grid and the auxiliary winding (0.2–0.3 p.u. rating) connected to the grid through a bidirectional AC–AC PWM converter is realized through the first space harmonics of airgap magnetic permeance (Λ_1). The constant component (Λ_0) results in a large leakage inductance (low power factor [0.7–0.8], in principle).
- Also, as $\Lambda_1/\Lambda_0 \leq 0.5$ [33], the coupling between the two windings yields an equivalent winding factor of max. 0.5, which excludes the claim of high torque density from principle.
- But the absence of brushes—present in standard doubly fed induction machines used for limited speed range variation in motors/generators—and the partial ratings of the PWM converter (20%–30%) makes their topology of some interest in niche applications (wind generators) and limited-speed range-control drives with light starting (large ventilators, etc.).

Note: The auxiliary (control) winding may be DC fed, and then the speed $n = f_1/p_r$, which is 50% of what a regular SM with same rotor would produce.

1.5.7 Magnetically-Geared Reluctance Electric Machine Drives

So far, we have treated FM-REMs with a single rotor where the flux modulator role was played by the stator or rotor itself.

However, to obtain more aggressive torque magnification, the magnetic gearing effect is used in FM-REMs with basically two rotors.

There are many topologies that fit this denomination, and they have been proposed recently to eliminate the mechanical gears in very-low-speed drives (less than, say, 100 rpm) with torque densities around/above 100 Nm/liter; Figure 1.26 illustrates a recent one [34].

PMs are present on the stator poles (with p_a poles) and on the high-speed rotor (with p pole pairs): the flux-modulator, which now is the low-speed main rotor, has q_m salient poles (in fact pole pairs). The FM principle still stands:

$$q_m = p + p_a \tag{1.25}$$

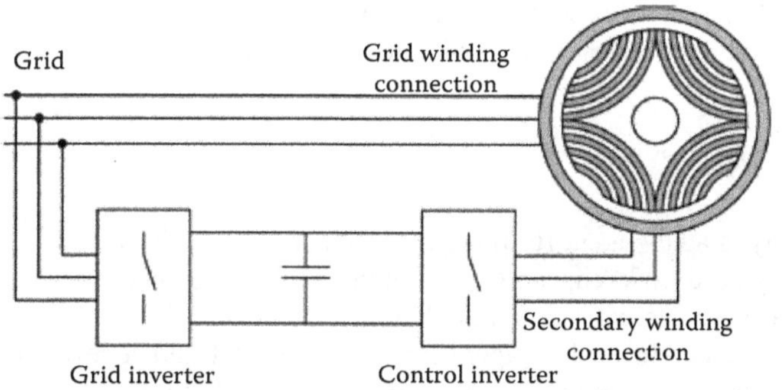

FIGURE 1.26 Brushless doubly fed reluctance machine–stator connection and variable reluctance rotor.

The outer rotor is integrated with the stator that contains both PMs and the AC winding, which has six coils like in a typical flux-reversal REM.

Here, the torque is the result of interactions between the high-speed PM rotor field (p pole pairs) and the stator PM fundamental field as modulated by the flux modulator and stator windings. The so-called gear ratio G_r is:

$$G_r = \frac{q_m}{p} = 1 + \frac{p_a}{p};\ p_a \gg p \tag{1.26}$$

- With $p = 2$ (Figure 1.27) pole pairs on the high-speed rotor, p_a is chosen for a given gear ratio G_r. A value of $p_a = 19$ pole pairs for the stator PM field is considered in Reference 34, the preferred option. Then $q_m = p + p_a = 2 + 19 = 21$ salient poles on the flux modulator. For a 50-Nm, 150-rpm motor, the efficiency was calculated at 90.8%.
- The NdFeB magnet weight was 0.834 kg, and the total motor active mass was calculated at roughly 5 kg.
- However, the measured efficiency was lower than 80%, showing that some losses have not been accounted for.
- The 10 Nm/kg is impressive, but the motor cost and complexity should be considered, too.
- For more on motor-integrated PM gears, see References 35 and 36.

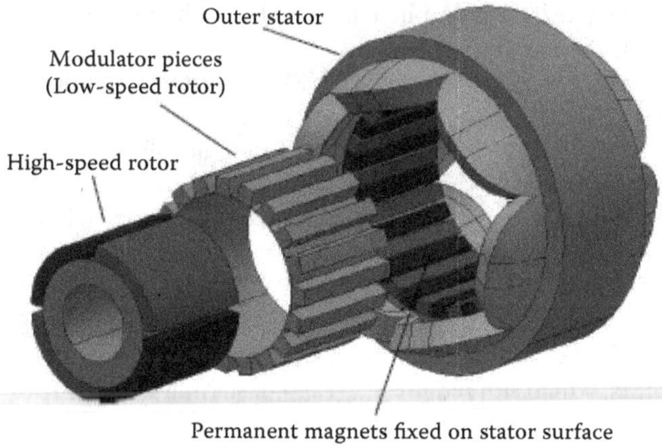

FIGURE 1.27 Magnetically geared FM-REM. (After P. M. Tlali et al., *IEEE Trans*, vol. MAG–52, no. 2, 2016, pp. 8100610. [34])

In Reference 36, for 10 MW at 9.65 rpm, 48.25 Hz, an efficiency of 99% is calculated for a total mass of 65 tons but with a total of 13 tons of sintered NdFeB magnets!

1.6 SUMMARY

- Electric energy is a key agent for prosperity in a cleaner environment.
- Electric energy needs in kWh/capita will continue to increase in the next decades, especially in emerging economies with large populations.
- More electric energy in an intelligent mix of fossil-fuel and renewable primary sources needs even better electric machines of moderate/affordable initial cost.
- Electric energy savings can be obtained by higher-efficiency line-start motors and variable-speed-drive electric machinery.
- Low speed–high torque applications (for direct motor and generator drives in vehicular propulsion and wind/hydro generator applications and robotics) prompt the need for new R&D efforts in the field.
- High-energy PM potential usage in large quantities in wind generators and electric vehicular propulsion has led recently to a large disequilibrium between demand and supply, with the consequence of a sharp increase in high-energy PM price (100 USD/kg or so).
- The high-energy ($B_r > 1.1$ T) PMs, high price should trigger a search for alternative electric machines that use fewer or no high-energy PMs per kW of motor/generator.
- This is how reluctance electric machines come into play again, but this time with improved torque density, efficiency, and power factors, at affordable costs for both line-start and variable-speed applications.
- REMs may use some PMs for assistance (preferably Ferrite magnets [6 USD/kg, $B_r = 0.45$ T] that should provide high magnetic saliency to guarantee high performance).
- In an effort to synthesize the extremely rich heritage of R&D efforts on REMs in the last two decades, we have divided them into two categories first:
  - Reluctance synchronous machines
  - Flux-modulation reluctance electric machines
- RSMs are traveling field machines with distributed stator windings and a variable reluctance rotor with strong magnetic saliency (high L_{dm}/L_{qm} ratio). The stator slot openings do not play any primary role in energy conversion.
- High-saliency rotors for RSMs include multiple flux-barrier and axially laminated anisotropic rotors. A small airgap is a precondition for high magnetic saliency. This limits the maximum speed, as the ratio pole pitch/airgap should be greater than 100/1 (preferably above 300/1 for ALA rotors).
- RSMs may be built for variable-speed PWM converter-controlled drives or for line-start applications.
- Adding PMs (even Ferrites) increases the power factor and allows for higher power or wider constant power speed range in variable-speed drives.
- Variable-speed RSMs have become commercial recently (up to 500 kW, 1500 kW occasionally), with 2%–4% more efficiency but 5%–8% less power factor than IMs (with the same stator) for an initial motor cost reduced by some 15%–20%.
- Line-start RSMs (split-phase or three-phase) of high class efficiency are needed and feasible, as they also maintain the starting current in the range of 6–6.5 p.u. value in contrast to high class efficiency IMs, which still require a 7–7.5 p.u. starting current (with the inconvenience of demanding the overrating of local power grids both in residential and factory applications).
- Claw-pole synchronous machines and switched reluctance machines are considered in the class of quasisynchronous machines.

- CP-SMs have a variable reluctance rotor to produce a multipolar airgap field with a single circular (or ring-shaped) DC excitation coil magnet in the CP rotor, offering the best utilization of the rotor-produced magnetic field.
- SRMs are reviewed, as recent progress has led to high performance (in efficiency [95%] and torque density [45 Nm/liter] at 60 kW and 3000 rpm) that may qualify them for electric propulsion on electric and hybrid electric vehicles.
- RSMs, CP-SMs, and SRMs are investigated in dedicated chapters in Part 1 of this book.
- Flux-modulation REM topologies are characterized by three (not two) main parts that contribute directly to electric energy conversion:
  - The exciter part
  - The flux-modulator part
  - The armature part

At least one part should be fixed (the armature), while one or both of the others may rotate. A kind of torque magnification occurs in such machines, which may be exploited for low/moderate-speed applications.

All FM-REMs require full (or partial) rating power PWM converter control for variable speed operation, and none has a cage on the rotor.

The key to the operation of FM-REMs is the contribution of both the constant and first space harmonic of airgap performance (produced by the variable reactance flux modulator) to the interaction between the exciter (with PMs, in general) and the armature mmf of the stator. The presence of two terms in torque production generates a kind of torque magnification (magnetic gearing effect), which is the main asset of FM-REMs.

From the rich literature on FM-REMs, a few main categories have been deciphered and characterized in this introductory chapter: Vernier PM machines, flux-switch (stator PM) machines, flux-reversal (stator PM) machines, transverse flux PM machines, dual-stator winding reluctance machines, and magnetically geared (dual rotor) machines.

Their main merits and demerits are emphasized, but their thorough investigation (modeling, design, and control) will be performed in dedicated chapters later in the second part of the book.

REFERENCES

1. I. Boldea, *Reluctance Synchronous Machines and Drives*, book, Oxford University Press, 1996.
2. G. Henneberger, I. A. Viorel, *Variable Reluctance Electrical Machines*, book, Shaker-Verlag, Aachen, Germany, 2001.
3. J. K. Kostko, "Polyphase reaction synchronous motor", *Journal of A.I.E.E.*, vol. 42, 1923, pp. 1162–1168.
4. P. F. Bauer, V. B. Honsinger, "Synchronous induction motor having a segmented rotor and a squirrel cage winding", U.S. Patent 2,733,362, 31 January, 1956.
5. ABB site.
6. A. J. O. Cruickshank, A. F. Anderson, R. W. Menzies, "Theory and performance of reluctance machines with ALA rotors", *Proc. of the Institution of Electrical Engineers*, vol. 118, no. 7, July 1971, pp. 887–894.
7. R. Mayer, H. Mosebach, U. Schroder, H. Weh, "Inverter fed multiphase reluctance machines with reduced armature reaction and improved power density", *Proc. of ICEM*, 1986, Munich, Germany, part 3, pp. 1138–1141.
8. I. Boldea, G. Papusoiu, S. A. Nasar, Z. Fu, "A novel series connected switched reluctance motor", *Proc. of ICEM*, 1986, Munich, Germany, Part 3, pp. 1212–1217.
9. J. D. Law, A. Chertok, T. A. Lipo, "Design and performance of field regulated reluctance machine", *Record of IEEE—IAS—1992 Meeting*, vol. 2, pp. 234–241.
10. D. Ursu, V. Gradinaru, B. Fahimi, I. Boldea, "Six-phase BLDC reluctance machines: FEM based characterization and four-quadrant control", *IEEE Trans*, vol. IA–51, no. 3, 2015, pp. 2105–2115.

11. L. Tutelea, D. Ursu, I. Boldea, S. Agarlita, "IPM claw-pole alternator system or more vehicle braking energy recuperation", 2012, www.jee.ro.
12. T. J. E. Miller, *Switched Reluctance Motors*, book, Oxford University Press, Clarendon Press, Oxford, England, 1993.
13. R. Krisnan, *Switched Reluctance Motor Drives*, book, CRC Press, Taylor and Francis Group, Boca Raton, Florida, 2000.
14. M. Takeno, A. Chiba, N. Hoshi, S. Ogasawara, M. Takemoto, "Test results and torque improvement of a 50 KW SRM designed for HEVs", *IEEE Trans*, vol. IA, 48, no. 4, 2012, pp. 1327–1334.
15. C. H. Lee, "Vernier motor and its design", *IEEE Trans*, vol. PAS–82, no. 66, 1963, pp. 343–349.
16. B. Kim, T. A. Lipo, "Operation and design principles of a PM Vernier motor", *IEEE Trans*, vol. IA–50, no. 6, 2014, pp. 3656–3663.
17. D. Li, R. Qu, T. A. Lipo, "High power factor Vernier PM machines", *IEEE Trans*, vol. IA–50, no. 6, 2014, pp. 3664–3674.
18. B. Kim, T.A. Lipo, "Analysis of a PM Vernier motor with spoke structure", *IEEE Trans*, vol. IA–52, no.1, 2016, pp. 217–225.
19. E. Binder, "Magnetoelectric rotating machine" (Magnetelektrische Schwungradmaschine), German Patent no. DE741163 C, 5 Nov. 1943.
20. S. E. Rauch, L. J. Johnson, "Design principle of flux switch alternator", *AIEE Trans*, vol. 74 III, 1955, pp. 1261–1268.
21. Z. Q. Zhu, J. Chen, "Advanced flux-switch PM brushless machines", *IEEE Trans*, vol. MAG–46, no. 6, 2010, pp. 1447–1453.
22. A. Fasolo, L. Alberti, N. Bianchi, "Performance comparison between switching-flux and PM machines with rare-earth and ferrite PMs", *IEEE Trans*, vol. IA–50, no. 6, 2014, pp. 3708–3716.
23. E. Sulaiman, T. Kosaka, N. Matsui, "A new structure of 12 slot/10 pole field-excitation flux-switching synchronous machine for HEVs", *Record of EPE—2011*, Birmingham, UK.
24. X. Liu, Z. Q. Zhu, "Winding configurations and performance investigations of 12-stator pole variable flux reluctance machines", *Record of IEEE—ECCE*, 2013.
25. R. P. Deodhar, S. Andersson, I. Boldea, T. J. E. Miller, "The flux reversal machine: A new brushless doubly-salient PM machine", *IEEE Trans*, vol. IE–33, no. 4, 1997, pp. 925–934.
26. I. Boldea, L. Tutelea, M. Topor, "Theoretical characterization of three phase flux reversal machine with rotor–PM flux concentration", *Record of OPTIM—2012*, Brasov, Romania (IEEEXplore).
27. C. H. T. Lee, K. T. Chau, Ch. Liu, "Design and analysis of a cost effective magnetless multiphase flux-reversal d.c. field machine for wind power conversion", *IEEE Trans*, vol. EC–30, no. 4, 2015, pp. 1565–1573.
28. Z. Q. Zhu, H. Hua, Di Wu, J. T. Shi, Z. Z. Wu, "Comparative study of partitioned stator machines with different PM excitation stators", *IEEE Trans*, vol. IA–52, no. 1, 2016, pp. 199–208.
29. H. Weh, H. May, "Achievable force densities for permanent magnet excited machines in new configurations", *Record of ICEM—1986*, Munchen, Germany, vol. 3, pp. 1101–1111.
30. J. Luo, S. Huang, S. Chen, T. A. Lipo, "Design and experiments of a novel axial circumferential current permanent magnet (AFCC) machine with radial airgap", *Record of IEEE—IAS—2001 Annual Meeting*, Chicago, IL.
31. I. Boldea, "Transverse flux and flux reversal permanent magnet generator systems introduction", *Variable Speed Generators*, book, 2nd edition, CRC Press, Taylor and Francis Group, New York, 2015.
32. D. Li, R. Qu, J. Li, "Topologies and analysis of flux modulation machines", *Record of IEEE—ECCE 2015*, pp. 2153–2160.
33. A. M. Knight, R. Betz, D. G. Dorrell, "Design and analysis of brushless doubly fed reluctance machines", *IEEE Trans*, vol. IA–49, no. 1, 2013, pp. 50–58.
34. P. M. Tlali, S. Gerber, R. J. Wang, "Optimal design of an outer stator magnetically geared PM machine", *IEEE Trans*, vol. MAG–52, no. 2, 2016, pp. 8100610.
35. P. O. Rasmussen, T. V. Frandsen, K. K. Jensen, K. Jessen, "Experimental evaluation of a motor - integrated PM gear", *IEEE Trans*, vol. IA–49, no. 2, 2013, pp. 850–859.
36. A. Penzkofer, K. Atallah, "Analytical modeling and optimization of pseudo – direct drive PM machines for large wind turbines", *IEEE Trans*, vol. MAG–51, no. 12, 2015, pp. 8700814.

2 Line-Start Three-Phase Reluctance Synchronous Machines

Modeling, Performance, and Design

2.1 INTRODUCTION

Line-start three-phase RSMs are built with a magnetically anisotropic rotor ($L_{dm} > L_{qm}$), host of a cage winding and sometimes of PMs, for better steady-state (synchronous) performance: efficiency and power factor.

They are proposed to replace induction motors in line-start constant-speed applications in the hope that they can produce high class efficiency in economical conditions (of initial cost and energy consumption reduction for given mechanical work, leading to an investment payback time of less than 3 years, in general).

The IEC 60034-30 standard (second edition) includes minimum efficiency requirements for IE 1, 2, 3, 4, and 5 efficiency classes of performance, according to Figure 2.1 [1].

These demanding requirements have to put the existing reality and trends in electric machinery R&D and fabrication face to face to assess the potential of meeting them.

Table 2.1 assesses the potential of various electric motor technologies in terms of efficiency [1].

Table 2.1 shows that, at least for efficiency classes IE4 and 5, there are very few candidates, and, among them, line-start PM motors are cited. To add more to the other challenges of IE2, 3, and 4 induction motors, the situation for the case of 7.5-kW four-pole motors is described in Table 2.2 [1].

From this table we may decipher a few hard facts:

- While the efficiency increases from 87% to 93%, the initial cost of the line start permanent magnet synchronous machine (LSPMSM) (class IE 4) is 233% of that of the IM of class IE 2. Class IE 1 is now out of fabrication in the E.U., but such existing motors will continue to operate for the next decade or so. Moreover, the starting current of LSPMSM is 7.8 p.u., lower than that of IM class IE3 (8.5 p.u.) but still too large to avoid oversizing local power grids that supply such motors without deep voltage surges during motor start.
- Still, the simple payback time for IE3-IM and IE4-LSPMSM 7.5-kW motors when replacing IE1-IMs has been calculated at 0.5 and, respectively, 2.8 years!
- Line-start PMSMs with strong PMs and small magnetic saliency have been most frequently considered so far. This is how efficiency was markedly increased, though with a notably higher-cost motor, which, in addition, is not yet capable of reducing the starting current to 6.2–6.8 p.u., as in IE1 class IMs (Table 2.2).
- So the argument for economical energy savings with reasonable payback time (less than 2–2.5 years) by line-start RSMs with strong magnetic saliency and eventually with lower-cost PM assistance is: Are they economically feasible, while also providing starting currents below 6.8 p.u. to avoid overrating the local power grids that supply such motors?

- As a step toward such a demanding goal, this chapter treats first line-start RSMs without any PMs and then ones with ferrite-PM assistance, while at the end, LSPMSMs with strong PMs and low magnetic saliency (1.8/1) are also investigated in terms of modeling, performance, and design, with representative case study results.

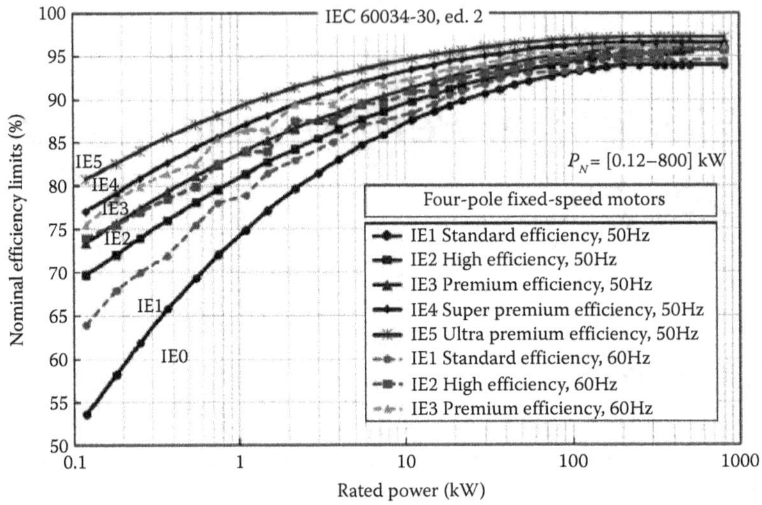

FIGURE 2.1 Efficiency class limits for four-pole three-phase AC motors (0.12–800 kW) according to IEC 60034-30 standard (second edition, 2014). (After A. T. Almeida, F. T. J. E. Ferreira, A. Q. Duarte, *IEEE Trans on*, vol. IA–50, no. 2, 2014, pp. 1274–1285. [1])

TABLE 2.1
Efficiency Potential of Various Electric Motors

Motor Type			Line-Start	IE1	IE2	IE3	IE4	IE5
Three-phase SCIM	Random- and form-wound windings:							
	IP2x (open motors)		Yes	Yes	Yes	Diff.	No	No
	IP4x and above (enclosed motors)	Random-wound windings	Yes	Yes	Yes	Yes	*Yes*	*Diff.*
		Form-wound windings	Yes	Yes	Yes	Yes	Diff.	No
Three-phase wound-rotor induction motors			Yes	Yes	Yes	*Yes*	*Diff.*	No
Single-phase SCIM	One capacitor		Yes	Yes	Dif.	No	No	No
	Two switchable capacitors		Yes	Yes	Yes	Diff.	No	No
Synchronous motors	VSD-fed PMSM		No	Yes	Yes	Yes	Yes	Diff.
	Wound-rotor		Some	Yes	Yes	Yes	Diff.	No
	LSPM		Yes[a]	Yes	Yes	Yes	*Yes*	*Diff.*
	Sinusoidal-field reluctance		Some	Yes	Yes	Yes	*Yes*	*Diff.*

Source: After A. T. Almeida, F. T. J. E. Ferreira, A. Q. Duarte, *IEEE Trans on*, vol. IA–50, no. 2, 2014, pp. 1274–1285. [1]

[a]LSPMs have some limitations on their line-start capabilities with respect to torque and external inertia.

Notes: The mark:

- "Yes" means that the motors are considered to be the state of present commercial technology and are therefore suitable for consideration in mandatory requirements in legislation (i.e., the efficiency class is achievable with present technology, although in some cases it may not be economical).
- "No" means the efficiency class is not achievable with present commercial technology.
- "Diff." (= "Difficult") means that the energy-efficiency level may be achieved with present commercial technology for some but not all power ratings, and the standardized frame-size may be exceeded.
- "Line-start" means the capability of the motor to start directly from the line without the need for a VSD (direct on-line [DOL] starting; Design N of IEC60034-12 for single-speed squirrel-cage induction motors [SCIMs]).

TABLE 2.2
Performance of 7.5-kW, Four-Pole IMs and LSPMSM in IE 2, 3, 4

IEC Class	Motor Technology	IEC Frame	Weight (kg)	Power Density (W/kg)	Moment of Inertia (kg m^2)	Rated Speed (r/min)	Full-Load Torque (Nm)	Locked Rotor to Full-Load Torque Ratio	Locked Rotor to Full-Load Current Ratio	Full-Load Efficiency (%)	Full-Load Power Factor	Price (%)
IE1	SCIM	132M	64.5	116.3	0.0465	1455	49.3	2.1	6.7	87.0	0.84	–
IE2	SCIM	132M	72.0	104.2	0.0528	1455	49.3	2.0	7.2	89.0	0.86	100
IE3	SCIM	132M	78.0	96.2	0.0642	1465	48.9	2.5	8.5	91.5	0.85	115
IE4	LSPM	132M	61.5	122.0	0.0500	1500	47.8	3.8	7.8	93.0	0.93	233

Source: After A. T. Almeida, F. T. J. E. Ferreira, A. Q. Duarte, *IEEE Trans on*, vol. IA–50, no. 2, 2014, pp. 1274–1285. [1]

2.2 THREE-PHASE LINE-START RELUCTANCE SYNCHRONOUS MACHINES: TOPOLOGIES, FIELD DISTRIBUTION, AND CIRCUIT PARAMETERS

2.2.1 Topologies

As implied in Chapter 1, three-phase line-start RSMs with cage rotors are built only with distributed symmetric three-phase windings, in general with q (slots/pole/phase) $> 2(3)$ and an integer number in order to reduce additional cage losses due to excessive stator mmf space harmonics and "harvest more completely the fruits of high magnetic saliency in the rotor."

The number of stator winding and rotor poles may be 2, 4..., but as experience shows, to get good torque density and an acceptable power factor, pole pitch/airgap > 150 and magnetic saliency $L_{dm}/L_{qm} > 5 \div 10$ [2] are required.

As the stator has, as for induction machines, uniform slotting and a distributed three-phase winding, Figure 2.2 illustrates a few representative rotor topologies for $2p = 2$ and $2p = 4$.

To assess performance, the flux distribution and then circuit parameters in the dq model are required:

$$L_d = L_{dm} + L_{sl},\ L_q = L_{qm} + L_{sl},\ R_s,\ R_{rd},\ R_{rq},\ L_{rld},\ L_{rlq} \text{ and } \mathrm{PM}_{emf} - E_{\mathrm{PM}} \text{ (if any).}$$

2.2.2 Analytical (Crude) Flux Distribution and Preliminary Circuit Parameters Formulae

To calculate even approximately the distribution of magnetic flux lines by analytical methods is helpful (Figure 2.3) if portrayed with the rotor axis d aligned to the stator mmf (axis d) or at 90° (axis q).

Typical FEM-derived flux lines in an RSM with no PMs, an ALA rotor (Figure 2.3a), and the airgap flux density variation with the rotor position for a given mmf in axes d and q (Figure 2.3b) reveal the complexity of the problem, as airgap flux densities have rich space harmonics content.

It is evident that, due to stator slotting and rotor multiflux barriers in the rotor, the airgap flux density in both the d and q axes shows high-frequency space harmonics. Also, in axis q, per half a pole, the airgap flux density changes polarity a few times, indicating that the magnetic potential of the rotor is not constant. Thus, the flux lines cross the airgap a few times, then the stator slot openings. This is good, as the fundamental flux density B$gq1$ is reduced, but the space harmonics may induce notable eddy currents in the rotor axial laminations.

They may be reduced by thin radial slits (three to four per total stack length).

On the contrary, in regular lamination (multiple flux barrier) rotors, the q axis airgap flux density does not, in general, show negative components under a pole, but its fundamental is reduced less, though still notably (Figure 2.4).

On top of that, the PM airgap flux density distribution (at zero stator currents)—Figure 2.5–shows space harmonics that will become time harmonics in the back emf (E_{PM}).

The results in Figures 2.3 through 2.5 have been obtained by FEM to completely illustrate the situation. For preliminary design (or performance calculations), simplified circuit constant parameter formulae are required. For that, only the fundamental in airgap flux density is considered. At a very preliminary stage, the expressions of magnetizing inductances L_{dm} and L_{qm} of an RSM (with distributed AC windings) are standard [2]:

$$L_{d,qm} = \frac{6 \cdot \mu_0 (W_1 K_{W1})^2 \tau \cdot l_{\text{stack}}}{\pi^2 p \cdot g_{d,q} \cdot \left(1 + K_{sd,q}\right)};\quad g_d = g \cdot K_{c1} \cdot K_{c2};\quad g_q = g_d + g_{\mathrm{FB}} \tag{2.1}$$

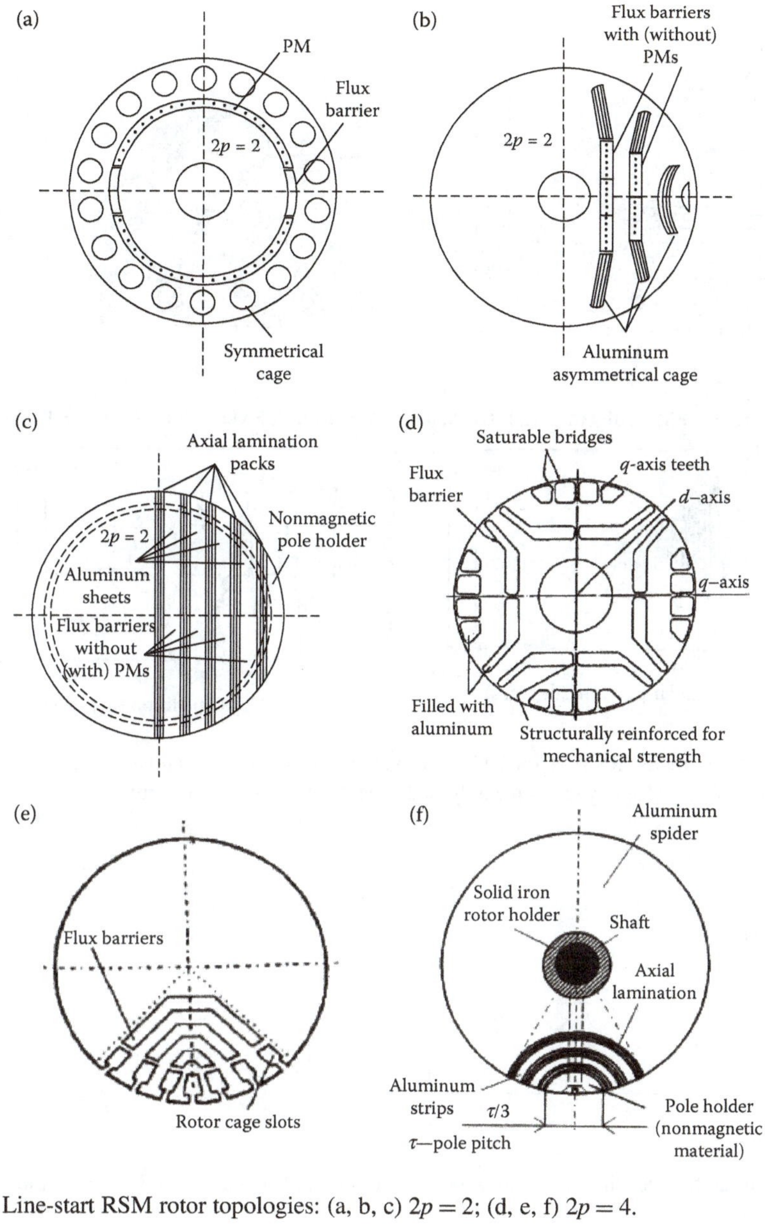

FIGURE 2.2 Line-start RSM rotor topologies: (a, b, c) $2p = 2$; (d, e, f) $2p = 4$.

where: W_1—turns per phase (one current path only); K_{W1}—fundamental winding factor; τ—pole pitch; l_{stack}–stack length; p—pole pairs; $K_{c1,2}$—Carter coefficient for stator and rotor; $K_{sd,q}$—equivalent saturation coefficients; g_{FB}—the equivalent additional airgap provided by the rotor flux barriers.

The average length of flux barriers along a q-axis flux line in the rotor makes a maximum of $2 \cdot g_{FB}$.

Note on Magnetic Saturation: Magnetic saturation is a complex phenomenon with local variations and, approximately, may be considered dependent on the i_d, i_q stator current and the PM flux linkage Ψ_{PM}, according to the concept of cross-coupling saturation. In the flux barrier rotor at very low currents, before the rotor flux bridges saturate, L_{qm} is large (almost equal to L_{dm}). As the stator currents increase, the rotor flux bridges saturate and thus L_{qm} decreases dramatically to produce reasonable torque (by magnetic saliency). For line-start RSMs, this variation of L_{qm} with current is beneficial for RSM self-synchronization.

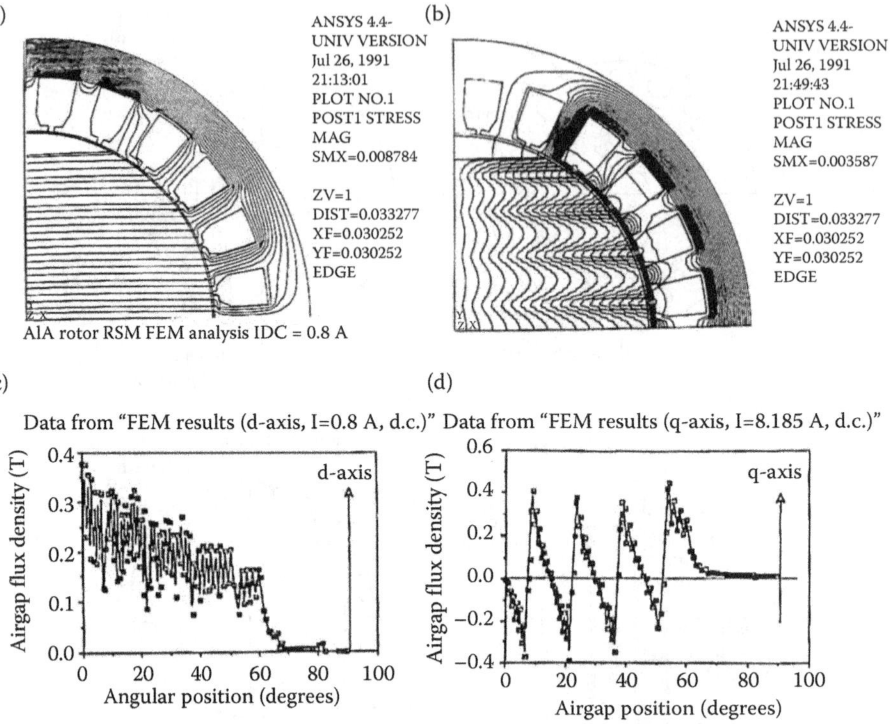

FIGURE 2.3 FEM-extracted RSM stator DC mmf magnetic flux lines at standstill in axes d and q (a, b); and airgap flux density in axes d and q (c, d) for a $2p = 2$ pole ALA rotor (as in Figure 2.1c).

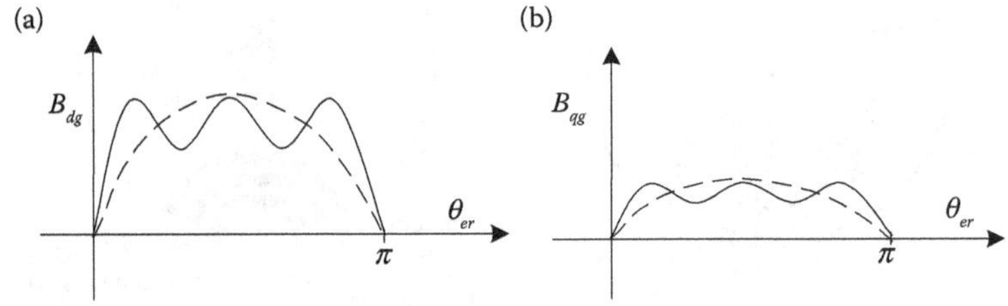

FIGURE 2.4 Airgap flux density versus rotor position in axis d, (a); and q, (b), for a multiple flux barrier rotor (as in Figure 2.2e).

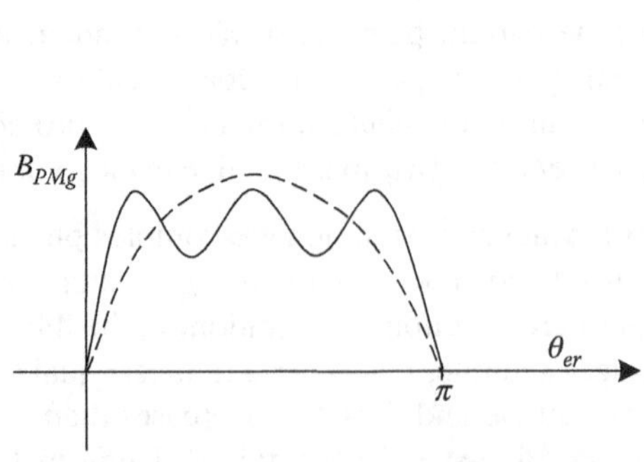

FIGURE 2.5 PM-produced airgap flux density in an RSM with flux-barrier rotor (as in Figure 2.1b).

Not so in ALA rotors, where L_{qm} is smaller but rather constant (independent of current): here, however, the larger magnetic saliency produces a higher peak reluctance torque that also helps self-synchronization under notable loads (1.2–1.3 p.u.).

For the simplified two-pole rotor in Figure 2.2a, the ratio of L_{dm}/L_{qm}

$$\frac{L_{dm}}{L_{qm}} = \frac{g_q}{g_d} \tag{2.2}$$

may be approximated by considering magnetic reluctances in axes d and q:

$$\frac{g_q}{g_d} \approx \frac{g + g_{\text{FB}} + g_{\text{cage}}}{g + g_{\text{cage}}}; \; g - \text{airgap} \tag{2.3}$$

where g_{FB}—the radial thickness of the flux barrier (only one flux barrier here), say, equal to $9 * g$, and g_{cage}—the equivalent radial airgap of the rotor cage bar region: with an equivalent height of the rotor cage (for infinite core permeability), $h_{\text{cage}} \approx 11 \cdot g$ and considering 60% core, 40% rotor slots $g_{\text{cage}} \approx h_{\text{cage}} * 0.6 = 6.6$ mm.

So, in this hypothetical cage [3],

$$\frac{g_q}{g_d} = \frac{L_{dm}}{L_{qm}} = \frac{g(1 + 9 + 6.6)}{g(1 + 6.6)} = 2.18 \tag{2.4}$$

This magnetic saliency ratio seems modest, but with high-energy PMs, it leads to outstanding performance, as seen later in this chapter.

The stator leakage inductance L_{sl} with its components is, as for IMs [2]:

$$L_{sl} = \frac{2\mu_0 W_1^2}{p g_1}\left(\lambda_{\text{slot}} + \lambda_z + \lambda_{\text{end}} + \lambda_{\text{diff}}\right) \tag{2.5}$$

Simple but still reliable expressions of nondimensional slot, airgap, end connection, and differential leakage permeances λ_{slot}, λ_z, λ_{end}, and λ_{diff} are [2]:

$$\begin{aligned}
\lambda_{\text{slot}} &\approx \frac{h_{su}}{3b_{sav}} + \frac{h_{s0}}{3b_{s0av}} < 1.5 - 2; \quad b_{s0av} = \frac{b_{s1} + b_{s0}}{2}; \quad b_{sav} = \frac{b_{s1} + b_{s2}}{2} \\
\lambda_z &= \frac{5gK_c/b_{s0} \times \left(3\beta_y + 1\right)/4}{5 + 4gK_c/b_{s0}} < 0.4 - 0.5; \quad K_c = K_{c1} \cdot K_{c2} \\
\lambda_{\text{end}} &\approx 0.34 \frac{g}{l_{\text{stack}}}(l_{\text{end}} - 0.64\tau) < 0.7 - 0.8 \text{ (for double layer windings)} \\
\lambda_{\text{diff}} &= \frac{3g}{\pi^2 g K_c} \sum^{\nu > 1} \frac{K_{w\nu}^2}{\nu^2}
\end{aligned} \tag{2.6}$$

with h_{su}, h_{sav}, h_{s0}, and b_{s0av} as useful (copper-filled) slot height, average useful slot width, slot neck height, and slot neck average width (Figure 2.6). The differential leakage inductance component (by λ_{diff}) refers to space harmonics of the stator airgap field and implies that all these harmonics cross the airgap; this is true for high-order harmonics if $\tau/\pi\nu > g$.

In a similar way, the PM emf (PMs in axis q here) may be written as:

$$E_{\text{PM}q1} = \omega_1\left(B_{g\text{PM}1} \cdot \frac{2}{\pi} \cdot \tau \cdot l_{\text{stack}}\right) \times W_1 \cdot k_{W1} \tag{2.7}$$

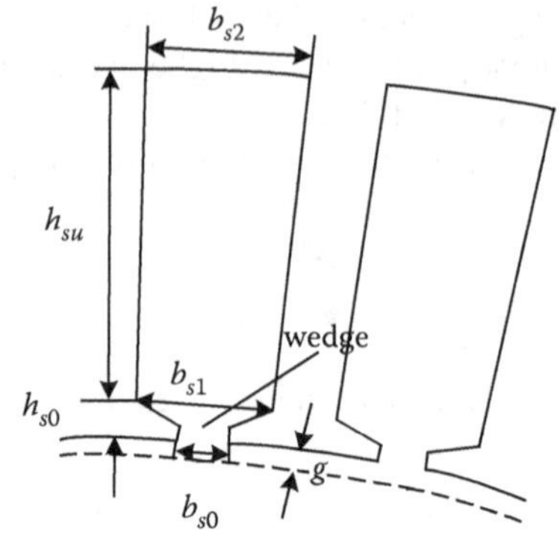

FIGURE 2.6 Typical stator semiclosed slots.

where

$$B_{gPM1} \approx \frac{4}{\pi} B_{gPM} \cdot \sin \alpha_{PM} \times \frac{2}{\pi} \tag{2.8}$$

$$B_{gPM} \approx B_r \cdot \frac{g_{FBPM}/\mu_{rec}}{\left(g_{FB}/\mu_{rec} + g + g_{cage}\right)\left(1 + K_{f\,\mathrm{ring}\,e}\right)} \tag{2.9}$$

where B_r is the remnant flux density of the PM, μ_{rec} is its recoil relative permeability ($\mu_{rec} =$ 1.05–1.11, in general), and $K_{f\,\mathrm{ring}\,e}$ accounts for the PM flux "lost" in the rotor; $K_{f\,\mathrm{ring}\,e}$, for the rotor in Figure 2.2a, is probably 0.15–0.25), but it may be calculated by FEM and then approximated better. The term g_{FBPM} is the equivalent thickness of flux barriers filled with PMs. We have to add the leakage inductance L_{rl} and resistance of the rotor cage R_r.

Considering again the symmetric cage of Figure 2.1a, we may simply use the standard formula [2]. But the presence of the flux bridge below the rotor slots notably reduces the slot leakage permeance λ_{slot}. For the simplified rotor slot in Figure 2.7, as shown in [4]:

$$\lambda_{slot\,r} \approx \frac{h_{r0}}{b_{r0}}(1 - C_W)^2 + \frac{h_{sr}}{3b_{sr}}[1 + 3C_W(C_W - 1)] \tag{2.10}$$

$$C_W = \frac{h_{r0}/b_{r0} + h_{sr}/2b_{sr}}{\mu_{ri} W_i/b_{sr} + h_{r0}/b_{r0} + h_{sr}/2b_{sr}} < 1 \tag{2.11}$$

with μ_{ri}—the relative permeability (10–20 p.u.) of the flux bridge W_i below the slots. When $W_i \approx \infty$, $C_w = 0$, $\lambda_{slot\,r}$ gets the standard formula:

$$\lambda_{slot\,r} = \frac{h_{r0}}{b_{r0}} + \frac{h_{sr}}{3b_{sr}} \tag{2.12}$$

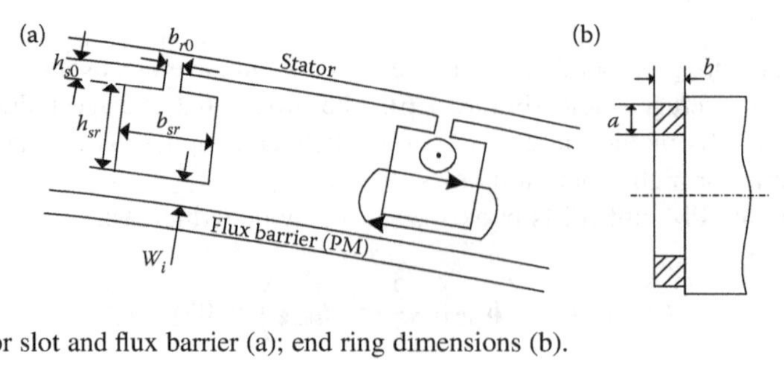

FIGURE 2.7 Rotor slot and flux barrier (a); end ring dimensions (b).

Also, the end connection (cage ring) permeance $\lambda_{e\,\text{ring}}$ is:

$$\lambda_{e\,\text{ring}} \approx \frac{3D_{\text{ring}}}{2N_r l_{\text{stack}} \sin\dfrac{\pi p}{N_r}} \cdot \ln 4.7 \cdot \frac{D_{\text{ring}}}{a+2b} < 0.7 - 1.0 \tag{2.13}$$

The combined cage bar and its end ring section resistance are approximated to:

$$R_{be} = R_{\text{bar}} + \frac{R_{\text{ring}}}{2 \cdot \sin^2\left(\dfrac{\pi p}{N_r}\right)} \tag{2.14}$$

where D_{ring} is the average diameter of the cage ring and a and b are its cross-section dimensions; N_r is the number of rotor slots, while R_{bar} and R_{ring} are the bar and its ring section (per both sides) electrical resistances.

Similarly, the bar + ring section leakage inductance L_{be} is:

$$L_{be} = \mu_0 l_{\text{stack}}(\lambda_{\text{slot}\,r} + \lambda_{zr} + \lambda_{e\,\text{ring}}) \tag{2.15}$$

The reduced values R_r, L_{rl} to the stator are [2]:

$$R_r = R_{be}\frac{N_r}{3K_i^2}; \quad L_{re} = L_{be} \cdot \frac{N_r}{3K_i^2} \tag{2.16}$$

$$K_i = \frac{N_r}{6 \cdot W_1 K_{W1}} \tag{2.17}$$

Note:

- The flux bridges in the rotor (close to flux barriers) modify (decrease) the leakage inductance of the rotor, which is beneficial for starting and self-synchronization under light load. Those bridges, however, saturate at small stator and rotor currents and give rise to additional torque pulsations during asynchronous operation.
- The calculation of L_{rl} and R_r was rather simple for the symmetric cage (Figure 2.2a), but it is more complicated when part of the flux barriers is occupied by the cage, which is now asymmetric, and, apparently, only FEM can be truly useful/practical in this case.

NUMERICAL EXAMPLE 2.1: CONSTANT CIRCUIT PARAMETERS

Let us consider a PM-assisted three-phase (delta connection) RSM with two poles and the rotor in Figure 2.2a, with a rectangular rotor bar as in Figure 2.7, with the data:

- Line voltage: $V_L = 380$ V (RMS)
- Frequency: $f_{1n} = 50$ Hz
- Inner stator diameter $D_{is} = 90$ mm
- Stator core length $l_{\text{stack}} = 90$ mm
- Airgap $g = 0.55$ mm
- Slots per pole per phase $q = 5$
- Rotor slots $N_r = 22$ with an alleged power of 5 kW at 3000 rpm
- PM radial thickness: $h_{\text{PM}} = 3$ mm
- The flux barrier (PM radial) thickness $g_{\text{FB}} = h_{\text{PM}} + 0.5 = 3.5$ mm
- Flux bridge thickness below-slots $W_i = 2$ mm

- Cage bar height $h_{sr} = 6$ mm
- Cage bar width $b_{sr} = 5$ mm
- Rotor slot opening $b_{r0} = 1$ mm;
- Height of rotor teeth tip $h_{r0} = 1$ mm
- Ring cross-section a (radial) $= h_{r0} + h_{sr} + w_i/2 = 8$ mm, b (axial) $= 25$ mm
- Ring average diameter $D_{\text{ring}} \approx D_{is} - 2g - a = 80.9$ mm
- Remnant flux density $B_r = 1.15$ T
- PM recoil relative permeability $\mu_{\text{rec}} = 1.07$ p.u.
- Chorded coils: $y/\tau = 13/15$
- Phase connection: D
- PM span: $135° = \alpha_{\text{PM}}$

Calculate

a. The number of stator turns if the no-load phase (line) voltage-PM emf is $E_{\text{PM}m} = 0.9\ V_L$ (RMS).

b. All circuit parameters considering the permeability of iron infinite all throughout, except in the rotor flux bridges below the slots, where $\mu_{ri} = 10$ (p.u.).

Solution

a. $E_{\text{PM}} = 0.9 \cdot V_{fn} = 342$ V (rms)

$$\psi_{\text{PM}} = \frac{E_{\text{PM}}}{\sqrt{2}\pi f_n} = 1.5395 \text{ Wb}$$

$$k_{w1} = \frac{\sin\left(\frac{pi}{6}\right)}{q\sin\left(\frac{pi}{6q}\right)} = 0.9567$$

$$k_{w2} = \sin\left(\frac{y}{\tau}\cdot\frac{\pi}{2}\right) = 0.9781$$

$$k_w = k_{w1}\cdot k_{w2} = 0.9358$$

$$B_{ag} = B_r \frac{\frac{h_{\text{PM}}}{\mu_{\text{rec}}}}{\left(\frac{h_{\text{PM}}}{\mu_{\text{rec}}} + g_{\text{FB}} - h_{\text{PM}}\right)\cdot k_{f\,\text{ring}\,e} + g} = 0.6597 \text{ T}$$

where $k_{f\,\text{ring}\,e}$ is approximated with:

$$k_{f\,\text{ring}\,e1} = 1 + \frac{2w_i\mu_{rb}}{\alpha_{\text{PM}}\frac{\pi}{90}D_{si}}\cdot\frac{g}{b_{sr}} = 1.0415$$

$$k_{f\,\text{ring}\,e} = k_{f\,\text{ring}\,e1}\cdot\frac{D_{si}}{D_{si} - 2(g + h_{r0} + h_{sr}) - g_{\text{FB}}} = 1.3128$$

$$\Phi_{\text{PM}} = \frac{2}{\pi}B_{ag1}\cdot\tau\cdot l_{\text{stack}}\cdot 10^{-6} = 0.0063 \text{ Wb}$$

where

$$\tau = \frac{\pi D_{si}}{2p}$$

$$W_1 = \frac{\Psi_{\text{PM}}}{k_w\Phi_{\text{PM}}} = 261.7187 \text{ turns}$$

W_1 is adjusted to an integer number multiple of the product of poles and number of slots per phase and pole, so, finally, $W_1 = 260$ turns.

b. $$g_1 = g + B_r \frac{\frac{h_{PM}}{\mu_{rec}} + g_{FB} - h_{PM}}{k_{f\,ring\,e1}} \cdot \frac{D_{si} - 2(g + h_{r0} + h_{sr}) - g_{FB}}{D_{si}} = 4.5484 \text{ mm}$$

$$g_d = \frac{1}{\frac{\sin\left(\frac{\alpha_{PM}}{2}\right)}{g_1} + \frac{1 - \sin\left(\frac{\alpha_{PM}}{2}\right)}{g}} = 2.9281 \text{ mm}$$

$$g_q = \frac{1}{\frac{1 - \cos\left(\frac{\alpha_{PM}}{2}\right)}{g_1} + \frac{\cos\left(\frac{\alpha_{PM}}{2}\right)}{g}} = 1.7855 \text{ mm}$$

$$L_{dm} = \frac{6\mu_0 (N_1 k_w)^2 \tau \cdot l_{stack}}{p\pi^2 g_d} = 0.1965 \text{ H}$$

$$L_{qm} = \frac{6\mu_0 (N_1 k_w)^2 \tau \cdot l_{stack}}{p\pi^2 g_d} = 0.3223 \text{ H}$$

$$R_{bar} = \rho_{Al} \frac{l_{stack}}{h_{sr} b_{sr}} = 9.69 \cdot 10^{-5}\ \Omega$$

$$R_{ring} = \rho_{Al} \frac{2\pi D_{ring}}{N_r a_{ring} b_{ring}} = 3.7315 \cdot 10^{-6}\ \Omega$$

$$R_{be} = R_{bar} + \frac{R_{ring}}{2\sin^2\left(\frac{p\pi}{N_r}\right)} = 1.8902 \cdot 10^{-4}\ \Omega$$

$$c_w = \frac{\frac{h_{r0}}{b_{r0}} + \frac{h_{sr}}{2b_{sr}}}{\frac{\mu_{rb} w_i}{b_{sr}} + \frac{h_{r0}}{b_{r0}} + \frac{h_{sr}}{2b_{sr}}} = 0.2857$$

$$\lambda_{sr} = \frac{h_{r0}}{b_{r0}} (1 - c_w)^2 + \frac{h_{sr}}{3b_{sr}} [1 + 3c_w (c_w - 1)] = 0.6653$$

$$\lambda_{rz} = \frac{2.5g}{5b_{r0} + 2g} = 0.2254$$

$$\lambda_{e\,ring} = \frac{3D_{ring}}{2N_r l_{stack} \sin\left(\frac{p\pi}{N_r}\right)} \log\left(4.7 \frac{D_{ring}}{a_{ring} + 2b_{ring}}\right) = 0.8098$$

$$L_{be} = \mu_0 l_{stack} (\lambda_{sr} + \lambda_{rz} + \lambda_{e\,ring}) = 1.9232 \cdot 10^{-7} \text{ H}$$

$$k_i = \frac{N_r}{6N_1 k_w} = 0.0147$$

$$R_r = R_{be} \frac{N_r}{3k_i^2} = 6.3788\ \Omega$$

$$L_{re} = L_{be} \frac{N_r}{3k_i^2} = 0.0065 \text{ H}$$

2.3 SYNCHRONOUS STEADY STATE BY THE CIRCUIT MODEL

At synchronous speed $n_1 = f_1/p$, ideally, the currents induced in the rotor cage are zero. Let us mention again that PMs are "planted" in axis q of the rotor (along L_q with $L_d > L_q$).

We may safely use the dq (space phasor) model of a cageless synchronous machine with L_d, L_q, and Ψ_{PMq1} under steady state [2]:

$$\bar{i}_s R_s - \overline{V}_s = -j\omega_r \overline{\Psi}_s; \quad \overline{\Psi}_s = L_d i_d + j(L_q i_q - \Psi_{PMq1}) \tag{2.18}$$

$$T_e = \frac{3}{2} p\left[\Psi_{PMq1} \cdot i_d + (L_d - L_q) \cdot i_d \cdot i_q\right] \tag{2.19}$$

$$\bar{i}_s = i_d + j i_q; \quad \overline{V}_s = V_d + jV_q; \quad \theta_{er} = p\theta_r$$

with

$$\overline{V}_s(\bar{i}_s) = \frac{2}{3}\left(V_a(i_a) + V_b(i_b)e^{j\frac{2\pi}{3}} + V_c(i_c)e^{-j\frac{2\pi}{3}}\right)e^{-\theta_{er}} \tag{2.20}$$

and V_a, V_b, V_c, i_a, i_b, i_c as stator phase voltages and currents and θ_r—rotor d axis position; $\theta_{er} = p\theta_r =$ electrical angle; p—pole pairs; $\overline{\Psi}_s$—stator flux linkage space vector, here in rotor coordinates.

The vector diagram based on the above equations, which here includes the iron losses (as a resistive circuit in parallel at terminals V_d, V_q), is presented in Figure 2.8.

$$V_d = i_{cd} R_{core}; \quad V_q = i_{cq} R_{core}; \quad \overline{V}_s = R_{core} \bar{i}_c$$

and (2.21)

$$i_{ds} = i_d + i_{cd}; \quad i_{qs} = i_q + i_{cq}; \quad \bar{i}_{ss} = i_{ds} + j i_{qs}$$

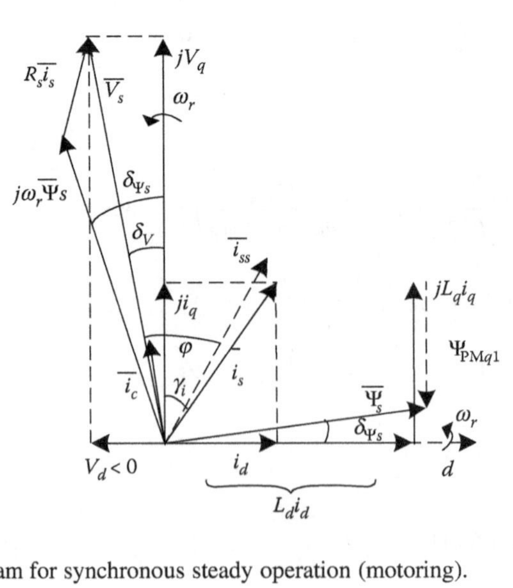

FIGURE 2.8 Vector diagram for synchronous steady operation (motoring).

A few remarks are in order:

- The assisting PMs (preferably of low cost) in axis q are producing some additional torque, but mainly they are increasing the power factor and thus, implicitly, the efficiency.
- The core loss (R_{core}), which in Equation 2.21 is related to total flux linkage (including PM contribution), leads to a decrease in efficiency, as expected, but also contributes to a slight increase in power factor (as stator winding losses do) in motor operation mode.

Equations 2.18 through 2.21 also lead to the dq-equivalent circuits in Figure 2.9.

Let us denote the power angle $\delta_{\Psi s}$ of this machine as the angle between total flux linkage $\overline{\Psi}_s$ and axis q.

Now, neglecting the core loss ($R_{core} \approx \infty$) and copper losses ($R_s = 0$), we may calculate the torque T_e versus power angle θ_Ψ characteristic, from Equations 2.18 through 2.21 as:

$$\Psi_s \sin\delta_{\Psi_s} = L_q i_q - \Psi_{PMq1}; \quad \Psi_s \cos\delta_{\Psi_s} = L_d i_d; \quad \Psi_s = \frac{V_s}{\omega_r} \tag{2.22}$$

$$(T_e)_{\substack{R_s=0 \\ R_{core}=\infty}} \approx \frac{3}{2} p \frac{V_s^2}{\omega_r^2}\left[\frac{\Psi_{PMq1}\omega_r}{V_s}\cdot\frac{\cos\delta_{\Psi_s}}{L_q} + \frac{1}{2}\left(\frac{1}{L_q}-\frac{1}{L_d}\right)\sin 2\delta_{\Psi_s}\right] \tag{2.23}$$

The problem with PMs is that there is no guarantee that the orientation of PM flux is such that its flux subtracts from $L_q i_q$ and, consequently, the PM torque component is positive in motoring. The power angle may take a value where the PM torque is negative and thus the efficiency is lower. This leads to the conclusion that for, say, rated torque to provide good efficiency, the PM flux should have such a value that only for positive PM torque is the former achieved with high efficiency (Figure 2.10) [6].

Figure 2.10 clearly shows that for no PMs does the rated torque obtained either in point A or A′ provide the same stable performance.

But for $\Psi_{PMq1} < \Psi_{PMq1}'$, there are still two stable operation points, B and B′. However, in B′, the performance (efficiency) will be lower because the PM torque is mild but negative.

On the contrary, for stronger PMs, only point C remains stable, and thus the machine, after self-synchronization, still settles for $\theta_{\Psi s}$ in point C.

This is an important design warning suggesting that the PM flux may not be too low in order to secure efficiency higher than without PMs.

The presence of losses, which are important in low-torque machines, will move points A, B, and C to the left for power, but the phenomenon remains the same.

Mathematically, we should use two more equations from the space phasor diagram:

$$\begin{aligned} &-\omega_r\Psi_s \sin\delta_\Psi + R_s\Psi_s\sin\delta_\Psi/L_d = V_d < 0; \\ &V_s^2 = V_d^2 + V_q^2\omega_r\Psi_s\cos\delta_\Psi + R_s(\Psi_s\sin\delta_\Psi + \Psi_{PMq1})/L_q = V_q \end{aligned} \tag{2.24}$$

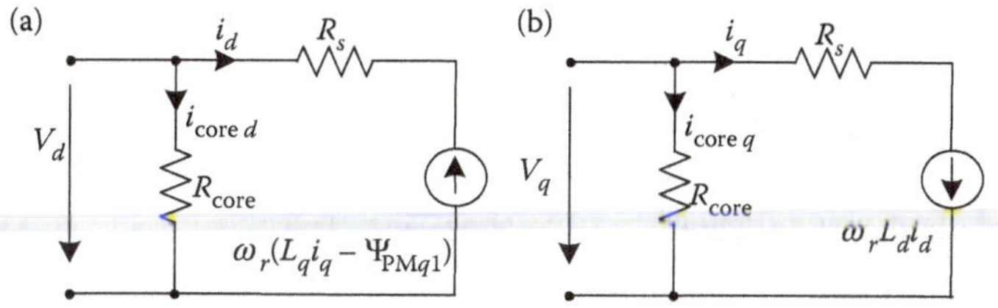

FIGURE 2.9 Steady-state equivalent circuits of PM-assisted RSM at synchronous speed: (a) in axis d; (b) in axis q.

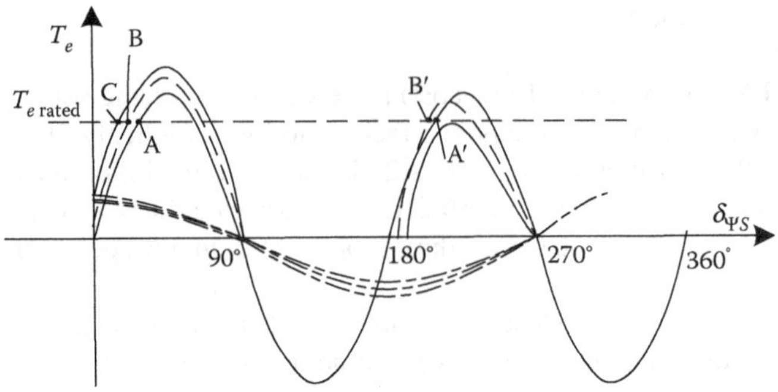

FIGURE 2.10 Torque versus power angle δ_Ψ (for zero losses). (After S. T. Boroujeni, M. Haghparast, N. Bianchi, *EPCS Journal*, vol. 43, no. 5, 2015, pp. 594–606. [6])

with

$$T_e = \frac{3}{2} \cdot p \cdot \Psi_s^2 \left[\frac{\Psi_{\text{PM}q1}}{L_q \Psi_s} \cdot \cos \delta_{\Psi_S} + \frac{1}{2} \left(\frac{1}{L_q} - \frac{1}{L_d} \right) \sin 2\delta_{\Psi_S} \right] \tag{2.25}$$

In general, the rated torque/peak torque ratio $T_{\text{rated}}/T_{\text{peak}} = 0.6–0.7$ for such PM-assisted RSMs, "to leave room" for self-stabilization after mild torque perturbations.

A simplified condition for maximum torque is obtained from Equation 2.25 as:

$$-\frac{\Psi_{\text{PM}q1}}{\Psi_s L_q} \sin \delta_{\Psi_{Sk}} + \left(\frac{1}{L_q} - \frac{1}{L_d} \right) \cdot \cos 2\delta_{\Psi_{Sk}} = 0 \tag{2.26}$$

for zero losses.

As expected, for $\Psi_{\text{PM}q1} = 0$, $\delta_{\Psi sk} \approx 45°$.

When PMs are absent, Equations 2.24 and 2.25 yield:

$$\Psi_s \left(-\omega_r \sin \delta_{\Psi_s} + \frac{R_s}{L_d} \cos \delta_{\Psi_s} \right) = V_d; \;\; \Psi_s \left(\omega_r \cos \delta_{\Psi_s} + \frac{R_s}{L_q} \sin \delta_{\Psi_s} \right) = V_q \tag{2.27}$$

$$V_s^2 = V_d^2 + V_q^2 \tag{2.28}$$

With δ_Ψ given, from Equation 2.27, Ψ_s is calculated. Then, from Equation 2.25 and $\Psi_{\text{PM}q1} = 0$:

$$T_{e\,\text{RSM}} = \frac{3}{2} p \Psi_s^2 \cdot \frac{1}{2} \left(\frac{1}{L_q} - \frac{1}{L_d} \right) \sin 2\delta_{\Psi_s} \tag{2.29}$$

Finally, from Equation 2.21, with $\Psi_{\text{PM}q1} = 0$,
i_d, i_q, and $i_s = \sqrt{i_d^2 + i_q^2}$ are determined.
The stator copper and core loss are:

$$p_{\text{copper}} = \frac{3}{2} R_s i_s^2; \;\; p_{\text{core}} = \frac{3}{2} \cdot \frac{V_s^2}{R_{\text{core}}} \tag{2.30}$$

Thus, the efficiency η may be calculated as:

$$\eta = \frac{T_e \cdot \omega_r/p}{T_e \cdot \omega_r/p + p_{\text{copper}} + p_{\text{core}} + p_{\text{mec}} + p_{\text{additional}}} \tag{2.31}$$

$$\bar{i}_{ss} = i_d + \frac{V_d}{R_{\text{core}}} + j\left(i_q + \frac{V_q}{R_{\text{core}}}\right) \tag{2.32}$$

So the power factor $\cos\varphi_1$ is:

$$\cos\varphi_1 = \frac{T_e\omega_r/(p\eta)}{\frac{3}{2}V_s i_{ss}} \tag{2.33}$$

If all losses are neglected, the ideal power factor $\cos\varphi_i$ is:

$$\cos\varphi_i = \frac{T_e\omega_r/p_1}{\frac{3}{2}V_s i_{ss}} = \frac{\left(1 - \frac{L_q}{L_d}\right)\sin 2\delta_{\Psi i}}{2\sqrt{\cos\delta_{\Psi i}^2 + \frac{L_d^2 \sin^2\delta_{\Psi i}}{L_q^2}}} \tag{2.34}$$

So, the maximum ideal power factor for the pure RSM is:

$$\cos\varphi_{i\max\text{RSM}} = \frac{1 - \frac{L_q}{L_d}}{1 + \frac{L_q}{L_d}} \tag{2.35}$$

Note: The losses in the machine improve the power factor in motoring while they worsen it in generating, but the former reduces efficiency in both cases.

As already alluded to in Chapter 1, it is feasible to produce, for the same stator, the same torque at better efficiency than in IM, but it is very difficult to have a better power factor unless L_q (RSM) $< L_{sc}$ (of IM). The latter condition apparently may be met only with ALA rotors.

EXAMPLE 2.2

For the PM-assisted RSM in Example 2.1, but with $\frac{\Psi_{PMq1}}{\Psi_s} = 0.1,\ 0.2,\ 0.3$, calculate the torque versus power angle $\delta_{\Psi 1}$ for zero losses in the machine. The stator leakage inductance is considered 10% from the main (d axis) inductance ($L_{dm} = 0.3223$ H, $L_{qm} = 0.1965$ H), and the stator resistance is $R_s = 2.7\ \Omega$.

Then, for $\frac{\Psi_{PMq1}}{\Psi_s} \approx 0.2$, calculate the steady-state machine performance (efficiency, power factor, torque, current) versus power angle δ.

Iron losses equivalent resistance $R_{\text{iron}} = 15000\ \Omega$, mechanical losses $p_{\text{mech}} = 2\%$ of ideal rated power obtained as 0.6 of peak ideal power.

Solution

$L_{s\sigma} = 0.1 * 0.3223 = 0.03223$ H

$L_d = L_{dm} + L_{s\sigma} = 0.3545$ H

$L_q = L_{qm} + L_{s\sigma} = 0.2287$ H

The ideal torque versus power angle is computed by Equation 2.23 and shown in Figure 2.11. The permanent magnet-induced voltage is:

$$E_{PM} = -\omega_1 \psi_{PM} = -76 \text{ V (rms)}$$

The currents in axes d and q are computed with following relation (reversing a 2×2 matrix is not a problem on modern computers).

$$\begin{pmatrix} I_d \\ I_q \end{pmatrix} = \begin{pmatrix} R_s & -\omega_1 L_q \\ \omega_1 L_q & R_s \end{pmatrix}^{-1} \begin{pmatrix} -V_{fn}\sin(\delta_{\psi s}) + E_{PM} \\ V_{fn}\cos(\delta_{\psi s}) \end{pmatrix}$$

The stator current components versus power angle are shown in Figure 2.12.

The core losses are not included in the previous equation, as they are computed separately and added to the input power.

$$P_{\text{iron}} = \frac{3 \cdot V_{fn}^2}{R_{\text{iron}}} = 28.88 \text{ W}$$

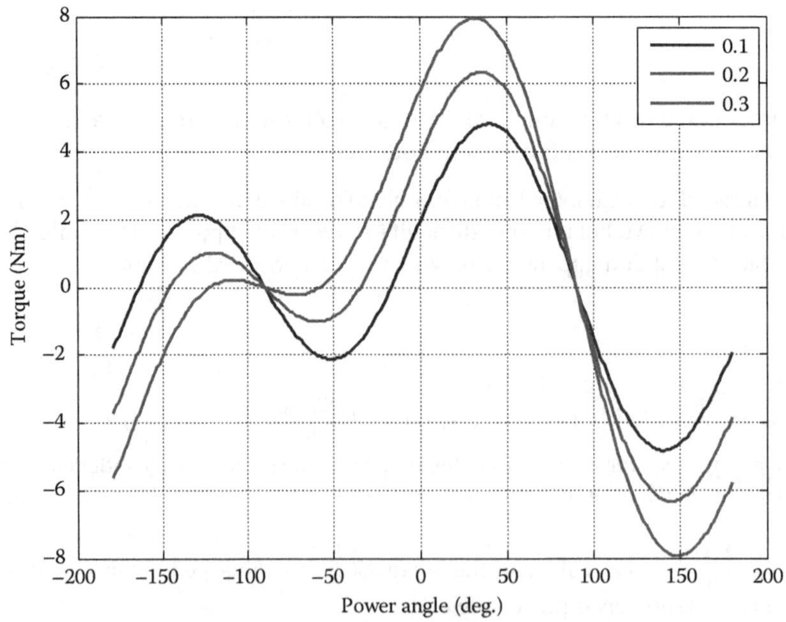

FIGURE 2.11 Ideal torque versus power angle.

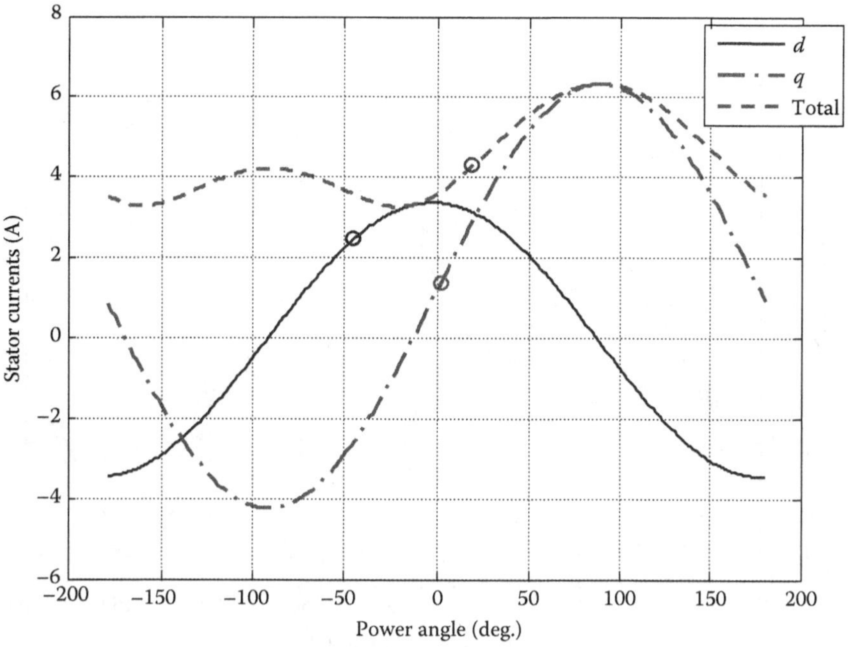

FIGURE 2.12 Stator current components versus power angle.

The complex values of the currents and voltage are:

$$\underline{I}_s = I_d + jI_q; \quad \underline{V}_s = -V_{fn} \cdot \sin(\delta_{\psi_s}) + jV_{fn} \cdot \cos(\delta_{\psi_s})$$

The input and output powers are computed:

$$P_{\text{grids}} = 3\text{Real}\left(\underline{V}_s \cdot \underline{I}_s^*\right) + P_{\text{iron}}$$

$$P_{\text{out}} = 3 \cdot \left[\omega_1 \cdot (L_d - L_q) \cdot I_q - E_{\text{PM}}\right] \cdot I_d - P_{\text{mec}}$$

The output torque is $T = \dfrac{pP_{\text{out}}}{\omega_1}$, and its dependence on the power angle in comparison with the ideal value is shown in Figure 2.13.

The efficiency is computed as the ratio between output and input power, considering that the negative values mean a generator regime, where the mechanical power means input power and the electrical power is the output power. For some angles, both mechanical and electrical powers are input powers and efficiency is considered zero, as shown in Figure 2.14.

The power factor is computed as the cosine angle between voltage and current.

The rated power angle is 2.0603 degrees; rated current values are: $I_{dn} = 3.3768$ A, $I_{qn} = 1.3747$ A, $I_n = 3.6459$ A, rated torque $T_n = 3.5015$ Nm, rated efficiency $\eta_n = 0.7552$, rated power factor 0.3435. This is only a didactic example where a PM machine is used with less PM than a PM-assisted reluctance machine, where the torque should be produced by a large difference between d and q axis inductance.

A practical PM-assisted machine has to have several flux barriers (as shown in Figure 2.2b to secure a large L_d/L_q ratio, while the topology in Figure 2.2a with results in Figures 2.11 through 2.14 has a small saliency [in this case only 1.5498]).

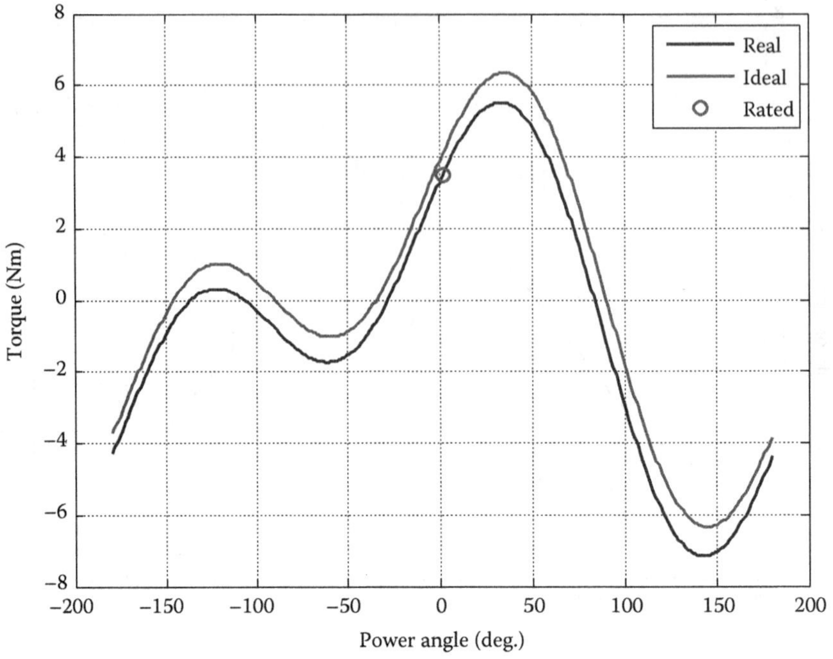

FIGURE 2.13 Output torque versus power angle.

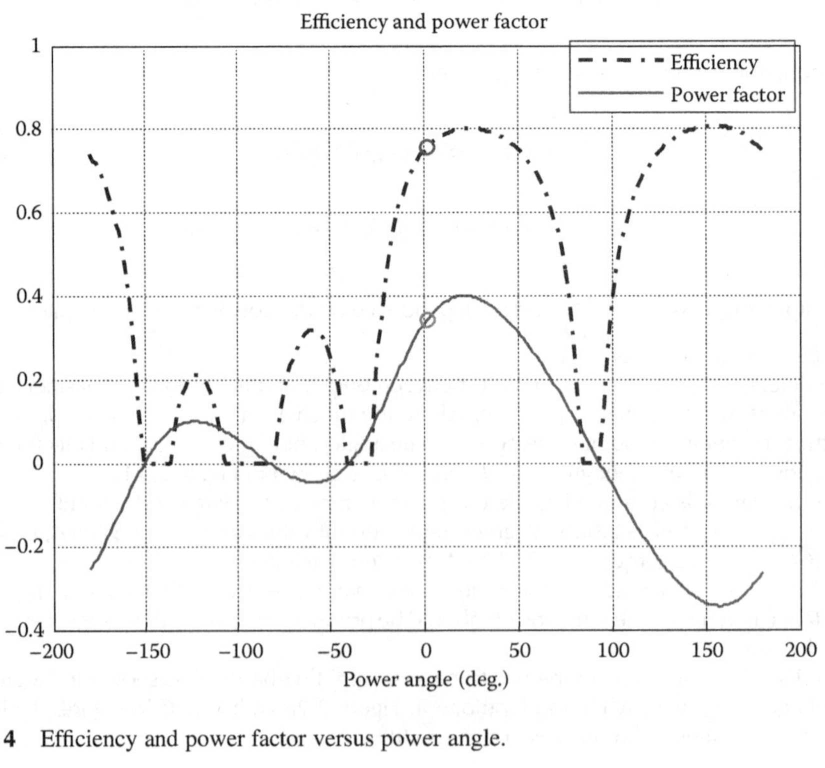

FIGURE 2.14 Efficiency and power factor versus power angle.

2.4 ASYNCHRONOUS TORQUE COMPONENTS

For asynchronous running transients and stability investigations, the dq model with cage rotor will be considered.

As the PM-RSM is investigated, the PMs will be considered in axis q ($L_q \ll L_d$), and thus the dq model (in rotor coordinates) is [4].

$$
\begin{aligned}
& V_d = R_s i_d + \frac{d\Psi_d}{dt} - \omega_r \Psi_q; \quad V_q = R_s i_q + \frac{d\Psi q}{dt} + \omega_r \Psi_d \\
& 0 = R_{dr} i_{dr} + \frac{d\Psi_{dr}}{dt}; \quad 0 = R_{qr} i_{qr} + \frac{d\Psi_{qr}}{dt} \\
& \Psi_d = (L_{dm} + L_{sl}) i_d + L_{dm} i_{dr}; \quad \Psi_q = (L_{qm} + L_{sl}) i_q - \Psi_{\mathrm{PM}q1} + L_{qm} i_{qr} \\
& \Psi_{dr} = L_{dm}(i_d + i_{dr}) + L_{drl} \cdot i_{dr}; \quad \Psi_{qr} = L_{qm}(i_q + i_{qr}) - \Psi_{\mathrm{PM}q1} + L_{qrl} \cdot i_q \\
& T_e = \frac{3}{2} p (\Psi_d i_q - \Psi_q i_d) \\
& \frac{J}{p}\frac{d\omega_r}{dt} = T_e - T_{\mathrm{load}}; \quad \frac{d\theta_{er}}{dt} = \omega_r \\
& \begin{vmatrix} V_d (or\ i_d) \\ V_q (or\ i_q) \end{vmatrix} = \frac{2}{3} \begin{bmatrix} \cos\theta_{er} & \cos\left(\theta_{er} - \frac{2\pi}{3}\right) & \cos\left(\theta_{er} + \frac{2\pi}{3}\right) \\ -\sin\theta_{er} & -\sin\left(\theta_{er} - \frac{2\pi}{3}\right) & -\sin\left(\theta_{er} + \frac{2\pi}{3}\right) \end{bmatrix} \cdot \begin{vmatrix} V_a (\text{or } i_a) \\ V_b (\text{or } i_b) \\ V_c (\text{or } i_c) \end{vmatrix} \\
& V_{a,b,c} = V_s \sin\left(\omega_1 t - (i-1)\frac{2\pi}{3}\right); \quad i = 1, 2, 3
\end{aligned}
\tag{2.36}
$$

Once the machine parameters—inductances, resistances, PM flux, and inertia J—are known, for given phase voltages V_a, V_b, V_c and given initial conditions, including initial speed and load torque versus speed, the system of Equation 2.36 may be solved numerically for a myriad of operation modes.

However, such a numerical approach would shed little light on the machine transients, say, during starting or asynchronous operation. Designing a PM-RSM with good starting, self-synchronization, and steady-state performance requires additional, intuitively obtained knowledge, as the torque has quite a few components and the magnetic saturation in the machine, both in the stator and rotor, varies notably during asynchronous acceleration and self-synchronization (L_{dm} and L_{qm} vary notably, but, L_{sl}, L_{drl}, and L_{qrl} also vary, though less).

It is rather straightforward that the machine torque during asynchronous running: $\omega_r < \omega_1$ [ω_1 is the frequency of stator voltages $V_a(t)$, $V_b(t)$, $V_c(t)$] is:

$$T_e = T_{\mathrm{PM}} + T_{\mathrm{cage}} \tag{2.37}$$

where:

$$T_{\mathrm{PM}} = \frac{3}{2} p \Psi_{\mathrm{PM}q1} \cdot i_d \tag{2.38}$$

$$
\begin{aligned}
T_{\mathrm{cage}} &= \frac{3}{2} p \left[L_{dm}(i_d + i_{dr}) i_q - L_{qm}\left(i_q + i_{qr}\right) i_d \right] \\
&= \left[\left(L_{dm} - L_{qm}\right) i_d i_q + L_{dm} i_q i_{dr} - L_{qm} i_d i_{qr} \right] \cdot \frac{3}{2} p
\end{aligned}
\tag{2.39}
$$

During asynchronous starting, the stator (and cage) currents are known to be notably larger than the rated current so that, immediately after start, the airgap flux (reflected by L_{dm}, L_{qm}) is around 50% of the rated stator flux $\left(\Psi_{\text{srated}} \approx \frac{V_1}{\omega_1}\right)$, but then it increases steadily during acceleration in asynchronous mode.

So, in fact, the leakage flux paths are saturated at start, while the main (airgap) flux paths are not, but the situation changes dramatically during asynchronous starting.

Only circuit and FEM (or time-step FEM) methods can capture these aspects with reasonable precision, although the computation time is still rather prohibitive for the complex case in point.

Even with such methods, the separation of different components of asynchronous torque is tedious, as only the frozen permeability methods may separate the real PM flux linkage, for example, in magnetic saturation conditions.

Let us consider here the cage asynchronous torque components in a PM-RSM at slip S:

$$S = 1 - \frac{n \cdot p}{f_1} \tag{2.40}$$

where n is the speed (in rps), p—pole pairs; $f_1 = \omega_1/2\pi$, the frequency of stator voltages.

2.4.1 The Cage Torque Components

The magnetic field of stator currents rotates at f_1/p, while the rotor rotates at speed n. Consequently, the stator currents i_d, i_q, with respect to the rotor, show the slip frequency $f_2 = S \cdot f_1$:

$$\begin{aligned} i_d &= i_S \sin \gamma_i \quad \gamma_i = S\omega_1 t \\ i_q &= i_S \cos \gamma_i \end{aligned} \tag{2.41}$$

γ_i (Figure 2.8) is the angle between $\bar{i}_S$ and axis q and is varying with f_2 frequency.

In a similar way, the rotor currents (in rotor coordinates) i_{dr} and i_{qr} may be written as (ω_r = const):

$$\begin{aligned} i_{dr} &= i_r \sin(\gamma_i - \Delta\gamma_i) \\ i_{qr} &= i_r \cos(\gamma_i - \Delta\gamma_i) \end{aligned} \tag{2.42}$$

Substituting Equations 2.41 and 2.42 into 2.39, the cage torque becomes (as in [3]).

$$T_{\text{cage}} = T_{\text{cage1}} + T_{\text{cage2}} \tag{2.43}$$

$$T_{\text{cage1}} = \frac{3}{2} p \cdot \frac{L_{dm} + L_{qm}}{2} \cdot i_s \cdot i_r \cdot \sin \Delta\gamma_i \tag{2.44}$$

$$T_{\text{cage2}} = \frac{3}{2} p \cdot \frac{L_{dm} - L_{qm}}{2} \cdot \left(i_s \cdot i_r \cdot \sin(2\gamma_i - \Delta\gamma_i) + i_s^2 \sin 2\gamma_i\right) \tag{2.45}$$

T_{cage1}, with $\Delta\gamma_i < 90°$, yields a positive (motoring) torque component influenced by the magnetic saturation produced periodically along axes d and q. This is an oscillating torque component at frequency Sf_1 (Figure 2.15) [5].

The PMs produce a distinct saliency that is not present either in IMs or pure reluctance synchronous machines by the variation in time of magnetic saturation, only during asynchronous operation. This PM-saturation–induced saliency creates slip frequency (Sf_1) pulsations in T_{cage1} in addition to the constant component. Both increase with machine inductances L_{dm}, L_{qm}. But T_{cage2}, in addition to the constant second term in Equation 2.45, shows torque pulsations at frequency $2 \cdot S \cdot f_1$. T_{cage2} occurs only if $L_{dm} \neq L_{qm}$, which is typical of all RSMs. In general, the amplitude of T_{cage2} is smaller than that of T_{cage1}, especially in small-saliency ($L_{dm}/L_{qm} \approx 2$), strong-magnet RSMs [5].

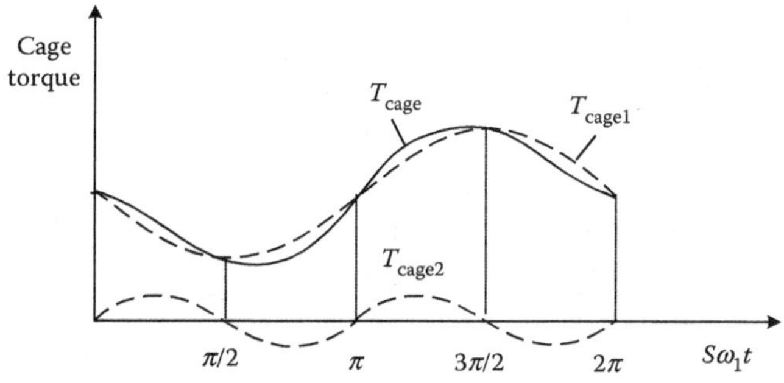

FIGURE 2.15 Cage torque components: $T_{cage1,\ cage2}$. (After A. Takahashi et al., *IEEE Trans on*, vol. EC–28, no. 4, 2013, pp. 805–814. [5])

2.4.2 Permanent Magnet Asynchronous Torque Components

In separating PM torque components, we may infer that the PM torque produces a braking torque of the rotor during asynchronous operation as the result of currents induced in the stator resistances by the traveling field at speed $n \neq f_1/p$.

At an arbitrary speed ω_r, the dq model of the PM-RSM in steady-state asynchronous mode would be:

$$\begin{aligned} V_d &= R_s i_d - \omega_r\left(L_q i_q - \Psi_{PMq1}\right); \quad E_{PM1} = -\omega_r \Psi_{PMq1} \\ V_q &= R_s i_d + \omega_r L_d i_d \end{aligned} \tag{2.46}$$

With respect to E_{PM1}, the stator voltage components V_d, V_q may be considered dephased by the angle $\gamma_v = S \cdot \omega_1 t$ in asynchronous operation mode (similar to [3]):

$$V_d = V_s \cos\gamma_V \quad V_q = V_s \sin\gamma_V \tag{2.47}$$

The PM torque is still obtained from Equation 2.38, but let us consider first the case of zero stator resistance ($R_s = 0$) and zero saliency ($L_d = L_q = L_s$).

$$(T_{PM})_{R_s=0} \approx \Psi_{PMq1} \cdot i_d \approx \frac{2}{3} p \Psi_{PMq1} \frac{V_s}{\omega_r L_s} \sin(S\omega_1 t) \tag{2.48}$$

So, for $R_s = 0$ and zero saliency, the PM torque pulsates with slip frequency but has a zero DC component.

In a similar way, for $R_s \neq 0$ [5]:

$$\begin{aligned} (T_{PM})_{R_s \neq 0} &\approx \frac{\Psi_{PMq1} \cdot V_S}{\sqrt{R_S^2 + \omega_r^2 L_S^2}} \sin(S\omega_1 t + \Phi) - \Psi_{PMq1}^2 \cdot \frac{\omega_r R_s}{R_s^2 + \omega_r^2 L_S^2} \\ \tan\Phi &= \frac{R_s}{\omega_r L_s} \end{aligned} \tag{2.49}$$

The second term in Equation 2.49 is the conventional constant, PM-produced braking torque in asynchronous operation (in fact at any speed).

But the first term in Equation 2.49 is pulsating, again, at slip frequency (as T_{cage1}).

Magnetic saturation of the rotor due to high cage currents (Figure 2.16) produces additional torque pulsations in T_{cage2} as follows [5].

When the stator current field is applied along the positive longitudinal axis, the PM torque reaches its positive maximum, but, at the same time, the main flux generated on the negative q axis is

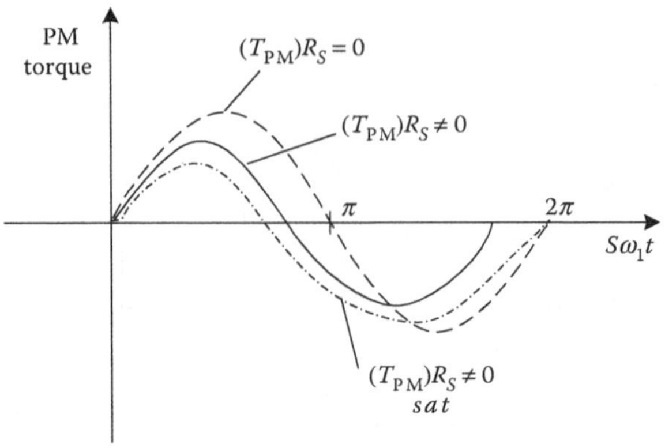

FIGURE 2.16 PM torque waveform during asynchronous operation. (After A. Takahashi et al., *IEEE Trans on*, vol. EC–28, no. 4, 2013, pp. 805–814. [5])

superimposed positively on the q axis PM flux, producing additional magnetic saturation. Thus, the positive maximum of PM torque is reduced (Figure 2.16). The opposite happens when the stator current field falls along the negative d axis, when the main flux further demagnetizes the magnets, leading to a large maximum negative PM torque [5].

Note: So this kind of magnetic saturation produces not only additional torque pulsations but a larger average PM braking torque, which has to be considered in a proper design.

Results on T_c obtained analytically (as above) and numerically by FEM are illustrated in [3].

The line-start PMSMs with strong magnets; $\Psi_{PM}/\Psi s = 0.896$ p.u.; and practically no initial saliency, $L_d \approx L_q \approx Ls$; $L_s/L_n = 0.53$ p.u., $R_s = 0.054$ p.u., $V_s = \sqrt{2} \cdot 115.5\,\text{V}$, $I_n\sqrt{2} = 14.5\sqrt{2}\,\text{A}$ are shown in Figure 2.17 [3].

These results show higher average negative values of PM torque versus speed, notably higher for all slips (but more aggressive at high slips, which may impede on motor starting if not considered in the design).

2.4.3 Reluctance Torque Benefits for Synchronization

Line-start RSMs without PMs, which have a high saliency, are known to provide for better self-synchronization; that is, to secure self-synchronization for larger load torques.

A two-pole LS-PMSM with moderate saliency ($L_{dm}/L_{qm} = 1.8$) and rather strong magnets (the configuration in Figure 2.2a) with a flux bridge between circular magnets of 13 degrees for a $\Psi_{PM}/\Psi_S = 0.896$ p.u. [5] increases the maximum load torque for safe self-synchronization from 1.25 to 1.42 p.u. The efficiency for steady-state synchronous operation in a 5-kW two-pole motor was also increased by more than 0.6% (4%–5% more than an IM), while the motor stator winding temperatures decreased by 37 K and the starting current was reduced by more than 25% [5]. This reduction of starting current, with a strong increase in efficiency, may qualify the line-start PMSM with mild saliency for high-efficiency standard motors at reasonably higher costs.

2.4.4 Asymmetric Cage May Benefit Self-Synchronization

In initial cost-sensitive line-start RSMs without or with weak (low-cost) PMs ($\Psi_{PM}/\Psi_S \approx 0.2$–$0.3$), the design of the rotor for both two and four poles tends to locate the cage in the flux barriers of the rotor (Figures 2.2b and 2.18).

In the absence of PMs (which anyway produce here much smaller torque than asynchronous torque), the average asynchronous torque may be obtained using the dq model of the machine

(a)

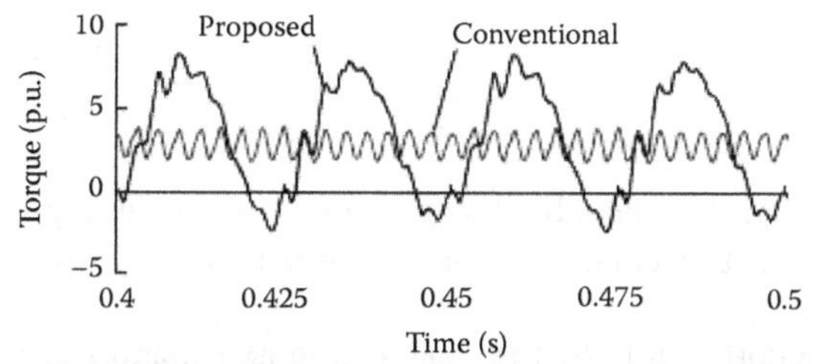

Pulsation frequency	Hz	0 (DC)	40 (sf_s)	80 ($2sf_s$)
Conventional method	p.u.	2.838	0.008	0.021
Proposed method	p.u.	3.230	4.642	0.256

(b)

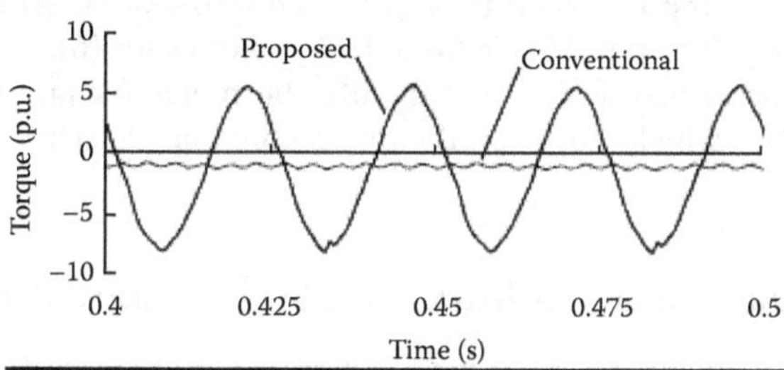

Pulsation frequency	Hz	0 (DC)	40 (sf_s)	80 ($2sf_s$)
Conventional method	p.u.	−0.964	0.003	0.001
Proposed method	p.u.	−1.356	6.673	0.271

FIGURE 2.17 Cage, (a) and PM, (b) torque components for a line-start PMSM. (After A. Takahashi et al., *IEEE Trans on*, vol. EC–30, no. 2, 2015, pp. 498–506. [3])

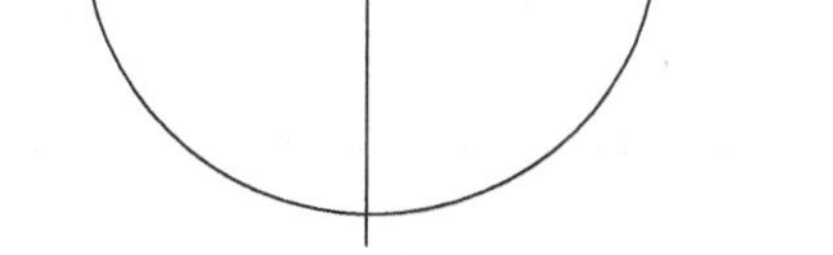

FIGURE 2.18 Four-pole asymmetric cage rotor line-start PM-RSM.

(Equation 2.36) with $d/dt = jS\omega_1$ and $\omega_r = ct$ to yield progressively [6]:

$$\underline{\Psi}_d = L_d(jS\omega_1)\cdot i_d(jS\omega_1);\;\; L_d(j\omega_1) = L_d - \frac{jS\omega_1 L_{dm}}{R_{rd} + jS\omega_1(L_{rdl} + L_{dm})}$$

$$\underline{\Psi}_q = L_q(jS\omega_1)\cdot i_q(jS\omega_1);\;\; L_q(j\omega_1) = L_q - \frac{jS\omega_1 L_{qm}}{R_{rq} + jS\omega_1(L_{rql} + L_{qm})} \quad (2.50)$$

$$\underline{\Psi}_d \approx \frac{\underline{V}_s}{\omega_1};\;\; \underline{\Psi}_q \approx -j\frac{\underline{V}_s}{\omega_1}$$

$$T_{\text{cage }av} \approx \frac{3}{2}p\frac{SV_s^2}{\omega_1}\left\{\frac{1}{R_{rd}}\cdot\left(\frac{L_{dm}}{L_d}\right)^2+\frac{1}{R_{rq}}\cdot\left(\frac{L_{qm}}{L_q}\right)^2\right\} \tag{2.51}$$

As $L_{md} \gg L_{mq}$, it is evident that the asynchronous torque in Equation 2.51, valid for small slip values (self-synchronization conditions), is increased more easily by reducing R_{rd} than by reducing R_{rq}.

But R_{rd} is defined predominantly by the cage area in axis q and R_{rq} by the cage area in axis d.

For a four-pole RSM of 10 Nm at 50 Hz and parameters as in Table 2.3 [6], the self-starting simulation with different values of rotor cage resistances R_{rd} and R_{rq} is shown in Figure 2.19 [6] to prove the beneficial influence on the process for $R_{rq} = 1.06\ \Omega$ and $R_{rd} = 1.81\ \Omega$ and high saliency ($L_{dm}/L_{qm} = 83$ mH/16 mH).

The addition of small NdFeB magnets ($\Psi_{\text{PM}q1} = 0.28$ Wb)—or bigger Ferrite magnets—in axis q has pushed the efficiency 3% up to 86.5% for a 0.77 power factor [6].

Though good, the performance does not qualify the motor for premium efficiency standards, mainly due to weak PMs, which, however, do not notably impede starting (Figure 2.20).

2.4.5 Transients and Stability of Reluctance Synchronous Machines

Considering $\delta_V > 0$, the voltage power angle (angle of axis q with voltage vector, Figure 2.8), for motoring:

$$V_d = -V\sqrt{2}\sin\delta_v \quad V_q = V\sqrt{2}\cos\delta_v \tag{2.52}$$

Now for the dq model of RSM (no PMs):

$$\frac{d\theta_{er}}{dt} \to \omega_1 - \frac{d\delta_v}{dt}; \quad \frac{d}{dt} \to s \tag{2.53}$$

where s is the Laplace operator.

From Equation 2.36 with 2.53, we first obtain $L_d(s)$ and $L_q(s)$:

$$L_d(s) = L_d\frac{(1+sT_d)}{(1+sT_0)}; \quad L_q(s) = L_q\frac{(1+sT_q)}{(1+sT_{q0})} \tag{2.54}$$

$$\Psi_d(s) = L_d(s)\cdot i_d(s); \quad \Psi_q(s) = L_q(s)\cdot i_q(s) \tag{2.55}$$

The linearized model in Equation 2.36 becomes:

$$\begin{aligned}
&-V\sqrt{2}(\cos\delta_{v0})\cdot\Delta\delta = s(\Delta\Psi_d) - \omega_1(\Delta\Psi_q) + \lambda_{q0}(s\Delta\delta_v) + R_s(\Delta i_d)\\
&-V\sqrt{2}(\sin\delta_{v0})\cdot\Delta\delta = s(\Delta\Psi_q) + \omega_1(\Delta\Psi_d) - \lambda_{d0}(s\Delta\delta_v) + R_s(\Delta i_q)\\
&\Delta\Psi_d = L_d(s)\Delta i_d\\
&\Delta\Psi_q = L_q(s)\Delta\\
&\Delta T_e + Js^2(\Delta\delta) = \Delta T_{\text{load}}\\
&\Delta T_e = \frac{3}{2}p(\Psi_{d0}\Delta i_q - \Psi_{q0}\Delta i_d + i_{q0}\Delta\Psi_d - i_{d0}\Delta\Psi_q)
\end{aligned} \tag{2.56}$$

TABLE 2.3
Geometrical Data of Considered Machine Stator

Motor poles	4
Motor stack length (mm)	80
Rotor radius (mm)	51
Stator slot number	36
Stator bore radius (mm)	51.5
Stator external radius (mm)	86.5
Stator winding turns number per phase; $g = 0.5$ mm	210

Parameters of Five Best Obtained Arc-Shape Geometries with Four FBs Per Pole; All Angles are in Degrees and α_1 is 41 For all Geometries

Geometry no.	t_1	t_2	t_3	t_4	α_2	α_3	α_4	β_1	γ_1	β_2	γ_2	β_3	γ_3	β_4	γ_4	L_d/L_q
1	6.5	5.5	7	8	30	19.5	9	53	53.12	52.54	50.72	47.45	38.43	30.90	4.01	13.35
2	6	6	4	12.5	30	22.5	13	56	56.61	56.48	55.11	54.46	51.89	45.27	2.53	13.31
3	6	5	4.5	12.5	30	22.5	13	53	53.15	52.54	50.96	49.64	46.53	39.31	2.01	13.14
4	6	5	5	10.5	30	20	11	56	56.61	56.48	55.46	53.02	48.21	41.51	2.53	13.06
5	4	5	6	12	35	25	12.5	53	53.19	53.15	52.54	50.96	47.00	38.43	2.01	12.59

Extent of the Used Variables in Arc-Shaped FBs; All Angles are in Degrees

Nr of FBs	α_1	α_2	α_3	α_4	t_1	t_2	t_3	t_4	β
1	20–43	–	–	–	7–20	–	–	–	50–60
2	15–27	30–43	–	–	4–9	7–14	–	–	50–60
3	10–19	22–31	34–43	–	4–7	4–7	7–14	–	50–60
4	9–16	19–25	28–35	37–43	4–7	4–7	4–7	7–14	50–60

Extent of Used Variables in Trapezoidal-Shaped FBs; all Angles are in Degrees

Nr of FBs	t_1	t_2	t_3	t_4	d_1	d_2	d_3	d_4	α	β
1	5–15	–	–	–	8–25	–	–	–	18–26	130–146
2	2–8	2–8	–	–	7–23	7–23	–	–	18–26	130–146
3	2–6	2–6	5–10	–	7–20	7–20	7–20	–	18–26	130–146
4	2–6	2–6	2–6	5–10	5–16	5–16	5–16	5–16	18–26	130–146

Estimated Parameters of Considered LS SynRel Motor

L_{md}	83 mH	R_s	2 Ω
L_{mq}	14 mH	R_{rd}	1.81 Ω
L_{ls}	4 mH	R_{rq}	1.06 Ω
L_{ldr}	8 mH	Vs	380 V (line–line)
L_{lqr}	4 mH	fn	50 Hz
J	$0.02\,\mathrm{kg}\cdot\mathrm{m}^2$	T_1	$11.3{\cdot}10{-}4\omega_r^2$

Source: After S. T. Boroujeni, M. Haghparast, N. Bianchi, *EPCS Journal*, vol. 43, no. 5, 2015, pp. 594–606. [6]

As instabilities have been shown experimentally to occur, especially close to no-load operation, the linearization will be operated around $i_{q0} = 0$ ($\Psi_{q0} = 0$, $\delta_{V0} \approx 0$, and i_{d0}, Ψ_{d0} nonzero, for machine magnetization):

$$i_{d0} = \frac{V\sqrt{2}}{Z_d}; \quad \Psi_{d0} = L_d i_{d0} \tag{2.57}$$

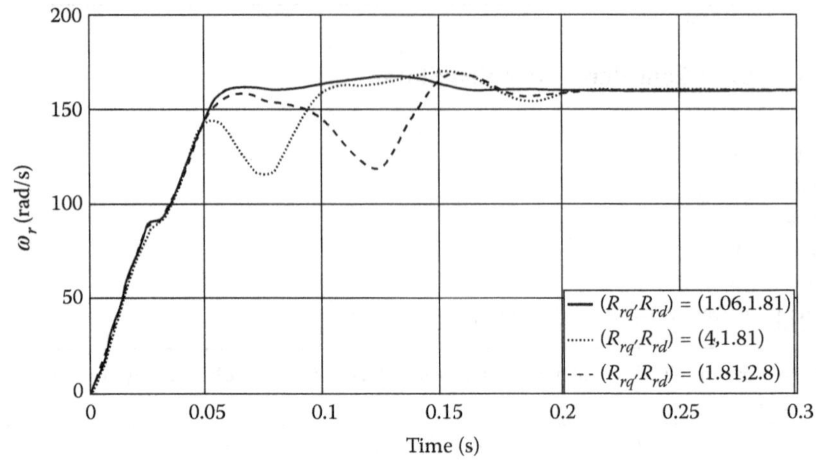

FIGURE 2.19 Self-starting speed transients for various asymmetric rotor cage resistance values. (After S. T. Boroujeni, M. Haghparast, N. Bianchi, *EPCS Journal*, vol. 43, no. 5, 2015, pp. 594–606. [6])

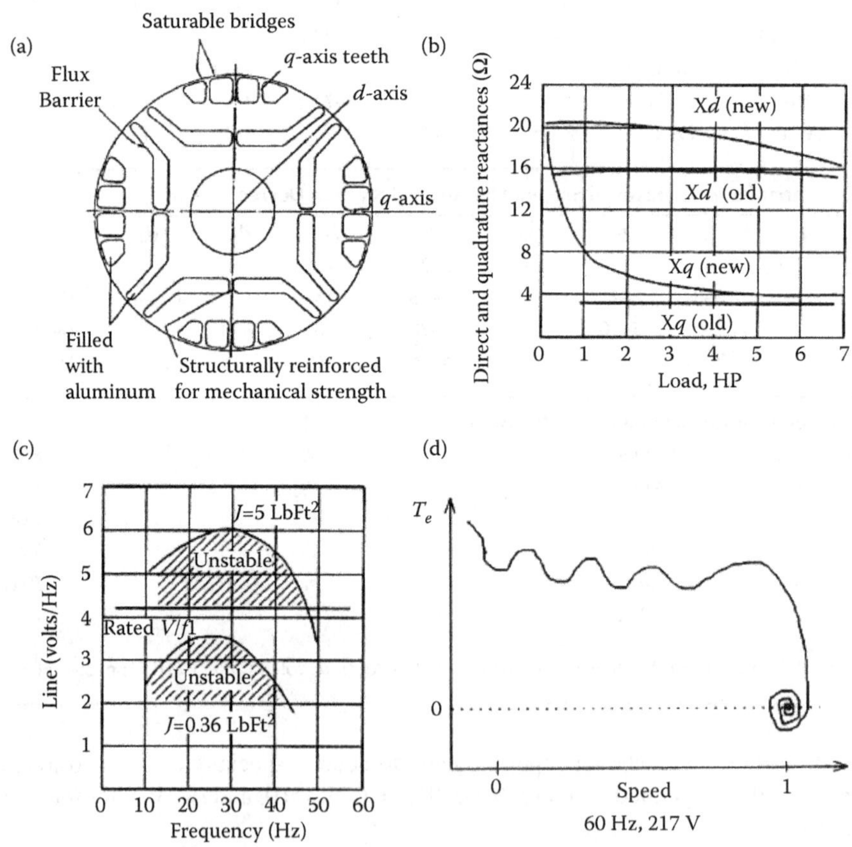

FIGURE 2.20 Inherently stable RSM transient performance: (a) the rotor topology; (b) the *d*, *q* reactances versus load; (c) stability boundaries; (d) transient torque during starting and self-synchronization.

From Equations 2.55 and 2.56, one obtains:

$$(G(s) + Ks^2)\Delta\delta_v = 0 \tag{2.58}$$

$$K = \frac{JL_q}{3p^2\left(\dfrac{V}{\omega_1}\right)^2} \cdot \left(\frac{Z_d}{\omega_1 L_q}\right)^2; \quad Z_d = \sqrt{R_s^2 + \omega_1^2 L_d^2} \tag{2.59}$$

$$G(s) = \frac{b_4 s^4 + b_3 s^3 + \cdots + b_0}{c_4 s^4 + c_3 s^3 + \cdots + c_0} \tag{2.60}$$

A high-order system is thus obtained whose characteristic Equation 2.60 has to be solved. Negative real part solutions of this equation mean stable operation, while zero real part solutions determine the margin of stability.

The results in Figure 2.20 [7] lead to conclusions such as:

- Inherently stable operation of RSMs requires notable saliency at full load (here, $L_d/L_q > 4$).
- $R_{dr} < R_{qr}$, which has been shown to add to stability, but the lower critical frequency f_c has to be kept low:

$$(f_{1c})_{\text{no load}} \approx \frac{K}{2\pi L_{mq}} \sqrt{R_{qr} R_s (1 - L_q/L_d)} \tag{2.61}$$

- Also, very low saliency at no load is required, and thus multiple flux barrier rotors with magnetic bridges that saturate notably below full load are suitable (Equation 2.61).
- The ALA rotor may allow lower values of R_{qr} for low f_{1c} but with a deep slot cage with a small skin effect around synchronous speed.
- The transient torque during acceleration shows small pulsations, which indicates the absence of a torque saddle at half-speed (due in general to cage asymmetry); also, the starting current is only 6 p.u.
- With $L_d/L_q = 4$ at full load, however, the maximum ideal power factor would be:

$$\cos\varphi_{i\max} = \frac{1 - \dfrac{L_q}{L_d}}{1 + \dfrac{L_q}{L_d}} = \frac{1 - \dfrac{1}{4}}{1 + \dfrac{1}{4}} = 0.6 \tag{2.62}$$

- This is how more sophisticated rotor topologies have to be found that mitigate starting torque, starting current with self-synchronization stability, and high steady state performance (with PM assistance).

2.5 ELECTROMAGNETIC DESIGN ISSUES

Electromagnetic design issues consist in general of:

- Machine specifications
- Machine modeling method
- Design optimization
- Optimal design (geometry) thorough validation by 3D FEM

2.5.1 Machine Specifications Are Closely Related to Application

As an example, typical specifications for an oil-pump application would be as follows:

Output power	20 kW
Speed	1000 rpm
Torque	200 Nm
Voltage	380 V
Frequency	50 Hz
Starting/rated current	<10/1
Starting/rated torque	>3/1
Rated efficiency	>0.94
Rated power factor	>0.93

The starting current is allowed to be higher than usual for IMs to provide unusually large starting torque, but it implies a dedicated—strong enough—local power grid.

The high efficiency, corroborated with large starting current and high power factor, implies the use of strong PMs (NdFeB or Sm_xCo_y), with a high $\Psi_{PM}/\Psi_S = E_{PM}/V_S > 0.9$ ratio and mild magnetic saliency, to add some reluctance torque. As expected, such specifications lead to a notably more expensive motor, but this is the price to pay for premium performance. In contrast, for 2%–4% higher efficiency but a 7%–12% lower power factor than an IM of the same stator and initial cost competitive design, for small refrigerator compressors, typical specifications would be a bit different. (Table 2.4).

The initial cost constraints impose a rotor topology with high magnetic saliency and weak PM flux $\Psi_{PM}/\Psi_s = EPM/V_s = 0.15$–$0.3$. The high magnetic saliency in the rotor may be obtained with a multiple-flux barrier topology where the rotor cage occupies the outer part. The smaller quantity of strong PMs (NdFeB or Sm_xCo_y) should be explored against low-cost Ferrite or bonded NdFeB magnets, mainly because the former are harder to demagnetize in critical conditions.

The starting current constraint is very important, as it allows the use of existing local power grids—especially in buildings—for supplying small refrigerator compressor loads at higher efficiency.

TABLE 2.4
Compressorlike Line-Start PM-RSM

Output power	1.5 kW
Speed	1500 rpm
Rated torque	10 Nm
Peak torque/rated torque	>1.6/1.0
Voltage	380 V
Frequency	50 Hz
Starting/rated current	6.5/1
Starting/rated torque	>1.5/1
Rated efficiency	>0.86
Rated power factor	0.77
Initial cost/IM cost	<1.25/1.00

2.5.2 Machine Modeling Methods

Models of RSMs starting with crude analytical approaches, as presented previously in this chapter, to calculate constant equivalent circuits allow the dq model to calculate the machine performance very approximately. Such models may be used for preliminary sizing of the machine, as done so far, and then more advanced field models for optimal design (sizing) around the preliminary (initial) design geometry can be used.

The magnetic equivalent circuit model [8,2,9] is used to account for magnetic saturation, slotting, and rotor flux barriers in the rotor by equivalent magnetic permeances of the latter.

However, for line-start RSMs, with and without PM assistance, this method becomes quite involved, and its computation time saving diminishes notably. It may be wiser to use FEM modeling of the machine directly (at least for steady state: asynchronous and synchronous modes), with only a few FEM runs per electrical period, to construct the sinusoidal mmf magnetic field distribution by regression methods and then calculate the machine circuit parameters by FEM, accounting for variable local magnetic saturation, slotting, flux barriers, PMs, and so on [10].

FEM, with limited computation time, has been proposed recently from the start for optimal design, though the total computation time may be over 50 hours, even with today's multicore desktop computers [11,12]. For line-start RSMs without or with PMs, with mild or strong magnetic saliency rotors, the direct use of FEM for optimal design is yet to be done.

2.5.3 Optimal Design Algorithms

Optimal design algorithms for electromagnetic devices is today a science in itself.

Deterministic and evolutionary methods, from the Monte Carlo method to the Hooke–Jeeves method, to particle swarm optimization (PSO), genetic algorithms (GAs), differential evolution (DE), ant colony, bee colony, imperialistic algorithms, and so on [2] have been proposed for electric machine optimal design.

They all have merits and demerits related to convergence time to a global (not local) optimum design. Some of them have been used against benchmarks to show comparative performance and have established "hierarchies" [10–12]. But from those referring to RSMs, most have been applied to cageless RSMs with or without PM assistance that are controlled by PWM converters for variable-speed drives.

For line-start RSMs, more basic (artlike) design algorithms have been applied so far, given the complexity and conflicts between starting, self-synchronization, and steady-state synchronous operation performance.

To illustrate this situation, we will show here sample results from representative recent literature case studies.

2.5.4 Sample Design Results for Oil-Pump-Like Applications

Traditionally, IMs have been used in most oil-pump applications, but recently, line-start PMSMs have been considered, too, in addition to variable speed drives which are only taking off marketwise.

The reason to replace IMs, especially when put in a limited-diameter pipe deep in the ground, is better efficiency. But the starting and self-synchronization performance also have to be kept close to those of the IM.

The specifications mentioned in Table 2.3 have been followed in [13] by investigating three different, rather rugged, rotor topologies (Figure 2.21).

For each rotor configuration and given stator (327-mm outer diameter, 115-mm inner diameter), with 54 slots and 6 poles (with $q_1 = 3$ slots/pole/phase, $y/\tau = 8/9$ chorded coils distributed winding) the emf, cogging torque, starting current and torque, and synchronous performance for variations in only a few variables, such as aluminum ring thickness (rotor a), PM span angle (rotor a), spoke magnet height (rotor b), stator, and rotor stack length, have been investigated heuristically to see the

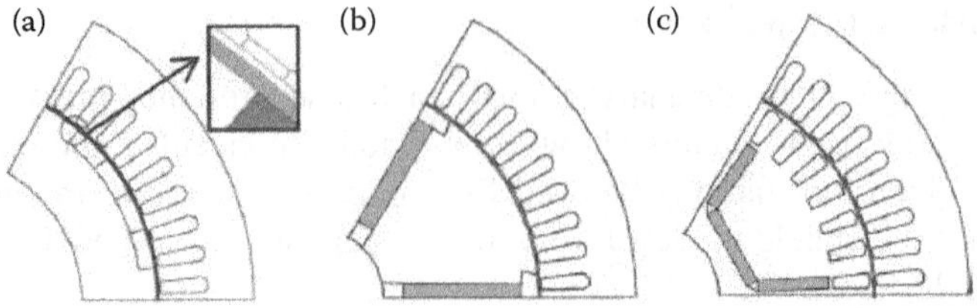

FIGURE 2.21 Three rotor configurations. (After A. Takahashi et al., *IEEE Trans on*, Vol. EC–30, No. 2, 2015, pp. 498–506. [3]): (a) with SPM and aluminum ring (very small magnetic saliency); (b) with spoke-type magnets and solid iron rotor; (c) with U-shaped PMs, regular cage and laminated core.

TABLE 2.5
Performance of LS-PM-RSMS, with Rotor 1, 2, 3 (Figure 2.22)

Parameters	Motor (a)	Motor (b)	Motor (c)
Efficiency	0.85	0.94	0.95
Power factor	0.86	0.99	0.99
Starting/rated current	9.86	9.78	9.98
Starting/rated torque	8.2	6.1	4.4

Source: After T. Ding et al., *IEEE Trans on*, vol. Mag.–45, no. 3, 2009, pp. 1816–1823. [13]

influence on performance. The investigation was performed by time-step FEM, which involves considerable computation time. The results are synthesized in Table 2.5 and Figure 2.22 [13].

While all topologies fulfill the starting current and torque strong demands, only motors (b) and (c) show the desired efficiency.

The best compromise between starting, steady-state, and self-synchronization performance is shown by motor c) (Figure 2.22), which is costlier. All three rotors use high-energy PMs whose cost is high. Another response surface-based methodology optimum design of a two-pole LS-PMSM with mild saliency and strong magnets led to an efficiency above 92% and an 83% power factor [14].

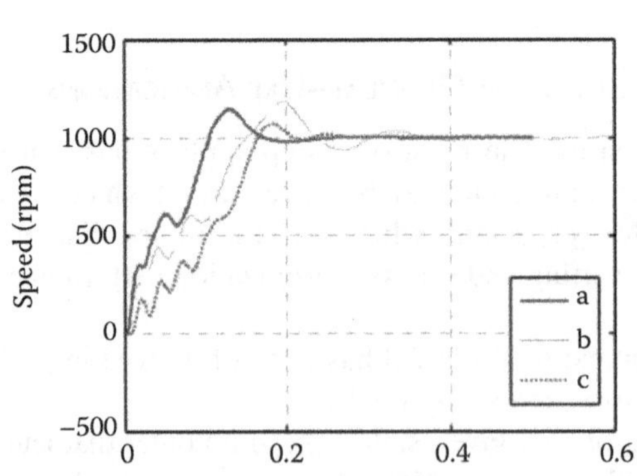

FIGURE 2.22 Speed time responses of the three LS-PMSMs. (After T. Ding et al., *IEEE Trans on*, vol. Mag.–45, no. 3, 2009, pp.1816–1823. [13])

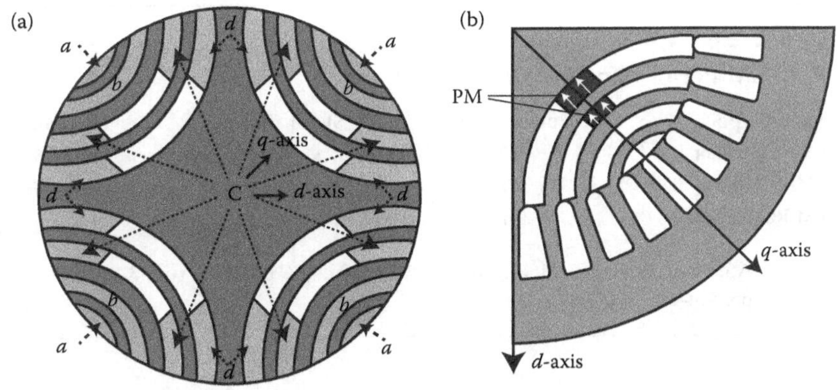

FIGURE 2.23 Rotor with multiple-flux-barrier and embedded cage: (a) no PMs; (b) with PMs. (After S. T. Boroujeni, M. Haghparast, N. Bianchi, *EPCS Journal*, vol. 43(5), 2015, pp. 594–606. [6])

2.5.5 Sample Design Results for a Small Compressorlike Application

As already implied, small compressorlike applications require high efficiency but are also very cost sensitive, while the starting is not very costly.

A thorough FEM-based investigation [6] for a motor with the specifications in Table 2.4 and the rotor as in Figure 2.23 was conducted.

For the PM-assisted RSM (Figure 2.23b), three PM flux linkage levels have been considered under the title motor i: motor 1 ($\Psi_{PM} = 0$), motor 2 ($\Psi_{PM} = 0.1$ Wb), and motor 3 ($\Psi_{PM} = 0.28$ Wb).

The calculated torque angle characteristics (Figure 2.24) [6] show that motors 1 and 3 guarantee stable operation at around the highest peak torque, where the PM flux is in the negative direction of axis q to produce positive (motoring) PM torque and thus better efficiency and power factor (operation in points A and B, but not C).

Results related to power factor, efficiency, and stator-rated current are summarized in Table 2.6.

The above results suggest that a four-pole LS-RSM at 10 Nm at 50 Hz has an efficiency better than existing IMs, but the power factor is still notably lower.

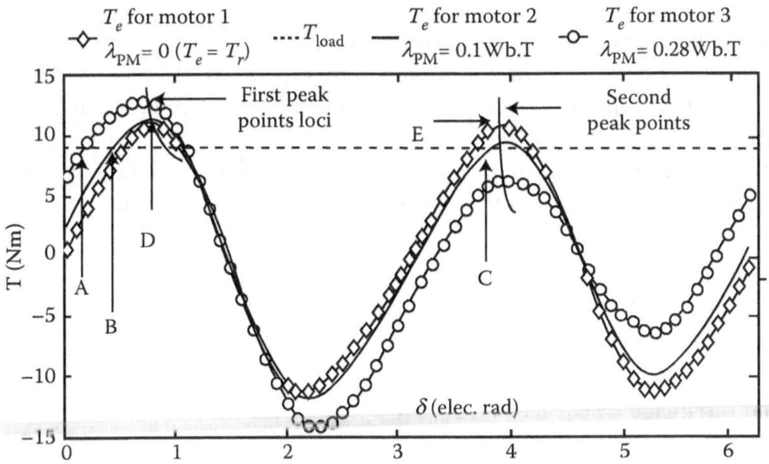

FIGURE 2.24 Torque angle characteristics for motors 1, 2, 3. (After S. T. Boroujeni, M. Haghparast, N. Bianchi, *EPCS Journal*, Taylor and Francis, New York, vol. 43(5), 2015, pp. 594–606. [6])

TABLE 2.6
PM-RSM Performance-Motors 1, 2, 3

Motor Type	Power Factor	Stator Current	Efficiency
LS-RSM-motor 1	0.71	10.70	83.80%
PM-RSM-motor 2	0.71	10.70	83.80%
PM-RSM-motor 3	0.77	9.5	86.50%

Source: After S. T. Boroujeni, M. Haghparast, N. Bianchi, *EPCS Journal*, vol. 43, no. 5, 2015, pp. 594–606. [6]

Moderately weak PM flux and reasonably large magnetic saliency: $L_d/L_q \approx 5.0$ (for $\cos\varphi = 0.71$ without PMs, but with losses considered), produce a 3% gain in efficiency of motor 3 (with weak but larger PMs). This is clearly a good justification to use PMs, but the starting and self-synchronization performance also have to be checked.

As super-high efficiency line-start motors should exhibit more than 90% efficiency at about 10 Nm and 1500 rpm (1.5 kW), more work is to be done to make the lower-cost LS-RSM with low-cost PMs suitable for the purpose.

Finally, in this section, only glimpses of the electromagnetic design of LS-PM-RSMs have been offered, while even a complete electromagnetic design, not to mention mechanical and thermal design, is beyond our scope here.

2.6 TESTING FOR PERFORMANCE AND PARAMETERS

Testing is indispensable in manufacturing any device, particularly an electric machine. The complex phenomena in LS-PM-RSMs, due to the simultaneous presence of distributed magnetic saliency, an aluminum cage, and, eventually, PMs, makes the testing of these machines even more suitable to get reliable information on performance and on circuit parameters. By "performance," we mean here:

- Synchronous steady-state-power angle, current, efficiency, power factor, efficiency versus torque (or output power) up to the maximum available.
- Asynchronous performance: current, torque versus speed (slip).
- Self-synchronization performance: speed versus time for acceleration for critical load torque application.
- Some of the discussed testing sequences for performance are included in the standards for line-start IMs but not for line-start PM-RSM machines yet, to the best of our knowledge.

Parameter estimation through testing is required to verify the various modeling (design) methodologies and theoretically investigate various transients and stability issues.

The parameters include:

- Stator and rotor cage resistances R_s, R_{dr}, R_{qr}
- Stator and rotor leakage inductances L_{sl}, L_{rdl}, L_{rql}
- Magnetization inductances $L_{dm}(i_d, i_q)$, $L_{qm}(i_d, i_q)$
- PM flux linkage Ψ_{PMq}
- Inertia J

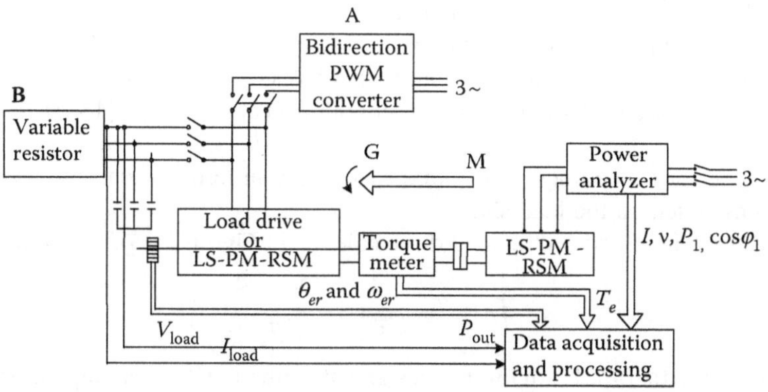

FIGURE 2.25 Testing platform for performance measurement: A: with load drive and bidirectional PWM converter for regenerative loading of LS-PM-RSM; B: back-to-back setup (M + G) with resistive load.

2.6.1 Testing for Synchronous Performance

The testing platform should contain a load drive or a back-to-back system (Figure 2.25).

For the back-to-back system, two identical LS-PM-RSMs are required; then, by measuring the output power P_{out} of the second identical machine, the efficiency may be directly (though approximately) calculated as:

$$\eta = \sqrt{\frac{P_{out}}{P_1}}; \quad P_{out} = 3 \cdot V_{load} \cdot i_{load} \tag{2.63}$$

For regenerative loading, adequate-for-temperature evolution investigation for various loads, a drive controlled in torque mode through a bidirectional PWM AC-DC, DC converter is required.

In this case, however, the drive torque has to be either measured or estimated (as in direct torque control (DTC) AC drives).

The rotor position is required for acquiring the load angle as the phase difference with the stator phase voltage or current to assess voltage or current power angles γ_V, γ_i.

Typical experimental results from synchronous performance testing are shown in Figure 2.26 [14].

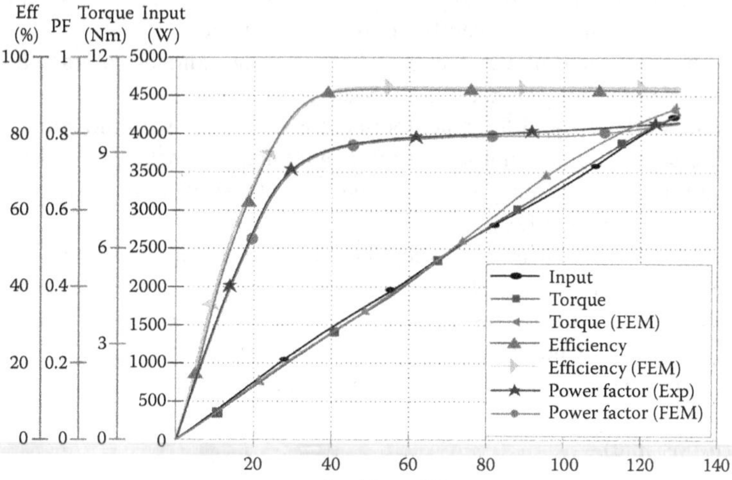

FIGURE 2.26 Typical steady-state synchronous performance experimental results from LS-PM-RSM. (After S. Saha, G. D. Choi, Y.-H. Cho, *IEEE Trans.*, vol. MAG–51, no. 11, 2015, pp. 8113104. [14])

The free acceleration and deceleration method introduced to offer both performance (efficiency) and all parameters, including inertia, for IMs may be adapted for LS-PM-RSMs as well.

Also, the virtual (synthetic) loading method, with a PWM converter to control the LS-PM-RSM alternatively as motor and generator (developed for both IMs and PMSMs [15–18]), may be adapted for the scope.

Asynchronous performance may be acquired easily only in terms of current, voltage, and speed (if an encoder is mounted on the load shaft).

If a torque-meter is mounted on the load drive side (as in Figure 2.25), the motor torque T_e is:

$$T_e = T_{\substack{\text{measured}\\ \text{load}}} + J_t \frac{d\omega_r}{dt} \tag{2.64}$$

where J_t is the total inertia of all moving masses and ω_r is the mechanical angular speed.

Observing the speed derivative from the encoder output is not a trivial task.

Self-synchronization tests for various loads may be performed with the platform in Figure 2.25, provided, again, the speed is calculated through a digital filter from the encoder (resolver) output.

2.6.2 Parameters from Tests

Typical methods to estimate all machine inductances and resistances are the flux decay and frequency response standstill tests, now standardized for IMs and cage-rotor SMs.

The free deceleration test may be used to measure the PM flux linkage Ψ_{PMq1} via no load voltage and its frequency measurement.

Reference 10 describes two methods, the single-phase standstill method and the dq model method, to calculate all resistances and inductances of LS-PM-RSMs from 2D FEM time-step calculations in DC and AC tests, considering the cage currents. Sample results [10], shown in Figure 2.27, should be similar to those obtained from standstill flux decay (L_{dm}, L_{qm}) and frequency tests [19].

2.7 SUMMARY

- Three-phase line-start (LS) reluctance synchronous motors, with an increased power factor (though with 2%–5% more efficiency) that is still lower than that of IMs with identical stators, have entered the markets recently in order to save energy.
- To increase efficiency and power, factor higher magnetic saliency rotors with multiple flux barriers per half-pole (3–4) have been introduced.
- Where premium efficiency [4,5] is the dominant goal, a rotor with moderate magnetic saliency $L_{dm}/L_{qm} = 1.6$–1.8 and high-energy PMs with $\Psi_{PM}/\Psi_S \approx 0.9$, placed behind the rotor cage, is used for two- and four-pole motors with good results. Even powers per unit of 250 kW have been investigated [20].
- To surpass the efficiency of IMs but keep the motor costs only up to 20% higher, multiple flux-barrier rotors with an embedded asymmetric rotor cage and weak low-cost PMs are used to produce high saliency and a better power factor, respectively.
- The presence of a rotor cage on the rotor has led to the exclusion of tooth-wound coil windings in the stator, where, additionally, to obtain higher magnetic saliency, distributed AC windings are used exclusively, with q (slots/pole/phase) ≥ 3 and high pole pitch/airgap: 150–200 (if possible).
- Reconciliation of reasonable starting/rating current and torque, self-synchronization (full acceleration time for a given maximum load torque [20]), and superior synchronous steady-state efficiency and power factor, all for a reasonable initial cost (materials plus fabrication), requires involved modeling and design tools.

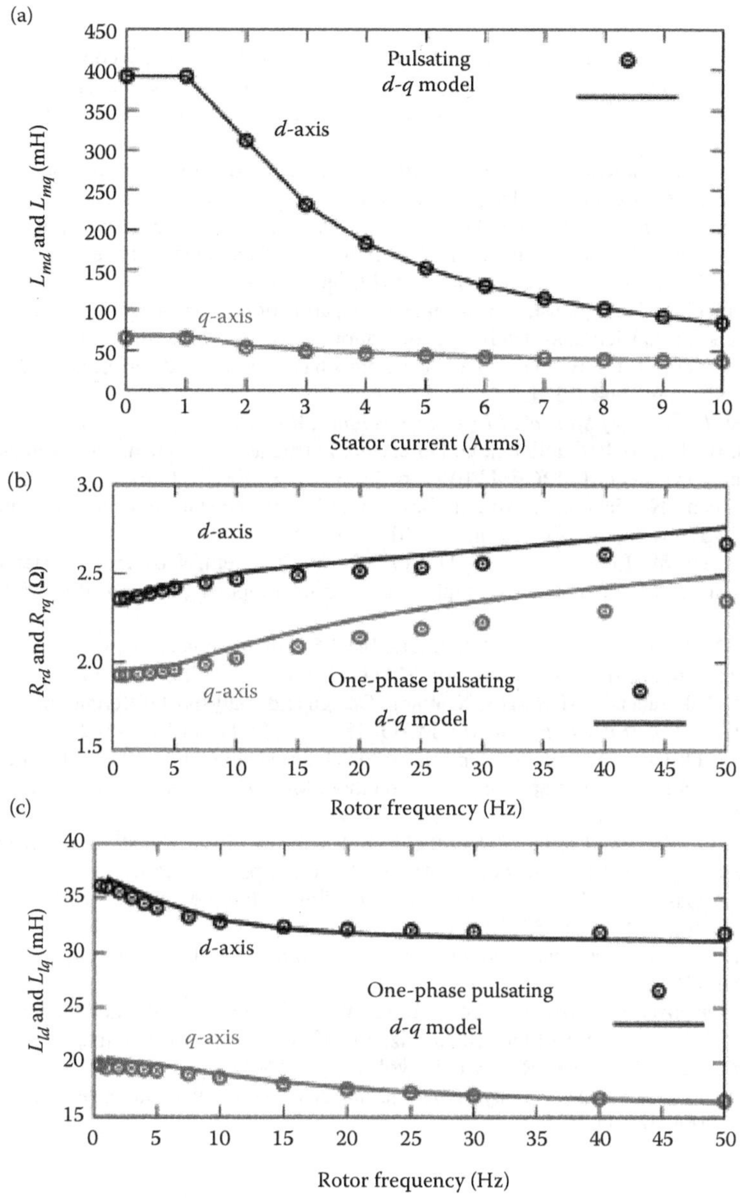

FIGURE 2.27 LS-RSM parameters from FEM (pulsating [one-phase] and dq [three-phase] model). (After S. T. Boroujeni, N. Bianchi, L. Alberti, *IEEE Trans on*, vol. EC–26, no. 1, 2011, pp. 1–8. [10]): (a) magnetization inductances L_{dm}, L_{qm}; (b) rotor resistances R_{rd}, R_{rq}; (c) rotor leakage inductances L_{rdl}, L_{rql}.

- They imply analytical, a magnetic equivalent circuit, and FEM modeling and design optimization techniques, not to mention structural and thermal design, which showed recently that LS-PM-RSMs are capable of running 31 K cooler for the same stator, torque, and speed with stator windings, a clear indication of higher efficiency, and so on.
- As premium line-start motors with reasonable starting current (6.2–6.8 p.u.) and torque (1.5–2 p.u.) are badly needed to provide for energy savings while keeping the existing local power grid ratings, LS-PM-RSMs are expected to remain a focus of R&D and industrial communities worldwide.

REFERENCES

1. A. T. Almeida, F. T. J. E. Ferreira, A. Q. Duarte, Technical and economical considerations on super high-efficiency three-phase motors, *IEEE Trans on*, vol. IA-50, no. 2, 2014, pp. 1274–1285.
2. I. Boldea, L. N. Tutelea, *Electric Machines*, book, CRC Press, Taylor and Francis, New York, 2011 (Part 2).
3. A. Takahashi, S. Kikuchi, K. Miyata, A. Binder, Asynchronous torque in line-starting PMSMs, *IEEE Trans on*, vol. EC–30, no. 2, 2015, pp. 498–506.
4. I. Boldea, *Reluctance Synchronous Machines and Drives*, book, Oxford University Press, 1996.
5. A. Takahashi, S. Kikuchi, K. Miyata, K. Ide, A. Binder, Reluctance torque utility in line-starting PM motors, *IEEE Trans on*, vol. EC–28, no. 4, 2013, pp. 805–814.
6. S. T. Boroujeni, M. Haghparast, N. Bianchi, Optimization of flux-barriers of line start synchronous reluctance motors for transient and steady-state operation, *EPCS Journal*, vol. 43, no. 5, 2015, pp. 594–606.
7. V. B. Honsinger, Inherently stable reluctance motors having improved performance, *IEEE Trans on*, vol. PAS–91, 1977, pp. 1544–1554.
8. V. Ostovic, *Dynamic of Saturated Electric Machines*, book, John Wiley, 1989.
9. A. Vagati, B. Boazzo, P. Guglielmi, G. Pellegrino, Ferrite assisted synchronous reluctance machines: a general approach, Record of ICEM 2012, pp. 1313–1319 (IEEEXplore).
10. S. T. Boroujeni, N. Bianchi, L. Alberti, Fast estimation of line start reluctance machine parameters by FEM, *IEEE Trans on*, vol. EC–26, no. 1, 2011, pp. 1–8.
11. A. Fatemi, D. M. Ionel, N. A. O. Demerdash, T. W. Nehl, Fast multi-objective CMODE-type optimization of PM machines using multicore desktop computers, *Record of IEEE-ECCE—2015*, pp. 5593–5600.
12. Y. Wang, D. M. Ionel, D. D. Dorrell, S. Stretz, Establishing the power factor limitation for synchronous reluctance machines, *IEEE Trans on*, vol. MAG–51, no. 11, 2015, pp. 8111704.
13. T. Ding, N. Takorabet, F.-M. Sargos, X. Wang, Design and analysis of different line-start PMSMs for oil pump applications, *IEEE Trans on*, vol. MAG–45, no. 3, 2009, pp.1816–1823.
14. S. Saha, G. D. Choi, Y.-H. Cho, Optimal rotor shape design of LSPM with efficiency and power factor improvement using response surface methodology, *IEEE Trans*, vol. MAG–51, no. 11, 2015, pp. 8113104.
15. L. Tutelea, I. Boldea, E. Ritchie, P. Sandholdt, F. Blaabjerg, Thermal testing for inverter fed IMs using mixed frequency method, *Record of ICEM—1998*, vol. 1, pp. 248–253.
16. A. Y. M. Abbas, J. E. Fletcher, The synthetic loading technique applied to PMSM, *IEEE Trans on*, vol. EC–26, no. 1, 2011, pp. 83–92.
17. I. Boldea, S. A. Nasar, *Induction Machine Design Handbook*, book, 2nd edition, CRC Press, Taylor and Francis, New York, 2010.
18. I. Boldea, *Synchronous Generators*, book, 2nd edition, CRC Press, Taylor and Francis, New York, 2015.
19. J. Soulard, H. P. Nee, Study of the synchronization of line-start permanent magnet synchronous motors, *Record of IEEE—IAS—2000*, Rome, Italy, vol. 1, pp. 424–431.
20. Q. F. Lu, Y. Y. Ye, Design and analysis of large capacity line start PM motor, *IEEE Trans on*, vol. MAG–44, no. 11, 2008, pp. 4417–4420.

3 Phase-Source Line-Start Cage Rotor Permanent Magnet–Reluctance Synchronous Machines
Modeling, Performance and Design

3.1 INTRODUCTION

Home (residential) appliances traditionally make use of the local AC one-phase power grid of standardized voltage (120 or 220 V and frequency 50 or 60 Hz). Also, traditionally, most home appliances (refrigerator compressors, washing machines, home heater water pumps, ventilators, etc.) use line-start cage rotor induction motor drives. These have a main and auxiliary stator winding to produce an airgap magnetic flux density with a strong (ideally pure) traveling wave character. A symmetric cage rotor is used (Figure 3.1a).

By using optimal design and increasing motor weight (and cost) at 100 W, two poles, and 50 Hz, such split-phase capacitor induction motors have reached an efficiency of 85%.

Efficiency, crucial in the cost of losses, is important countrywide or worldwide, as there are so many such motors (a few in most houses).

Introducing variable-speed drives for many/most home appliances is a way to decrease total losses in the drive, if and only if the drive efficiency $\eta_{drive} = \eta_{motor}{}^{*}\eta_{converter}$ is larger than the efficiency of the constant-speed line-start motor, unless, by variable speed, the working machine (appliance)—compressor—efficiency itself increases more, and then the initial cost of the variable-speed drive (motor + PWM converter) has to be compared with the line-start motor alone.

As making $\eta_{drive} > \eta_{line\ start\ motor}$ is not an easy task, still better efficiency line-start motors, despite their high starting current and problems of acceleration and self-synchronization, in the form of cage–rotor reluctance synchronous motor (without and with PMs), are investigated.

As for three-phase RSMs (Chapter 2), when higher efficiency is a critical goal, high-energy (NdFeB) magnets that produce $E_{PM}/V_1 \approx 0.9$ and a low magnetic saliency ($L_{qm}/L_{dm} = 1.6$–1.8) rotor represent the solution (Figure 3.1b). When initial cost and higher efficiency have to be investigated, high-saliency anisotropic rotors ($L_{qm}/L_{dm} > 3$–4) with weak assisting PMs ($E_{PM}/V_1 \approx 0.3$–0.4) seem the way to go (Figure 3.1c).

For the sake of generality, the spatial angle ξ of the main and auxiliary windings is different from 90°. Though it was shown that $\xi = 100° - 110°$ may help in asynchronous starting of IM and reducing the PM braking torque during asynchronous starting of PM-RSMs, in general, $\xi = 90°$ in almost all such commercial motors.

The windings of all three configurations in Figure 3.1 are similar, and to reduce additional losses in the rotor cage and stator and rotor cores, the mmf space harmonics of the stator windings are reduced

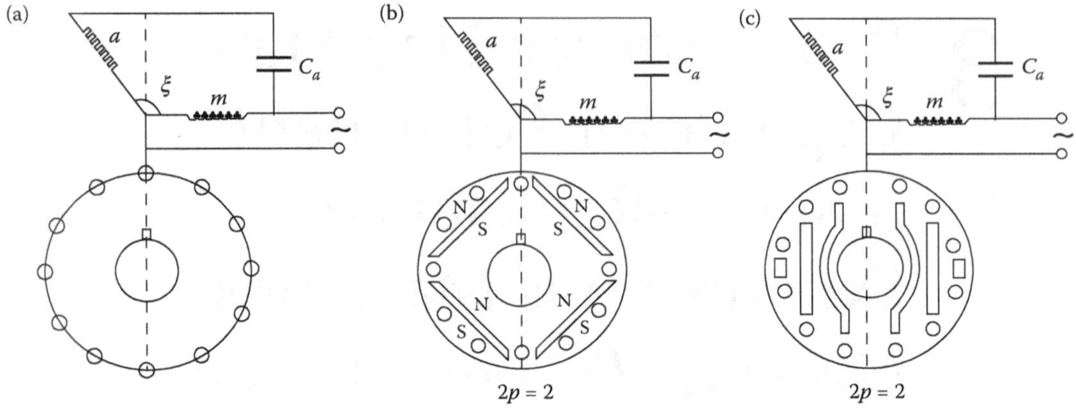

FIGURE 3.1 Line-start 1-phase source AC two-pole motors: (a) IM; (b) strong PM reluctance synchronous motor; (c) weak PM-RSM.

by using coils of different numbers of turns per coil for the main and auxiliary winding in various slots.

A typical such distribution per half a pole for two poles (24 slots in all) is:

$$\begin{aligned} W_{1\text{main}} &= \text{Integer}\big([123 \quad 168 \quad 95 \quad 55 \quad 24 \quad 0] \times K_{\text{main}}\big) \\ W_{1\text{aux}} &= \text{Integer}\big([0 \quad 0 \quad 45 \quad 81 \quad 73 \quad 63] \times K_{\text{aux}}\big) \end{aligned} \tag{3.1}$$

Coefficients K_{main} and K_{aux} may be varied from 1.0 as required in the design, but maintaining the same (small) harmonics content. The distribution of mmf for the main and auxiliary winding for a motor, as shown in Figure 3.2a and b, shows space harmonics contents as in Figure 3.2a and b,

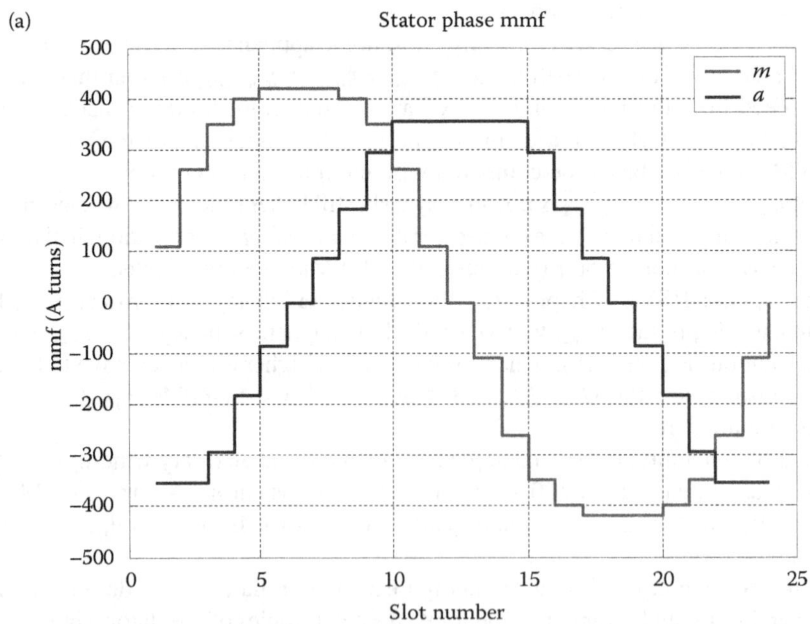

FIGURE 3.2 Stator winding mmf distribution (a). *(Continued)*

(b)

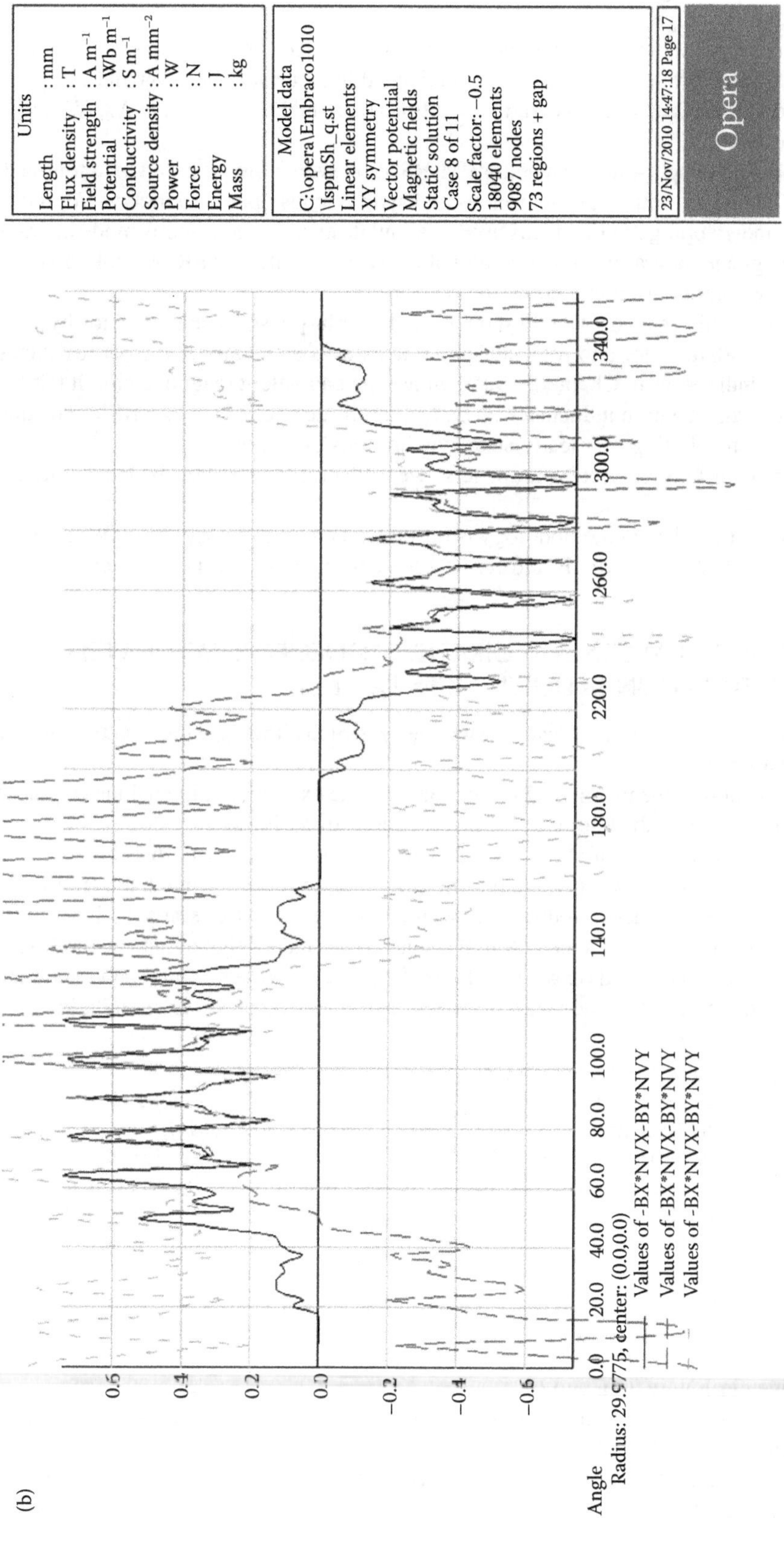

FIGURE 3.2 (Continued) space harmonics content (b).

illustrating a rather strong fundamental with small space harmonics to guarantee low additional losses in the rotor core and cage.

The slot openings on both the rotor and stator will introduce additional space harmonics in the airgap flux density and torque pulsations. To complicate things further, the two windings produce a pure traveling field in the airgap only in certain conditions (a certain turn ratio $a = N_a/N_m$ and close to the most frequent load).

So the modeling of split-phase capacitor PM-RSMs is even more involved than for the three-phase PM-RSMs (Chapter 2), both in asynchronous and synchronous operation modes. For the three-phase PM-RSM, a time-stepping 2D FEM was used to simulate even asynchronous mode operation considering local magnetic saturation, but for the split-phase capacitor PM-RSM, this goal is still to be accomplished within a reasonable computation effort.

This is why, in this chapter, we will model first the split-phase capacitor motor by analytical formulae for the standard stator and rotor leakage inductances and stator and rotor resistances of main and auxiliary windings, as in Chapter 2, but then use the equivalent magnetic circuit (EMC) model to investigate the magnetization inductances L_{dm}, L_{qm} for the dq axes of the PM-RSM, considering magnetic saturation and slotting for steady-state synchronous operation.

Further on, with known circuit parameters, we will investigate asynchronous and synchronous performance.

Finally, an optimal design methodology is introduced on a case study at 100 W, 50 Hz, 3000 rpm, two poles, and 220 V, with NdFeB and ferrite PMs and medium magnetic saliency.

3.2 EQUIVALENT MAGNETIC CIRCUIT MODEL FOR SATURATED MAGNETIZATION INDUCTANCES L_{dm}, L_{qm}

Let us start with the description of machine geometry, which allows the introduction of main geometric (design) variables.

The cross-section of the machine (Figure 3.3a) and the axial main dimensions (Figure 3.3b) illustrate a two flux-barrier/half-pole rotor with embedded strong PMs.

Additional geometry data of the rotor and stator slots is offered in Figure 3.4.

The magnetic circuit model [1] is simplified here by considering the stator and rotor magnetic saturation separately in order to reduce the computation effort. Also, sinusoidal airgap flux density distribution is considered when calculating the stator yoke and tooth magnetic saturation. The airgap flux density is calculated based on given voltage, frequency, number of turns, and geometry. Typical MATLAB® code to calculate the stator magnetic saturation factor K_{ss} is given below:

```
fi1=Vn*sqrt(2)*kvm/(w1*Nm*kmwp);                    % polar flux
Bmag1=fi1/(lstack*tau)*1e6*pi/2;                    % airgap flux density
Bst=Bmag1*tauSslot/wst;                              % stator tooth flux density
Bsy=fi1/(2*shy*lstack)*1e6;                          % rotor yoke flux density
Hmst=getHx(Bref,Href,Bst,sw3/wst,ssf);              % stator magnetic field strength
Hmsy=getHx(Bref,Href,Bsy*cos(pi*(0:ns1-1)/(2*ns1)),mean(shl(1,:))/shy,ssf)
mmf=[kC*hag*Bmag1/mu_0 shlmax*Hmst lmsy*mean(Hmsy)]*1e-3;
                                                     % stator mmf
kss=sum(mmf)/mmf(1);                                 % stator saturation factor
```

The magnetic equivalent circuits for rotor axis d (along PMs—minimum inductance) and along axis q are given in Figure 3.5a and b.

The rotor magnetic saturation is computed iteratively, with its initial value calculated considering the already-determined stator flux level.

As visible in Figure 3.5a, the R_{mrMs} reluctance (of the rotor bridge above the rotor slots) is much larger than the rotor tooth reluctance $R_{\mathrm{mrMs}}(1)$ in parallel with $R_{\mathrm{mrb}}(1)$ and $R_{\mathrm{mrMs}}(2)$. Consequently,

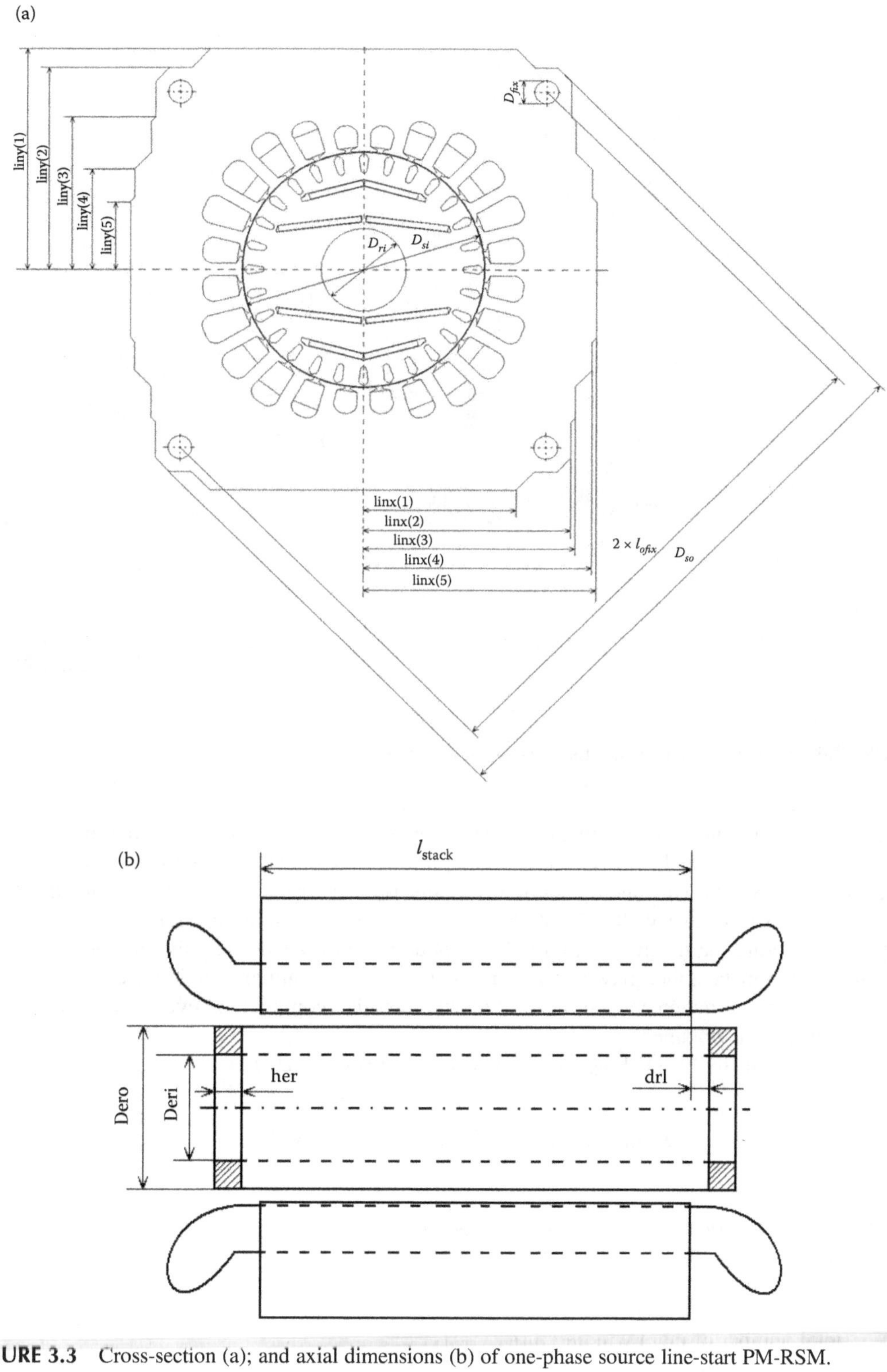

FIGURE 3.3 Cross-section (a); and axial dimensions (b) of one-phase source line-start PM-RSM.

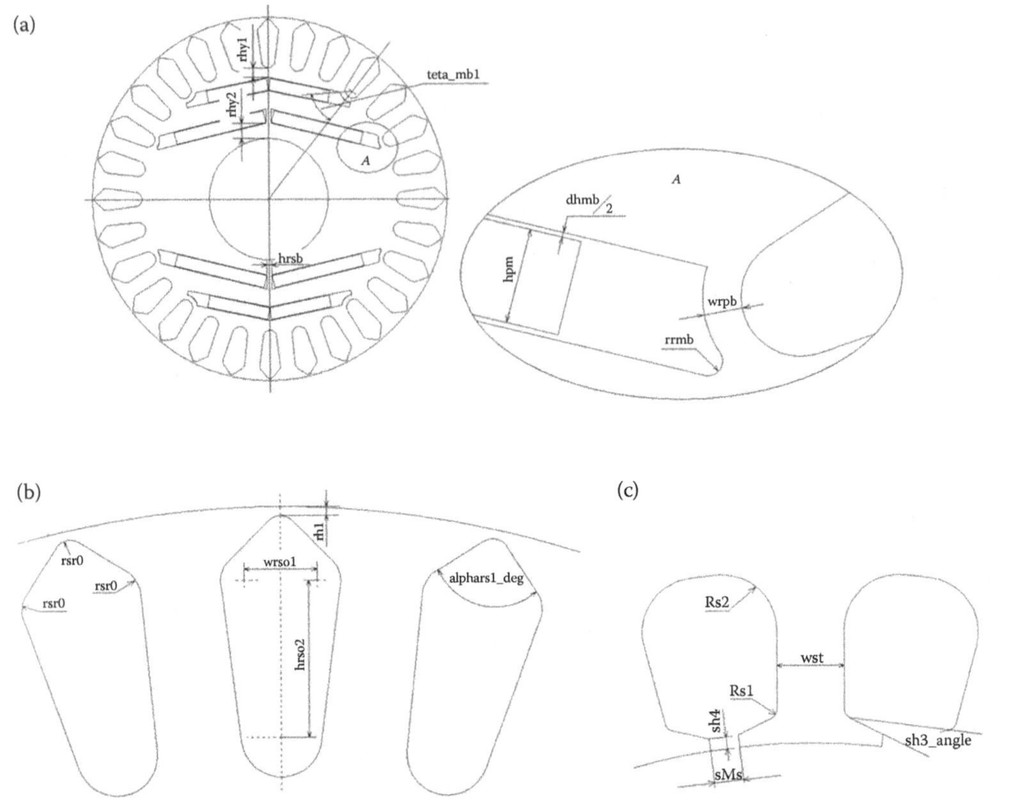

FIGURE 3.4 Flux barrier data (a); rotor slots (b); stator slots (c).

the number of nodes with nonzero magnetic scalar potential is reduced from eight to four to cut down the computation time. The PM flux is computed in the absence of stator currents. As good efficiency means positive PM and reluctance synchronous torque ($i_d < 0$ for $L_{dm} < L_{qm}$), the stator currents are not expected to spectacularly alter the stator's already-assumed saturation level.

After calculating the magnetic flux in all magnetic circuit branches and the flux densities in all rotor and iron branches, then having compared them with their initial (previous) values and using over-relaxation to introduce new permeability values, another computation cycle is done until sufficient convergence is obtained.

Finally, the emf (by PMs) E_{1PM} in the main winding is calculated as:

$$E_{1PM} = \omega_1 B_{agPM1} \cdot \frac{2}{\pi} \cdot tau \times l_{stack} \times (N_m K_{wme}) \tag{3.2}$$

B_{agPM1}—*PM-produced airgap flux density fundamental*
ω_1—stator voltage angular frequency
tau—pole pitch
l_{stack}—stator core stack length
N_m—total number of main winding turns
K_{wme}—equivalent main winding factor for the fundamental mmf (considering coils with different numbers of turns)

The d axis magnetization inductance L_{dm} is calculated first at the PM level of magnetization.

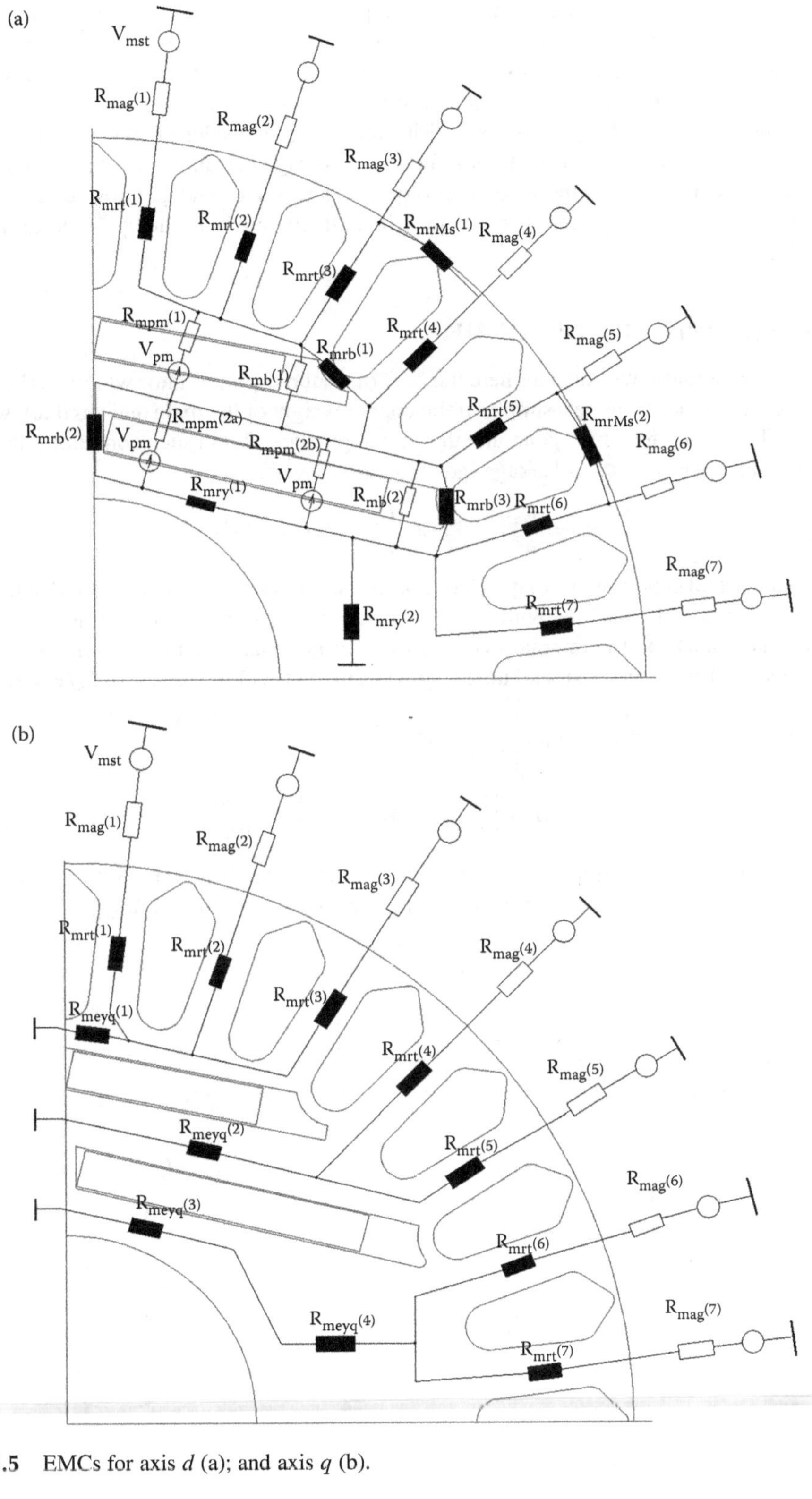

FIGURE 3.5 EMCs for axis *d* (a); and axis *q* (b).

A similar procedure has to be used in axis q, starting with a given current value I_q in the main winding. The q axis flux is now calculated in the presence of PM flux and thus the magnetic permeability of the core is calculated for the resultant flux density: $\sqrt{B^2_{mq\_old} + B^2_{PM}}$. Other than that, the computation procedure in axis q is the same as in axis d, and finally, the value of L_{qm} is calculated.

Having calculated L_{dm}, L_{qm}, E_{PM} from EMC modeling, considering the magnetic saturation approximately, but in synchronous operation mode—zero cage currents—and having the analytical expressions (as in Chapter 2) for the stator and rotor resistances and leakage inductances (R_s, R_{rd}, R_{rq}, L_{rdl}, L_{rql}), the electric circuit of the machine may be defined. We may thus proceed to develop the circuit model.

3.3 THE ELECTRIC CIRCUIT MODEL

For the sake of generality, we consider here the case of nonorthogonal stator windings (Figure 3.1b). Also, electric circuit modeling gets simpler if the copper weight of the main and auxiliary windings is the same [3]. This is not, however, generally the case in practice. But in such conditions, the main and auxiliary winding resistances and leakage inductances are related as:

$$R_m = \frac{R_a}{a^2}; \quad L_{ml} = \frac{L_{al}}{a^2}; \quad a = \frac{N_a k_{wa}}{N_m k_{wm}} \tag{3.3}$$

To solve the actual situation, we may externalize the stator resistances and leakage inductances by switching to an "internal" machine with V'_m, V'_a voltages instead of V_m and V_a (Figure 3.6).

Now we may transform the internal machine to an orthogonal winding machine (α, β), with the actual number of turns in the main and auxiliary winding, but with R_m, R_a, L_{ml}, L_{al} externalized [2]:

$$\begin{vmatrix} V'_m \\ V'_a \end{vmatrix} = M_{\nu 02p} \begin{vmatrix} V_\alpha \\ V_\beta \end{vmatrix}; \quad \left|M_{\nu 02p}\right| = \begin{vmatrix} 1 & 0 \\ -a\cos\xi & a\sin\xi \end{vmatrix} \tag{3.4}$$

$$\begin{vmatrix} i_\alpha \\ i_\beta \end{vmatrix} = M_{ip20} \begin{vmatrix} i_m \\ i_a \end{vmatrix}; \quad \left|M_{ip20}\right| = \begin{vmatrix} 1 & a\cos\xi \\ 1 & a\sin\xi \end{vmatrix} \tag{3.5}$$

The orthogonal (α, β) machine does not run in symmetric conditions and thus we may apply the direct and inverse [1(d), 2(i)] decomposition by the transformation matrix M_{di20}:

$$\left|M_{di20}\right| = \frac{1}{\sqrt{2}} \begin{vmatrix} 1 & 1 \\ j & -j \end{vmatrix} \tag{3.6}$$

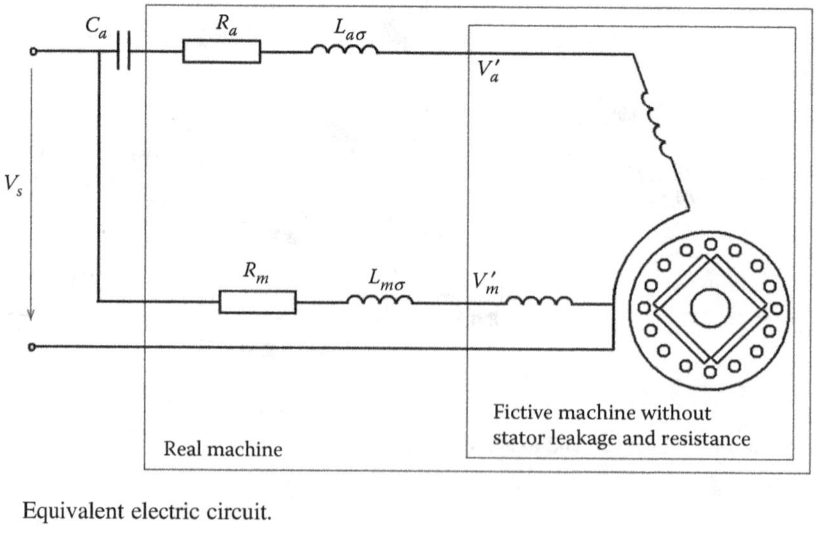

FIGURE 3.6 Equivalent electric circuit.

3.4 ASYNCHRONOUS MODE CIRCUIT MODEL

The direct and inverse impedances of the machine in axis d and axis q, valid in a synchronous operation mode [S (slip) $\neq 1$], $\underline{Z}_{1d}$, $\underline{Z}_{2d}$, $\underline{Z}_{1q}$, $\underline{Z}_{2q}$ are defined as below:

```
Z1d= (j*xdm*Rrd./slip-xrds*xdm)./(Rrd./slip+ j*(xrds+xdm));
Z1q= (j*xqm*Rrq./slip-xrqs*xqm)./(Rrq./slip+j*(xrqs+xqm));
Z1d= Z1d*Rfe./(Z1d+Rfe);          % add iron losses
Z1q= Z1q*Rfe./(Z1q+Rfe);          % add iron losses
Z1= (Z1d+Z1q)/2;
Z2d= (j*xdm*Rrd./(2-slip)-xrds*xdm)./(Rrd./(2-slip)+j*(xrds+xdm));
Z2q= (j*xqm*Rrq./(2-slip)-xrqs*xqm)./(Rrq./(2-slip)+j*(xrqs+xqm));
Z2= (Z2d+Z2q)/2;
```

We may now arrange the direct and inverse (d, i) component equations in matrix form:

$$|\underline{V}_{d,i}| = |\underline{Z}_{di}||\underline{I}_{di}|; \quad \underline{Z}_{di} = \begin{vmatrix} \underline{Z}_1 & 0 \\ 0 & \underline{Z}_2 \end{vmatrix} \tag{3.7}$$

Using the above transform matrixes $M_{\gamma d12p}$ and M_{ip2i} yields:

$$\begin{vmatrix} V'_m \\ V'_d \end{vmatrix} = M_{\nu di2p}|Z_{di}||M_{ip2i}| \cdot \begin{vmatrix} I_m \\ I_a \end{vmatrix} = |\underline{Z}_0| \begin{vmatrix} \underline{I}_m \\ \underline{I}_a \end{vmatrix} \tag{3.8}$$

Adding the “external circuit”: $R_{\text{sm}} + jX_{\text{sml}}$, $R_{\text{sa}} + jX_{\text{sal}}$, one obtains:

$$\begin{vmatrix} \underline{V}_s \\ \underline{V}_s \end{vmatrix} = \left| |\underline{Z}_0| + \begin{vmatrix} R_{\text{sm}} + jX_{\text{sml}} & 0 \\ 0 & R_{\text{sa}} + jX_{\text{cal}} - jX_{\text{cap}} \end{vmatrix} \right| \cdot \begin{vmatrix} \underline{I}_m \\ \underline{I}_a \end{vmatrix} = |Z_p| \begin{vmatrix} \underline{I}_m \\ \underline{I}_a \end{vmatrix} \tag{3.9}$$

and

$$\begin{vmatrix} \underline{I}_m \\ \underline{I}_a \end{vmatrix} = |Z_p^{-1}| \cdot \begin{vmatrix} \underline{V}_s \\ \underline{V}_s \end{vmatrix} \tag{3.10}$$

Given the values of parameters and slip frequency ($S\omega_1$), the above electric circuit model produces the stator currents I_m and I_a and, on the way, the direct and inverse asynchronous torque average components:

$$\begin{aligned} T_{\text{eas}} &= T_{\text{das}} + T_{\text{ias}} \\ T_{\text{das}} &\approx \text{Re}\left[\underline{Z}_1 |\underline{I}_d^2|\right] p/\omega_1 \\ T_{\text{ias}} &\approx \text{Re}\left[\underline{Z}_2 |\underline{I}_i^2|\right] p/\omega_1 \end{aligned} \tag{3.11}$$

3.5 PERMANENT MAGNET AVERAGE BRAKING TORQUE

A simplified expression of average PM braking torque—due to stator current components $i_{d\text{PM}}$, $i_{q\text{PM}}$ induced in the stator (considered short-circuited) by the PM flux by motion at frequency ω_r [4]—may be obtained rather easily, as it corresponds to the PM-induced stator winding current Joule losses.

$$\begin{aligned}
T_{bPMav} &= p\sin\xi\left[\frac{1}{a}\Psi_{dPM}\cdot i_{qPM} - \Psi_{qPM}\cdot i_{dPM}\right] \\
i_{dPM} &= \frac{-(1-S)^2(X_q - X_c)}{R_s^2 + X_d(X_q - X_c)(1-S)^2}E_{PM};\quad E_{PM} = \omega_r\Psi_{PM}; \\
i_{qPM} &= \frac{-(1-S)R_s}{R_s^2 + X_d(X_q - X_c)(1-S)^2}E_{PM}; \\
\Psi_{dPM} &= \frac{X_d I_{dPM} + E_0}{\omega_r};\quad \Psi_{qPM} = \frac{(X_q - X_c)I_{qPM}}{\omega_r}
\end{aligned} \tag{3.12}$$

Note: As investigated in Chapter 2 [5], magnetic saturation due to interaction of PM flux–stator and rotor cage currents produces, during asynchronous running, additional average and pulsating braking PM torque even for three-phase PM-RSMs, and more so for split-phase capacitor PM-RSMs where the stator currents $i_m(t)$, $i_a(t)$ are inherently asymmetric. However, due to the utter complexity, only a computation time–consuming time-step FEM seems the way to go on this issue, other than experimental effort. For small motors, the experimental way may be less expensive and faster in producing fully reliable results.

Typical results on asynchronous running of a 100-W, 50-Hz, 3000-rpm, 220-V line-start PM-RSM with dual flux barriers per pole rotor (Figure 3.3), based on the above-described method, are given in what follows (Figures 3.7 and 3.8).

The asynchronous performance illustrates the good starting properties (above 0.6 Nm starting torque [$S=1$]) and sufficient (rated) torque, for self-synchronization at small slip ($S=0.04$).

Note: The circuit parameters for the machine performance in Figures 3.7 and 3.8 are:

$L_{md}=0.511$ H, $L_{mq}=1.532$ H, $L_{ml}=0.09$ H, $L_{al}=0.0477$ H, $R_m=23.29\ \Omega$, $R_a=5.03\ \Omega$, $a=0.8429$, $\Psi_{PM}=1.147$ Wb; running capacitor $C_r=4\ \mu$F (starting capacitor $C_s=8C_a$)

The main geometry is characterized by:

$D_{is}=57.3$ mm; inner stator diameter
$L_{stack}=44.7$ mm, stator stack length
$h_{PM}=1.4$ mm, PM height (NeFeB)
$D_{ro}=56.8$ mm, rotor outer diameter
$D_{so}=134$ mm, stator outer diameter

Now, using the EMC model described above, the synchronous performance of the same motor (with the given geometry, total number of turns in the two phases [Equation 3.1], and wire diameters) is calculated.

3.6 STEADY STATE SYNCHRONOUS PERFORMANCE/SAMPLE RESULTS

Following the data of the machine above, the EMC model for extracting L_{dm} and L_{qm} was used to calculate the latter for steady-state synchronous operation using the standard equations [2] where only the PM torque and reluctance torque components are present. This time, performance is calculated versus rotor power angle.

The main and auxiliary winding currents I_m, I_a (Figure 3.9a and b) and their direct and inverse components (I_1 and I_2) (Figure 3.9c) show that the machine operates in most cases away from symmetrical conditions ($I_2=0$).

Voltages and powers are illustrated in Figure 3.10, where negative power angles are to be considered for motoring.

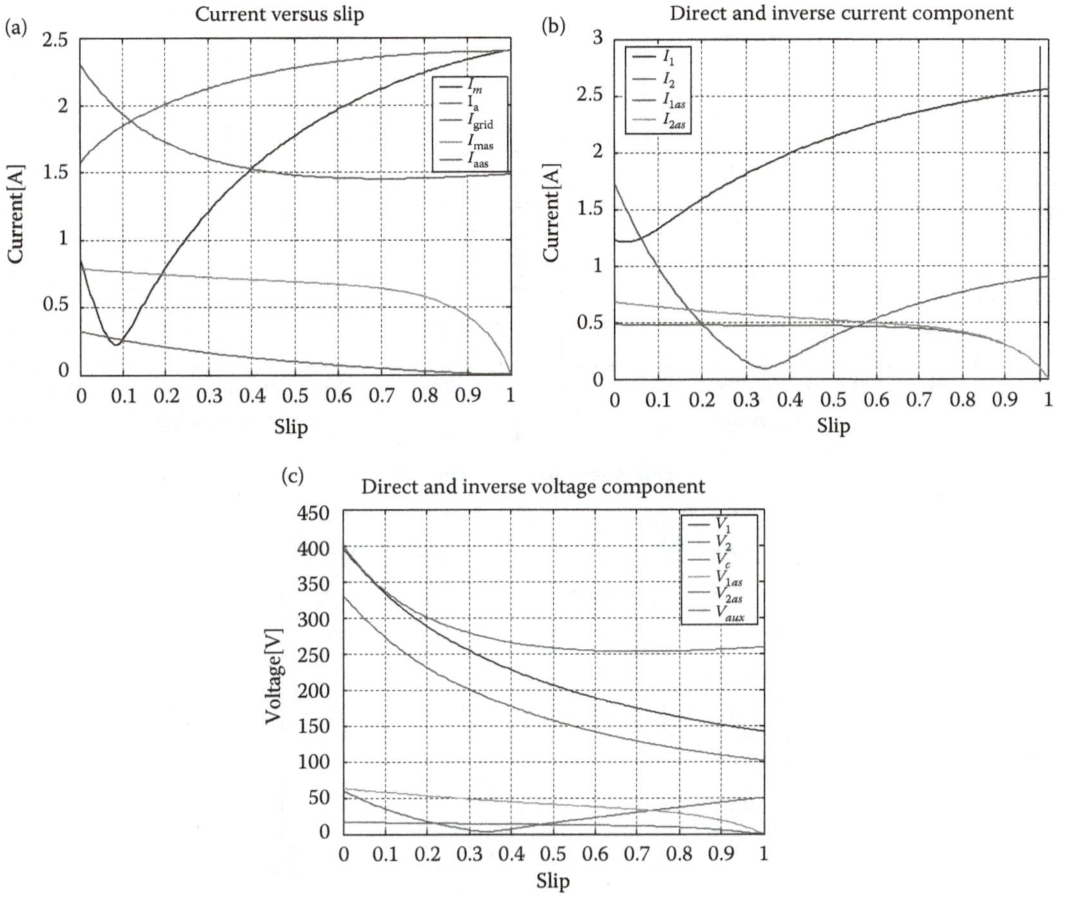

FIGURE 3.7 Asynchronous operation: (a) stator currents: I_m, I_a, I_{grid}; (b) direct and inverse sequence currents; (c) voltages.

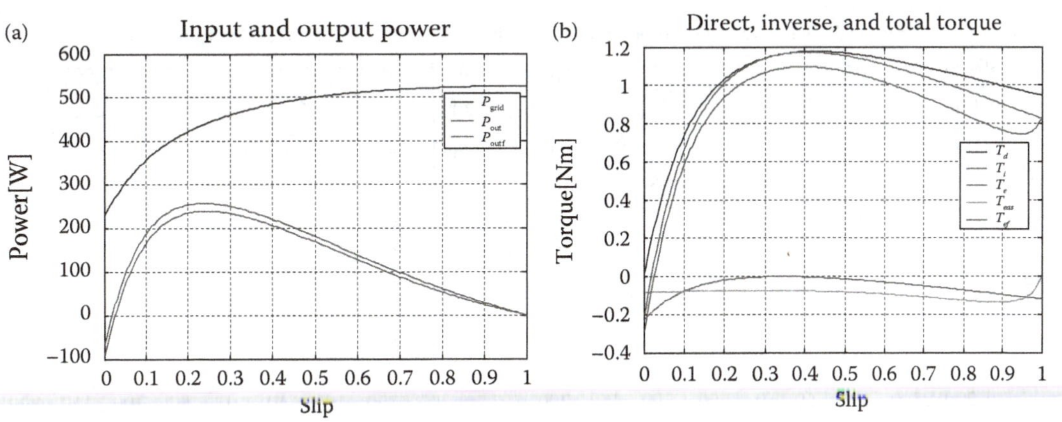

FIGURE 3.8 Asynchronous operation: (a) input and output power versus slip; (b) average torque components T_{das}, T_{ias}, T_{ePM}, versus slip. T_{ePM}—average PM braking torque; T_{ef}—total torque considering the asymmetry of stator windings ($R_a \neq R_m a^2$).

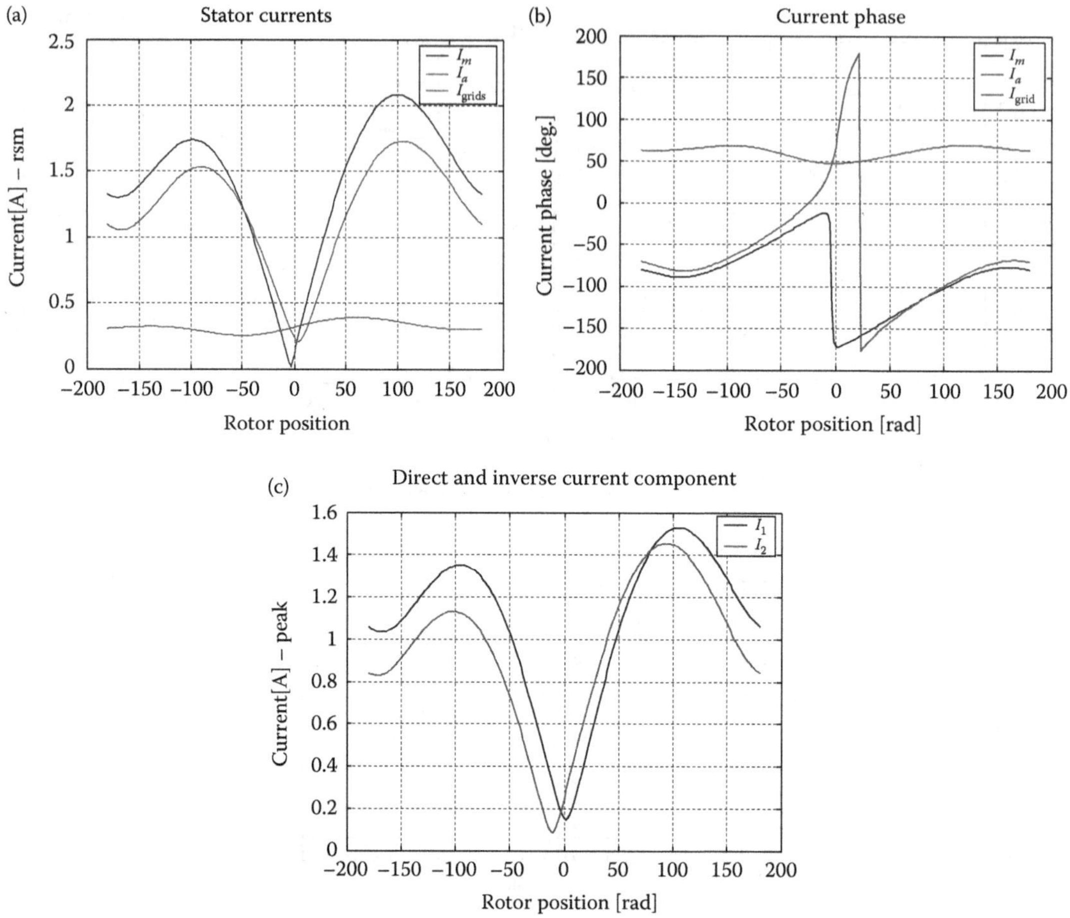

FIGURE 3.9 Current components at synchronous operation versus rotor power angle: (a) I_m, I_a; (b) $\underline{I}_m$, $\underline{I}_a$; (c) I_1, I_2.

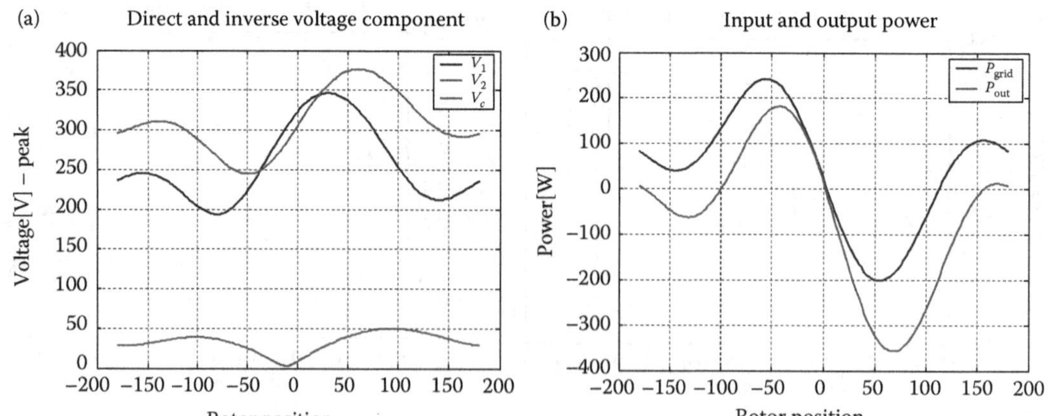

FIGURE 3.10 Voltages and powers for synchronous operation versus power angle, (a) V_1, V_2, V_c (capacitor voltage); (b) input and output power.

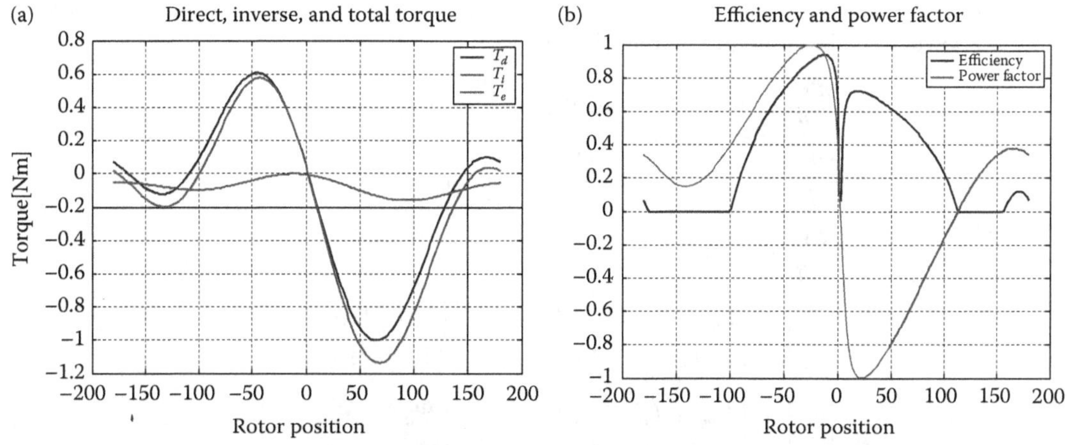

FIGURE 3.11 Synchronous performance: (a) torque; (b) efficiency and power factor.

Torque, efficiency, and power factor, as illustrated in Figure 3.11, show that at ~0.3 Nm (rated torque) and power angle ~15°, 100 W, the electrical efficiency is about 0.9 and the power factor is 0.953.

A simplified treatment of split-phase LS-PM-RSM or LS-RSM steady state is given in [2].

Note: Torque in asynchronous mode may be decomposed into six components (two for PM influence and four for cage torque), which also show large torque pulsations that impede on asynchronous acceleration and self-synchronization [6]. Given the complexity of the subject, we will lump all such components into the model for transients.

3.7 THE *dq* MODEL FOR TRANSIENTS

For the study of transients, the dq model is instrumental. However, as the rotor is not symmetric along axes d and q, the dq axes should be fixed to the rotor. In this case, however, the stator has to be symmetric in the sense that, even with a different number of turns in the main and auxiliary windings N_m, $N_a(a = N_a/N_m)$, the quantity of copper has to be roughly the same in the two windings; that is, the resistances and leakage inductances are related as:

$$R_m a^2 = R_a \quad L_{ml} a^2 = L_{al} \tag{3.13}$$

If this is not the case, phase coordinates may be used; however, coupling inductances depend on rotor position, and the computation time is rather prohibitive. For the steady state, we examined this case in previous sections. Starting with nonorthogonal but symmetric stator windings (condition [Equation 3.13] is fulfilled), the dq model is:

$$\begin{aligned}&\begin{vmatrix} V_d \\ V_q \end{vmatrix} = \begin{vmatrix} \cos\theta_{er} & \sin\theta_{er} \\ -\sin\theta_{er} & \cos\theta_{er} \end{vmatrix} \cdot \begin{vmatrix} V_\alpha \\ V_\beta \end{vmatrix}; \quad \begin{vmatrix} V_\alpha \\ V_\beta \end{vmatrix} = \begin{vmatrix} 1 & 0 \\ K_{V_m} & K_{V_a} \end{vmatrix} \cdot \begin{vmatrix} V_m \\ V_a \end{vmatrix} \\ &K_{V_m} = \frac{1}{\tan\xi}; \quad K_{V_a} = \frac{1}{a\sin\xi}; \quad V_a = V_m - V_c\end{aligned} \tag{3.14}$$

For currents, the same transformation holds, but with $K_{im} = \cos\xi$ and $K_{ia} = \sin\xi/a$ in order to conserve power; ξ is the angle of winding ($\xi \geq 90°$).

$$\frac{d\Psi_d}{dt} = V_d - R_s i_d + \omega_r \Psi_q; \quad \frac{d\Psi_q}{dt} = V_q - R_s i_q - \omega_r \Psi_d$$

$$\frac{d\Psi_{dr}}{dt} = -R_{rd} \cdot i_{dr}; \quad \frac{d\Psi_{qr}}{dt} = -R_{rq} i_{qr}$$

$$C\frac{dV_c}{dt} = i_a$$

$$\begin{vmatrix} i_d \\ i_{dr} \end{vmatrix} = \begin{vmatrix} L_{md} + L_{ml} & L_{md} \\ L_{md} & L_{md} + L_{rdl} \end{vmatrix}^{-1} \begin{vmatrix} \Psi_d - \Psi_{PM} \\ \Psi_{dr} - \Psi_{PM} \end{vmatrix} \tag{3.15}$$

$$\begin{vmatrix} i_q \\ i_{qr} \end{vmatrix} = \begin{vmatrix} L_{mq} + L_{ml} & L_{mq} \\ L_{mq} & L_{mq} + L_{rdl} \end{vmatrix}^{-1} \begin{vmatrix} \Psi_q \\ \Psi_{qr} \end{vmatrix}$$

$$T_e = p(\Psi_d i_q - \Psi_q i_d)$$

$$\frac{J}{p}\frac{d\omega_r}{dt} = T_e - T_{load}; \quad \frac{d\theta_{er}}{dt} = \omega_r$$

A seventh-order system in Equation 3.15, with Ψ_d, Ψ_q, Ψ_{dr}, Ψ_{qr}, V_c, ω_r, θ_{er} as variables, could be solved numerically. Magnetic saturation of the main field may be considered by dedicated functions $\Psi_{dm}(i_{dm}, i_{qm})$ and $\Psi_{qm}(i_{dm}, i_{qm})$, which may be calculated by the EMC or by FEM and then "inverted" by using polynomial approximations that allow derivatives. Similarly, $L_{drl}(i_{dr})$, $L_{qrl}(i_{qr})$, might be considered. Sample digital simulations of acceleration and then steady-state operation of the same 100-W, two-pole, 50-Hz (3000-rpm) motor in previous paragraphs, the dq model, is given in the following:

- By the dq model, Figure 3.12a. The block diagram
- Figure 3.12b. Capacitor and capacitor switch model
- Figure 3.13a. The single-phase LS-PM-RSM circuit model for transients
- Figure 3.13b. The model input circuit parameters
- Figure 3.14a. Grid current during motor acceleration
- Figure 3.14b. Speed during motor acceleration
- Figure 3.14c. Torque during motor acceleration
- Figure 3.15a. Grid current: steady state
- Figure 3.15b. Speed: steady state
- Figure 3.15a. Torque: steady state
- Figure 3.16. Motor grid input power and input and output energy during motor acceleration

The case study reveals that the machine investigated here for transient performance fulfills the specifications and self-synchronizes at a load torque of 0.4 Nm, larger than the rated torque (0.3 Nm). The next step is to introduce an optimal design methodology and, finally, show some FEM results for its validation.

3.8 OPTIMAL DESIGN METHODOLOGY BY A CASE STUDY

Here, we will briefly describe a rather involved optimal design methodology and code for the same 100-W, two-pole, 50-Hz (3000-rpm), 220-V, one-phase LS-PM-RSM. The optimal design methodology has to deal with the following issues:

- Specifications
- Machine (device) model and optimization variables

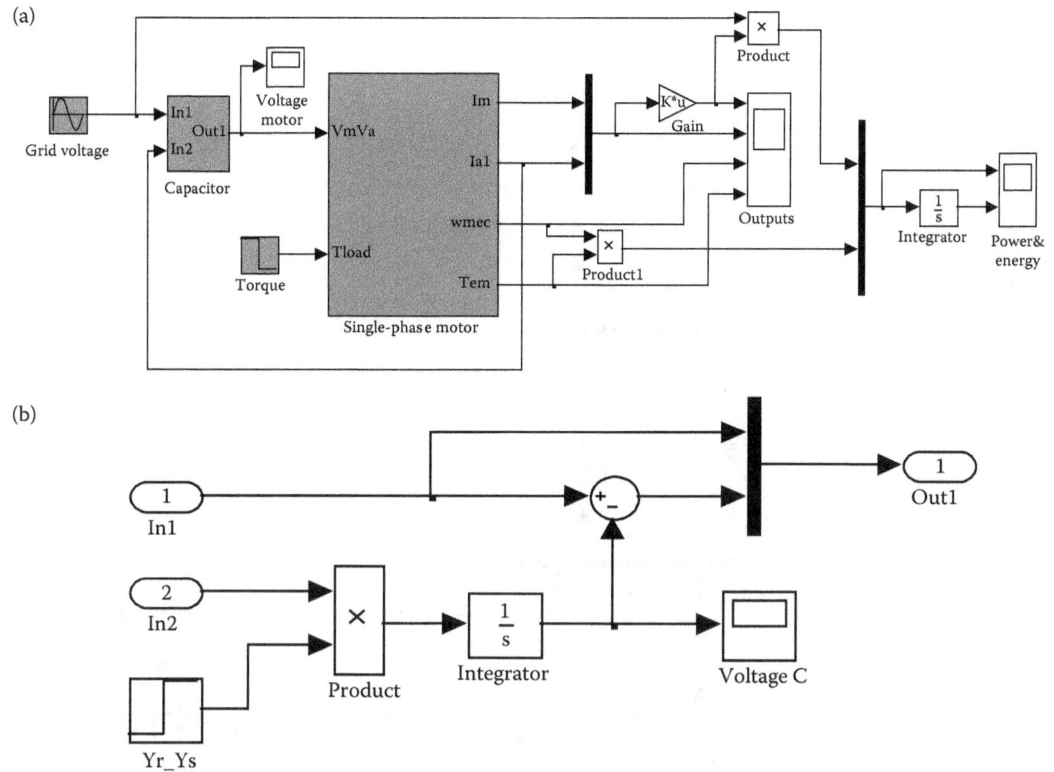

FIGURE 3.12 LS-PM-RSM model for transients, (a) block diagram; (b) capacitor and capacitor switch model.

- Optimization objective (fitting) function(s)
- Optimization mathematical algorithm
- Complete optimal design code
- Input and output file numerical and graphical data
- FEM validation of optimal design geometry (machine)

The specifications are those already mentioned related to power, frequency, speed, and voltage, but they could be given new constraints, depending on application.

The machine model includes the crude analytical expressions of resistances and leakage inductances, PM flux, and the EMC-derived magnetization inductances L_{dm}, L_{qm} valid for synchronous steady state. For asynchronous operation (already illustrated above), L_{dm} and L_{qm} have been considered constant for the sake of simplicity.

The design code is able to run in "evaluation (e) mode" or in "optimal"(o) mode.

Finite element method magnetics (FEMM) 4.2 is embedded in the design code at a few levels (for validation):

- No FEM validation
- Only PM flux in the main winding at zero currents
- PM and stator current flux linkages in axes d and q for main and auxiliary windings
- Synchronous torque versus rotor position γ

The entire optimal design code in MATLAB–Simulink® has a structure as in Figure 3.17.

Some details of the motor models have been presented above. Here, we will focus on the quasi-multiobjective optimal design implemented using the Hooke–Jeeves modified method [7].

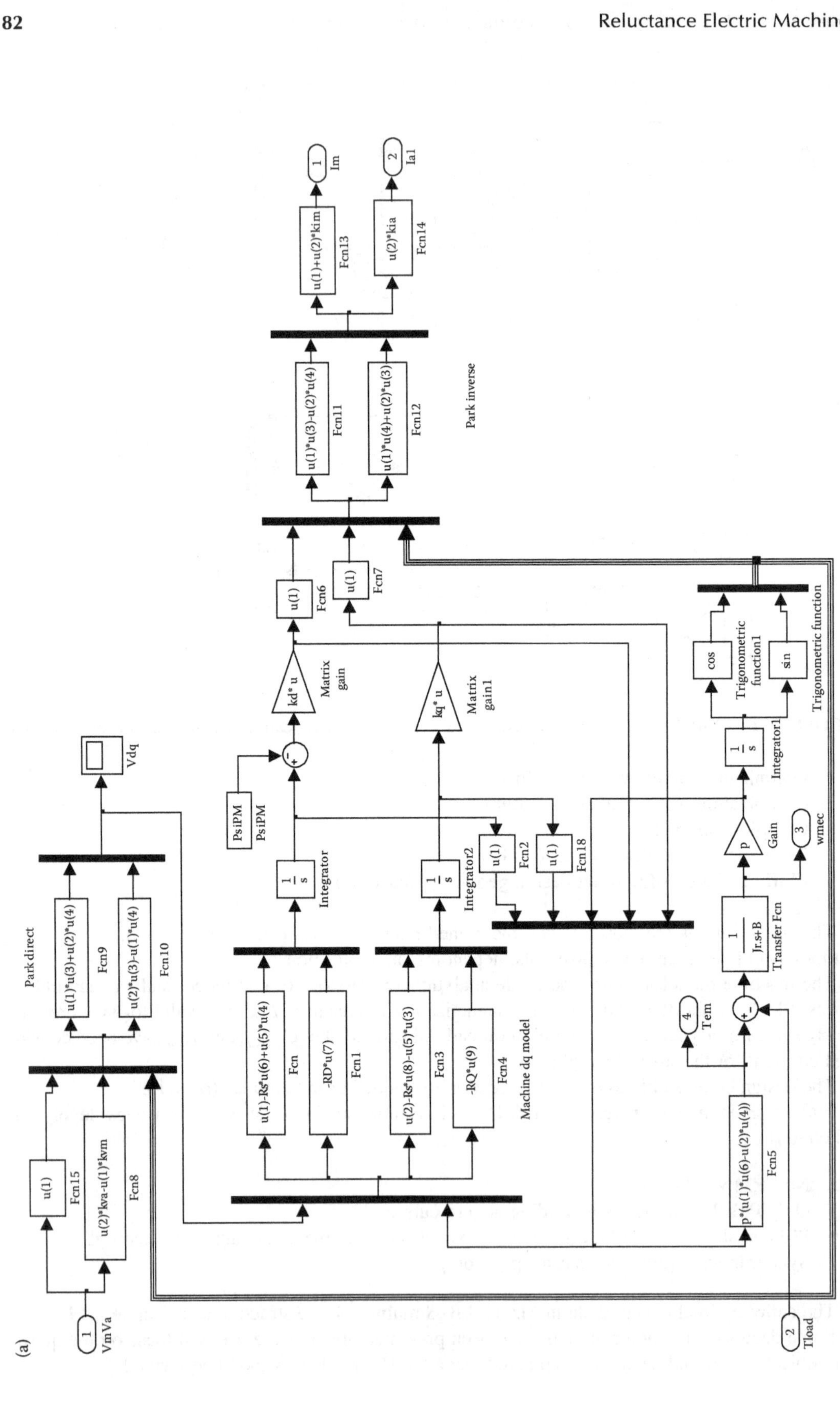

FIGURE 3.13 The single-phase LS-PM-RSM, (a) the complete model.

(Continued)

(b)

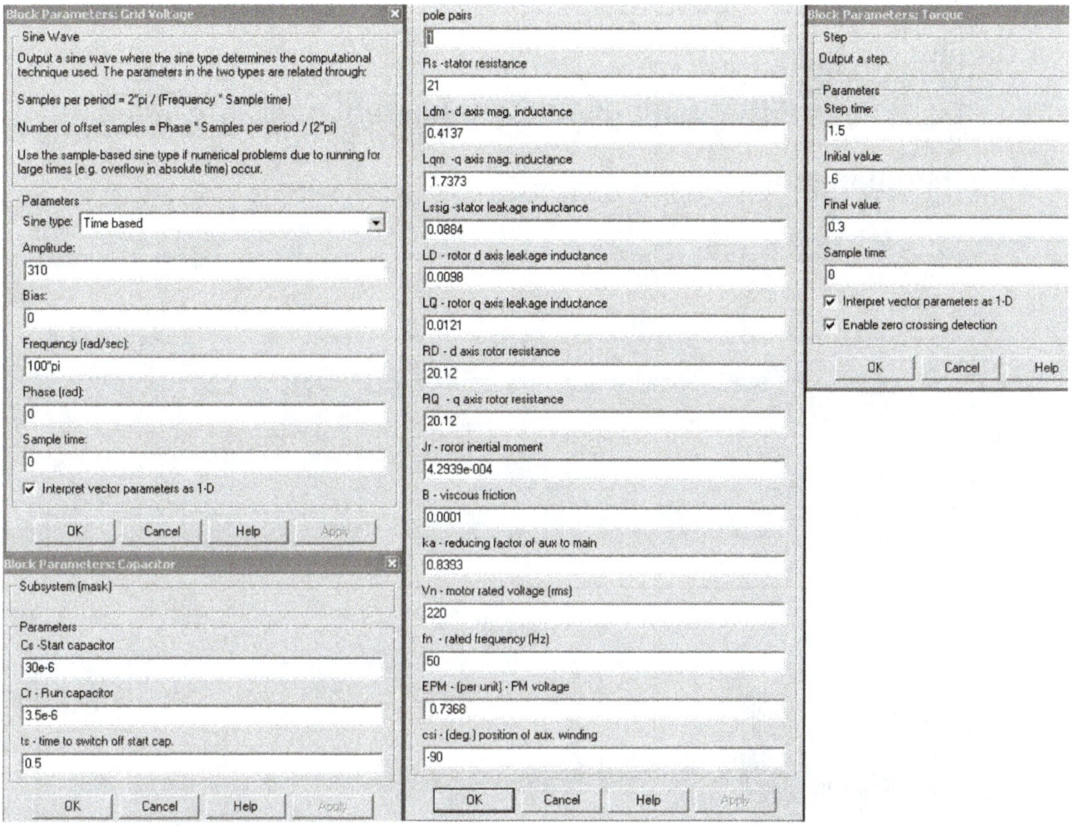

FIGURE 3.13 (Continued) The single-phase LS-PM-RSM, (b) the model input parameters.

The optimization objectives are:

- Reducing the initial motor cost (including capacitor cost)
- Increasing the efficiency
- Satisfying the starting and synchronization conditions for rated load torque

All these objectives are aggregated in a single objective cost function t_cost:

```
cu_c=cu_pr*mcu;                         %USD copper cost
lam_c=lam_pr*msiron;                    %USD lamination cost
PM_c=PM_pr*mpm;                         %USD PM cost
rotIron_c=mriron*rotIron_pr;            %USD rotor iron cost
rc_c=mrw*Al_pr;                         %USD rotor cage cost
pmw_c=pmw_pr*mmot;                      %USD passive material cost
mot_cost=cu_c+lam_c+PM_c+rotIron_c+rc_c+pmw_c; %USD initial cost
c_cost=2*pc0+kpc*Cap*1e6++kpc*Ca*1e6; %USD capacitor costs
i_cost=mot_cost+c_cost;
energy_c=energy_pr*Pn*(1/etan-1)/1000*hpy*ny; %USD/kWh energy price
if(Tmax>kstt*Tn)
st_cost=(max((sn-ssmax)/ssmax,0)+max(1-Tmin/(kstt*Tn),0)+      +2*max
(Pn*ktk/Pom-1,0))*i_cost;      %startability penalty cost
    else
```

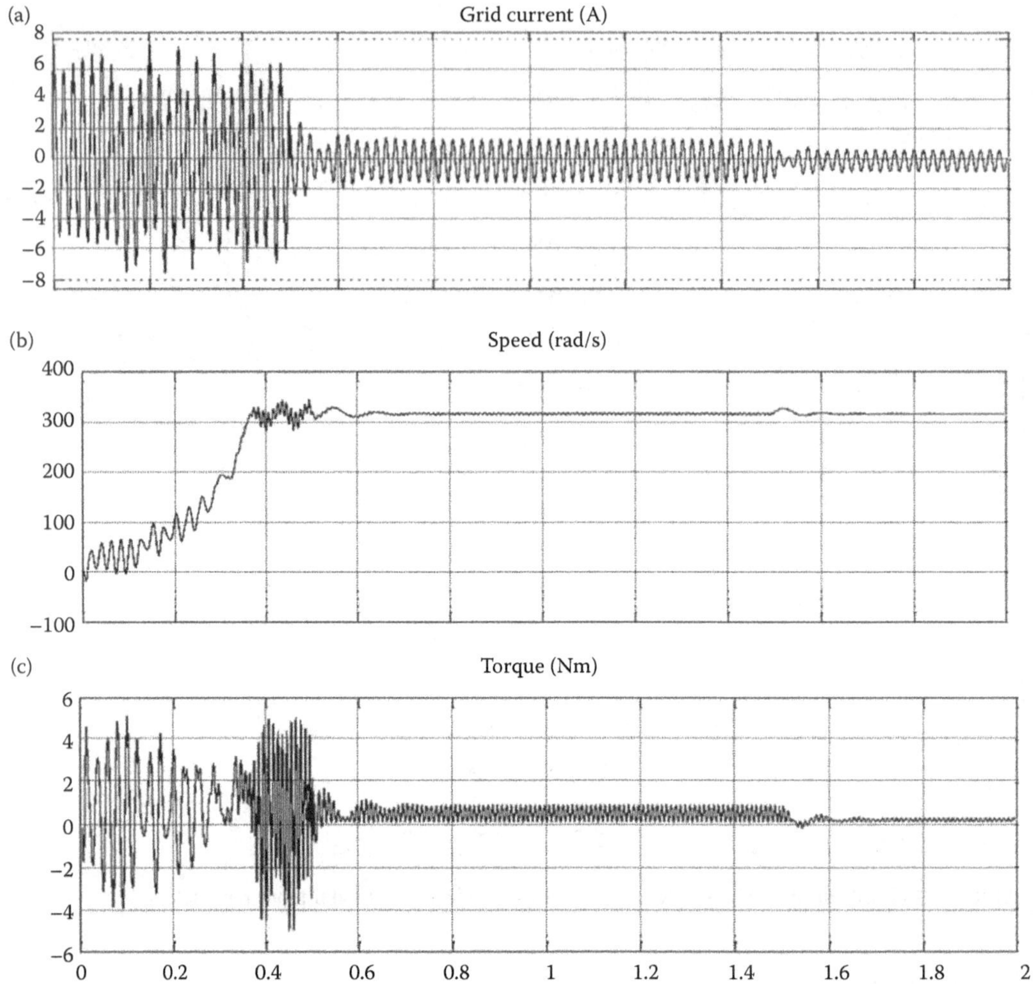

FIGURE 3.14 Machine grid current (a); speed (b); and torque (c) during running up to speed from standstill.

```
st_cost=(max((sn-ssmax)/ssmax,1)*kstt*Tn/Tmax+max(1-Tmin/(kstt*Tn),0)
+max(Pn*ktk/Pom-1,0))*i_cost;      %startability penalty cost
    end
t_cost=i_cost*max(1,Pn/Pom)+energy_c+st_cost; %USD total cost
```

The cost of the copper, stator laminations, rotor cage, rotor laminations, PMs, frame, and capacitors are all computed. If the motor is not capable of delivering the rated output power, an over-unity power penalty factor (P_n/P_{om}) multiplies the initial cost.

The startability penalty cost st_cost is added if the starting requirements are not met. We refer to starting torque at zero speed and torque at $S_{smax} = 0.04$ (slip value). The latter should be smaller than k times the rated torque (k_{ts} for starting, k_{tk} for synchronization).

Only sample data from input and output table files are shown in Tables 3.1 through 3.4.

The results show an electrical efficiency around 0.929 and power factor of 0.939 for a 4.22-kg motor with NdFeB magnets (0.125 kg) at an initial cost of the motor of 26.6 USD. Efficiency, mass, and cost (objective) function evolution during optimization in Figure 3.18 shows that 30 optimal design cycles are sufficient to secure solid results.

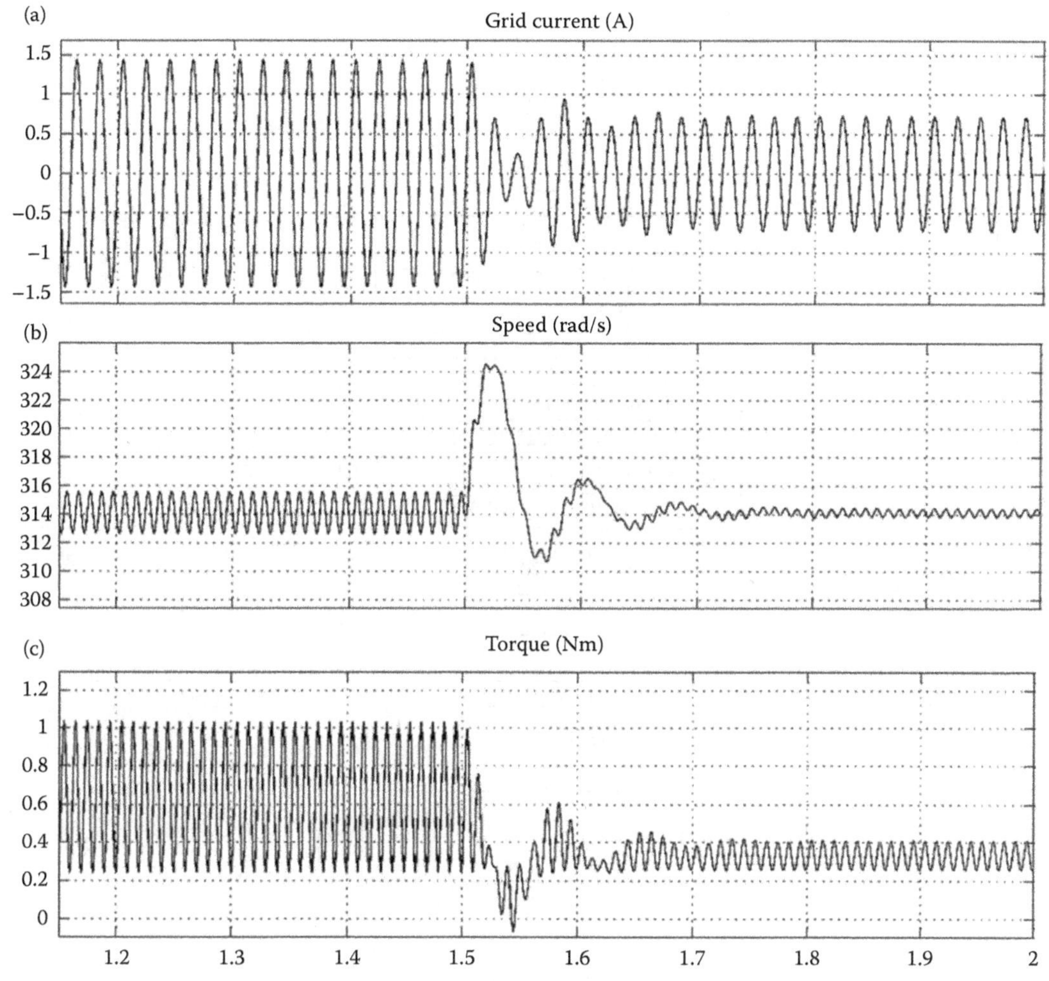

FIGURE 3.15 Machine synchronous steady state: grid current (a); speed (b); torque (c).

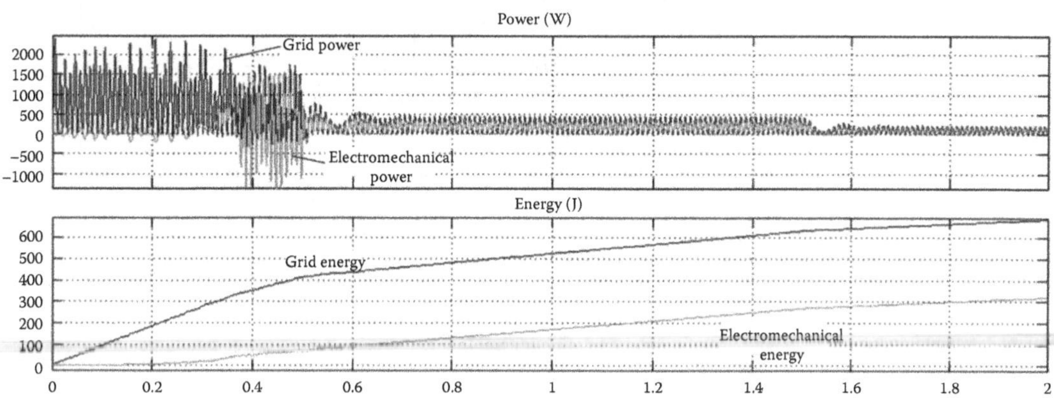

FIGURE 3.16 Motor input power and input and output energy during motor acceleration to speed (load torque during starting was 0.6 Nm [rated torque: 0.3 Nm] and was then reduced to 0.4 Nm).

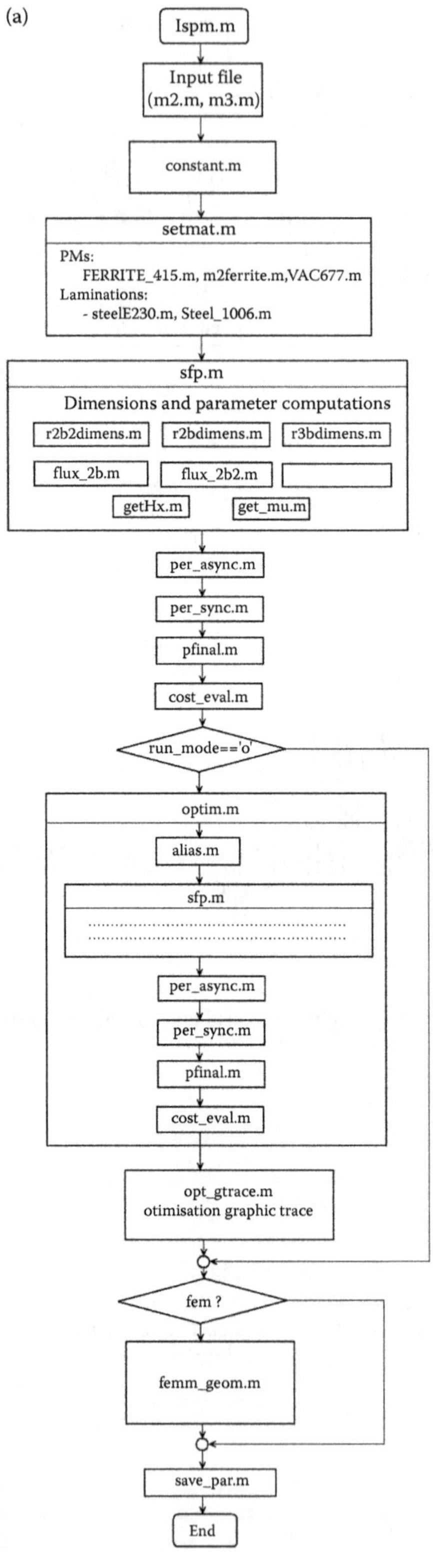

FIGURE 3.17 The optimal design code block diagram (a). *(Continued)*

(b)

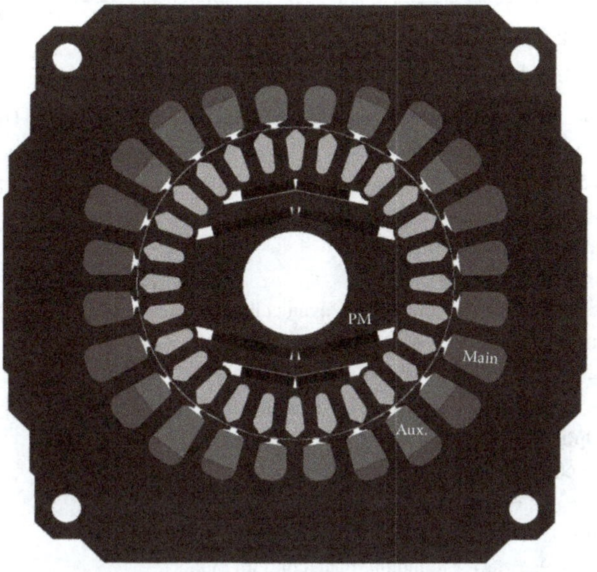

FIGURE 3.17 (Continued) cross-section of motor (b).

As the performance during asynchronous running, steady-state operation, starting, and synchronization transient have already been exposed for the same motor as in the optimal design, here we will only show some FEM validation results of the optimally designed motor.

3.9 FINITE-ELEMENT MODELING VALIDATION

The following items related to FEM validation are illustrated here:

- PM flux density in the airgap (Figure 3.19)

The magnetic field line map for the resultant magnetic field produced concurrently by PMs, demagnetization $I_{md} = -0.398$ A, and $I_{aq} = +0.392$ A for rated torque conditions is shown in Figure 3.20.

The FEM-calculated magnetization inductances for rated torque are: $L_{md\text{FEM}} = 0.578$ H, $L_{aq\text{FEM}} =$ 2.30 H (total), with $L_{mq\text{FEM}} = 3.009$ H.

These values show good magnetic saliency. The $L_{mq\text{FEM}}/L_{md\text{FEM}} = 3.009/0.578$ ratio is sufficiently high to secure superior electrical efficiency (zero mechanical losses) for reasonable initial cost and motor weight.

Note: A ferrite PM motor, designed for the same specifications, yields an electrical efficiency of 0.8984, power factor: 0.955, initial (motor) cost: 34.44 USD with 7.895 kg of motor weight. The ferrite PM motor ends up costlier at lower efficiency, as the quantities of iron and copper are notably larger to secure a good-enough efficiency (above the value of 0.86 for the IM).

3.10 PARAMETER ESTIMATION AND SEGREGATION OF LOSSES IN SINGLE-PHASE CAPACITOR PERMANENT MAGNET-RELUCTANCE SYNCHRONOUS MACHINES BY TESTS

3.10.1 Introduction

Segregation of losses—performed through special no-load tests—has been a rather general method for getting load performance (efficiency) without actually loading various types of commercial electrical machines.

TABLE 3.1
The Input File

```
%Line Start Interior permanent magnet machine (PMM)—Input Data File
Pn = 100; % W—rated power
fn = 50; % Hz—base speed
Vn = 220; % V—rated voltage
%Primary Dimension
pp = 2; % numbers of poles
Nmc0 = [123 168 95 55 24 0]; % main coil turns
Nac0 = round(1.5*[0 0 45 81 73 63]); % auxiliary coil turns
Nr = 28; % rotor bars—should be multiple of 4
Ca = 3.5e-6; %F—Capacity on auxiliary phase
% Optimization Variable
Dsi = 60.08; % mm—Stator inner diameter
lstack = 48; % mm—Core stack length
ksext = 1; % factor to modify external stator dimensions
knm = 0.9; % factor to modify main coil turn number
kna = 0.9; % factor to auxiliary coil turn number
dmw_i = 11; % main wire diameter index
daw_i = 12; % auxiliary wire diameter index
hpm = 2; % mm—PM height
rhy1 = 1.5; % mm—distance between first barrier and rotor bar
rhy3 = 2.5; % mm—distance between second barrier and shaft
wst_pu = 0.4; % stator tooth relative width
wpm_pu = 0.85; % width of PM relative to the field barrier
wrso1 = 2.6; % mm—width of rotor slot—between trim centers
hrso2 = 4.97; % mm—rotor slot height—distance between trin = m center
teta_mb1 = pi/4;
iCa = 4; % running capacitor index
iCap = 10; % start capacitor—index
```

As early as 1935, C. Veinott [8] introduced such a method for single-phase capacitor induction motors, and little has been added to it since then.

As single-phase capacitor PM-RSMs have been investigated for low-power high-efficiency applications [9], it seems timely to try to develop a complete but comfortable-to-use methodology for parameter estimation and loss segregation for this machine.

It is almost needless to say that loss segregation implies parameter estimation first. The tasks will be performed as follows:

- Present the revolving (+−) theory (model) characteristic equation of a one-phase capacitor PM-RSM.
- Define specific tests and apply the theory to estimate the motor parameters (resistances and inductances) and perform loss segregation without driving or loading the machine.
- Validate the above methodologies through load testing.

3.10.2 Theory in Short

To reduce complexity, in essence, we use here the revolving field (+−) model for a two-phase motor with orthogonal stator windings and PMs and a cage on the anisotropic rotor (Figure 3.21) [2].

TABLE 3.2
The Output File

% Electrical Parameters
Rm = 23.289969; % Ohm—main phase resistance
Ra = 25.032505; % Ohm—auxiliary phase resistance
Rr = 57.878568; % Ohm—rotor equivalent resistance
Lsm = 0.090244; % H—leakage inductance of main winding
Lsa = 0.047708; % H—leakage inductance of aux. winding
Lsrd = 0.079865; % H—rotor equivalent *d* axis leakage inductance
Lsrq = 0.079865; % H—rotor equivalent *q* axis leakage inductance
Lmd = 0.511171; % H—magnetization inductance on *d* axis (around 0 d axis current)
Lmq = 1.532661; % H—magnetization inductance on *d* axis at $I_q = 0.642824$ A
Nm = 876.000000; %—turns on main winding
Na = 742.000000; %—turns on auxiliary winding
kmwp = 0.851851; %—main winding factor
kawp = 0.847687; %—auxiliary winding factor
a = 0.842891; %—factor of reducing aux. winding to main winding
lpm = 1.147396; % Wb—linkage permanent magnet flux in main phase
Elpm = 360.465002; % V—main phase emf produced by PM in main phase (peak value)
Imn = 0.289350; % A—main phase rated current
Ian = 0.290508; % A—auxiliary phase rated current
Igridn = 0.498713; % A—grid rated current
Irn = 0.088533; % A—rotor current
Pcum = 1.949919; % W—main phase copper losses
Pcua = 2.112610; % W—auxiliary phase copper losses
Pcurn = 0.453658; % W—rotor copper losses
Pfe = 1.920377; % W—iron losses
etan = 0.938945; %—rated efficiency
cosphip = 0.953201; %—power factor
Ca = 3.3; % uF—running capacitor
Cap = 15.0; % uF—auxiliary starting capacitor

The two stator windings need not have identical distribution or the same copper weight. First, we reduce the auxiliary to the main winding:

$$I'_a = aI_a; \quad a = \frac{N_a K_{wa}}{N_m K_{wn}} \tag{3.16}$$

$$V'_{a0} = \frac{V_{a0}}{a} \tag{3.17}$$

Again, N_a, N_m are turns per phase and K_{wa}, K_{wm} their winding coefficients. To accommodate the case of $R_m \neq R_a/a^2$, we will define internal fictitious voltages for the main and auxiliary phases: V_{a0} and V_{m0} instead of V_a, V_m.

So, from Figures 3.3 through 3.5, we have:

$$\underline{V}_m = \underline{V}_a + \underline{V}_{ca} = V_{a0} - jX'_{ca}\underline{I}_a = \underline{V}_{m0} - jX'_{cm}\underline{I}_m \tag{3.18}$$

$$X'_{ca} = X_{ca} - \left(X_{al} - jR_a\right)(1 - m) \tag{3.19}$$

TABLE 3.3
Winding Details and Rotor Dimensions

Nm = 876; % turns per main phase
Na = 742; % turns per auxiliary phase
Nmc = [116 158 89 52 23 0]; % main coil turns
Nac = [0 0 64 115 103 89]; % auxiliary coil turns
dmw = 0.510000; % mm—diameter of main winding wire
daw = 0.450000; % mm—diameter of auxiliary winding wire
Dro = 56.800000; % mm—rotor outer diameter
Dri = 20.000000; % mm—rotor inner diameter
wrt = 3.000221; % rotor tooth width
rh1 = 0.275000; % mm—height of rotor slot tip
rsr0 = 0.500000; % mm—corner radius for rotor slot
alphars1_deg = 90.000000; % degree—angle between rotor slot edge
hrso1 = 0.550000; % mm—rotor slot height—distance between trim centers
wrso1 = 1.100000; % mm—width of rotor slot—between trim centers
hrso2 = 2.900000; % mm—rotor slot height—distance between trim centers
rsr2 = 0.721845; % mm—rotor slot top radius
her = 7.633211; % mm—rotor end ring height—used only in e1 run_mode
Deri = 45.306311; % mm—rotor end ring inner diameter—used only in e1 run_mode
Dero = 56.800000; % mm—rotor end ring outer diameter—used only in e1 run_mode
wpm = [13.000000 19.500000]; % mm—permanent magnet width
fi_rmb = [−14.818843−6.815769]; %—flux barrier angle
hmb = 1.600000 % mm—flux barrier height
rmb = 1.521845; % mm—flux barrier radius

$$X'_{cm} = -(X_{ml} - jR_m)(1 - m); \quad X_{ca} = \frac{1}{\omega_1 C_a} \tag{3.20}$$

with $m = 1$ for symmetric (equivalent) windings and $m = 0$ otherwise. The $+-$ transformations are:

$$\begin{vmatrix} \underline{V}_1 \\ \underline{V}_2 \end{vmatrix} = \frac{1}{\sqrt{2}} \begin{vmatrix} 1 & -j \\ 1 & +j \end{vmatrix} \cdot \begin{vmatrix} V_{m0} \\ V'_{a0} \end{vmatrix} \tag{3.21}$$

$$\begin{vmatrix} V_{m0} \\ V'_{a0} \end{vmatrix} = \frac{1}{\sqrt{2}} \begin{vmatrix} 1 & 1 \\ j & -j \end{vmatrix} \cdot \begin{vmatrix} \underline{V}_1 \\ \underline{V}_2 \end{vmatrix} \tag{3.22}$$

Making use of Equations 3.16 and 3.17 and Equations 3.21 and 3.22 in Equations 3.18 and 3.19 yields:

$$\underline{V}_m\sqrt{2} = aj(\underline{V}_1 - \underline{V}_2) + X'_{ca}(\underline{I}_1 - \underline{I}_2)/a \tag{3.23}$$

$$\underline{V}_m\sqrt{2} = \underline{V}_1 + \underline{V}_2 - jX'_{cm}(\underline{I}_1 + \underline{I}_2) \tag{3.24}$$

The direct field voltage equation (Figure 3.22) is:

$$\underline{V}_1 - \underline{E}_1 = \underline{Z}_1\underline{I}_1; \quad \underline{Z}_1 = R_{1e} + jX_{1e} \tag{3.25}$$

TABLE 3.4
Objective Function and FEM Validation

cu_c = 8.176390; % USD—copper cost
lam_c = 3.144986; % USD—lamination cost
PM_c = 6.264258; % USD—PM cost
rotIron_c = 0.804382; % USD—rotor iron cost
rc_c = 0.328975; % USD—rotor cage cost
pmw_c = 5.318692; % USD—weight penalty cost (frame)
mot_cost = 24.037683; % USD—motor cost
c_cost = 2.555400; % USD—capacitor cost
i_cost = 26.593083; % USD—initial cost
energy_c = 19.507677; % USD—energy loss penalty cost
st_cost = 1.163339; % USD—starting penalty cost
t_cost = 47.264099; % USD—total cost—objective function
psi_pm = −0.977362; % Wb FEM computed flux
psi_md = −0.746750; % Wb total *d* axis flux from FEM at main current 0.398456 A peak
Lmd_fem = 0.578764; % H main winding *d* axis inductance
psi_aq = −0.902739; % Wb total *q* axis flux from FEM at auxiliary current −0.392182 A peak
Laq_fem = 2.301840; % H auxiliary winding *q* axis inductance
psi_mda = −0.768317 % Wb total *d* axis flux, main current (d) 0.398456 A and aux. (q) −0.392182 A peak
psi_mq = 1.199161; % Wb total *q* axis flux from FEM at main current 0.398456 A peak
Lmq_fem = 3.009515; % H auxiliary winding *q* axis inductance
psi_pma = −0.811229 % Wb total *d* axis flux, auxiliary winding (d) with main current 0.398456 A peak

with

$$R_{1e} = mR_m + \omega_1\left(L_{md} - L_{mq}\right)\sin 2\gamma \tag{3.26}$$

$$X_{1e} = mX_{ml} + \frac{\omega_1}{2}\left[\left(L_{md} + L_{mq}\right) - \left(L_{md} - L_{mq}\right)\cos 2\gamma\right] \tag{3.27}$$

The current dq angle γ is the variable for load representation. For the inverse field (sequence), we use the known approximation:

$$\underline{V}_2 = \underline{Z}_2\underline{I}_2 \tag{3.28}$$

$$\underline{Z}_2 \approx \frac{1}{2}\left(\underline{Z}_{2d} + \underline{Z}_{2q}\right) \tag{3.29}$$

$$\underline{Z}_{2d} = \frac{\left(R_{rd}/(2-S) + jX_{rdl}\right)jX_{md}}{R_{rd}/(2-S) + j(X_{md} + X_{rdl})} + \left(R_m + jX_{ml}\right)m \tag{3.30}$$

$$\underline{Z}_{2q} = \frac{\left(R_{rq}/(2-S) + jX_{rql}\right)jX_{mq}}{R_{rq}/(2-S) + j\left(X_{mq} + X_{rql}\right)} + \left(R_m + jX_{ml}\right)m \tag{3.31}$$

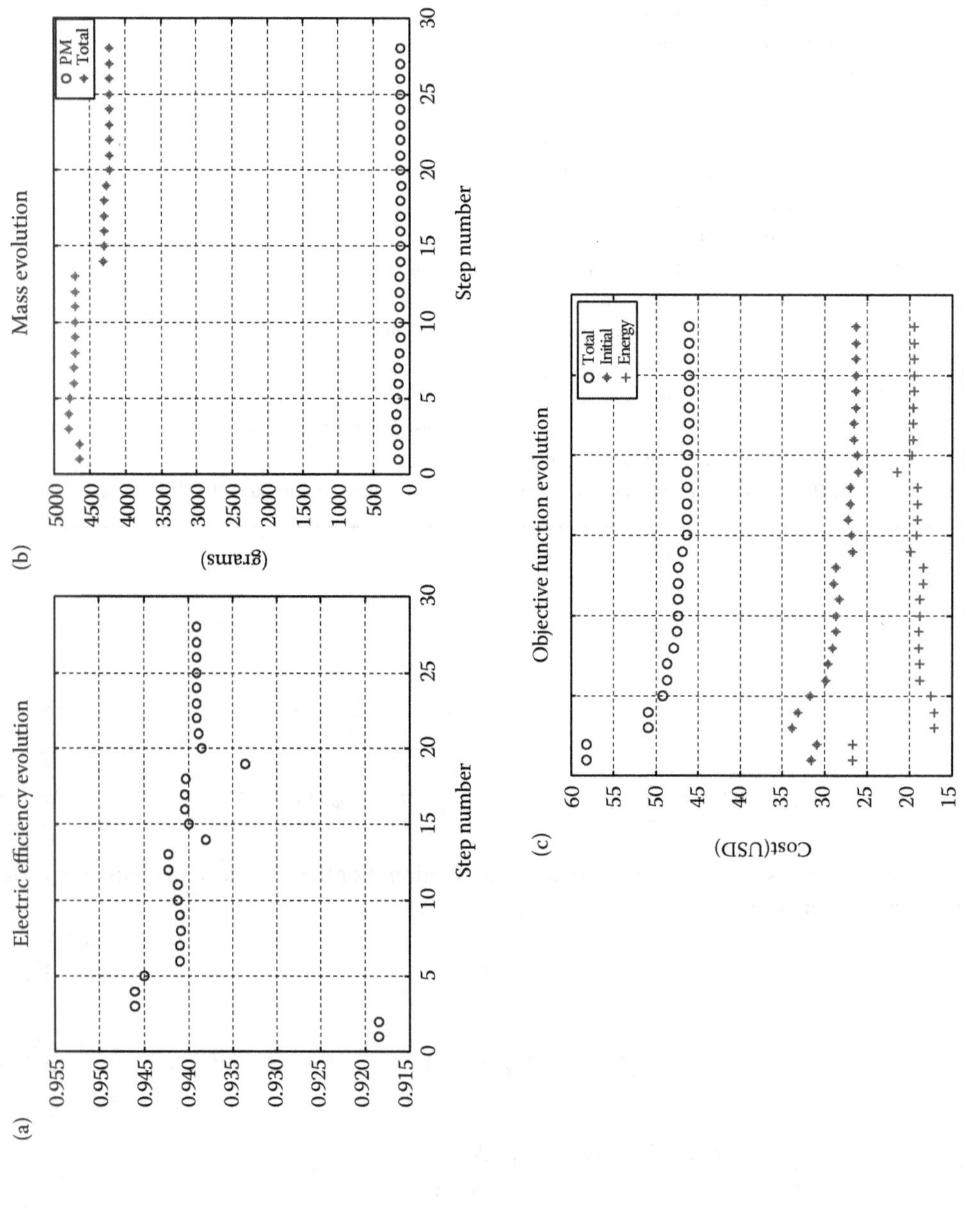

FIGURE 3.18 Electrical efficiency (a); mass (b); and objective function evolution (c) during optimal design process.

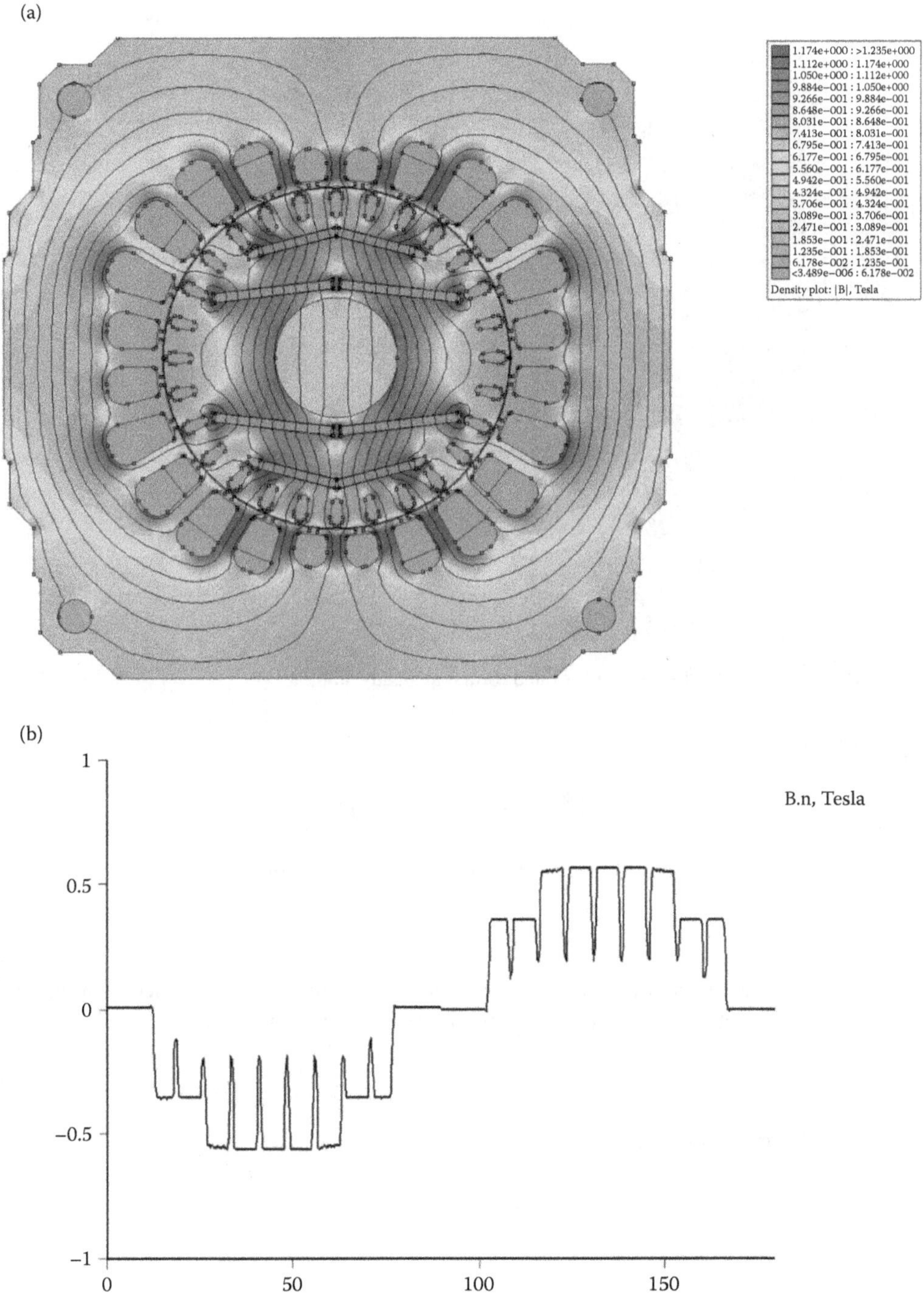

FIGURE 3.19 PM flux density map (a); and its distribution in the airgap (b).

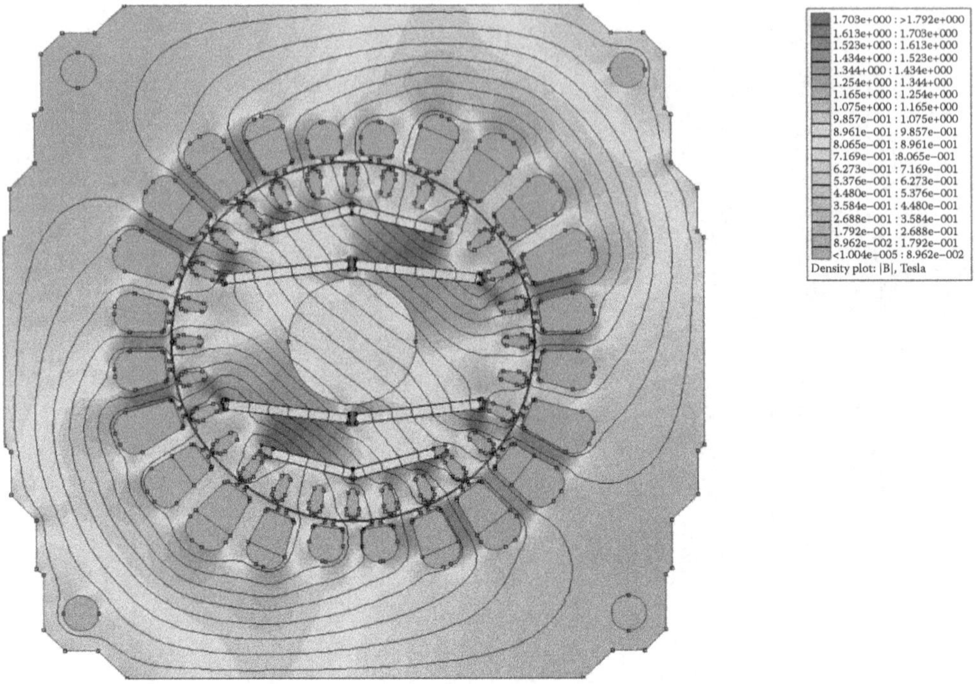

FIGURE 3.20 Magnetic field line map for rated torque at steady state.

The electromagnetic torque is:

$$T_{e+} = \left[\mathrm{Re}\left(\underline{V}_1\underline{I}_1^*\right) - R_m I_1^2 \cdot m\right] \cdot p_1/\omega_1 \tag{3.32}$$

$$T_{e-} = -\left[\mathrm{Re}\left(\underline{V}_2\underline{I}_2^*\right) - R_m I_2^2 \cdot m\right] \cdot p_1/\omega_1 \tag{3.33}$$

$$T_e = T_{e+} + T_{e-} \tag{3.34}$$

Core losses. As the inverse field travels mostly along leakage paths, we neglect here the inverse field core losses and consider the direct field core losses proportional to V_1:

$$P_{\text{iron1}} = V_1^2/R_{\text{iron1}} \tag{3.35}$$

where R_{iron1} is determined by tests and considered constant.

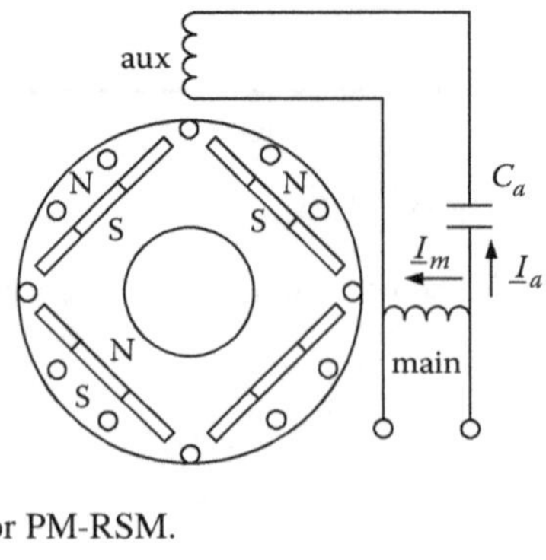

FIGURE 3.21 One-phase capacitor PM-RSM.

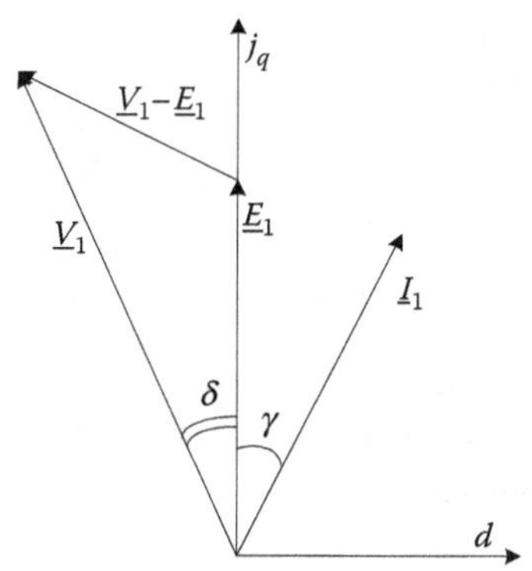

FIGURE 3.22 Phasor diagram for direct field.

Magnetic saturation. In most such machines, saturation effects are moderate. However, the parameter estimation methodology to be developed here will produce the dependence of main synchronous inductances L_{md}, L_{mq} on I_d and I_q currents.

The above theory will now be applied to the tests chosen for parameter estimation and loss segregation.

3.10.3 Parameter Estimation through Standstill Tests

At standstill, PMs will not produce any induced voltages, so the machine will behave like an induction motor with nonsymmetric stator and rotor windings for AC tests and as a typical PMSM for DC-type tests. The parameters to be estimated are those appearing in Z_1 and Z_2.

Step voltage tests will be used to estimate L_d and L_q with one phase turned on (or off) at a time.

a. DC current decay tests

   Supplying the main winding with DC current and then turning it off (Figure 3.23) will make the current decay to zero through the freewheeling diode.

   The corresponding equation is:

$$(\lambda_{PMi} + L_d \cdot i_{d0}) - \lambda_{PMf} = \int R_m i_d \, dt + \int V_{diode} \, dt \tag{3.36}$$

   If saturation is negligible, the initial and final values of PM flux linkage are the same: $\lambda_{PMi} = \lambda_{PMf}$; I_{d0}—initial current.

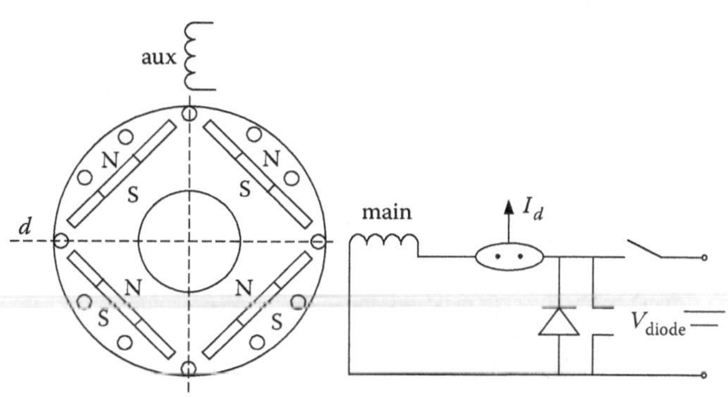

FIGURE 3.23 DC current decay test in axis d at standstill.

One test suffices to produce L_d (unsaturated value, though) with R_m previously measured in a steady DC small voltage–current test.

If the saturation influence is to be explored, tests at various initial currents are done and the PM flux λ_{PMf} is determined from a zero-current generator test:

$$\lambda_{PMf} = V_m\sqrt{2}/\omega_1 \tag{3.37}$$

Now, we only have to find $\lambda_d = \lambda_{PMi} + L_d \cdot I_{d0}$ as a function of I_{d0}, from Eq. 3.37. Performing the same test once for the auxiliary winding, where, again, the rotor will naturally align along the d axis, yields:

$$\left(\lambda_{PMai} + L_d^a \cdot i_{d0}^a\right) - \lambda_{PMaf} = \int R_a i_d^a\, dt + \int V_{\text{diode}}\, dt \tag{3.38}$$

The same reasoning as above applies, and (with saturation neglected) we find L_d^a. The turn ratio a will be:

$$a = \left(\frac{L_d^a}{L_d}\right)^{\frac{1}{2}} \tag{3.39}$$

As the resistances are known already, a new value of a is:

$$a' = \left(\frac{R_a}{R_m}\right)^{\frac{1}{2}} \tag{3.40}$$

If $a = a'$, we may assume that the two windings are equivalent and $m = 1$ in Equations 3.18 through 3.34.

To produce DC current decay tests in axis q, for the main winding, we have to fix (latch) the rotor in the position for auxiliary phase d axis tests and then proceed as above (Figure 3.24).

This time, we get:

$$L_q \cdot i_{q0} = \int R_m i_q\, dt + \int V_{\text{diode}}\, dt \tag{3.41}$$

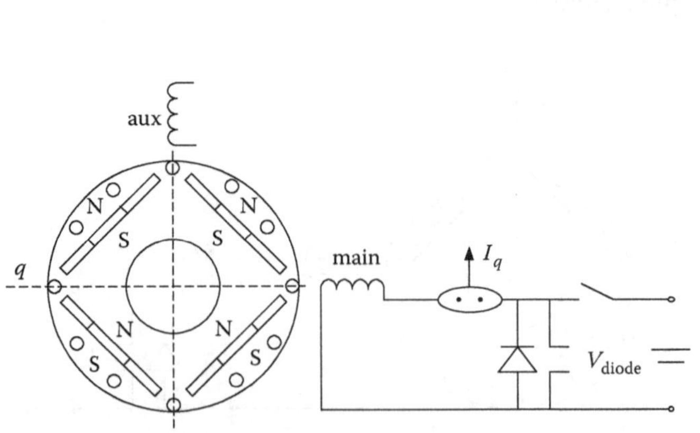

FIGURE 3.24 q-axis current decay tests (main phase) at standstill.

Here, we may repeat the test for various initial currents to determine the saturation effect as $L_q(I_{q0})$. To complete the estimation of parameters occurring in Z_1, we should also find the main phase leakage inductance L_{ml}. The AC standstill tests that follow will deliver L_{ml}, L_{dl}', R_{rd}, R_{rq}, L_{rdl}, L_{rql} (reduced to the main winding).

b. AC standstill tests

If frequency effects on the rotor cage are negligible, a test at grid frequency suffices. The AC tests in axes d and q are done only for the main winding (Figure 3.25).

We measure I_m, V_m, $P_m(\varphi_m)$ and calculate:

$$R_{ed} = \frac{P_m}{I_m^2} \tag{3.42}$$

$$X_{ed} = \sqrt{\left(\frac{V_m}{I_m}\right)^2 - R_{ed}^2} \tag{3.43}$$

where

$$Z_{ed} = R_{ed} + jX_{ed} = \underline{Z}_{2d}(S = 1) \tag{3.44}$$

Note that

$$Z_{ed} = \frac{(R_{rd} + jX_{rdl})jX_{dm}}{R_{rd} + j(X_{rdl} + X_{dm})} + (R_m + jX_{ml}) \tag{3.45}$$

We now have two equations, know L_d, and have to calculate X_{rdl}, X_{ml}, R_{rd}.

As one more equation is needed, we return to the step voltage test.

c. DC voltage main winding turn-off on a known DC current in the main phase (Figure 3.26). The pertinent equations for a wheeling diode at standstill for the stator and rotor are:

$$V_d(t) = -L_{dm}\frac{di_{rd}}{dt} \tag{3.46}$$

$$R_{rd}i_{rd} + (L_{dm} + L_{rdP})\frac{di_{rd}}{dt} = 0 \tag{3.47}$$

with the solution:

$$i_{rd} = Ae^{-t(R_{rd}/(L_{md}+L_{rdl}))} \tag{3.48}$$

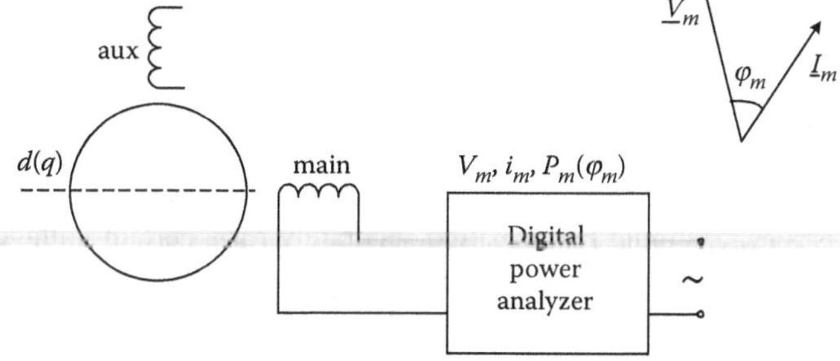

FIGURE 3.25 AC standstill test in axis d for the main phase.

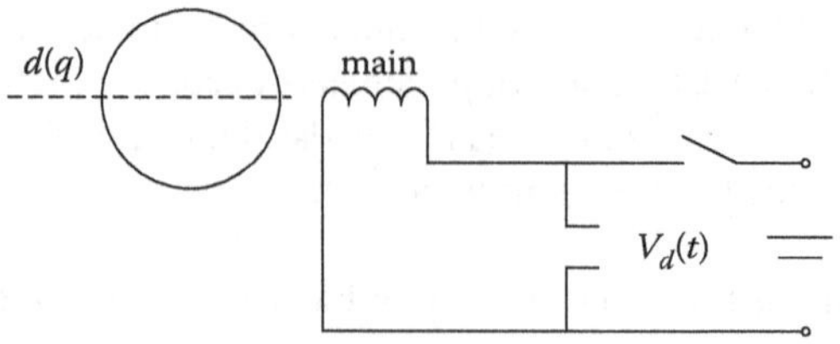

FIGURE 3.26 DC voltage turn-off in the main phase in axis *d*.

At time zero, we may consider that the flux does not change and thus:

$$L_d i_{d0} = L_{dm} i_{rd0} = A L_{dm} \tag{3.49}$$

$$i_{rd} = \frac{L_d i_{d0}}{L_{dm}} e^{-t(R_{rd}/(L_{dm}+L_{rdl}))} \tag{3.50}$$

$$V_d(t) = \frac{L_d i_{d0} \cdot R_{dr}}{L_{dm} + L_{rdl}} e^{-t \cdot R_{rd}/(L_{dm}+L_{rdl})} \tag{3.51}$$

The initial residual voltage V_{d0} is:

$$V_{d0} = \frac{L_d \cdot R_{rd}}{L_{dm} + L_{rdP}} \cdot i_{d0} \tag{3.52}$$

This is the third equation sought. An iterative procedure is required to solve Equations 3.42 and 3.43 or 3.45. An initial solution may correspond to $X_{rdl} = X_{ml}$, when Equation 3.52 is used to recalculate the rotor resistance R_{rd0} and thus start the iterative process. Note that neglecting X_{dm} in Equation 3.43 is not acceptable in a machine with PMs. But, alternatively, $V_d(t)$ may also be used with V_{d0} measured to find L_d [use a logarithmic time scale for $V_d(t)$].

$$\ln\ V_d(t) = -\frac{V_{d0} \cdot R_{rd} \cdot t}{L_{dm} + L_{rdl}} \tag{3.53}$$

The same AC standstill tests will be performed in axis q for the main winding to determine R_{rq}, X_{qrl}.

So far, we clarified the parameters' estimation from standstill tests, so we may now proceed to loss segregation, where we are going to use all the parameters estimated above.

3.10.4 Loss Segregation Tests

The losses in the machine (Figure 3.27) occur in the capacitor P_{cap}, stator windings P_{co}, rotor cage P_{al}, stator core P_{iron}, and as mechanical P_{mec}. Stray losses are small and are included (though partially) in P_{co}, as the stator current does not vary much from no load to full load for low-power motors.

The capacitor losses P_{capn} should be measured separately for a certain voltage V_n; thus:

$$P_{cap} = P_{capn}\left(V_{ca}/V_{cn}\right)^2 \tag{3.54}$$

To segregate the losses, the single-phase no-load motoring test is proposed (to start the motor, the auxiliary phase with capacitor is connected, but after synchronization, the latter is turned off;

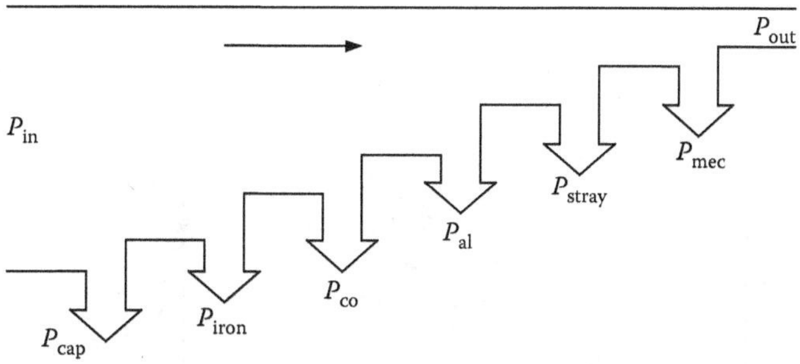

FIGURE 3.27 Power balance in line-star capacitor permanent magnet synchronous machines (C-PMSMs).

Figure 3.28). This test will be performed for various voltage levels. The input power is:

$$P_{m0} = R_m(I_m)^2 + R_{2r}(I_2)^2 + P_{iron} + P_{mec} \tag{3.55}$$

Now, as $I_a = 0$, according to Equation 3.21, applied to currents,

$$\underline{I}_2 = \underline{I}_1 = \underline{I}_m/.707 \tag{3.56}$$

R_{2r} is the rotor equivalent resistance for $S = 2$:

$$R_{2r} = \mathrm{Re}(\underline{Z}_2) - R_m * \mathrm{m} \tag{3.57}$$

Making use of Equations 3.28–3.31 with all parameters known, Equation 3.57 yields R_{2r} and thus the term $R_{2r} * (I_2)^2$ in Equation 3.55 is calculated. The first term in Equation 3.55 is straightforward, as I_m is measured. So, in fact, we have to further segregate only the core loss from the mechanical loss.

To do so, we assume that the core loss is proportional to V_m squared and obtain the characteristics of Figure 3.29.

Notice that we get approximately a straight line whose intersection with the vertical axis produces the mechanical losses, as the latter are independent of voltage.

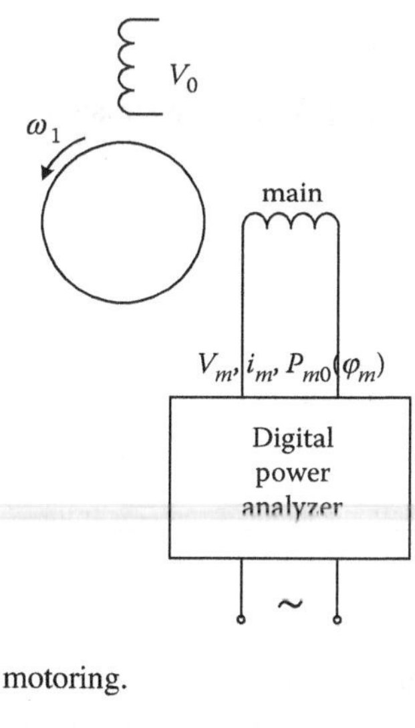

FIGURE 3.28 No-load single-phase motoring.

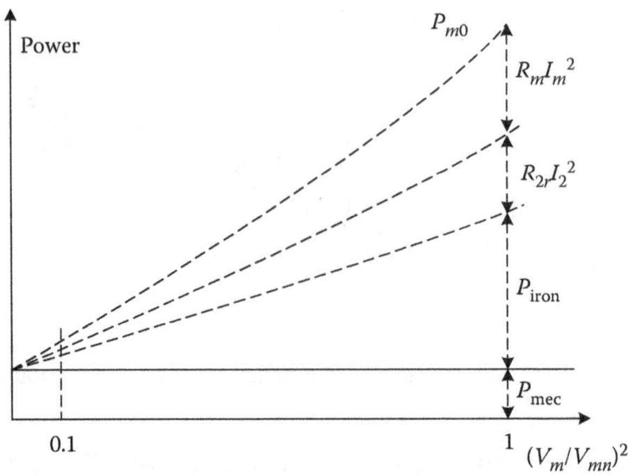

FIGURE 3.29 Single-phase no-load power versus $(V_m)^2$.

Note: A fast verification of some parameter estimations is available, provided we also measure the voltage across auxiliary phase V_{a0} in the one-phase no-load motoring steady state (Figure 3.28):

$$V_{a0} = a * \left|\underline{V}_1 - \underline{V}_2\right| / \sqrt{2} \tag{3.58}$$

Alternatively, the same Equation 3.58 may be used to find a, sparing the DC decay test for the auxiliary phase.

For the rated voltage, we may also assume that the dq current angle for the direct field component $\gamma \approx 90°$; that is, $I_q = 0$, thus obtaining the phasor diagram on Figure 3.30 and Equation 3.59 based on Equation 3.25:

$$E_1 \approx \sqrt{V_1^2 - (R_m I_1)^2} - \omega_1 L_d \underline{I}_1 \tag{3.59}$$

This way, we may also calculate E_1 (PM-induced voltage); thus, the PM flux λ_{PM} required to explore the magnetic saturation effects through DC decay standstill tests becomes available without a zero-current generator test, which would require a driving motor.

3.10.5 Validation Tests

To validate the above methodologies for parameter estimation and loss segregation, two basic tests are used:

- No-load motoring test
- Load test

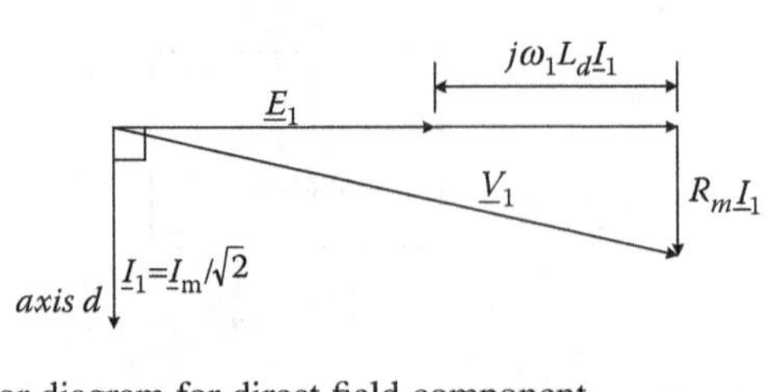

FIGURE 3.30 No-load phasor diagram for direct field component.

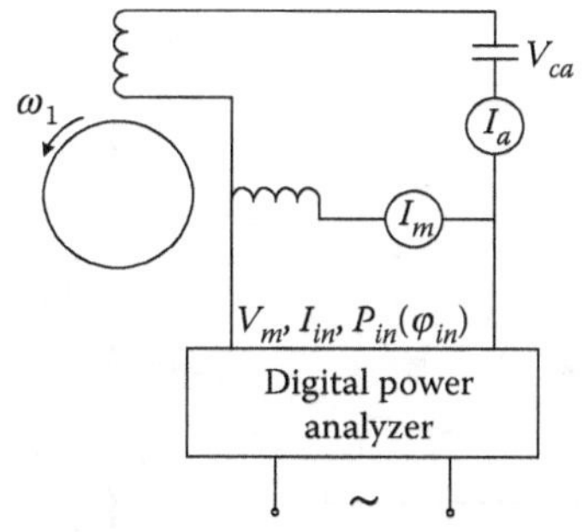

FIGURE 3.31 No-load capacitor motoring.

a. No-load motoring

The arrangement is as in Figure 3.31, and we directly measure all three currents (I_m, I_a, I_{in}), the input power and voltage (P_{in} and V_m), and the capacitor voltage V_{ca}.

Based on Figure 3.32, we may calculate the angles between I_m, I_a, and V_m and thus prepare for calculating I_1 and I_2:

$$\varphi_{am} = -\varphi_{in} + \cos^{-1}\left(\left(I_a^2 + I_{in}^2 - I_m^2\right)/2I_aI_{in}\right) \tag{3.60}$$

$$\varphi_m = +\varphi_{in} + \cos^{-1}\left(\left(I_m^2 + I_{in}^2 - I_a^2\right)/2I_mI_{in}\right) \tag{3.61}$$

The power loss division will be similar to the single-phase no-load case but for different currents and V_1 and V_2. The capacitor losses are first subtracted from the loss division in Figure 3.33.

From Equation 3.21 with Equations 3.60–3.61, we get I_1 and I_2:

$$\underline{I}_1 = \left(\underline{I}_m - j\underline{I}_a \cdot a\right)/\sqrt{2} \tag{3.62}$$

$$\underline{I}_2 = \left(\underline{I}_m + j\underline{I}_a \cdot a\right)/\sqrt{2} \tag{3.63}$$

Also, from Equation 3.28, we get V_2, with Z_2 calculated from Equation 3.60 for $S = 2$. Again:

$$V_m\sqrt{2} = \underline{V}_1 + \underline{V}_2 \tag{3.64}$$

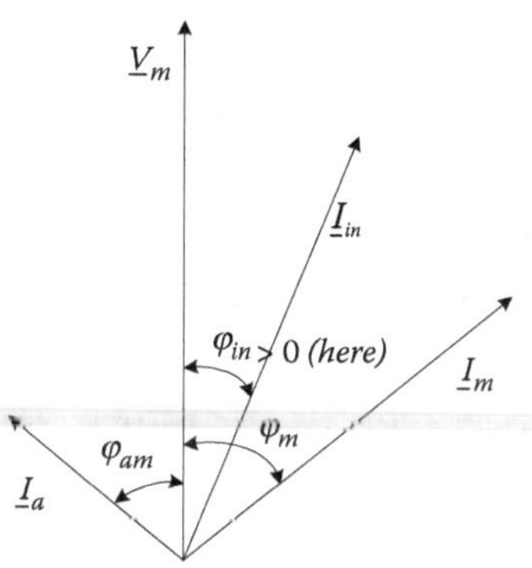

FIGURE 3.32 Phasor angles.

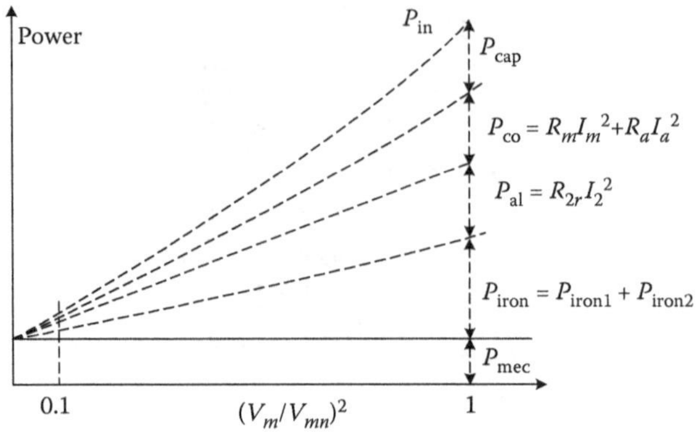

FIGURE 3.33 No-load capacitor motoring power losses versus V_m squared.

serves to find V_1. The mechanical losses should be the same as for the single-phase no-load test. Now, we can separate the core loss from Figure 3.25 and, as we know V_1, we may calculate (validate) R_{iron}. Again, E_1 can be calculated (validated) from Equation 3.59.

b. Load tests

The same set of measurements as above is performed in a load test (Figure 3.34).

From the calibrated load machine, we get the output power P_{out} of our motor, and as the input motor power P_{in} is measured, the efficiency η_m is:

$$\eta_m = \frac{P_{out}}{P_{in}} \tag{3.65}$$

The validation consists of proving the measured efficiency by calculating the output power P_{out} from the measured input power P_{in} minus the sum of segregated losses:

$$P_{outc} = P_{in} - R_m I_m^2 - R_a I_a^2 - R_{2r} I_2^2 - p_{iron} - P_{cap} - P_{mec} \tag{3.66}$$

With R_m and R_a known and I_m and I_a measured, the first two terms in Equation 3.66 are calculated.

Capacitor losses are calculated easily, as V_{ca} is measured. Now, using the same procedure as for no-load capacitor tests, we calculate $\underline{I}_1$, $\underline{I}_2$, $\underline{V}_1$, $\underline{V}_2$, and thus $\underline{R_2 r}$ * $(\underline{I}_2)^2$ and P_{iron} are

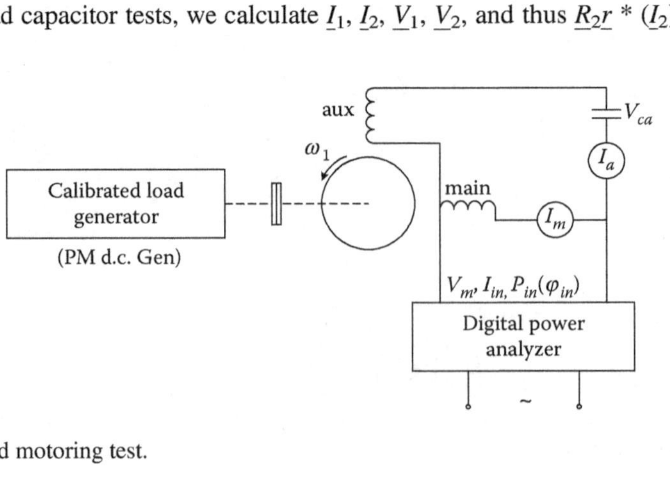

FIGURE 3.34 Load motoring test.

determined. Finally, the calculated efficiency is:

$$\eta_c = \frac{P_{\text{outc}}}{P_{\text{in}}} \tag{3.67}$$

Efficiency is verified for various loads.

Note: To avoid using a calibrated DC machine, two identical one-phase capacitor PM-RSMs may be mounted back to back with their rotor d-axes shifted by twice the estimated rated direct torque angle (2*δ, Figure 3.22). One of them will act as a motor, while the other will be generating. The input power to the motor P_{in} and the total power from the grid ($2^*\sum P_{\text{loss}}$) will be measured; the losses in the two machines are considered equal to each other.

The efficiency η'_m is:

$$\eta'_m = \sqrt{1 - \frac{2\sum P_{\text{loss}}}{P_{\text{in motor}}}} \tag{3.68}$$

An updated approach to the same problem as here is unfolded in [10] for one-phase capacitor IMs.

3.11 SUMMARY

- One-phase source line-start PM-RSMs are being proposed to replace their induction motor counterparts, mainly for home appliances, for notably better efficiency.
- When notably better efficiency is paramount, 1.6–1.8 magnetic saliency cage rotors with strong PMs ($E_{\text{PM}}/V_1 = 0.9$) are used.
- When better efficiency and initial costs have been mitigated, multiple flux barrier (higher saliency: above 2.5–3) cage rotors with low-cost magnets ($E_{\text{PM}}/V_1 = 0.3$—0.4) are to be investigated. In this case, the rotor cage will be embedded in the outer parts of the rotor flux barriers.
- Analytical standard expressions for the stator and rotor resistances and leakage inductances may be used (as for IMs), but if the rotor slots are closed (flux barrier bridges), magnetic saturation of leakage rotor flux paths has to be considered for transients and stability, especially at low loads.
- For the magnetic inductances L_{dm}, L_{qm}, the equivalent magnetic circuit model including magnetic saturation may be used to accomplish this complex goal within lower computation time than when using FEM, especially for the investigation of transients or for optimal design.
- Self-starting (asynchronous operation) and self-synchronization are crucial to line-start machines, together with steady-state (here synchronous) operation performance.
- Steady-state operation design asynchronous operation may be approached best by the dq model using the circuit parameters calculated analytically and based on the EMC model, as mentioned above, but the dq model is instrumental for the investigation of transients and stability.
- During acceleration to speed, the one-phase source line-start PM-RSM exhibits quite large asynchronous and PM-produced torque oscillations and a PM braking torque: this makes the starting performance of this machine inferior to its IM counterpart; this aspect should be carefully investigated in any optimal design methodology.

- Optimal design methodologies, such as the one presented briefly in this chapter with rather extensive sample results, show that an efficiency of 93% for a 100-W, 3000-rpm one-phase source line-start PM (NeFeB)-RSM for a 4.5-kg motor may be obtained; torque is two times the rated torque (0.3 Nm), while the asynchronous torque at 0.05 slip is above the rated torque to secure safe self-synchronization. The objective function contains initial motor cost, capacitor cost and efficiency, starting torque, and 0.05 slip torque as penalty functions. Only a few tens of iterations are required with a modified Hooke–Jeeves optimization algorithm.
- Finally, a complete testing sequence is introduced to estimate (and validate) both the machine parameters and steady-state efficiency for various load levels.
- The use of a weak (Ferrites) and high magnetic saliency (flux barrier) rotor with embedded cage in one-phase source line-start PM-RSMs is still in the early stages of investigation, but important new progress might occur soon.

REFERENCES

1. V. Ostovic, *Dynamics of Saturated Electric Machines*, book, Springer Verlag, New York, 1989.
2. I. Boldea, T. Dumitrescu, S. A. Nasar, Unified analysis of 1-phase a.c. motors having capacitors in auxiliary windings, *IEEE Trans*, vol. EC–14, no. 3, 1999, pp. 577–582.
3. I. Boldea, S. A. Nasar, *Induction Machine Design*, handbook, 2nd edition, Chapter 23, pp. 725, Boca Raton, Florida, CRC Press, Taylor & Francis, 2010.
4. V. B. Honsinger, Permanent magnet machine: Asynchronous operation, *IEEE Trans*, vol. PAS–99, 1980, no. 4, pp. 1503–1509.
5. A. Takahashi, S. Kikuchi, K. Miyata, A. Binder, Asynchronous torque in line-start permanent magnet synchronous motors, *IEEE Trans*, vol. EC–30, no. 2, 2015, pp. 498–506.
6. M. Popescu, T. J. E. Miller, M. I. McGilp, G. Strappazzon, N. Trivillin, R. Santarossa. Line-start permanent magnet motor: Single phase, *IEEE Trans*, vol. IA–39, no. 4, 2003, pp. 1021–1030.
7. I. Boldea, L. N. Tutelea, *Electric Machines,* book, Chapters 13–14, CRC Press, Boca Raton, Florida, Taylor and Francis Group, New York, 2010.
8. C. G. Veinott, Segregation of losses in single phase induction motors, *A.I.E.E. Trans*, vol. 54, no. 2, 1935, pp. 1302–1306.
9. T. J. E. Miller, Single phase permanent magnet motor analysis, *IEEE Trans*, vol. IA–21, no. 3, 1985, pp. 651–658.
10. B. Tekgun, Y. Sozer, I. Tsukerman, Modeling and parameter estimation of split-phase induction motors, *IEEE Trans*, vol. IA–52, no. 2, 2016, pp. 1431–1440.

4 Three-Phase Variable-Speed Reluctance Synchronous Motors

Modeling, Performance, and Design

4.1 INTRODUCTION

Variable-speed drives are required for variable output processes in order to increase productivity and for energy savings. Pumps, ventilators, refrigerator compressors, and so on can all benefit from variable speed.

PMSMs with high-energy magnets have been proposed recently to replace some induction motor variable-speed drives for the scope in order to increase the power factor and efficiency. However, the high price of high-energy PMs puts this solution in question, especially for initial cost constraint applications.

This is how reluctance synchronous motors [1], with or without assisting PMs (with low-cost Ferrites or even NdFeB magnets in small quantities), where the reluctance torque (produced through high enough magnetic rotor saliency) is predominant, come into play. In general, in such motor drives, $e_{PM}/V_s < 0.3$ and thus a total magnetic saliency $L_d/L_q > 3.5$–5 is required for reasonable performance (with respect to IMs) and initial cost; yes, still subject to optimal design. Especially for a wide constant power speed range CPSR > 2.5–3 and power factor PF > 0.8, assisting PMs are a must.

Another great divide is between distributed stator windings ($q_1 \geq 2$–3, slots/pole/phase) and tooth-wound (concentrated) stator windings ($q_1 < 1$). By now it has been demonstrated that for high enough magnetic saliency, distributed windings are preferable [2] but not exclusive when fabrication cost is to be reduced.

In addition to high ratios of pole pitch τ/airgap g: $\tau/g > 100$–150 and $q_1 \geq 3$, for high saliency, distributed anisotropy rotors with either multiple flux barriers per pole (multiple flux barrier anisotropic [MFBA]-rotor) or with axial laminations interspersed with insulation layers (ALA-rotor) (Figure 4.1a and b) [2] are to be used.

The merits and demerits of MFBA- and ALA-rotors are summarized below:

MFBA-rotor

- It uses regular (transverse) laminations and stamping to produce the rotor flux barriers (easy fabrication).
- It produces, for given q_1 and τ/g ratios, a mild saliency ratio L_d/L_q, which leads to similar torque densities with IMs (of the same stator), at a few percent (+4%–1.5%) higher efficiency but at a smaller power factor (−10%–15% [3]. Only with assisting PMs can the power factor and wide CPSR (>2.5–3) of IMs be reached or surpassed.
- The total cost of initial materials and fabrication is estimated to be 12%–13% smaller than for variable-speed IMs.

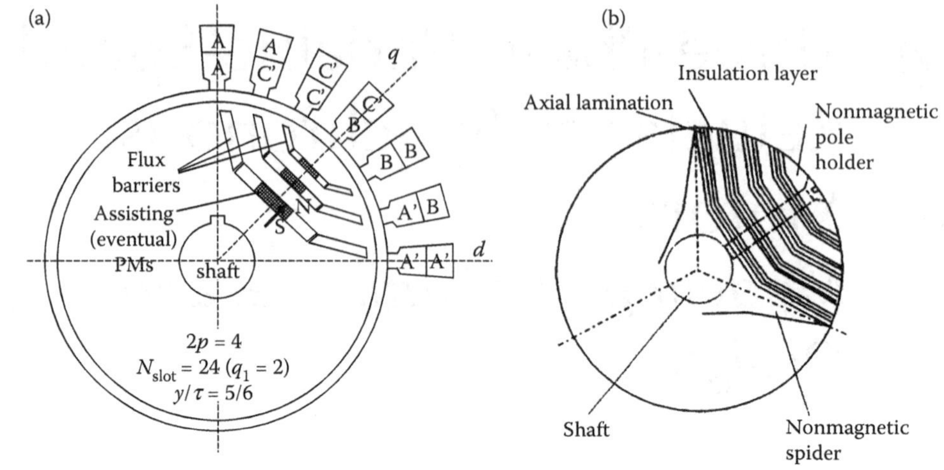

FIGURE 4.1 Typical RSM topologies: (a) with MFBA-rotor; (b) with ALA-rotor and distributed three-phase stator winding.

- MFBA-rotor RSMs have recently reached industrial fabrication, from 10 kW to 500 (1500) kW for variable-speed drives [3], so far without assisting PMs and mostly with four poles.

ALA-rotor

- It needs axial (transformer) laminations cut out in a dedicated fabrication rig and, after assembly on rotor poles, mechanical machining of the rotor is required.
- It may produce higher torque density, higher efficiency, and a higher power factor than the IM (of the same stator) due to higher magnetic saliency, especially for $2p = 2$, 4 poles, even without assisting PMs.
- The total initial cost is smaller, but the fabrication cost seems higher than for IMs.
- The q axis stator-current flux space harmonics produce eddy current losses (even on no load) in the rotor axial laminations; they may be, however, reduced by lamination thin slits placed at three to four axial positions.
- The two-pole ALA-rotor 1.5-kW RSM has been proven to produce a peak power factor of 0.91 (with no assisting PMs) in a non-through-shaft rotor configuration [4].

Due to the presence of real (or virtual, by flux bridges) slot openings in the rotor, in addition to the stator and with magnetic saturation, the flux distribution and inductance calculation with high precision should be approached by 2D FEM rather than by analytical methods. Also, for short stack length (l_{stack})/pole pitch (τ) ratios, say, $l_{stack}/\tau < 1.2$–1.5, even 3D (axial) field components should be considered to further improve results, especially if skewing is applied for torque ripple reduction.

However, to grasp the key factors (variables) that influence flux distribution and demagnetization inductance L_{dm}, L_{qm} variation, approximate analytical methods are still useful, especially if magnetic saturation is also considered [5]. The latter methods prove useful to drastically reduce the computation effort (time) in optimal design algorithms, though computationally efficient FEM models have recently begun to penetrate optimal design methodologies from the start [6].

The above arguments led us to the treatment of both analytical and FEM modeling of RSM in order to calculate flux distribution, L_{dm}, L_{qm}, inductances, torque, losses, and performance, and then to optimal design algorithms, with quite a few representative case study results.

4.2 ANALYTICAL FIELD DISTRIBUTION AND $L_{dm}(I_d)$, $L_{qm}(I_q)$ INDUCTANCE CALCULATION

Here, we will examine first the MFBA-rotor topology, as it reached industrial fabrication up to 500 (1500) kW/unit without assisting PMs, then the PM-assisted MFBA-rotor. Finally, the ALA-rotor topology without assisting PMs will be tackled. We mentioned magnetization inductances in the title above since they have different expressions from those of IMs, while the leakage inductances have expressions similar to those of IMs and line-start three-phase RSMs in the previous chapters.

4.2.1 THE MFBA-ROTOR

The extended magnetic circuit, accounting for slotting on both the stator and rotor and for magnetic saturation, may be used for the purpose [7], with a computation effort more than 10 times smaller than that for 2D-FEM, but still large and, inevitably, obscuring a few key parameters that predominantly influence q axis magnetization inductance, which, in turn, is crucial for torque density, losses, and power factor calculations.

Hereby, first, a simple two-pole MFBA non-through-shaft rotor is considered (Figure 4.2a), where the thin flux bridges are considered by equivalent slot openings b_{sr}, a hypothesis that is easy to implement when the latter saturate early (at small load already) and have a tangential length l_b, a thickness t_b, and a low relative permeability (corresponding to a given $B_s = 2$ T): $\mu_{br} = 10$–50.

$$b_{sr} \approx \frac{l_b}{\mu_{br}} \tag{4.1}$$

So, in fact, the rotor becomes, for calculus, segmented (as the ALA-rotor) but still uses regular transverse laminations. Consequently, this method [8] may also be safely used for the ALA-rotor. Stator slotting is still disregarded.

For the kth rotor segment in Figure 4.2b and c, the mean value of the q axis stator mmf is f_k, while the former's magnetic potential is r_k. While the d axis field distribution and inductance L_{dm} are straightforward, for axis q, there are two main flux inductance components:

- One flux that goes through the rotor segments and across flux barriers L_{qmf}
- A circulating component across the airgap (due to the shaded area mmf variation in Figure 4.2c) L_{qmc}

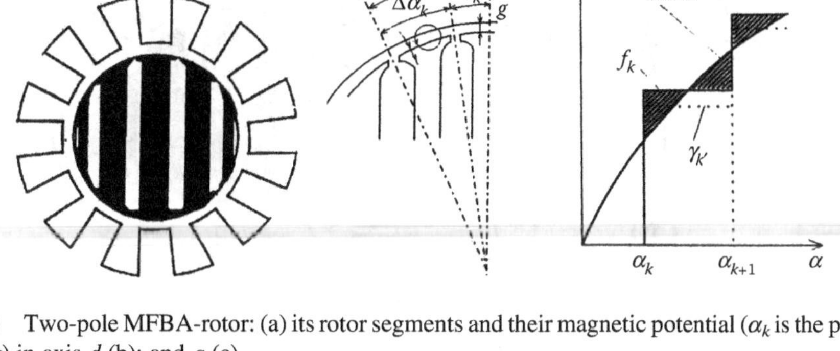

FIGURE 4.2 Two-pole MFBA-rotor: (a) its rotor segments and their magnetic potential (α_k is the positive angle of segment k) in axis d (b); and q (c).

According to the above assumptions, the magnetization inductances are:

$$L_{dm} \approx A \int_{-\pi}^{\pi} \sin^2 \alpha d\alpha = A\pi$$

$$L_{qmc} = A \int_{-\pi}^{\pi} [\sin \alpha - f(\alpha)] \sin \alpha d\alpha \tag{4.2}$$

$$L_{qmf} = A \int_{-\pi}^{\pi} [f(\alpha) - \gamma(\alpha)] \sin \alpha d\alpha$$

Using the staircase approximation (Figure 4.2c):

$$\frac{L_{qmc}}{L_{dm}} = 1 - \frac{4}{\pi} \sum f_k^2 \Delta\alpha_k$$
$$\frac{L_{qmf}}{L_{dm}} = \frac{4}{\pi} \sum f_k \cdot (f_k - \gamma_k) \Delta\alpha_k \tag{4.3}$$

Apparently, the ratio L_{qmc}/L_{dm} decreases with an increasing number of rotor flux barriers (segments) such that, for four segments/pole (Figure 4.3), the ratio $L_{qmc}/L_{dm} \approx 0.027$ [8].

Now, to calculate L_{qmf}/L_{dm}, the calculation of magnetic potentials of rotor segments γ_k is needed.

The number of stator slots (though considered approximately by Carter's coefficient) also influences the magnetic potential of rotor segments. For simplification, let us consider here that the number of stator slots and rotor segments is the same (though, in reality, this situation has to be avoided to reduce torque ripple).

The magnetic equivalent circuit in axis q (Figure 4.3a) allows us to calculate the p.u. magnetic potentials γ_1, γ_2 of rotor segments (l_1, l_2 are flux barrier widths and S_i is their length per half pole pitch; S_s—stator slot pitch). With $\Delta f_k = f_k - f_{k-1}$, L_{qmf} in Equation 4.3 becomes

$$\frac{L_{qmf}}{L_{dm}} = \frac{4}{\pi} p \frac{k_c g}{\gamma} \sum \Delta f_k^2 \frac{s_k}{l_k}; \tag{4.4}$$

where: g—airgap, γ—rotor radius, and p—pole pairs of the machine.

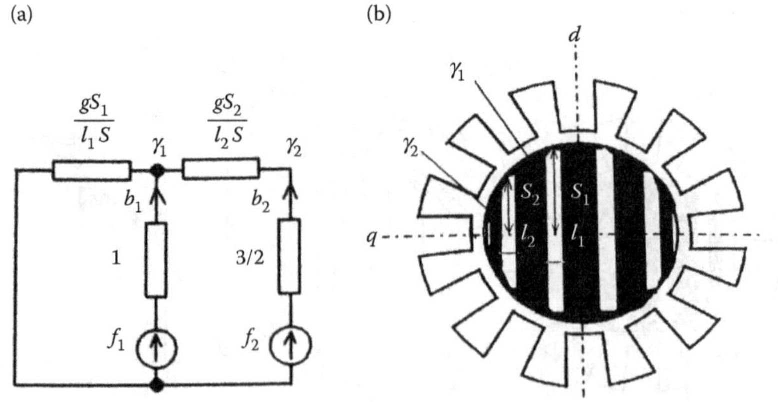

FIGURE 4.3 Equivalent magnetic circuit in axis q to calculate rotor segment magnetic potentials γ_1 and γ_2, (a); and the machine rotor and stator slotting, (b).

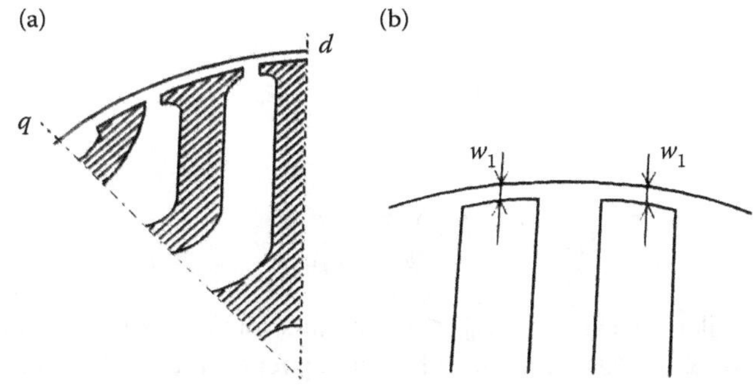

FIGURE 4.4 Four-pole rotor optimized geometry (a) and rotor saturated flux bridges (ribs) (b).

If a shaft (of radius a) is present, the radial dimension along axis q (Figure 4.4a) may be divided into iron l_{iq} and air l_{aq}:

$$l_{iq} + l_{aq} = r - a \tag{4.5}$$

Considering l_{iq} constant (as the d axis pole flux may be fixed), with r and a given, l_{aq} is given. With n rotor slots per pole, $n - 1$ partial derivatives of L_{qmf}/L_{dm} with respect to l_k will be zero:

$$l_1 = l_{aq} - \sum_{k=2}^{n} l_k; \ \frac{\partial l_1}{\partial l_k} = -1; \ \frac{\partial(L_{fqm}/L_{dm})}{\partial l_k} = 0 \tag{4.6}$$

$$\text{Finally:} \frac{l_k}{l_1} = \frac{\Delta f_k}{\Delta f_1}\sqrt{\frac{s_k}{s_1}} \tag{4.7}$$

Consequently, the thickness l_k of a flux barrier is proportional to the square root of its depth per half-pole pitch s_k.

With s_k decreasing toward the q axis, so will l_k; thus, for a four-pole rotor, $L_{qmf}/L_{dm} \approx 0.04$ [8].

From Equations 4.4 and 4.7, L_{qmf}/L_{dm} becomes:

$$\frac{L_{qmf}}{L_{dm}} = \frac{4}{\pi} p \frac{k_c g}{k_c g + l_{aq}} \sum \Delta f_k^2 \frac{s_k}{l_k} \tag{4.8}$$

The rotor flux bridges (ribs) introduce an additional component for the q axis magnetization inductance L_{qmribs}, which diminishes both average torque and power factor:

$$L_{\text{qmribs}} \approx \frac{4}{\sqrt{3}} l_{\text{stack}} \frac{N_1 K_{w_1}}{i_q} B_s \tag{4.9}$$

where N_1—turns/phase, K_{w1}—stator winding factor.

As B_s (flux density in the rotor flux bridges) varies with i_q (stator current in axis q), an approximate ratio $B_s/i_q \approx$ ct for saturated conditions may be calculated:

$$B_s/i_q = \mu_{rb}\mu_0 H_{\text{bridge}}/i_q; \ H_{\text{bridge}} \cdot n \cdot l_b \approx \frac{3\sqrt{2}N_1 k_{w1} i_q}{p \cdot q_1} \tag{4.10}$$

Again, μ_{rb} is the average iron p.u. permeability in the rotor flux bridges.

Thus, the total magnetization inductance L_{qm} is:

$$L_{qm} = L_{qmf} + L_{qmc} + L_{\text{qmribs}}, \tag{4.11}$$

while for axis d:

$$L_{dm} \approx \frac{6 \cdot \mu_0 (N_1 k_{w1})^2 \tau \cdot l_{\text{stack}}}{\pi^2 k_c \cdot g \cdot p_1} \cdot k_{dm} \tag{4.12}$$

Equation 4.12 is similar for IMs, but $k_{dm} < 1$ accounts for the case when the airgap is increased notably close to axis q, as in ALA-rotors with nonmagnetic pole holders, or as in regular salient pole synchronous machines:

$$k_{dm} \approx \frac{\tau_s}{\tau} + \frac{1}{\pi}\sin\frac{\tau_p}{\tau}\pi \leq 1;\ \tau_p - \text{pole shoe};\ \tau - \text{pole pitch} \tag{4.13}$$

Still, magnetic saturation has not been considered in Equation 4.13. In general, to secure good performance but still reduce the stator outer diameter (yoke depth), the stator is designed more saturated than the rotor. As an equivalent airgap g_e may be used instead of g: $g_e = g \cdot (1 + K_{ss})$, where K_{ss} defines the machine saturation equivalent level, values of K_{ss} up to 0.3–0.5 may be accepted, as RSMs, in general, have a small airgap to reduce magnetization current and thus secure a satisfactory power factor.

It is recognized here that K_{ss} may be calculated analytically but, for preliminary designs, it may be assigned different values; later on, by FEM, K_{ss} may be calculated and the design adjusted accordingly.

4.2.2 The MFBA-Rotor with Assisting Permanent Magnets

The temptation to develop analytical approaches for field distribution, magnetization inductances L_{dm}, L_{qm}, and $e_{PM\alpha}$ (the PM emf with PMs in axis q), which are considered small, e_{PMd} ($V_s < 0.3$), to secure lower motor initial cost, but a satisfactory power factor and CPSR, is strong, at least for preliminary designs.

But the optimal design of a PM-RSM also has to consider PM demagnetization avoidance (calculated in general for 2 p.u. current). For Ferrite-PMs, to avoid demagnetization, the thickness of flux barriers is recommended to be the same along their length (Figure 4.5a) [9].

Alternatively, short NdFeB magnets may be placed on the bottom of rotor flux barriers (Figure 4.5b) [10].

For the case in Figure 4.5a, a complete analytical design is available in [9].

It is based, again, on a per-unit equivalent circuit (Figure 4.6a and b).

When all angles (positions) of flux barriers—$\Delta\xi_i$—are equal, the rotor (called "complete") can be easily defined if the number n of equivalent rotor slots per pole pair is given. For the sake of simplicity, this case is treated here. The f_1, f_2, f_3 in Figure 4.6 refer to mmfs; p_1, p_2, p_3 refer to flux barrier permeances; r_1, r_2, r_3 refer to magnetic potentials of rotor flux segments; and p_g is airgap magnetic permeance. The fluxes are normalized with respect to $\mu_0 F_q l_{\text{stack}}$, as F_q is divided by $\mu_0 l_{\text{stack}}$. Thus, the PM p.u. mmf m_k is:

$$m_k = \frac{B_r l_k}{\mu_0 F_q};\ B_r - \text{PM remnant flux density} \tag{4.14}$$

$$p_k = \frac{s_k}{l_k} \approx ct;\ p_g = \frac{\gamma \Delta\xi_r}{\pi k_c g} \tag{4.15}$$

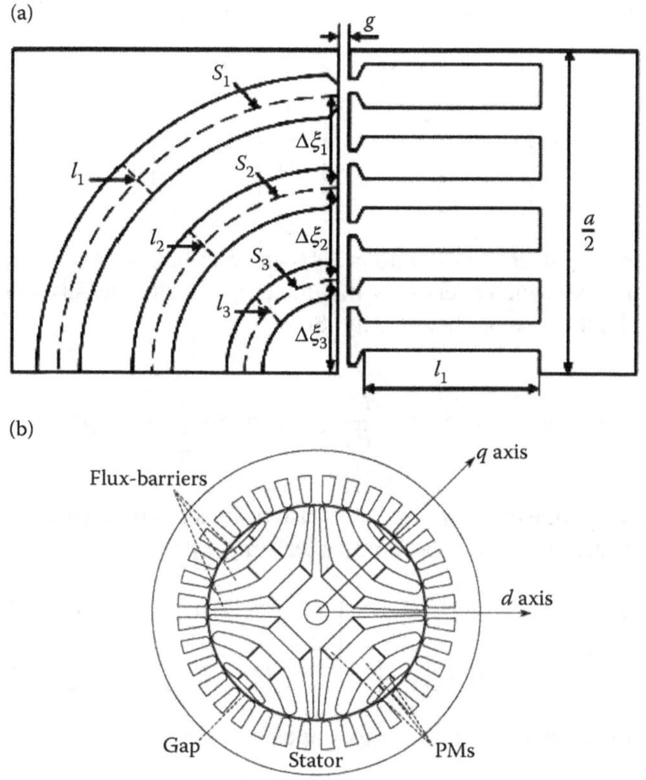

FIGURE 4.5 Circular MFBA-rotor, (a); boat-shaped MFB-rotor, (b) with assisting PMs.

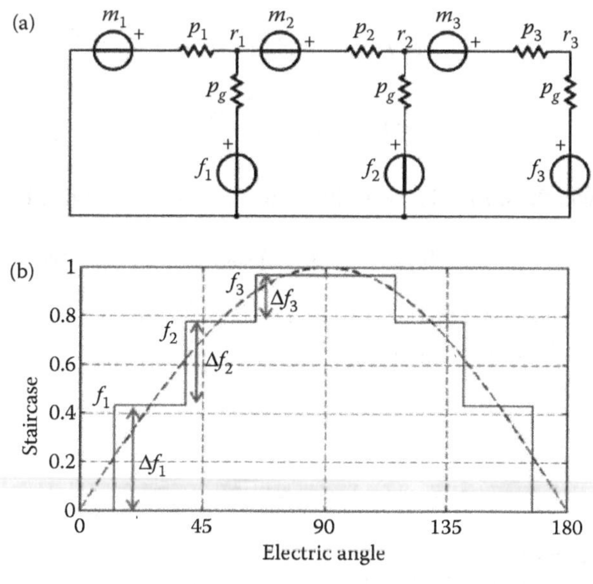

FIGURE 4.6 EMC for three flux circular barriers per rotor pole, (a); and the stator mmf step function, (b).

The EMC in Figure 4.6 has the standard solution matrix equation from [9]:

$$A_1 \Delta\gamma = Bm + C\Delta f \tag{4.16}$$

$$\Delta\gamma = \begin{bmatrix} \gamma_1 \\ \gamma_2 - \gamma_1 \\ \gamma_3 - \gamma_2 \end{bmatrix}; \quad m = \begin{bmatrix} m_1 \\ m_2 \\ m_3 \end{bmatrix}; \quad \Delta f = \begin{bmatrix} f_1 \\ f_2 - f_1 \\ f_3 - f_2 \end{bmatrix} \tag{4.17}$$

To simplify the matrices A, B, C in Equation 4.16, the flux barrier permeances p_k are designed to make $\Delta\gamma$ proportional to Δf, which eventually minimizes q axis flux density space harmonics, while all magnets work at about the same flux density. So:

$$m_1 = m_2 = m_3 = m = \frac{M}{F_q}\Delta f; \; l_k = \Delta f_k \frac{\sum_n l_k}{\sum_n \Delta f_k} = \Delta f_k \frac{l_{aq}}{F_q} \tag{4.18}$$

But equal flux barrier permeance ($p_k = p_b = \text{ct}$) also insures minimum q-axis flux density harmonics and, in this case, matrices A_1, B, C become rather simple [9]:

$$A_1 = \begin{bmatrix} \beta+1 & -\beta & 0 \\ 1 & \beta+1 & -\beta \\ 1 & 1 & \beta+1 \end{bmatrix}; \; B = \begin{bmatrix} \beta & -\beta & 0 \\ 0 & \beta & -\beta \\ 0 & 0 & \beta \end{bmatrix}; \; C = \begin{bmatrix} 1 & 0 & 0 \\ 1 & 1 & 0 \\ 1 & 1 & 1 \end{bmatrix} \tag{4.19}$$

Finally, the PM flux density B_m (with i_q flux opposite to PM flux) is:

$$\left(B_{mpu}\right)_k = \frac{B_m}{B_r} = 1 - \frac{\Delta r_k}{m_k} \tag{4.20}$$

Equation 4.20 notably eases the assessment of the magnetization stage of PMs, as all magnets suffer the demagnetization effect of large i_q stator currents equally (PMs are in axis q).

For no load ($\Delta f = 0$), Equations 4.16–4.19 yield the p.u. PM flux density $B_{m0\text{p.u.}}$[9]:

$$B_{m0p.u.} = \frac{1}{1 + ((s_1/l_{aq}) \cdot (g/\tau) \cdot (2\pi/\Delta\xi_r)) \sin(\Delta\xi_r/2)} \approx \frac{1}{1 + ((\pi^2/2l_{aqp.u.}) \cdot (g/\tau))} \tag{4.21}$$

$$\text{with } l_{aqp.u.} = \frac{2 \cdot l_{aq}}{\tau}$$

Once $B_{m0p.u.}$ is known, the no-load airgap flux density may be calculated. Consequently, $E_{\text{PM}d}$ is:

$$E_{\text{PM}d} = \omega_1 \cdot \Psi_{\text{PM}q01}; \quad \Psi_{\text{PM}q01} - \text{peak pole airgap flux linkage by PMs} \tag{4.22}$$

with stator resistance R_s, leakage inductance L_{sl} is easy to calculate, and with $L_d = L_{dm} + L_{sl}$ and $L_q = L_{qm} + L_{sl}$, we now have all parameters needed for the circuit model of the PM-RSM.

4.3 THE AXIALLY LAMINATED ANISOTROPIC–ROTOR

As mentioned above, the analytical methods applied to the MFBA-rotor are basically also valid for the ALA-rotor, with $L_{\text{qmribs}} = 0$.

The saturated rotor iron ribs have been proved to reduce the torque by as much as 10%–15% with respect to their absence.

Also, the ALA-rotor basically allows making the rotor poles by alternating thin laminations (say, 0.35 mm thick) with insulation layers (say, 0.2 mm thick).

It is like the number of multiple flux barriers has been increased notably.

A high-order harmonic occurs due to this "multislot" rotor both in the d and q axis airgap flux densities, but the magnetic saliency is proved larger than for MFBA-rotors [4,10].

The analytical method based on defining a variable equivalent airgap was also applied simply to an ALA-rotor [4,11–12], while a more sophisticated approach was developed in [13]. Reference 4 investigates a two-pole ALA-rotor RSM with its topology in Figure 4.7.

Also, adopting an equivalent stator saturation factor K_{ss}, the equivalent airgap is augmented by Δg, dependent on K_{ss} and on the power angle δ [13]:

$$\Delta g(sat, \delta) = \frac{3(K_{ss}+1)}{(K_{ss}+1)+2+2K_{ss}\cos 2\delta} \tag{4.23}$$

Thus, the inverted airgap function $g^{-1}(\delta)$ may be written as:

$$g^{-1}(\delta) = (g^{-1} - k_m) + k_m \cos 2\delta; \; k_m = \frac{2}{3}\frac{K_{ss}}{g(K_{ss}+1)} \tag{4.24}$$

Again, K_{ss} is only the contribution of iron and is thus $K_{ss} \approx 0.2$–0.5 in RSMs.

Starting with the calculation of phase self and mutual inductances $|L_{abc}|$ and using the Park transformation $P(\theta_{er})$, the $dq0$ inductance matrix L_{dq0} is obtained:

$$L_{dq0} = \begin{vmatrix} L_d & L_{dq} & L_{d0} \\ L_{dq} & L_q & L_{q0} \\ L_{d0} & L_{q0} & L_{00} \end{vmatrix}; \; |L_{dq0}| = |P(\theta_{er})| \cdot |L_{abc}| \cdot |P(\theta_{er})|^{-1} \tag{4.25}$$

Figure 4.8a and b [4] shows typical analytical results of k_{dm} and k_{qm} (p.u.) with respect to constant airgap inductances showing some remarkable peculiarities.

FEM and experimental verifications have shown that the average L_d, L_q are calculated with acceptable errors by the analytical method (Figure 4.9).

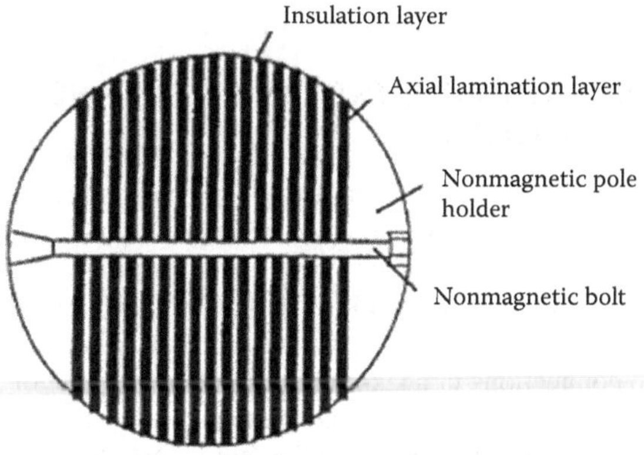

FIGURE 4.7 Two-pole ALA-rotor RSM with no through-shaft. (After I. Boldea et al., *Conference Record of the 1992 IEEE Industry Applications Society Annual Meeting*, Part I, pp. 212–218. [4])

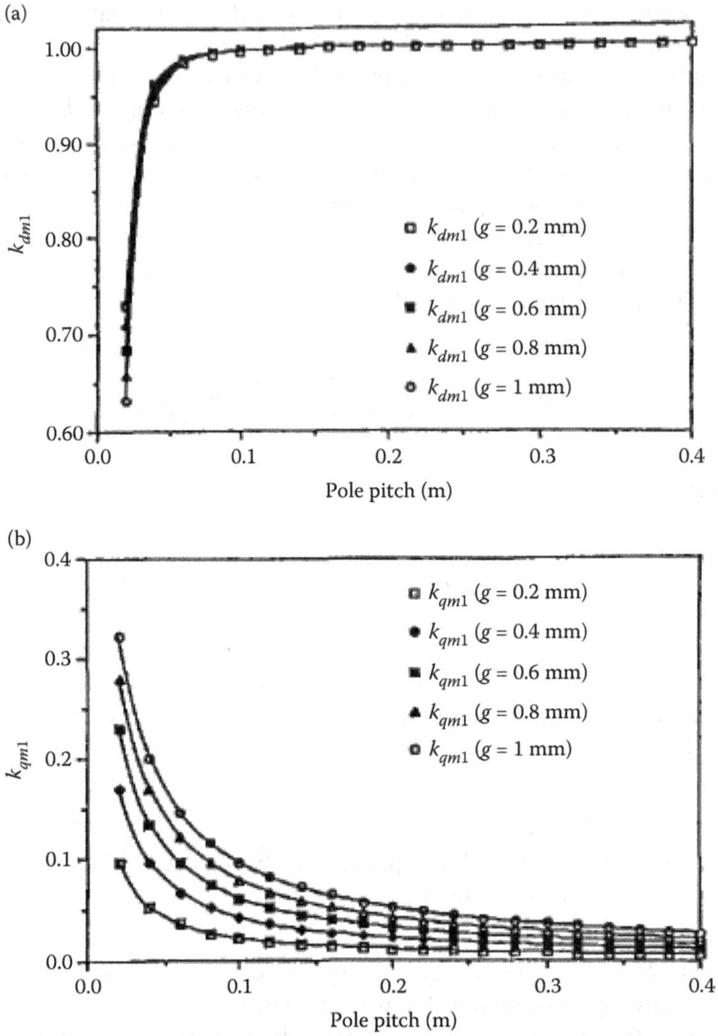

FIGURE 4.8 Inductances (analytical): (a) k_{dm}; (b) k_{qm}. (After I. Boldea et al., *Conference Record of the 1992 IEEE Industry Applications Society Annual Meeting*, Part I, pp. 212–218. [4])

4.4 TOOTH-WOUND COIL WINDINGS IN RELUCTANCE SYNCHRONOUS MOTORS

Earlier in this chapter, it was implied that distributed ($q \geq 2$) windings are more suitable for RSMs because they exploit the magnetic saliency better, mainly because the machine leakage inductance is smaller due to lower space harmonics content of the stator mmf.

An attempt to compare tooth-wound coil (TWC) and distributed (D) windings for same stator has shown [15] that by allowing a larger slot fill factor (58%), TWC windings may provide similar efficiency due to shorter end connections in a six-slot/four-pole configuration (Figure 4.10). However, the magnetic saliency is notably smaller than for D windings, and so is the power factor (from 0.62 to 0.483!). As the inverter cost is usually higher than the motor cost, the lower power factor, due to the additional (leakage) inductance of multiple stator mmf space harmonics, means more inverter current and kVA and thus more drive initial costs for the TWC-RSM.

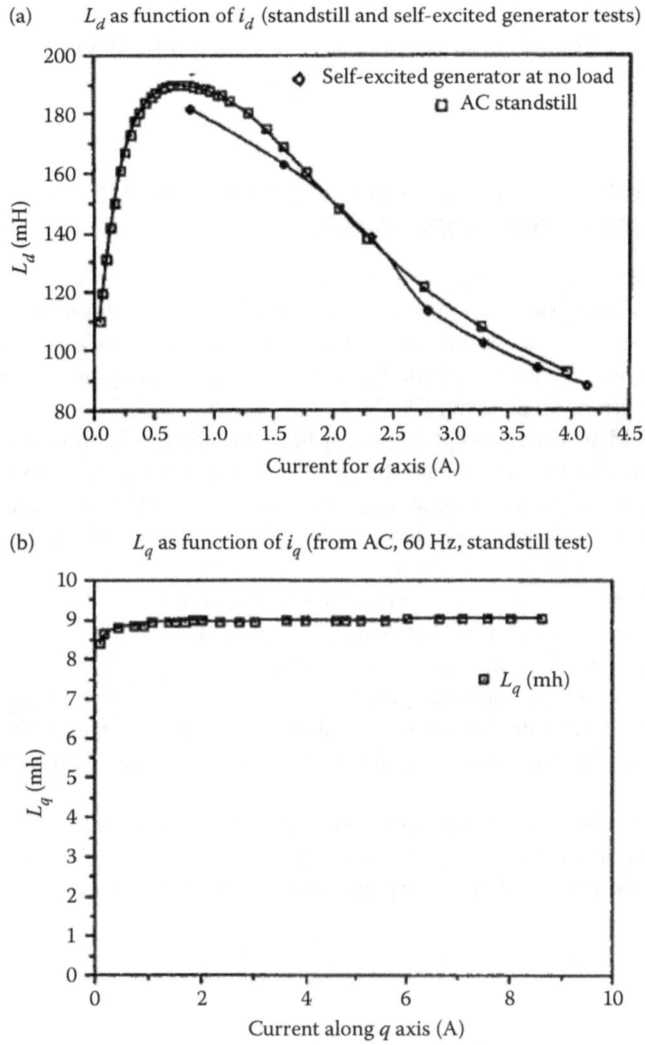

FIGURE 4.9 L_d, (a); L_q, (b) versus phase current (experimental). *Note*: The presence of PMs (instead of insulation layers) has to be treated as for the MFBA-rotor. (After I. Boldea et al., *Conference Record of the 1992 IEEE Industry Applications Society Annual Meeting*, Part I, pp. 212–218. [4])

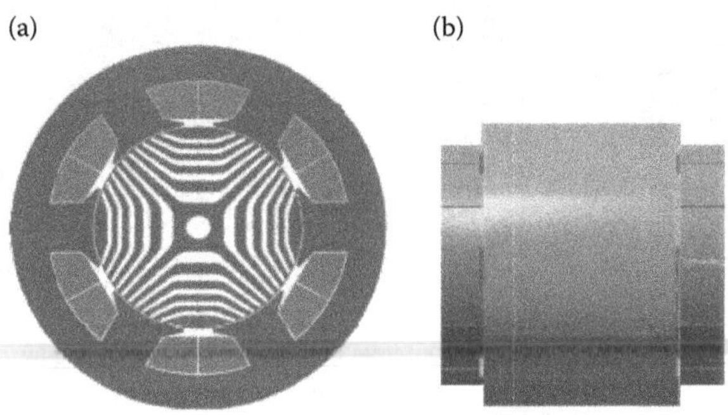

FIGURE 4.10 Six-slot/four-pole TWC-RSM: (a) cross-section; (b) stator prototype. (After C. M. Spargo et al., *2013 International Electric Machines & Drives Conference*, pp. 618–625. [15])

Still, in initial cost–critical applications, the simplicity of stator winding and the shorter frame (due to the shorter end connections of stator coils) may lead to lower fabrication costs that may result in an overall beneficial solution, especially if Ferrite-PMs are added for power factor correction.

4.5 FINITE-ELEMENT APPROACH TO FIELD DISTRIBUTION, TORQUE, INDUCTANCES, AND CORE LOSSES

The rather "crude" analytical methods so far have produced expressions for main circuit parameters: R_s, L_d, L_q, Ψ_{PM01q} (E_{PMd}). However, the presence of slotting, the multiple flux barrier rotor, and magnetic saturation lead to space (time) harmonics in flux density in various parts of the machine, accompanied by torque pulsations and additional core losses and cogging torque (when PM assistance is available), warranting the use of FEM.

In general, 2D-FEM suffices, except for axis q field distribution in short stator stacks ($l_{stack}/\tau <$ 1.5) when some axial flux fringing occurs (because L_{qm} is small due to high magnetic reluctance) and should be accounted for by an increase in L_{qm} that may go up to 10% and deteriorate performance. FEM analysis of electric machines is a technique in itself now treated in dedicated books [16,17], with a rich literature [18].

Here, only sample results are given to illustrate various aspects.

Early FEM results on field distribution and airgap flux density variation in an ALA-rotor two-pole RSM in axes d and q are shown in Figure 4.11a and b.

They have been obtained at standstill, placing the rotor in axis d with respect to phase a by supplying phase a in series with phases b and c in parallel ($i_a = -2i_b = -2i_c$); to simulate the behavior in axis q, only phases b and c in series are supplied with ($i_b = -i_c$), while the rotor is stalled in the position for axis d.

The airgap flux densities in axes d and q (Figure 4.12) exhibit high pulsations (due to rotor lamination/insulation multiple layers) in axis d and positive/negative pulsation due to stator slotting in axis q. This way, the ratio of flux density fundamentals is high, and so is L_{dm}/L_{qm}:

$$L_{dm}/L_{qm} = \frac{B_{qd1}}{B_{gq1}} \cdot \frac{i_b}{i_a} \frac{2}{\sqrt{3}} \quad (4.26)$$

$i_d = i_a$ (axis d): three phases are active; $i_q = i_b \cdot \dfrac{2}{\sqrt{3}}$ (axis q): two phases are active.

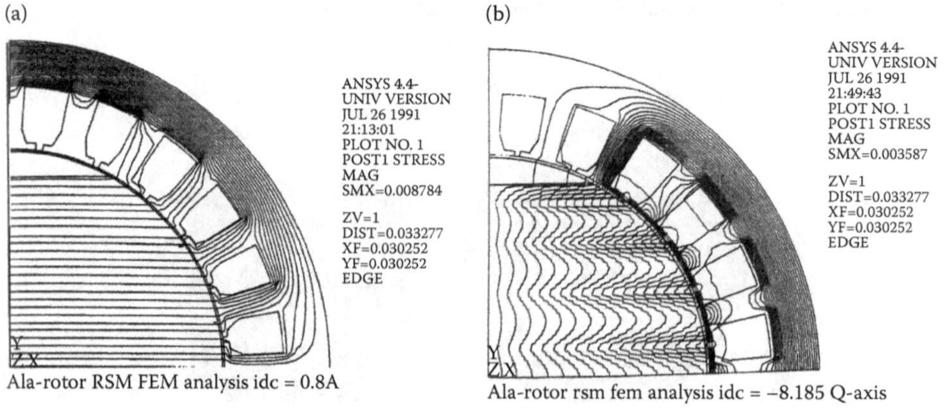

FIGURE 4.11 ALA-rotor two-pole RSM (1.5 kW, 50 Hz): FEM field lines: (a) in axis d; (b) in axis q. (After I. Boldea, *Reluctance Synchronous Machines and Drives*, book, Oxford University Press, New York, 1996. [2])

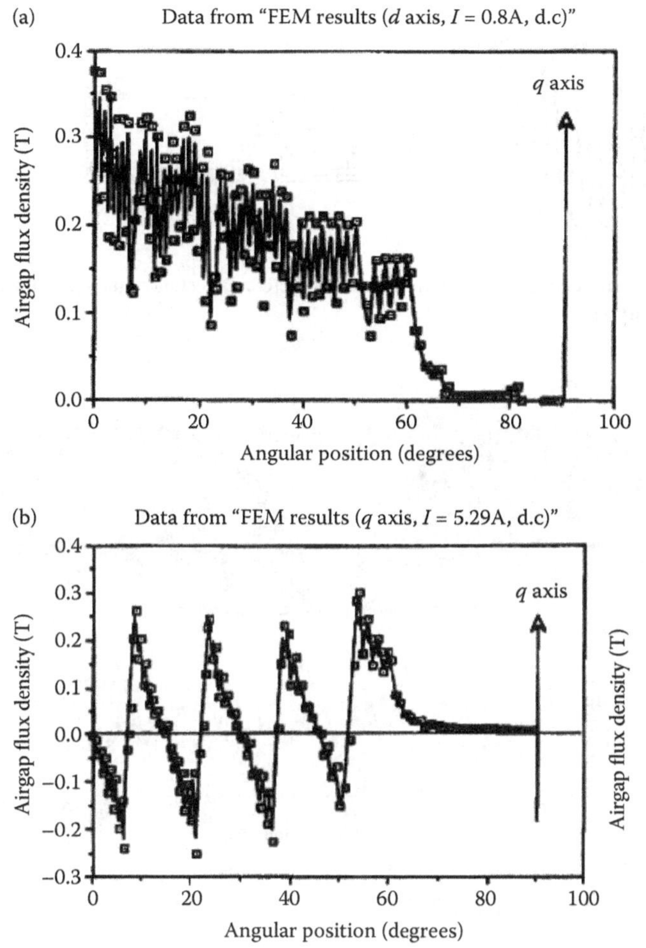

FIGURE 4.12 ALA-rotor two-pole RSM (1.5 kW, 50 Hz): FEM airgap flux density: (a) in axis d; (b) in axis q.

A value L_{dm}/L_{qm} for a full load of 30 was FEM-calculated (and measured) in [4] for an ALA-rotor, two-pole RSM (1.5 kW, 50 Hz).

However, the q axis flux density plus/minus pulsations within the pole span lead to high eddy current losses in the rotor axial laminations (because the flux lines go perpendicularly to the rotor lamination plane). To reduce them to reasonable values, a few (three to five) thin radial slits are machined in the finished rotor (Figure 4.13).

Typical FEM calculation results referring to torque pulsations for MFBA-rotor RSMs with and without assisting PMs are shown in Figure 4.14 [19].

A few remarks are in order:

- The geometry of rotor flux barriers is different, since the placement of PM requires simplifications.
- As expected, the average torque for the same stator and same currents is smaller for RSMs, because the assisting Ferrite PMs add some torque (though their main role is to improve the power factor and thus enlarge CPSR).
- There are torque pulsations due to "slotting" and magnetic saturation (RSM) and PMs (cogging torque) in PM-RSMs: torque ripple in both cases is around 30% and thus optimal design efforts should reduce it when the application requires it, to, perhaps, 8%–10%.

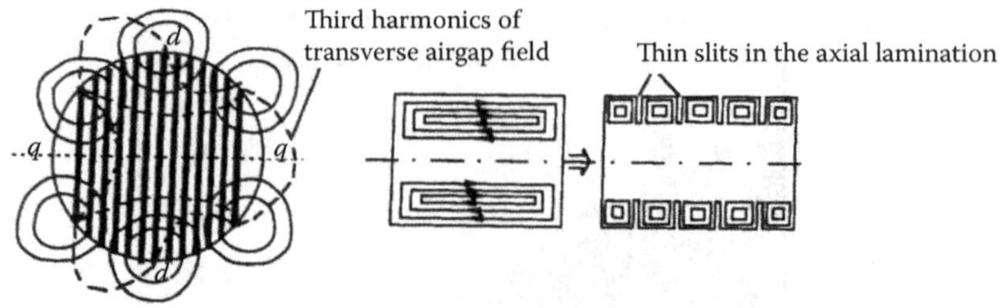

FIGURE 4.13 Eddy currents in rotor axial laminations due to q axis stator flux space harmonics (Figure 4.12b): (a) without slits; (b) with slits.

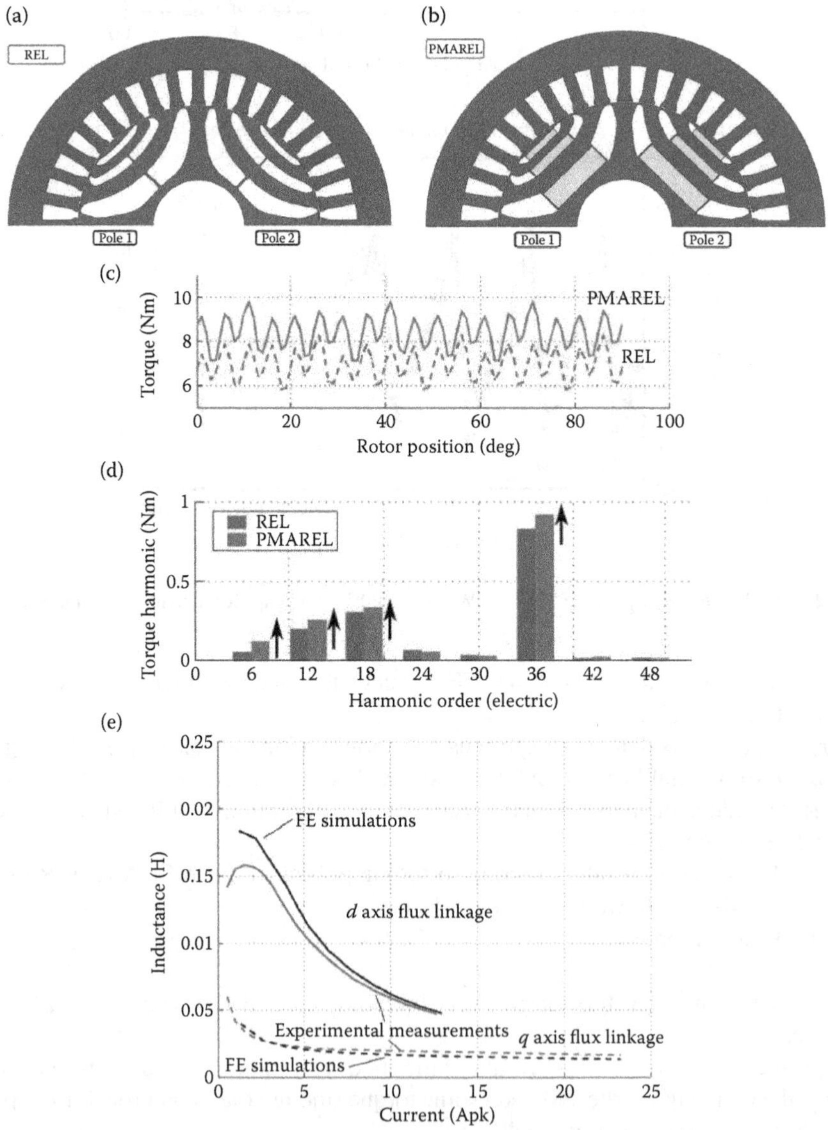

FIGURE 4.14 MFBA-rotor four-pole RSM and PM-RSM by FEM: (a) RSM cross-section, (b) PM-RSM cross-section, (c) torque versus rotor position for both motors for sinusoidal currents, (d) torque ripple harmonics, (e) d and q axis inductances. (After M. Ferrari et al., *IEEE Trans on*, vol. IA-51, no. 1, 2015, pp. 169–177. [19])

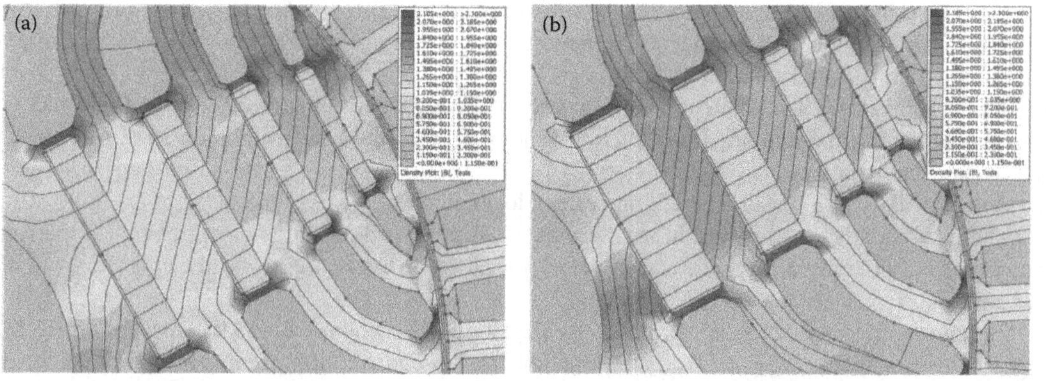

FIGURE 4.15 Flux lines by FEM in a Ferrite-PM-RSM at rated load: (a) demagnetization free design at overload; (b) initial (large PM weight) design. (After P. Guglielmi et al., *IEEE Trans on*, vol. IA-49, no. 1, 2013, pp. 31–41. [20])

A key factor in designing PM-RSMs is PM demagnetization avoidance, especially with Ferrite-PM rotors [20], where the optimal design for minimum PM weight endangers the magnets at overload while the demagnetization-free design still allows 30% less PM weight (from initial design)—Figure 4.15a and b.

4.6 THE CIRCUIT *dq* (SPACE PHASOR) MODEL AND STEADY STATE PERFORMANCE

The space phasor (*dq*) model of PM-RSM, with magnets in axis *q* of the rotor ($L_d > L_q$), is straightforward [14], in rotor coordinates:

$$\begin{aligned}
&\overline{V}_s = R_s\overline{i}_s + \frac{d\overline{\Psi}_s}{dt} + j\omega_r\overline{\Psi}_s;\ \overline{\Psi}_s = \Psi_d + j\Psi_q;\ \overline{i}_s = i_d + ji_q;\\
&\Psi_d = L_d i_d;\ \Psi_q = L_q i_q - \Psi_{\mathrm{PM}q};\ \overline{V}_s = V_d + jV_q\\
&T_e \approx \frac{3}{2}p_1(\Psi_d i_q - \Psi_q i_d) = \frac{3}{2}p_1\left[\Psi_{\mathrm{PM}q} + (L_d - L_q)\cdot i_q\right]\cdot i_d\\
&\frac{J}{p_1}\frac{d\omega_r}{dt} = T_e - T_{\mathrm{load}};\ \frac{d\theta_{er}}{dt} = \omega_r
\end{aligned} \tag{4.27}$$

The torque expression refers to the case of sinusoidal emf, and phase inductances are dependent on sinus or cosine of 2 θ_{er}.

To include torque pulsations due to stator mmf space harmonics, slot openings, rotor flux barriers, and magnetic saturation, new terms should be added:

$$T_{ef}(\theta_{er}) = \frac{3}{2}p\left[\Psi_{\mathrm{PM}q} + (L_d - L_q)\cdot i_q\right]\cdot i_d + p\frac{\partial W_{mc}(i_d, i_q, \theta_{er})}{\partial\theta_{er}} + T_{\mathrm{cogging}} \tag{4.28}$$

$W_{mc}(i_d, i_q, \theta_{er})$ is the magnetic co-energy in the machine expressed in *dq* coordinates and yields all torque pulsations except for cogging torque (PM torque at zero current), the last term in Equation 4.28.

In PM-RSMs, much of the torque is not produced by the PMs and thus cogging torque is less notable (with respect to strong PM flux synchronous machines: surface permanent magnet synchronous motors (SPMSMs) and IPMSMs).

In general, only FEM could produce effective results for $T_{et}(\theta_{er})$. The first term, however, was shown to calculate the average electromagnetic torque of the machine rather well. For control purposes, in general, T_e in Equation 4.27 suffices.

But, in general, Ψ_d and Ψ_q depend nonlinearly on both dq current components, and thus a cross-coupling effect occurs. The functions $\Psi_d(i_d, i_q)$ and $\Psi_q(i_d, i_q)$ may be FEM-computed or measured in flux decay standstill tests; see Figure 4.16 [21].

For an MFBA-rotor, the cross-coupling effect dependence of Ψ_d of i_q and of Ψ_q of i_d is even stronger and may endanger estimation in such sensorless drives by a too-large error position if the rotor saturation is not limited or the position error caused by cross-coupling is not considered in the control design.

Let us now investigate the performance by Equation 4.27 where we add the core losses by 1(2) equivalent resistances R_{ironS} and R_{ironR} (in relation to rotor contribution) to obtain the equivalent circuits in Figure 4.17.

The terms L_{df} and L_{qf} are the transient inductances along d and q axes:

$$L_{dt} \approx L_d + \frac{\partial L_d}{\partial i_d} i_d \le L_d;\ L_{qt} \approx L_q + \frac{\partial L_q}{\partial i_q} i_q \le L_q, \tag{4.29}$$

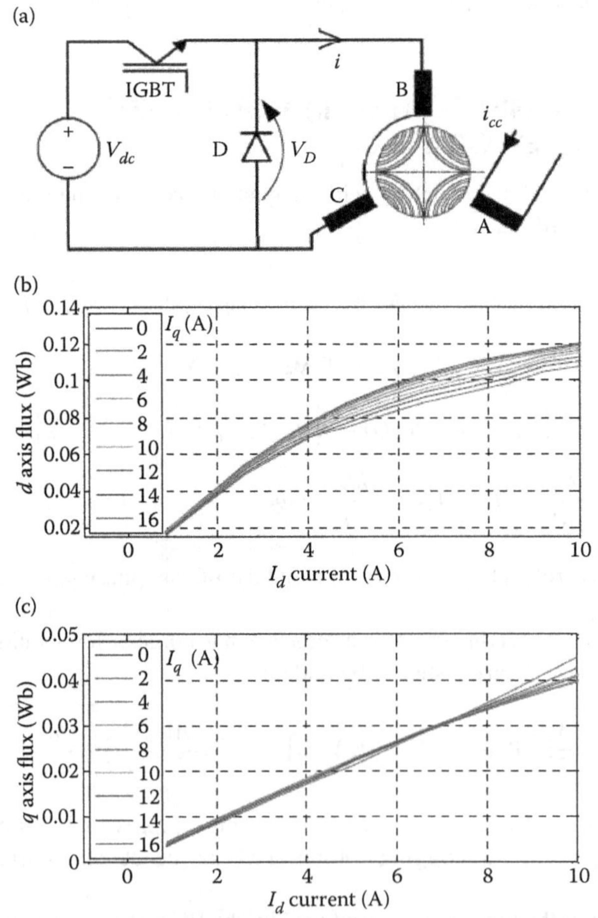

FIGURE 4.16 ALA-rotor four-pole small (1 Nm) RSM flux linkage current curves: (a) experimental arrangement; (b) axis d; (c) axis q.

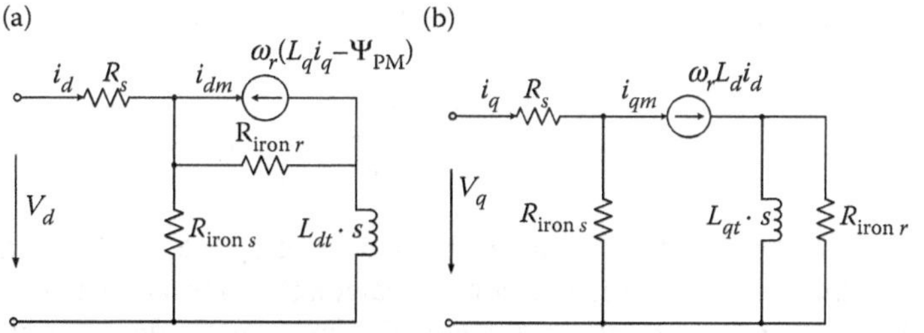

FIGURE 4.17 Equivalent dq circuits of PM-RSM (with core losses considered).

relevant in transient processes in iron. Core losses may be calculated from static FEM field variation calculations in each FEM element by summation using 2(3) term analytical core loss standard formulae, or from AC standstill FEM field calculations. L_d, L_q differ from L_{dt} and L_{qt} only if magnetic saturation is advanced, when, for approximate calculations above 10% load, $L_q \approx$ const, while the real $L_d(i_d)$ function is considered through a second-order polynomial.

With Equation 4.27, vector diagrams could be built. For steady state $d/dt = s = 0$ and for RSM and PM-RSM, the vector diagrams are as in Figure 4.18.

From the vector diagrams, we may first notice that the assisting PMs reduce the power factor angle φ and also decrease the flux amplitude Ψ_{s0} for given stator currents i_d, i_q, for better torque at the same speed for the same stator.

For a given speed and currents, lower voltage V_{s0} is required and thus wider CPSR is feasible.

But, also, Figure 4.18 allows us to simply calculate the performance such as efficiency η and power factor angle φ:

$$\eta = \frac{\frac{3}{2}(V_{d0}i_{d0} + V_{q0}i_{q0}) - \frac{3}{2}R_s i_s^2 - p_{\text{iron}} - p_{\text{mec}}}{(3/2)(V_d i_d + V_q i_q)} \tag{4.30}$$

$$\tan\varphi \approx \frac{V_{q0}i_{d0} - V_{d0}i_{q0}}{V_{d0}i_{d0} + V_{q0}i_{q0}}$$

with

$$\begin{aligned} &\omega_{r0}L_d i_{d0} + R_s i_{q0} = V_{s0}\cos\delta_V, \quad \delta_V > 0\ (\text{for motoring}) \\ &\omega_{r0}(L_q i_{q0} - \Psi_{PMq}) - R_s i_d = V_{s0}\sin\delta_V \\ &\tan\gamma_i = \frac{i_{d0}}{i_{q0}};\ \varphi = \delta_V + \delta_i;\ i_{s0} = \sqrt{i_{d0}^2 + i_{q0}^2} \end{aligned} \tag{4.31}$$

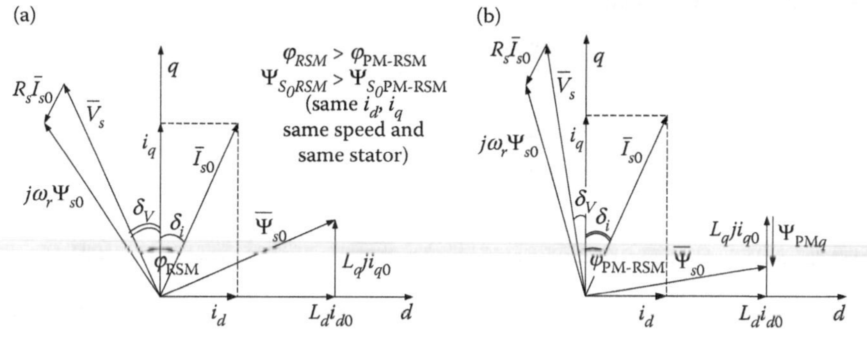

FIGURE 4.18 Vector diagrams at steady state (motoring) with zero core losses: (a) RSM; (b) PM-RSM.

Approximately (neglecting the rotor core losses), the core loss p_{iron} is added:

$$p_{\text{iron}} = \frac{3}{2}\frac{V_{s0}^2}{R_{\text{iron }s}} \tag{4.32}$$

Now, by giving values to ω_r and V_s for increasing values of δ_v, we may calculate i_{d0} and i_{q0} from Equation 4.31, with a known (from no-load testing or calculations) $\Psi_{\text{PM}q}$. Then, i_{s0} is also calculated from Equation 4.31 together with γ_{i} and finally φ (which may be verified by Equation 4.30). Then, with $R_{\text{iron }s}$ known, p_{iron} is computed from Equation 4.32 and, with mechanical losses given as a function of speed $\omega_r(\omega_r = p_1\Omega_r = p_1 2\pi n)$, efficiency may be calculated.

The real power factor is slightly "improved" (1%–2%) by considering core and mechanical losses (but neglected in Equation 4.31) in motoring and worsened in generating when $\delta v < 0$.

In the absence of PMs, the RSM ideal maximum power factor (losses neglected) from Equation 4.30 becomes:

$$\cos\varphi_{\max 0} = \frac{1 - L_q/L_d}{1 + L_q/L_d} \tag{4.33}$$

This is similar to IM, where L_d, L_q are replaced by L_s (no load inductance) and, respectively, L_{sc} (short-circuit inductance). As in two- or four-pole IMs, in general, $L_s/L_{sc} > 10$ even at 1 kW, 50 Hz, designing a competitive RSM in the same stator is a daunting task; still, the ALA-rotor may achieve this goal. Yes, in a rotor topology that is still considered less manufacturable.

The power factor is improved notably by assisting PMs (Figure 4.27), and no apparent maximum of it occurs. Experiments on load with an inverter-fed drive may be run, and according to the above simple performance calculation routine, results as shown in Table 4.1 have been obtained [22] under no load torque (when $i_d \gg i_q$) [2].

δ_i is the current angle, and the test also yields the $L_d(i_d)$ curve and core loss, if mechanical losses are known (or neglected).

Progressive load tests could be run to calculate efficiency, power factor, i_d, i_q, γ, φ versus voltage angle γ_v (or output power) and thus completely characterize the RSM or PM-RSM.

The machine may also be run as a capacitor-excited AC generator with variable resistive load to extract all its characteristics if a drive with torque, speed, and rotor position sensor is available. A measure of such a load motoring performance with an ALA-rotor 1.5 kW, two-pole RSM

TABLE 4.1
***d* Axis, Zero Load Torque Tests**

$\delta°$	i_s (A)	V_s (V)	Input Power (w)	i_{dm} (A)	λ_d (W_b)	R_{m1} (Ω)	L_d (mH)
26.15°	3.15	18.62	45	2.831	0.114	12.65	39.208
20.52°	8.27	52.17	267	7.75	0.3114	17.02	40.18
18.49°	12.84	75.07	504	12.18	0.4480	19.27	35.00
13.65°	21.37	91.28	775	20.77	0.5447	21.04	26.225
12.24°	25.31	43.90	862	24.74	0.5603	21.74	22.647
11.36°	28.61	97.00	950	28.04	0.8788	22.56	20.64
						$L_q = 5.55$ mH	

Source: After L. Xu et al., *IEEE Trans on*, vol. IA-27, no. 5, 1991, pp. 1977–1985. [22]

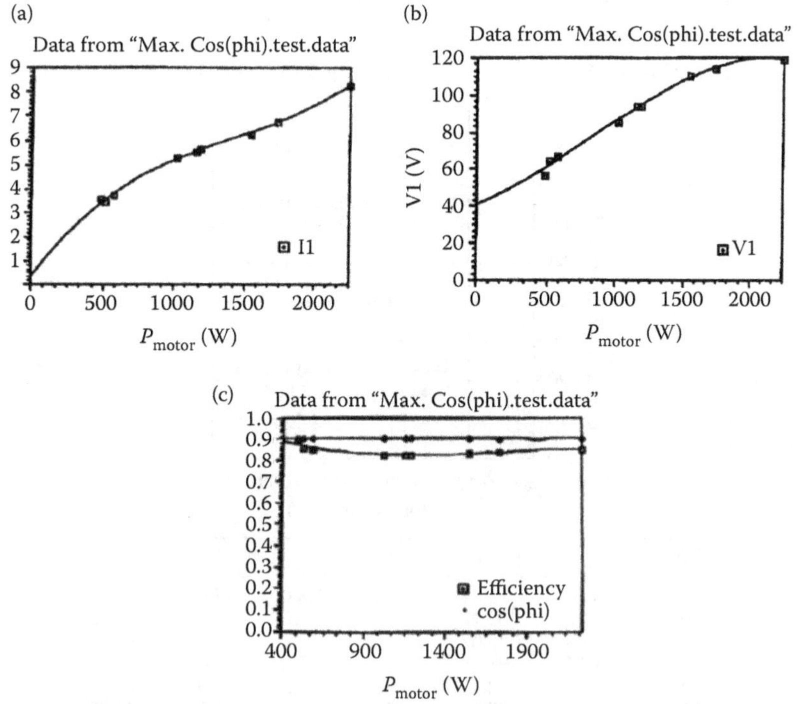

FIGURE 4.19 Measured load motoring performance of an ALA-rotor two-pole RSM of 1.5 kW at 60 Hz: (a) current versus output power; (b) voltage (RMS, phase voltage) versus output power for maximum power factor; (c) efficiency and power factor.

connected to the grid (220 V, 60 Hz) and loaded slowly by a variable alternating current device (VARIAC) (transformer) to maintain the maximum power factor is shown in Figure 4.19.

The performance is very good, but, again, it is an ALA-rotor RSM, which is still considered hard to fabricate at the industrial scale.

Load tests on an industrial 1500-rpm, four-pole, 14-kW MFBA-rotor RSM show comparative performance as in Table 4.2.

While a +2% improvement in efficiency over IM was obtained, the power factor decreased from 0.82 to 0.77. Considering the inverter losses, the system efficiency was increased by 2.1%, which, by using a less expensive motor and the same inverter, leads to an overall better variable speed drive, not to mention that it runs at lower temperatures (average and in critical spots inside the machine).

4.7 DESIGN METHODOLOGIES BY CASE STUDIES

4.7.1 Preliminary Analytical Design Sequence by Example

Let us design an MFBA-rotor two-pole, three-phase RSM that, when inverter-fed at 380 Vdc, produces $P_n = 1.5$ kW and a max speed of 4500 rpm for an efficiency above 90%. No CPSR is necessary (air-compressor-like load).

4.7.1.1 Solution

For an estimated efficiency $\eta = 90\%$, the rated electromagnetic torque T_{en} is:

$$T_{en} = \frac{P_n}{\eta \cdot 2\pi n_{\max}} = \frac{1500}{0.9 \cdot 2\pi \cdot 4500/60} = 3.538 \text{ Nm} \tag{4.34}$$

TABLE 4.2
Inverter-Fed Load Testing of Four-Pole MFBA-Rotor RSM and IM (Same Stator)

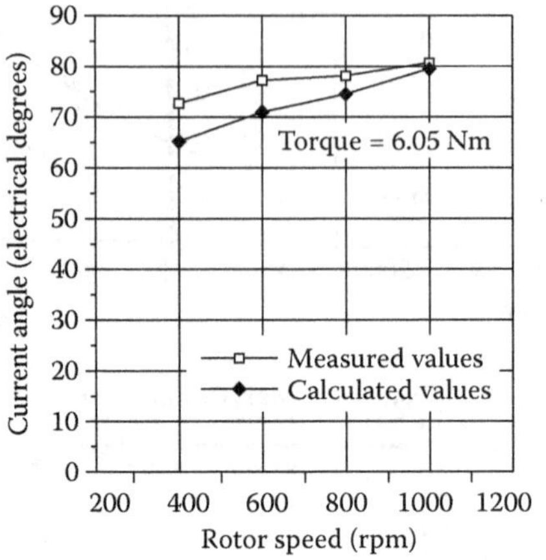

$\theta°$	I_{spk}	Fund. V_s	P_{in}	I_{dm}	λ_{ds}	R_m
26.15	3.15	18.62	45	2.831	0.1111	12.65
20.52	8.27	52.17	267	7.75	0.3114	17.02
18.49	12.84	75.07	504	12.18	0.4480	19.27
13.65	21.37	91.28	775	20.77	0.5447	21.04
12.24	25.31	93.90	862	24.74	0.5603	21.74
11.36	28.61	97.00	950	28.05	0.5788	22.55

Source: R. R. Moghaddam, F. Gyllensten, *IEEE Trans*, vol. IE–61, no. 9, 2014, pp. 5058–5065 [23].

Adopting a specific shear stress on the rotor $f_t = 0.7\ \text{N/cm}^2$ and an interior stator diameter $D_{is} =$ 60 mm, the stator core (stack) length l_{stack} is obtained:

$$T_{en} = f_t \pi D_{is} l_{\text{stack}} \frac{D_{is}}{2}; \; l_{\text{stack}} = \frac{2 \times 3.538}{\pi \cdot 0.7 \times 10^4 \times 0.06^2} \approx 0.100\,\text{m} = 100\,\text{mm} \tag{4.35}$$

The tentative rotor configuration is shown in Figure 4.20.

The aspect ratio of the stator is given by $l_{\text{stack}}/\tau = 2 \cdot 100/(60\pi) = 1.05$, which is rather favorable for shorter p.u. coil end connections.

Let us now approximately express the dq axis inductances: $L_{dm} = L_m k_{dm}$ with L_m and k_{dm} yields from Equations 4.12 and 4.13.

From Equation 4.11, approximately, $L_{qmc}\ (4.3) \approx 0.03 L_{dm}$.

$$L_{qmf} \approx \frac{L_m g k_c}{g k_c + (n \cdot l_b/\pi)}; \; n \cdot l_b \approx n \cdot l_i \approx \frac{\tau}{2}\left(\frac{1}{2} - \frac{1}{3}\right) \tag{4.36}$$

$$\begin{gathered} L_{qm\,\text{bridges}} \approx \frac{L_m g k_c}{g k_c + ((n l_i/\mu_{sat\,\text{Re}\,l}) \cdot (h_{sc}/h_{\text{bridge}})) \times 1/(1 + (1/2))}; \; \mu_{sat\,\text{Re}\,l} = (30 - 100) \\ h_{sc} = \frac{\tau}{\pi} \cdot \frac{1}{2} \end{gathered} \tag{4.37}$$

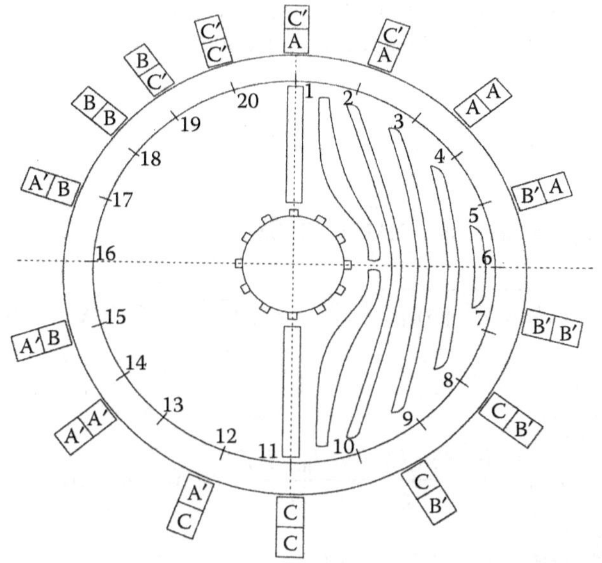

FIGURE 4.20 Typical two-pole MFBA rotor RSM with 15 stator slots and 20 equivalent rotor slots (flux barriers).

with $h_{\text{bridge}} = 0.5$ mm, $g = 0.3$ mm, $D_{ir} \approx 60$ mm, $\mu_{sat\,\mathrm{Re}\,l} = 90$, $k_c = 1.25$, we get:

$$nl_b = \frac{\tau}{2} \times \frac{1}{3} = 15\,\text{mm} \tag{4.38}$$

$$\frac{L_{qm\,\text{bridge}}}{L_m} = L_m 0.025;\ \frac{L_{qmf}}{L_m} \approx 0.023;\ L_{qm0} = 0.03 \cdot L_m;\ L_{dm} = 0.95 \cdot L_m \tag{4.39}$$

For 60% iron, 40% air proportions in the rotor core, finally:

$$\frac{L_{dm}}{L_{qm}} \approx \frac{0.05}{0.023 + 0.03 + 0.025} \approx \frac{12}{1} \tag{4.40}$$

Now, if we adopt $L_{sl} = 0.05\,L_{dm}$ (leakage inductance):

$$\frac{L_d}{L_q} = \frac{L_{dm} + L_{sl}}{L_{qm} + L_{sl}} = \frac{0.95 + 0.050}{\frac{1}{12} + 0.050} \approx 7.00 \tag{4.41}$$

Magnetic saturation will be kept low (especially in the rotor) to preserve this total saliency under full load. Also, hopefully, L_{sl} will be smaller than above when calculated, as shown below.

The current i_d provides machine magnetization, and for an 18-slot two-pole stator (two-layer–chorded-coil winding with $k_{w1} \approx 0.925$), the slot mmf along axis d for magnetization $(n_s I_d)_{\text{RMS}}$ is:

$$(n_s I_d)_{\text{RMS}} = \frac{B_{g1dm}\pi g \cdot K_{c1}(1 + K_{ss})}{\mu_0 \cdot 3\sqrt{2}q_1 \cdot k_{w1} \cdot k_{dm}} = \frac{0.7 \cdot \pi \cdot 0.3 \cdot 10^{-3} \cdot 1.25 \cdot 1.56}{1.256 \cdot 10^{-6} \cdot 3\sqrt{2} \cdot 3 \cdot 0.95 \cdot 0.97}$$
$$= 80\ \text{Aturns per slot} \tag{4.42}$$

B_{g1d}, the peak airgap no-load flux density in the airgap, is chosen as $B_{g1dm} = 0.7\ T$. Now, we design the machine for the maximum power factor, for which, ideally:

$$i_q = i_d\sqrt{\frac{L_d}{L_q}} \tag{4.43}$$

For this case, the ampere turns/slot (RMS) will be:

$$n_s i_s = \sqrt{(n_s i_d)^2 + (n_s i_q)^2} = 80 \times \sqrt{1+7} \approx 215 \text{ Aturns/slot} \tag{4.44}$$

Now

$$L_{dm} = \frac{6\mu_0(3n_s k_{\omega 1})^2\tau \cdot l_{\text{stack}} k_{dm}}{\pi^2 gkc(1+k_s)} = 0.09 \cdot n_s^2 \cdot 10^{-3} \tag{4.45}$$

The torque expression (Equation 4.27) is still:

$$\begin{aligned} T_e &= 3p(L_d - L_q) i_{d\text{ phase RMS}} \cdot i_{q\text{ phase RMS}} \\ &= 3 \cdot 1 \cdot 0.09 \cdot 10^{-3}(1 - 1/7) \cdot 80 \cdot 200 = 3.7 \text{ Nm} > 3.5 \text{ Nm} \end{aligned} \tag{4.46}$$

So the machine produces the required torque with a reserve.

The maximum power factor $\cos\varphi_{\max i}$ is:

$$\cos\varphi_{\max i} = \frac{1 - L_q/L_d}{1 + L_q/L_d} = \frac{1 - 1/7}{1 + 1/7} = \frac{6}{8} = 0.75 \tag{4.47}$$

It could be 0.77 when losses are considered.

The stator slot geometry is determined easily, as we know the slot mmf $n_s i_s = 215$ Aturns/slot and may adopt the current density (say, $j_{\text{con}} = 4.77\ \text{A/mm}^2$) and the slot fill factor $k_{\text{fill}} = 0.45$. The slot useful area $A_{\text{slot}\,n}$ is:

$$A_{\text{slot}\,n} = \frac{n_s i_s}{j_{\text{con}} K_{\text{fill}}} = \frac{215}{4.77 \times 0.45} \approx 100 \text{ mm}^2 \tag{4.48}$$

But the slot pitch $\tau_s = \tau/9 = 94.2/9 = 10.46$ mm.

For a tooth width $w_s = 5$ mm, the slot width $w_{s1} = 5.46$ mm.

With a slot depth of only $h_{sa} = 13$ mm, the recalculated useful area of the trapezoidal slot (Figure 4.21) $A_{\text{slot}\,nf}$ is

$$A_{\text{slot}nf} = \frac{w_{s1} + w_{s2}}{2} h_{sa} = \frac{5.46 + 10.42}{2} \times 13 = 103.906 \text{ mm}^2 > 100 \text{ mm}^2$$

The stator yoke h_{cs} is:

$$h_{cs} \approx \frac{\tau}{\pi} \cdot \frac{B_{g1\max}}{B_{cs\max}} \approx \frac{94}{3.14} \cdot \frac{0.7}{1.4} \approx 15 \text{ mm} \tag{4.49}$$

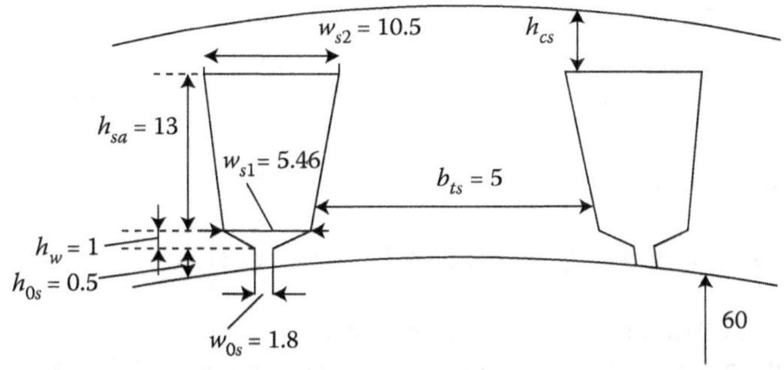

FIGURE 4.21 Stator slot geometry.

And now, the leakage inductance L_{sl} may be computed [17]:

$$L_{sl} = \frac{2\mu_0 (n_s pq)^2}{pq} (\lambda_s + \lambda_z + \lambda_f) \cdot l_{\text{stack}} \tag{4.50}$$

$$\lambda_s = \frac{2h_{sa}}{3(w_{s1} + w_{s2})} + \frac{w_{0s}}{h_{0s}} + \frac{2h}{w_{s1} + w_{0s}} = \frac{13 \cdot 2}{3(10.52 + 5.46)} + \frac{0.5}{1.8} + \frac{2 \times 1}{5.46 + 1.8} = 1.095 \tag{4.51}$$

The airgap leakage coefficient λ_z is:

$$\lambda_z = \frac{5g}{5w_0 + 4g} = \frac{5 \times 0.3}{5 \cdot 1.8 + 0.3} = 0.15 \tag{4.52}$$

Similarly, for end coils:

$$\lambda_f \approx 0.34 \cdot \frac{q_1}{l_{\text{stack}}} (l_f - 0.64 \cdot y) = \frac{0.34 \cdot 3}{0.1} \cdot 0.7 \cdot \frac{8}{9} \cdot 0.094 = 0.596, \tag{4.53}$$

$$\text{with } y = \frac{8}{9}\tau \tag{4.54}$$

Finally, L_{sl} (Equation 4.50) is:

$$L_{sl} = \frac{2 \cdot 1.256 \times 10^{-6} (n_s \cdot 1 \cdot 3)^2}{1 \cdot 3} (1.095 + 0.15 + 0.596) \cdot 0.1$$
$$= 1.382 \times 10^{-6} n_s^2 \tag{4.55}$$

and now the ratio $L_{sl}/L_{dm} = 1.387 \cdot 10^{-6} \, {n_s}^2 / (90 \cdot 10^{-6} \, n_s^2) < 0.02$ (it was assigned as 0.05).

Thus, the chosen MFBA rotor provides a better than 7/1 total saliency:

$$\frac{L_d}{L_q} = 8.5 > 7 \tag{4.56}$$

As this is only a preliminary result, let us keep the initial design so far for safety and continue with stator resistance R_s:

$$R_s = \rho_{Co} \frac{l_c \cdot p_1 \cdot q \cdot n_s^2}{i_s \cdot n_s} j_{\text{con}} = 0.060 \cdot n_s^2 \times 10^{-2} \text{at } 100°\text{C} \tag{4.57}$$

where $\rho_{Co} = 2.3 \times 10^{-8}\ \Omega$m for 100°C; the stator turn length l_c is:

$$l_c = 2 \cdot l_{\text{stack}} + 4 \times 5 + 2 \cdot 1.34 \cdot \left(\frac{8}{9}\right) \cdot \tau = 444\ \text{mm} \tag{4.58}$$

So the rated copper losses are:

$$P_{\cos} = 3 \cdot R_s \cdot I_s^2 = 3 \times 0.060 \times 10^{-2} \times (215)^2 = 94.3\ \text{W} \tag{4.59}$$

The frequency at 4500 rpm is 75 Hz ($p = 1$ pole pairs).

Approximately at 1.5 T and 50 Hz, the core losses are considered 3 W/kg; thus, they are $3 \times 1.5^2 = 6.75$ W/kg at 1.5 T and 75 Hz.

The stator core weight, on the other hand, is:

$$W_{\text{iron}} \approx \frac{\pi(D_{\text{sout}}^2 - D_{is}^2)}{4} \times 0.4 \cdot \gamma_{Fe} \cdot l_{\text{stack}} \approx 3.1\ \text{kg} \tag{4.60}$$

Thus, the iron losses are approximately $p_{\text{iron}} = 3.1 \times 6.75 \approx 21$ W.

Adopting 2% mechanical losses ($p_{\text{mec}} = 30$ W), the estimated machine efficiency is:

$$\eta_n = \frac{P_n}{P_n + p_{\cos} + p_{\text{core}} + p_{\text{mec}}} = \frac{1500}{1500 + 94.3 + 21 + 30} = 91\% \tag{4.61}$$

what now remains to be done is to calculate the number of turns per slot n_s that will satisfy the limit AC-phase voltage $V_s = 152$ V for 380 Vdc. With $L_q = 11.8 \times 10^{-6} \cdot n_s{}^2$; $L_d = 91.8 \times 10^{-6} \cdot n_s{}^2$; $R_s = 6.8 \cdot 10^{-4} \cdot n_s{}^2$; $n_s i_d = 80$ Aturns; $n_s i_q = 200$ Aturns, the dq voltage equations:

$$\begin{aligned} V_d &= -\omega_r L_q i_q + R_s i_d \\ V_q &= \omega_r L_d i_d + R_s i_d \end{aligned} \tag{4.62}$$

yield $V_{d\ \text{phase RMS}} = 0.98 \cdot n_s$; $V_{q\ \text{phase RMS}} = 3.5578 \cdot n_s$

$$\text{Finally, } n_s = \frac{152}{\sqrt{0.98^2 + 3.5578^2}} = 42\ \text{turns/slot} \tag{4.63}$$

As the winding has two layers, each coil has 21 conductors. The conductor diameter:

$$d_{\text{copper}} = \sqrt{\frac{4}{\pi} \cdot \frac{I_s}{j_{con}}} = \sqrt{\frac{4}{\pi} \cdot \frac{215/42}{4.77}} = 1.1\ \text{mm} \tag{4.64}$$

with $i_{sn} = n_s \cdot I_{sn}/n_s = 215/42 = 5.112$ A (RMS per phase),

We may now recalculate the power factor:

$$\cos\varphi \approx \frac{P_n}{3 V_s I_s \eta_n} = \frac{1500}{3 \times 152 \times 5 \cdot 112 \times 0.912} \approx 0.705 \tag{4.65}$$

This value is lower than the initial approximately calculated one, leaving room for improvement in the design or adopted for design safety.

Finally, the machine circuit parameters are: $L_d = 0.161$ H, $L_q = 0.0208$ H, $R_s = 1.2\ \Omega$, $R_{\text{iron}} = 3 \cdot V_s{}^2/P_{\text{iron}} = 3465\ \Omega$. With this data, the investigation of transient and control design may be approached after a thorough verification of analytically forecasted performance by FEM inquires. The above preliminary design may also serve as a solid start for optimal design.

4.7.2 MFBA-Rotor Design in Reluctance Synchronous Motor Drives with Torque Ripple Limitation

There is an extremely rich literature on RSM modeling and design with MFBA-rotors [24].

There are quite a few specifications to comply with in an RSM drive design, such as:

- Average torque
- Torque ripple (in p.u. [%])
- Efficiency (motor + converter)
- Power factor
- A certain CPSR
- Initial motor + converter cost

For the time being, no PM assistance (to increase average torque, power factor, and CPSR) is considered.

In any case, high $L_d - L_q$ is required for large average torque and a large L_d/L_q ratio is needed for a high power factor.

High airgap saliency L_{dm}/L_{qm} and low leakage inductance ratio L_{sl}/L_{dm} sum up the above requirements even better. To yield a small L_{qm}, however, implies at least an MFBA-rotor: the thicker the flux barriers, the lower the L_{qm}, while, unfortunately, due to heavy magnetic saturation, L_{dm} decreases more than L_{qm} and thus average torque is, in fact, reduced [19]. Saturating the rotor heavily is also not suitable for AC saliency-based rotor position estimation by signal injection in sensorless drives.

So, "there should be more iron than air" in the MFBA-rotor. The insulation ratio per MFBA-rotor was defined [25] as K_{ms}:

$$K_{a,d,q} = \frac{\text{radial air length}}{\text{radial iron length}} \tag{4.66}$$

There is still a debate if $K_{ms,d,q}$ should be smaller than 1.0, but this is the trend to avoid excessive rotor magnetic saturation.

But it is also important to have a large total radial air length per airgap:

$$l_{ag} = \frac{\text{radial air length}}{\text{airgap}} \tag{4.67}$$

to yield low L_{qm}/L_{dm} ratio [26].

Two main questions arise:

- Should the air ratio ($K_{a,d,q}$) be the same in axes d and q?
- Should the flux barrier angles be equal or not, and should they be spread over all the pole span or not?

Strong worldwide previous R&D efforts are integrated in [27] in an articulated optimal industrial design methodology to reply to the above questions for securing high average torque and limited torque ripple.

Note that there are still two other important issues:

- How to reduce harmonics-produced core losses in the rotor
- How to reduce radial force peaks to reduce noise and vibration

We believe that direct FEM geometrical optimization of rotor flux barriers after their width, length, and placement angles have been assigned by "macroscopic" optimal design, as suggested above, is a practical approach in industry.

But let us go back to the optimal macroscopic design of MFBA-rotor. First, a rotor structure as in Figure 4.22a is adopted [23].

The optimal design strategy in [23–24], in essence, targets the separation of high average torque from low ripple torque in the rotor flux barrier design by:

- Keeping the rotor slot (barrier) pitch α_m constant mainly in the d axis, while 2β is allowed to vary around $2\beta = \alpha_m$ for a certain number of flux barriers. A virtual slot (point B, β) is introduced, and by varying β with respect to α_m, only a few FEM calculations are needed to mitigate torque ripple. Also, additional torque ripple reduction may be obtained by changing the radial positions of flux barriers in axis $q(y_q)$; see Figure 4.22a.
- q-axis flux is minimized to produce high average torque (and power factor); this is accomplished by intelligently relating the flux barrier lengths s_k and their thickness w_{li}, as already done in Equation 4.7 with symbols as in Figure 4.22a:

$$\frac{w_{li}}{w_{lj}} = \frac{\Delta f_i}{\Delta f_j}\sqrt{\frac{s_{bi}}{s_{bj}}};\ \frac{s_i}{s_j} = \frac{f_{di}}{f_{dj}}, \tag{4.68}$$

where the flux barrier permeances have been considered equal to each other.

Sensitivity studies have shown [23] that this way, the solving of the two main problems, high average torque and low torque ripple, may be separated.

Consequently, a straightforward design methodology was developed [23]:

- Assume that the stator geometry is given: $l_a + l_y = \text{ct}$.
- Then, for a given pole number ($2p$), barrier number per pole (k), insulation ratio in axes d and $q(k_{wd}, k_{wq}$ in Figure 4.22a), and β angle values, the rotor flux barrier dimensions are calculated by Equation 4.68.
- For simplicity, the stator slotting and magnetic saturation are disregarded, as the resultant geometry is analyzed later by FEM anyway.
- With β, the number of flux barriers k and the positions of flux barriers at airgap (points D_i in Figure 4.22b and c) are obtained easily, and:

$$\alpha_m\left(k + \frac{1}{2}\right) = \frac{\pi}{2p_1} - \beta \tag{4.69}$$

- "The iron" may first be distributed in the rotor pole according to Equation 4.68 and $l_a + l_y = \text{ct}$ (given):

$$\frac{f_{d_1}}{f_{d_2}} = \frac{2s_1}{s_2};\ \frac{f_{d_{i+1}}}{f_{d_2}} = \frac{s_{i+1}}{s_2};\ i = 2, \ldots k;\ \sum_1^k s_i = \frac{l_a + l_y}{1 + k_{aq}} \tag{4.70}$$

- Now "the air" in the rotor pole may be distributed again according to Equation 4.68:

$$\left(\frac{\Delta f_i}{\Delta f_1}\right)^2 = \frac{w_{li}}{w_{l1}};\ i = 2, \ldots k;\ \sum_1^k w_{li} = l_a = \frac{l_y + l_a}{1 + k_{aq}} \times k_{aq} \tag{4.71}$$

- Similarly, in axis d, each barrier dimension can be calculated if the sizes of flux barriers in axis $q(w_i, s_i)$ are given (as calculated above) and the insulation (air) ratio in axis d is given:

$$\left|\frac{w_{id}}{w_{1d}}\right|_{d\ \text{axis}} = \left|\frac{w_{li}}{w_{l1}}\right|_{q_{\text{axis}}};\ i = 2, \ldots, k \tag{4.72}$$

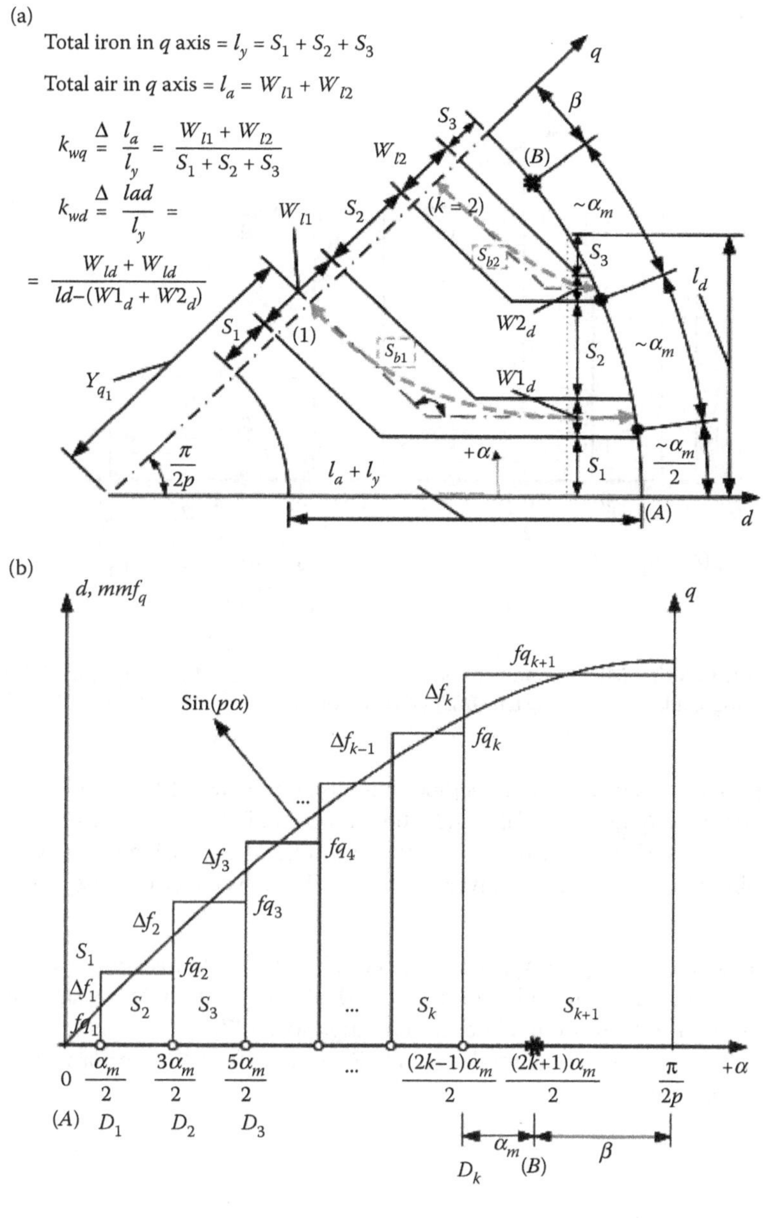

FIGURE 4.22 MFBA-rotor geometry, (a) d and q axis mmf p.u. segmented distribution (b) and (c). (After R. R. Moghaddam, F. Gyllensten, *IEEE Trans*, vol. IE–61, no. 9, 2014, pp. 5058–5065. [23]) *(Continued)*

- By modifying y_{qi} (the radial positions of flux barriers in axis q) such that the dimensions calculated above meet the constant rotor slot pitch condition (α_m), further torque ripple reduction may be obtained.
- Sensitivity studies could clarify the influence of the k_{ad}/k_{aq} ratio; its optimum seems to be in the interval 0.6–0.7, [24]; k_{aq} seems to have an optimum around 0.6.
- For constant slot pitch and $2\beta = \alpha_m$ high torque ripple for $k = 4$ flux barriers/pole is obtained and thus has to be avoided.
- The effect of β on average torque is negligible but on torque ripple is important, as expected; a value of $\beta = 9°$ for a four-pole, four–flux-barrier rotor-RSM with $k_{ad} = 0.3$, $k_{ad} = 0.7$ is found to offer minimum torque ripple (Figure 4.23) [23]: 12%.

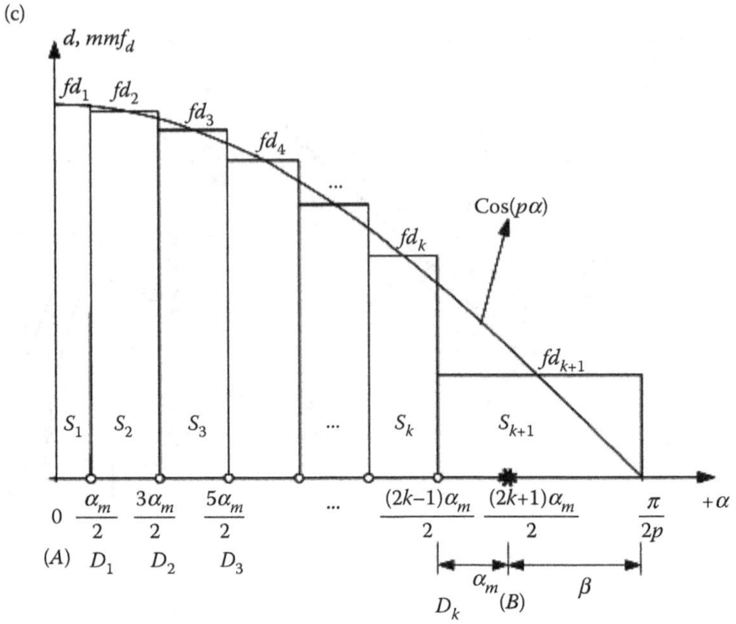

FIGURE 4.22 (Continued) MFBA-rotor geometry, (a) d and q axis mmf p.u. segmented distribution (b) and (c). (After R. R. Moghaddam, F. Gyllensten, *IEEE Trans*, vol. IE–61, no. 9, 2014, pp. 5058–5065. [23])

The final results of the above design methodology as applied to a 14-kW, four-pole, 50-Hz RSM with MFBA-rotor, already given in Table 4.2, show remarkable improvements over the existing IM drive: +2.1% in system efficiency (motor + converter) in a less expensive motor with lower temperature operation in the same stator and inverter, though the power factor is 3% smaller than for the IM drive.

Note: Further reduction of torque ripple and rotor core loss has been investigated by stator (or rotor) skewing, rotor pole displacement, and asymmetric ("Machaon") rotor poles by shaping each flux barrier by direct geometrical FEM optimal design investigations [27], and so on. It was also found that small fabrication tolerances can notably deteriorate RSM and PM-RSM performance, mainly due to the mandatory small airgap.

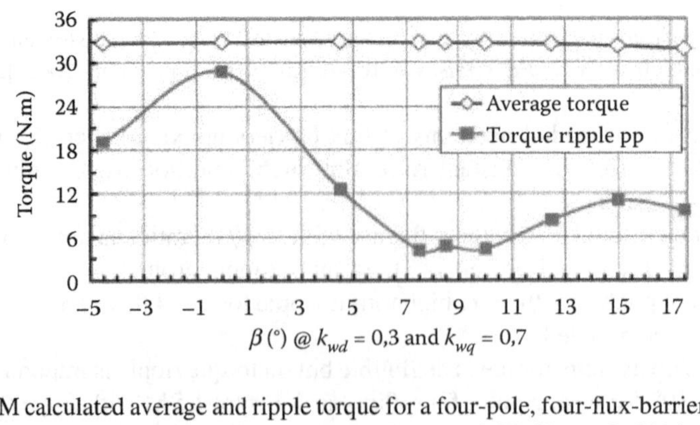

FIGURE 4.23 FEM calculated average and ripple torque for a four-pole, four-flux-barrier rotor with $k_{ad} = 0.3$, $k_{aq} = 0.7$. (After R. R. Moghaddam, F. Gyllensten, *IEEE Trans*, vol. IE-61, no. 9, 2014, pp. 5058–5065. [23])

4.8 MULTIPOLAR FERRITE-PERMANENT MAGNET RELUCTANCE SYNCHRONOUS MACHINE DESIGN

Using low-cost (Ferrite) PMs in RSM with notable saliency to produce a good power factor and increased efficiency is a clear way to further improve RSM performance in variable speed motor/generator drives.

For Ferrite-PMs, the highest danger of demagnetization, which occurs at low temperatures, may be reduced by proper shaping of rotor flux barriers filled with magnets.

The complete design of Ferrite-PM RSMs with multiple poles, presented in [28], is synthesized here. The modeling of this machine was treated summarily earlier in this chapter.

A relationship between the p.u. PM volume V_m p.u. and the airgap PM flux density B_{gaPM} is also established [28]:

$$B_{gaPM} = \frac{V_m, p.u.}{l_{a,p.u.}} \cdot \frac{\Delta\xi \cdot \cos(\Delta\xi/2)}{\tan(\Delta\xi/2)} B_{ma} \approx 2\frac{V_m, p.u.}{l_{a,p.u.}}\left(1 - \frac{\pi^2}{\eta_r^2}\right)B_{mr} \tag{4.73}$$

with

$$\frac{B_{ma}}{B_{mr}} = 1 \bigg/ \left(1 + \frac{4\pi S_1}{l_{a,p.u.} \cdot \tau} \cdot \frac{\sin(\Delta\xi/2)}{\Delta\xi} \cdot \frac{g}{\tau}\right); \quad \Delta\xi = \frac{2\pi}{\eta_r} \tag{4.74}$$

For:

- Constant-thickness flux barriers per their length
- Uniform rotor slotting (n_r slots per periphery)
- Barrier thicknesses proportional to Δf_{qi} (current loading)
- Barrier length proportional to same p.u. quantities (as previously) admitted in this chapter, to provide uniform flux density in the Ferrite-PMs that fill the flux barriers

A complete ("natural") compensation of flux in axis q is considered in the design for rated power.

$$L_q i_q - \Psi_{PMq} = 0; \quad \Psi_{PMq} = N_1 k_{w1} \frac{2}{\pi} B_{gaPM \cdot \tau \cdot l_{stack}}, \tag{4.75}$$

The cross-section of the machine is shown in Figure 4.24 [28].

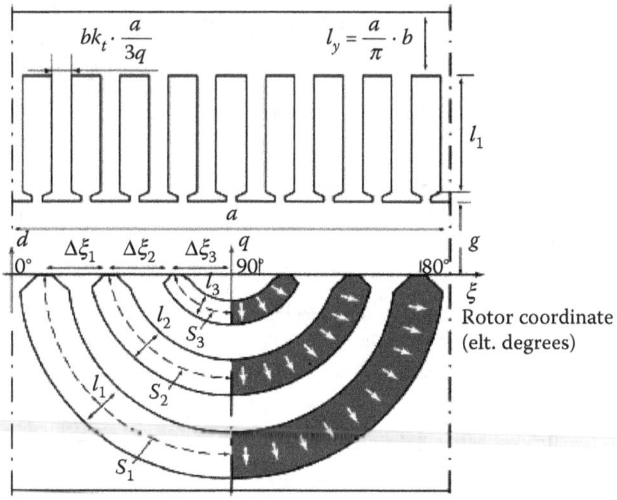

FIGURE 4.24 Circular MFBA-rotor PM-RSM geometry. (After B. Boazzo et al., 2015, *IEEE Trans on*, vol. IE-62, no. 2, pp. 832–845. [28])

All variables in Equations 4.73 and 4.74 are presented in Figure 4.24, while B_{ma} is the uniform flux density in the magnets and B_{gaPM} is the peak flux density in the airgap for zero stator currents.

From the condition (Equation 4.75) and L_{qm} and L_{sl} expressions developed previously in the chapter, the characterizing linear current loading A_{q0} (Aturns/m):

$$A_{q0} = \frac{3\sqrt{3}(N_1 k_{w1})}{p\tau} I_q \tag{4.76}$$

A high ratio $\tau/g \approx 50$–200 is recommended for a good design, and the PM p.u. volume $V_{m,\text{p.u.}}$ should be about 40%–45% with Ferrite-PMs.

Then the shear rotor stress is calculated:

$$\begin{aligned} \sigma &= B_{\text{gap},d} A_{q0} - B_{\text{gap},q} A_d \text{ in } (\text{N/m}^2) \\ &\approx b B_{Fe} A_{q0}, \end{aligned} \tag{4.77}$$

with the second term negligible. The d-axis magnetic flux density $B_{\text{gap},d}$ is produced by i_d current and may be imposed by design (as in the previous preliminary design paragraph). Typical design sensitivity studies of q-axis characteristic current loading A_{q0} and of shear stress versus τ/g ratio (Figures 4.25 and 4.26) emphasize:

- The notable influence of stator tooth depth per airgap ratio, l_t/g
- The mild influence of τ/g and of remnant ferrite flux density B_r
- The important influence on shear stress of both B_r and relative air (PM) length per pole in axis q

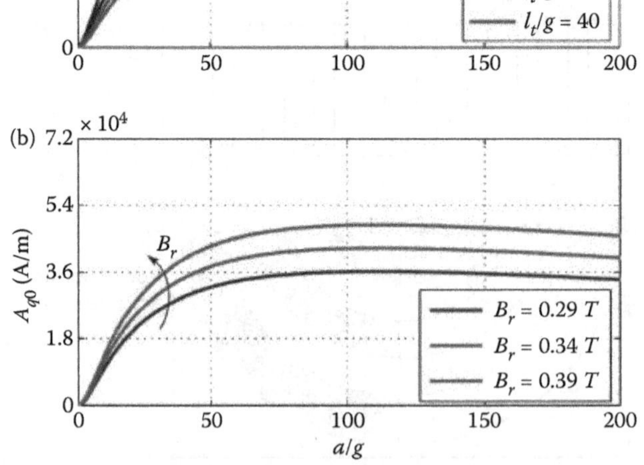

FIGURE 4.25 Characteristic electric q-axis loading A_{q0} versus τ/g for $q_1 = 3$, $n = 3$, $n_r = 14$, $B_r = 0{,}34$ T. (After B. Boazzo et al., 2015, *IEEE Trans on*, vol. IE-62, no. 2, pp. 832–845. [28])

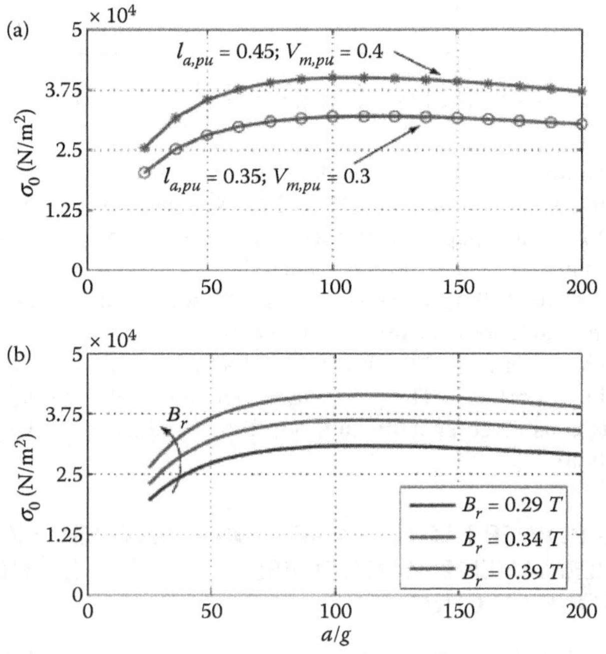

FIGURE 4.26 Shear stress versus τ/g.

By condition (Equation 4.75), the ideal power factor expression becomes simply:

$$\tan\varphi \approx \frac{A_d}{A_{q0}} = \frac{i_d}{i_{q0}} \tag{4.78}$$

A large power factor is needed, but both A_d and A_{q0} (linear current loadings) contribute to it.

However, as τ/g increases above 50/1, the power factor is above 0.8 and tends to unity for large τ/g values (200/1) [28].

The demagnetization current loading $A_{q\,\mathrm{demag}}$ may be defined as:

$$A_{q\,\mathrm{demag}} = \frac{\pi B_r l_{a,p.u.}}{\mu_0 f_{qn}}\left(1 - \frac{B_{m\,\mathrm{demag\,p.u.}}}{B_{m0\,\mathrm{p.u.}}}\right) \tag{4.79}$$

with $B_{\mathrm{m\,demag\,p.u.}}$ imposed. As the most critical situation occurs at low temperature, even preheating the magnets by AC stator currents before operation may be used to avoid the situation.

An upper limitation of pole pairs $p_{\max}$ is found for given τ, g, l_t (tooth depth) [28], compliant with small core loss:

$$p_{\max} = \frac{\pi \cdot D_{is}}{2g}\left(\left.\frac{\tau}{g}\right|_{\min}\right)^{-1}\left(1 - \frac{2l_t}{D_{is}}\right) - b \tag{4.80}$$

with D_{is} and g given; (l_t/D_{is}) and $b(B_{gp,d}/B_{Fe})$ have to be adopted in the design.

There is also a minimum number of poles, as the yoke thickness increases with p (pole pairs) decreasing.

Finally, an optimal number of pole pairs p_0 for Joule loss minimization was found [28].

$$p_0 \approx \sqrt[3]{\frac{3}{2}\left(b + \frac{\pi}{3}k_{ch}\frac{D_{is}}{2l_{stack}}\right)\left(\frac{m_0}{bB_{Fe}}\frac{D_{is}}{2g}TV\right)^2} \tag{4.81}$$

where k_{ch} is the coil chording factor; TV—torque/stator volume.

In the two case studies investigated in [28], a 14-kW, 168-rpm, 0.38-m outer stator diameter, 0.25-m stack length, 0.75-mm airgap, and, respectively, a 2-MW, 15-rpm, 4-m outer stator diameter, 1.5-m stack length, 4-mm airgap have been considered.

A power factor of 0.75 for 120 kg active weight in a 10-pole configuration for the first (800- Nm) machine with $10^4\,W/m^2$ Joule loss density was obtained.

For the second machine, for $2p = 22$ poles, 1.273 MNm, the Joule loss density was 7600 W/m^2, while the active weight was 50 tons. The power factor should be above 0.85, as $\tau/g = 145/1$ [28].

The performance is considered encouraging, but further reduction in weight is required before eventual industrialization.

4.9 IMPROVING POWER FACTOR AND CONSTANT POWER SPEED RANGE BY PERMANENT MAGNET ASSISTANCE IN RELUCTANCE SYNCHRONOUS MACHINES

In some applications (spindles, traction, etc.) a wide constant power speed range is required (CPSR $\geq\geq 3$).

In such cases, the design should include assisting PMs (preferably Ferrites). Also, the design is performed for base speed (peak torque, maximum inverter voltage: $k_{s\,max}$) and then checked for wide CPSR by a constraint (in the optimal design) for torque "unrealization" at max. speed.

A light traction drive example with RSMs and PM-RSMs is investigated in [19] for a two-pole pair configuration for 7000-rpm/1500-rpm CPSR.

The topologies (Figure 4.27a), torque and power (Figure 4.27b), and power factor and efficiency (Figure 4.27c) lead to remarks such as:

- The addition of Ferrite-PMs increases the torque envelope notably (more than 25%) and allows for a 7000-rpm/1500-rpm CPSR at more than the specified 1.5 kW.
- At the same time, as expected, the power factor was increased dramatically above base speed.
- Even the efficiency was increased by 3%–4%, to a maximum of 92%.
- The no-load voltage at maximum speed (7000 rpm) should be less than 150% p.u., as the torque contribution of PMs at low speed is about 25%; consequently, no overvoltage protection against uncontrolled generator (UCG) faulty operation is required.

Note: A sequence of studies investigated PM-RSMs for traction for electric and hybrid electric vehicles and found them capable of 75% of torque in the same volume, but similar performance with IPMSMs [29].

4.10 RELUCTANCE SYNCHRONOUS MACHINE AND PERMANENT MAGNET-RELUCTANCE SYNCHRONOUS MACHINE OPTIMAL DESIGN BASED ON FINITE-ELEMENT METHOD ONLY

Optimal design codes based exclusively on computationally efficient FEM of the RSM [6] and PM-RSM [30,31] have been put in place lately, but the total computation effort, with multiframe hardware, is still in the range of 50 hours.

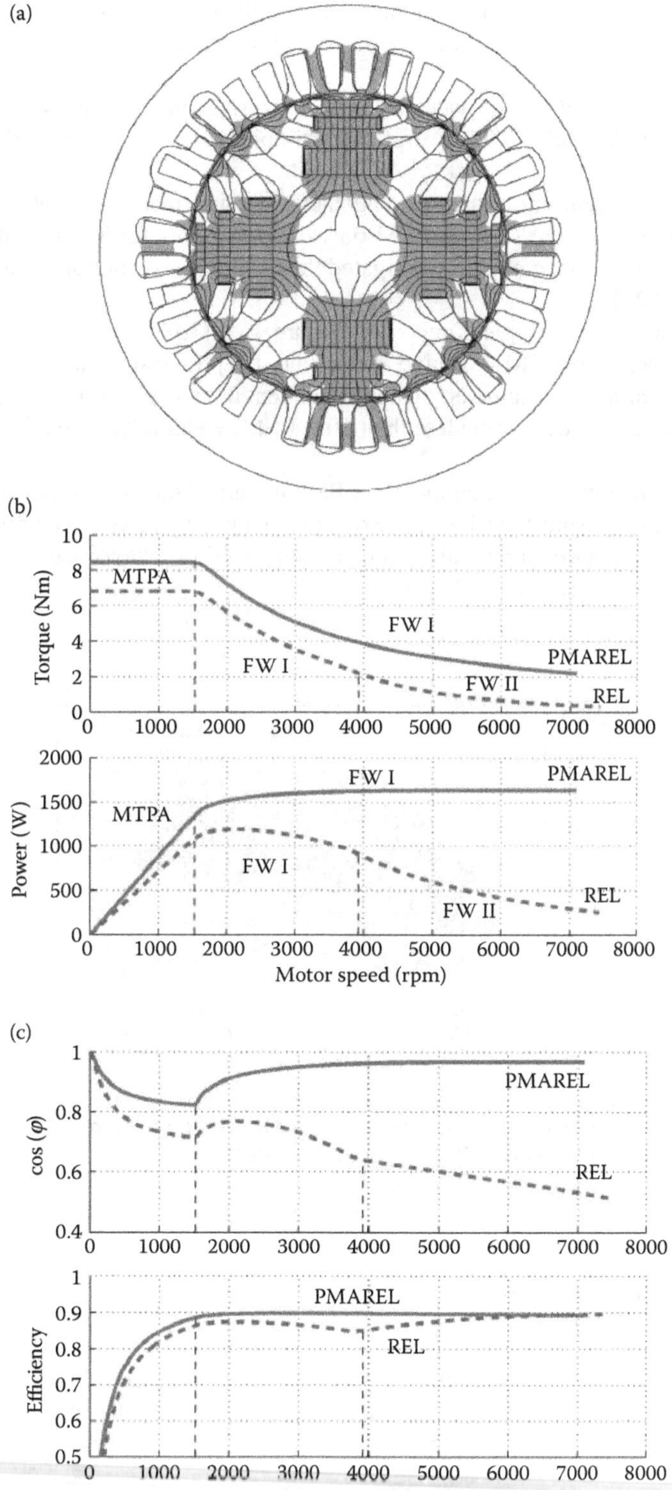

FIGURE 4.27 RSM and PM-RSM for wide CPSR: (a) motor topologies; (b) torque and power versus speed; (c) power factor (coil) and efficiency versus speed. (After M. Ferrari et al., *IEEE Trans on*, vol. IA-51, no. 1, 2015, pp. 169–177. [19])

A differential evolution algorithm [6] has been applied to RSMs and finally compared for PM-RSMs in terms of p.u. torque, losses, and power factor, mainly for maximum torque per ampere (MTPA) conditions.

Sample Pareto clouds of the power factor against machine "badness" ($\sqrt{p_{\text{copper}}}$/torque)—Figure 4.28—show a dramatic (0.95) power factor improvement with PM-RSM; also, more than 25% more power is available.

Rated speed and rated current test results satisfactorily confirmed the FEM predictions [6].

In another multiobjective GA optimal FEM-only design attempt for RSMs, with three algorithms for comparisons, the torque ripple was investigated thoroughly for reduction for a 4.5-Nm, 5.0-krpm case study (Figure 4.29) [30].

An extremely small torque ripple is reported (Figure 4.29 [30]).

Again, based solely on FEM machine modeling, [31] investigated a 1.6-Nm, 5–50-krpm, obtaining first an 80-mm outer diameter, 50-mm stack length, two-pole configuration with an airgap $g = 0.25$ mm, torque and torque ripple less than 10% RSM, yielding good performance except for a 0.51 power factor.

Also, by adding high energy and temperature PMs in part of the flux barrier area, notable power and power factor improvements have been, as expected, obtained ($\cos\varphi = 0.85$ for 40% PM fill). The rotor did not allow (or include) a retaining ring, so the flux bridges have also been verified mechanically by FEM, up to a maximum of 70 krpm.

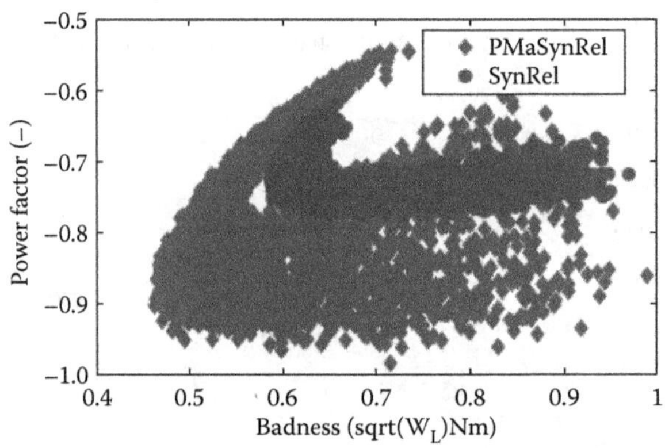

FIGURE 4.28 Pareto clouds with DE-FEM optimization for RSM and PM-RSM. (After Y. Wang et al., *IEEE Trans on*, vol. IA-52, no. 4, pp. 2971–2978, July-Aug. 2016. [6])

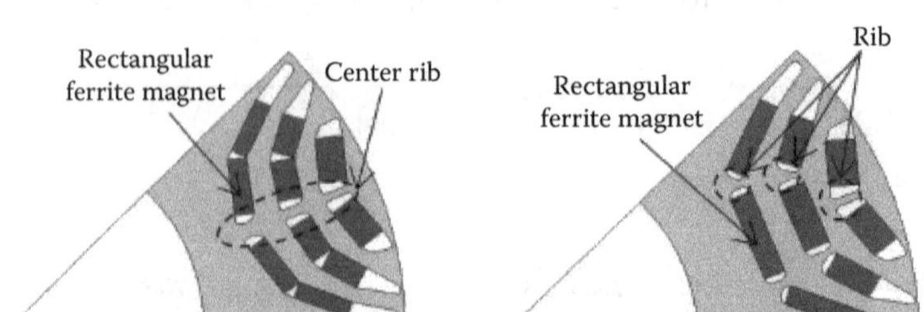

FIGURE 4.29 Pareto clouds of 10 optimization runs stopped after 3000 function calls for torque ripple versus torque by three algorithms (MODE, MOGA, MOSA.) (Adapted from F. Cupertino et al., *IEEE Trans*, vol. IA-50, no. 6, 2014, pp. 3617–3627. [30])

4.11 SUMMARY

- Three-phase RSMs, proposed in 1927, have been sporadically used in variable-speed drives ever since, but have only recently reached industrialization in 10 kW to 500 (1500) kW per unit in four-pole topologies, where they yield 2%–2.2% more system (motor + converter) efficiency in the same stator (and converter) than for IMs, though at 5%–8% less motor power factor; the motor initial cost is also about 10% less than that of an IM.
- Multiple flux-barrier or axially laminated rotors with distributed magnetic anisotropy are needed for competitive performance ($L_d/L_q > 5$–8); to secure such saliency, pole pitch/airgap ratios above 100/1 are recommended.
- PM assistance (preferably by low-cost Ferrite-PMs) is required when a wide constant power speed range (CPSR > 3) is required, as in spindle or traction applications; implicitly, the power factor is drastically increased, while $e_{PMd}/V_s < 0.25$–0.3; this leads to a less than 150% no-load voltage at maximum speed in uncontrolled generator mode for a CPSR = 3–4; consequently, no overvoltage protection provision is required.
- Simple analytical methods have been proposed to calculate the magnetizing inductances L_{dm}, L_{qm} and e_{PMd} for both MFBA- and ALA-rotors; they include circulating and flux bridge additional (fringing) flux contributions to increase L_{qm}.
- More advanced equivalent magnetic circuit models include the above aspects but also magnetic saturation and the presence of assisting PMs.
- The above methods are very useful for preliminary design that can serve as initial design in optimal design algorithms.
- The circuit model of RSMs and PM-RSMs satisfactorily portrays both steady-state and transient performance for control design, as illustrated in the chapter by a case study.
- To consider not only a large average torque (large L_{dm}, L_{qm}) but also the reduction of total torque pulsations with sinusoidal current control in variable-speed drives, a more elaborate design methodology of rotor multiple flux barriers is required.
- According to such an industrial design approach, the uniform angle placement (α_m) of rotor flux barriers in axis d in equal magnetic reluctance flux barriers provides for large average torque with three to four flux barriers per half-pole.
- To reduce torque pulsations, a fictitious rotor slot is introduced at angle β from axis q and thus, along the latter axis, the flux barriers are not placed uniformly. An optimum value of β is found for four poles and four flux barriers per pole as $\beta_{optim} = 9°$; thus, the goals of high average torque and low torque pulsations are separated in the design.
- 2D-FEM is mandatory in the above industrial design methodology; 3D-FEM may be used to calculate L_{qm} more precisely, especially in short stator core stacks ($l_{stack}/\tau < 1.5$).
- Ferrite-PM assistance in RSMs for circular uniform flux barrier rotors and boat-shaped flux barrier rotors are treated by case studies to show dramatic improvement in power factor and some efficiency improvement over RSMs for applications as diverse as directly driven elevator drives, light traction drives [32], and direct-driven wind generators (2 MW, 15 rpm).
- Computationally efficient FEM-only optimization algorithms have been introduced recently in the optimal design of RSMs and PM-RSMs with promising performance results, especially with torque ripple below 10%. Still, the computation time for such advanced methods—on multiframe hardware—is on the order of 50 hours; drastic progress in reducing this computation time is needed while these methods are here to stay, as they allow sensitivity studies that document the robustness of the design to fabrication tolerances and material property variations.
- The ALA-rotor produces better efficiency and power factor for the same stator than the MFBA-rotor but is still considered hard to manufacture. As the high no-load rotor iron core loss problem has been solved (see above in the chapter), it is hoped that competitive fabrication technologies will evolve to soon move the ALA-rotor in RSM (PM-RSM) to the mass production stage.

REFERENCES

1. J. Kotsko, Polyphase reaction synchronous motors, *Journal of the American Institute of Electrical Engineers*, vol. 42, no. 11, pp. 1162–1168 Nov. 1923.
2. I. Boldea, *Reluctance Synchronous Machines and Drives*, book, Oxford University Press, New York, 1996.
3. ABB, http://search-ext.abb.com/library/Download.aspx?DocumentID=3AUA0000120962&LanguageCode=en&DocumentPartId=1&Action=Launch.
4. I. Boldea, Z. X. Fu, S. A. Nasar, Performance evaluation of axially-laminated anisotropic (ALA) rotor reluctance synchronous motors, *Conference Record of the 1992 IEEE Industry Applications Society Annual Meeting*, Part I, pp. 212–218.
5. I. Boldea, T. Fukao, T. A. Lipo, L. Malesani, T. J. E. Miller, A. Vagati, Synchronous reluctance motors and drives: A new alternative, *IEEE Tutorial Course at IEEE Industry Applications Society. Annual Meeting*, 1994.
6. Y. Wang, D. M. Ionel, M. Jiang, S. J. Stretz, Establishing the relative merits of synchronous reluctance and PM-assisted technology through systematic design optimization, *IEEE Trans on*, vol. IA-52, no. 4, pp. 2971–2978, July-Aug. 2016.
7. V. Ostovic, *Dynamics of Saturated Electric Machines*, book, John Wiley, Springer Verlag New York Inc., 1989.
8. A. Vagati, G. Franceschini, I. Marongiu, G. P. Troglia, Design criteria of high performance synchronous reluctance motors, *Conference Record of the 1992 IEEE Industry Applications Society Annual Meeting*, vol. 1, pp. 66–73.
9. A. Vagati, B. Boazzo, P. Guglielmi, G. Pellegrino, Ferrite assisted synchronous reluctance machines: a general approach, *2012 XXth International Conference on Electrical Machines*, pp. 1315–1321.
10. C. J. Heyne, A. M. El-Antably, Reluctance and doubly-excited reluctance motors, ORNL/SUB—81–95013/1, Report.
11. N. Bianchi, B. J. Chalmers, Axially laminated reluctance motor: Analytical and finite element methods for magnetic analysis, *IEEE Transactions on Magnetics*, vol. 38, no. 1, Jan 2002.
12. I. Torac, A. Argeseanu, Analytical model of the synchronous reluctance motor with axially laminated rotor for optimization purpose, *Record of OPTIM—2008*, pp. 27–32 (IEEEXplore).
13. E. S. Obe, Calculation of inductances and torque of an axially-laminated synchronous reluctance motor, *IET Electric Power Applications*, 2010, vol. 4, no. 9, pp. 783–792.
14. I. Boldea, S. A. Nasar, *Electric Drives*, book, 3rd edition, CRC Press, Taylor and Francis Group, New York, 2016.
15. C. M. Spargo, B. C. Mecrow, J. D. Widmer, Application of fractional slot—concentrated windings to synchronous reluctance machines, *2013 International Electric Machines & Drives Conference*, pp. 618–625.
16. N. Bianchi, *Electric Machines Analysis Using Finite Elements*, book, CRC Press, Taylor and Francis Group, New York, 2005, pp. 239–245.
17. I. Boldea, L. N. Tutelea, *Electric Machines: Steady State, Transients and Design with MATLAB*, book, part 3, CRC Press, Taylor and Francis Group, New York, 2010.
18. S. T. Boroujeni, N. Bianchi, L. Alberti, Fast estimation of line-start reluctance machine parameters by finite element analysis, *IEEE Transactions on Energy Conversion*, 2011, vol. 26, no. 1, pp. 1–8.
19. M. Ferrari, N. Bianchi, E. Fornasiero, Analysis of rotor saturation in synchronous reluctance and PM assisted reluctance motors, *IEEE Trans on*, vol. IA-51, no. 1, 2015, pp. 169–177.
20. P. Guglielmi, B. Boazzo, E. Armando, G. Pellegrino, A. Vagati, Permanent-magnet minimization in PM-assisted synchronous reluctance motors for wide speed range, *IEEE Trans on*, vol. IA-49, no. 1, 2013, pp. 31–41.
21. S. Agarlita, Ion Boldea; Frede Blaabjerg, High-frequency-injection-assisted "active-flux"-based sensorless vector control of reluctance synchronous motors, with experiments from zero speed, *IEEE Trans on*, vol. IA-48, no. 6, 2012, pp. 1931–1939.
22. L. Xu, X. Xu, T. A. Lipo, D. W. Novotny, Vector control of a synchronous reluctance motor including saturation and iron loss, *IEEE Trans on*, vol. IA-27, no. 5, 1991, pp. 1977–1985.
23. R. R. Moghaddam, F. Gyllensten, Novel high-performance SynRM design method: an easy approach for a complicated rotor topology, *IEEE Trans*, vol. IE-61, no. 9, 2014, pp. 5058–5065.
24. R. R. Moghaddam, F. Magnussen, C. Sadarangani, Novel rotor design optimization of synchronous reluctance machine for low torque ripple, *2012 XXth International Conference on Electrical Machines*, pp. 720–724.

25. D. A. Staton, T. J. E. Miller, S. E. Wood, Maximizing the saliency ratio of the synchronous reluctance motor, *IEE Proceedings B—Electric Power Applications*, 1993, vol. 140, no. 4, pp. 249–259.
26. A. Vagati, G. Franceschini, I. Marongiu, G. P. Troglia, Design criteria of high performance synchronous reluctance motors, *Conference Record of the 1992 IEEE Industry Applications Society Annual Meeting*, vol. 1, pp. 66–73.
27. N. Bianchi, M. Degano, E. Fornasiero, Sensitivity analysis of torque ripple reduction of synchronous reluctance and IPM motors, *2013 IEEE Energy Conversion Congress and Exposition*, pp. 1842–1849.
28. B. Boazzo, A. Vagati, G. Pellegrino, E. Armando, P. Guglielmi, Multipolar ferrite assisted synchronous reluctance machines: a general design approach, *IEEE Trans on*, vol. IE-62, no. 2, 2015, pp. 832–845.
29. M. Obata, S. Morimoto, M. Sanada, Y. Inoue, Performance of PMASynRM with ferrite magnets for EV/HEV applications considering productivity, *IEEE Trans*, vol. IA-50, no. 4, 2014, pp. 2427–2435.
30. F. Cupertino, G. Pellegrino, C. Gerada, Design of synchronous reluctance motors with multiobjective optimization algorithms, *IEEE Trans*, vol. IA-50, no. 6, 2014, pp. 3617–3627.
31. F. Cupertino, M. Palmieri, G. Pellegrino, Design of high speed synchronous reluctance machines, *2015 IEEE Energy Conversion Congress and Exposition (ECCE)*, pp. 4828–4834.
32. E. Carraro, M. Morandin, N. Bianchi, Traction PMASR motor optimization according to a given driving cycle, *IEEE Trans*, vol. IA-52, no. 1, 2016, pp. 209–216.

5 Control of Three-Phase Reluctance Synchronous Machine and Permanent Magnet–Reluctance Synchronous Machine Drives

5.1 INTRODUCTION

Variable-speed drives (VSDs) now cover 30–40% of all drives used in association with power electronics to increase productivity and save energy in variable-output energy conversion processes.

VSDs imply motion control—position, speed, torque control—in motoring and output voltage amplitude (frequency) and power control in generating. Generating will not be considered in this chapter. However, regenerative braking is inherent to VSDs and thus will be investigated here. Control of VSDs, RSMs and PM-RSMs included, may be implemented:

- With position sensors: encoder-VSDs
- Without position sensors (sensorless): encoderless-VSDs

Position sensors are planted in safety-critical and servo-drives when precise control of absolute position, speed, and torque is required down to zero speed.

Also, in many applications when the maximum to minimum speed control range is above 200 (1000 and more)/1, and torque response ($\pm$ rated torque) is required within 1–3 milliseconds, an encoder is mandatory. Sensorless (encoderless) control implies using online (and offline) information and computation to estimate—observe—the rotor position within three to four electrical degrees of current fundamental maximum frequency (speed).

The speed control range for an encoderless drive has increased notably in the last decades and is now above 200/1 with 2–3 millisecond $\pm$ full torque response, a steady-state speed error of 2–3 rpm, and up to high speeds (of say, 10 krpm and more).

5.2 PERFORMANCE INDEXES OF VARIABLE-SPEED DRIVES

Performance of VSDs is gauged by performance indexes. Though not standardized yet, the latter, introduced in [1], may be an useful guide and is summarized as:

- Energy conversion indexes:
  - Drive (motor + converter) power efficiency at rated (base) speed: η_P = output power/input power
  - Motor + converter energy efficiency in highly dynamic applications: η_e = output energy/input energy

  - RMS kW/kVA output kW/input kVA: refers to the product of efficiency and power factor in sinusoidal current drives
  - Losses/torque or $\sqrt{\text{losses}}$/torque: the "badness" torque factor
- Drive response performance

Peak torque (T_{ek})/inertia (J), which leads to ideal acceleration

$$a_{i\max} = \frac{T_{ek}}{J}\left(\text{rad}/s^2\right) \tag{5.1}$$

The ideal acceleration time t_{ai} to ω_b (electrical angular speed: $\omega_b = p_1\Omega_b$).

$$t_{ai} = \frac{\omega_b/p_1}{a_{i\max}} = \frac{\omega_b}{p_1}\cdot\left(\frac{T_{ek}}{J}\right)^{-1}; \quad p_1 - \text{pole pairs} \tag{5.2}$$

Field weakening range; to increase speed above ω_b (where full inverter voltage $V_{s\max}$ full magnetic flux in the machine produces the accepted full [rated] power in a continuous [or specified] duty cycle), the magnetic flux in the machine is reduced to allow operation at $V_{s\max}$ such that to conserve base power P_{bc}:

$$P_{bc} = T_{eb}\frac{\omega_{rb}}{p_1} \tag{5.3}$$

up to ω_{rmax}.

This is called the continuous (or specified duty cycle) constant power speed range; there are applications where no flux weakening is required and where $\omega_{rb} = \omega_{r\max}$ (say, refrigerator compressor drives); in spindlelike or tractionlike applications, CPSR $\geq 3/1$ is generally required.

Note: Here in flux-weakening or wide CPSR drives, the RSM and especially PM-RSM come into play, as the flux weakening is basically done by decreasing the id stator current component in contrast to strong magnet ($e_{PM} > 0.7\ V_{s\max}$) IPMSMs. Consequently, the efficiency and power factor in PM-RSMs is superior to IPMSM (with strong PM emf) in the flux-weakening speed range ($\omega_{rb} < \omega_r < \omega_{r\max}$).

- Variable speed ratio $\omega_{r\max}/\omega_{r\min}$:
  1. $\omega_{r\max}/\omega_{r\min} > 200/1$ for servo-drives, with encoder (1–3 ms torque response)
  2. $100 < \omega_{r\max}/\omega_{r\min} < 200$ for advanced (closed-loop) sensorless (encoderless) drives ($\pm$full torque response in 2–6 ms)
  3. $\omega_{r\max}/\omega_{r\min} < 20$ open loop sensorless drives based on V/f (I-f) control (scalar drives): torque response is slower (in tens of milliseconds range)
- Torque rise time at zero speed: $t_{T_{ek}}$, which is in the millisecond range for servo drives and advanced sensorless (encoderless) drives and in the tens of milliseconds for open loop (V/f- or I-f–based) controlled drives
- Torque ripple range $\Delta T_e/T_{eb}$

The torque ripple compounds cogging torque (at zero currents when assisting PMs are used) and torque pulsations due to current control (PWM type), motor stator and rotor slotting, magnetic saturation, and stator winding type.

Note: In general, applications such as pumps, fans, compressors, and traction, a 10% torque ripple may be acceptable to reduce noise and vibration, but for, say, car-steering-assist (or servo) drives, $\Delta T_e/T_b < 2\%$.

- Motion control precision and robustness: ΔT_e, $\Delta\omega_r$, $\Delta\theta_r$ for torque, speed, and position control, respectively, in the presence of torque perturbation, inertia, and machine parameter variations (ΔP_{ar}) defined as $\Delta T_e/\Delta P_{ar}$, $\Delta\omega_r/\Delta P_{ar}$, $\Delta\theta_r/\Delta P_{ar}$.

- Dynamic stiffness: $\Delta S = \Delta T_{\text{perturbation}}/\Delta x$; Δx-controlled variable variation. The load torque frequency and amplitude are varied and the response error is measured to determine maximum acceptable torque perturbation frequency and amplitude for a given minimum accepted control precision at critical operation points.
- Temperature, noise weight, and total cost
  - Direct drives are in contact with the load machine and thus the motor temperature T_{motor} has to be limited: $T_{\text{motor}} < 20°\text{C} + T_{\text{ambient}}$.

    For drives with a mechanical transmission, the motor temperature restrictions are milder and given by the stator winding insulation class (B, E, F).
  - The noise of a VSD is produced by both the motor and the static power converter. The accepted radiated noise level (at 1 m from the motor, in general) depends on application and, for machine tools, is:

$$L_{\text{noise}} \approx 70 + 20\log\left(\frac{P_n}{P_{n0}}\right);\ P_{n0} = 1\,\text{kW},\ P_n = (1-10)\,\text{kW} \tag{5.4}$$

  - The overall (total) cost of a VSD, C_t, maybe defined as:

$$C_t = C_{\text{mot+conv}} + C_{\text{loss}} + C_{\text{maint}} \tag{5.5}$$

with

$C_{\text{mot + conv}}$—cost of motor + converter, sensors, control, protection
C_{loss}—capitalized loss cost (motor + converter) for a given equivalent/average duty cycle over the entire drive life for a given application
C_{maint}—maintenance repair costs over VSD life

For a better assessment of C_t, the net present worth cost is considered, by allowing inflation, part of electric energy price evolution and premium in investment dynamics, over the VSD life. Based on C_t, the payback time may be calculated to justify the introduction of a VSD in a certain industrial process. With variable output, only payback times from energy bill reduction alone of less than 3–5 years, in general, are accepted as practical by users, especially in the kW, tens of kW, and hundreds of kWs per unit applications.

- Motor specific weight: peak torque/kg or peak torque per outer-diameter stator stack volume: Nm/liter. Torque densities of 80 Nm/liter have recently been reached in hybrid electric vehicles for power up to 100 kW and torque up to 400 Nm, where equipment volume (weight) is strongly restricted. A small electric airplane also has strong weight restrictions for its VSDs.
- Converter-specific volume; converter kVA/volume: kVA/liter; for HEVs in general, 5–10 and more kVA/liter is necessary, but for efficiency above 0.97–0.98 in the 150-kVA range, this is how silicon carbide (SiC) power switches come into play; in other applications, the kVA/liter requirement is somewhat relaxed.

5.3 RELUCTANCE SYNCHRONOUS MACHINE AND PERMANENT MAGNET–RELUCTANCE SYNCHRONOUS MACHINE CONTROL PRINCIPLES

In principle, it would be possible to start directly with the PM-RSM and then consider the RSM as a particular case for $\Psi_{\text{PM}q} = 0$.

However, "simple to complex" seems here the best method to more easily grasp the essentials.

Besides the encoder and encoderless VSD divide, we may classify the wide plethora of RSM control strategies proposed and applied so far as:

- Fast torque response (vectorial) drives:
- Scalar (V/f- or I-f–based) control without and with stabilizing closed loops where, in general, no speed loop control is used; the speed range is smaller and the torque response is slower (in the tens of milliseconds range).

  Ventilator, compressor, and pump loads are typical for many scalar drives that are simpler in terms of control and finally less costly, but still competitive in terms of energy conversion by special measures that essentially reduce flux with torque reduction.

5.4 FIELD-ORIENTED CONTROL PRINCIPLES

Core loss plays some role in control, not only in the sense of lower efficiency but also by creating a certain coupling between the d-q model axes and thus delaying the response in the first milliseconds of transients [2].

However, for the time being, in stating the FOC principle, core loss, stator mmf space harmonics, slotting, and airgap flux harmonics due to magnetic saturation are all neglected. Again, as there is no cage effect on the rotor, the space-phasor (d-q) model of the RSM in rotor coordinates is [1] copied here for convenience (with more details):

$$\begin{aligned}
&\overline{V}_S = R_S\overline{I}_S + \frac{d\overline{\Psi}_S}{dt} + j\omega_r\overline{\Psi}_S; \quad \overline{\Psi}_S = L_d i_d + jL_q i_q = \Psi_d + j\Psi_q \\
&\overline{I}_S = i_d + ji_q; \quad \overline{V}_S = V_d + jV_q; \quad \Psi_d = L_{sl}i_d + \Psi_{dm}, \quad \Psi_q = L_{sl}i_q + \Psi_{qm} \\
&T_e = \frac{3}{2}p_1(\Psi_d i_q - \Psi_q i_d) = \frac{3}{2}p_1(L_d - L_q)i_d i_q \\
&J\frac{d\omega_r}{dt} = T_e - T_{\text{load}}; \quad \frac{d\theta_{er}}{dt} = \frac{\omega_r}{p_1}; \quad \Psi_{dm} = L_{dm}i_d, \quad \Psi_{qm} = L_{qm}i_q \\
&\overline{V}_S = \frac{2}{3}\left(V_a + e^{j\frac{2\pi}{3}}V_b + e^{-j\frac{2\pi}{3}}V_c\right)e^{-j\theta_{er}}; \quad i_m = \sqrt{i_d^2 + i_q^2}
\end{aligned} \tag{5.6}$$

There is a cross-coupling magnetic saturation effect (as discussed in the previous chapter) [2] such that:

$$\begin{aligned}
&\Psi_{dm} = L_{dm}(i_m)\cdot i_d, \quad \Psi_{qm} = L_{qm}(i_m)\cdot i_q \\
&L_{dm}(i_m) = \frac{\Psi^*_{dm}(i_m)}{i_m}; \quad L_{qm}(i_m) = \frac{\Psi^*_{qm}(i_m)}{i_m}
\end{aligned} \tag{5.7}$$

The time derivatives of Ψ_{dm} and Ψ_{qm} are:

$$\begin{aligned}
\frac{d\Psi_{dm}}{dt} &= L_{ddm}\frac{di_d}{dt} + L_{dqm}\frac{di_q}{dt} \\
\frac{d\Psi_{qm}}{dt} &= L_{qdm}\frac{di_d}{dt} + L_{qqm}\frac{di_q}{dt}
\end{aligned} \tag{5.8}$$

$$L_{dqm} = L_{dm} = (L_{dmt} - L_{dm})\frac{i_d i_q}{i_m^2}; \quad L_{dmt} - L_{dm} = L_{qmt} - L_{qm}$$

$$L_{dmt} = \frac{d\Psi^*_{dm}}{di_m}; \quad L_{qmt} = \frac{d\Psi^*_{qm}}{di_m}; \quad L_{ddm} = L_{dmt} \cdot \frac{i^2_{dm}}{I^2_m} + L_{dm} \cdot \frac{I^2_{qm}}{I^2_m} \tag{5.9}$$

$$L_{qqm} = L_{qmt} \cdot \frac{I^2_{qm}}{I^2_m} + L_{qm} \cdot \frac{i^2_{dm}}{I^2_m}$$

$\Psi^*_{dm}(i_m)$, $\Psi^*_{qm}(i_m)$ are unique curves (Figure 5.1) obtained by FEM or in standstill DC flux decay tests with d and q current components present. This is why $i_m = \sqrt{i_d^2 + i_q^2}$ is used as the unique variable.

The above model of cross-coupling saturation [2] allows us to compute steady-state and transient performance with fluxes, currents, or a combination of them as variables but with Ψ_{dm} and Ψ_{qm} as distinct variables.

Note: Ψ^*_{dm} and Ψ^*_{qm}, which occur with both i_d and i_q present in Equation 5.7 and Figure 5.1, degenerate into Ψ_{dm} and Ψ_{qm} only when the operation mode (or test) includes either i_d or i_q.

Transient inductances $L^{(im)}_{dmt}$, $L^{(im)}_{qmt}$ (Equation 5.9) are in general smaller than $L^{(im)}_{dm}$, $L^{(im)}_{qm}$ due to saturation. On the other hand, at very low AC currents, the first part of the magnetization curve (hysteresis cycles) is active, where the incremental inductances L^i_{dm}, L^i_{qm} are in place:

$$L^i_{dm} = \frac{\Delta\Psi^*_{dm}}{\Delta i_m}, \quad L^i_{qm} = \frac{\Delta\Psi^*_{qm}}{\Delta i_m} \tag{5.10}$$

L^i_{dm}, L^i_{qm} are smaller than both L_{dm}, L_{qm} and L^t_{dm}, L^t_{qm}; see Figure 5.2. They occur at very small currents, say, in standstill frequency FEM calculations or tests.

The curves $\Psi^*_{dm}(i_m)$ and $\Psi^*_{qm}(i_m)$ may be approximated many ways by analytical expressions. For example:

$$\Psi^*_{dm} = L_{dm0} i_m - K_{dm0} i_m^2; \quad \Psi^*_{qm} = L_{qm0} i_m - K_{qm0} i_m^2 \tag{5.11}$$

$$\text{Or even } L^{(i_d)}_{dm} \approx L_{dm0} - K_{dm0} i_d \text{ and } L_{qm} \approx L_{qm0} - K_{qm0} i_m, \tag{5.12}$$

when (however) the cross-coupling effect is neglected. Yet another approximation of cross-coupling saturation is simply:

$$\begin{aligned} \Psi_d &\approx L_d i_d + L_{dq} i_q \\ \Psi_q &\approx L_q i_q + L_{dq} i_d \end{aligned} \tag{5.13}$$

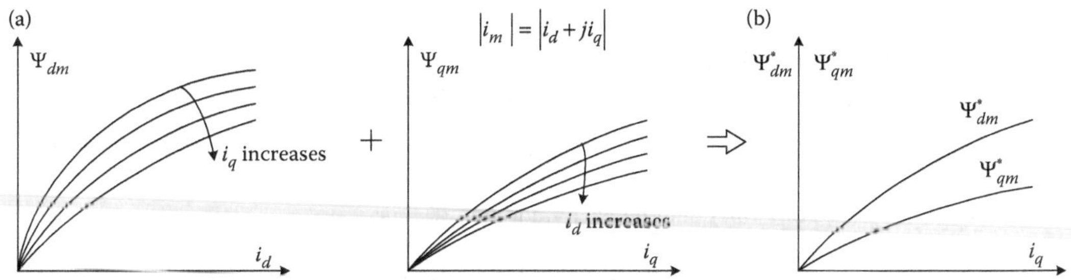

FIGURE 5.1 Magnetization curves in axis i_m, d, q: real, (a); and equivalent, (b). (From I. Boldea, S. A. Nasar, *Proc. IEE*, vol. 134–B, no. 6, 1987, pp. 355–363. [2])

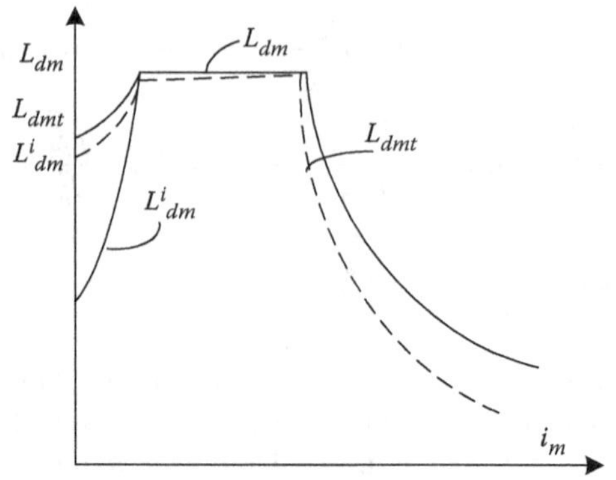

FIGURE 5.2 L_{dm}, L^t_{dm}, L^i_{dm} versus i_m.

Let us now return to FOC, neglecting magnetic saturation first. Controlling i_d and i_q separately is called FOC.

As the rotor is cageless (windingless), there is no decoupling current circuit (in contrast to cage-rotor IMs) and thus, if i_d^* is referenced independently (constant or dependent on torque or speed), i_q^* can be calculated from the reference torque expression:

$$i_q^* = \frac{2}{3p_1}\frac{T_e^*}{(L_d - L_q)i_d^*} \tag{5.14}$$

But defining a new flux, called "active flux" [3], Ψ_d^a as:

$$\Psi_d^a = (L_d - L_q)i_d^*;\quad \overline{\Psi}_d^a = \overline{\Psi}_s - L_q\bar{i}_s, \tag{5.15}$$

$$i_q^* = \frac{2}{3p_1}\frac{T_e^*}{\Psi_d^{a*}} \tag{5.16}$$

The reference torque is obtained as the output of the speed closed loop or separately in the torque mode control (in tractionlike applications).

It turns out that the active flux is aligned along axis d (if cross-coupling saturation is neglected) and the d-q model has only an L_q inductance:

$$\overline{V}_S = R_S\bar{I}_S + j\omega_r L_q\bar{i}_S + L_q\frac{d\bar{i}_S}{dt} + \frac{d\overline{\Psi}_d^a}{dt} + j\omega_r\overline{\Psi}_d^a \tag{5.17}$$

$$T_e = \frac{3}{2}p_1\overline{\Psi}_d^a \cdot i_q \tag{5.18}$$

The FOC of an RSM may be expressed in active flux terms as in Figure 5.3, after the motion-induced voltage compensations $V_{q\text{comp}}^*$, $V_{d\text{comp}}^*$ are added in:

$$V_{q\text{comp}}^* = \omega_r\Psi_d^{a*};\quad V_{d\text{comp}}^* = -\omega_r L_q i_q^* \tag{5.19}$$

The function $\Psi_d^{a*}(\omega_r)$ or $i_d^*(\omega_r)$ may be easily adopted if i_d^* is:

$$\begin{aligned} i_d^* &= i_{dn}\ \text{const for}\ \omega_r \le \omega_b \\ i_d^* &= i_{dn}\cdot\frac{\omega_b}{\omega_r} < i_{dn} \end{aligned} \tag{5.20}$$

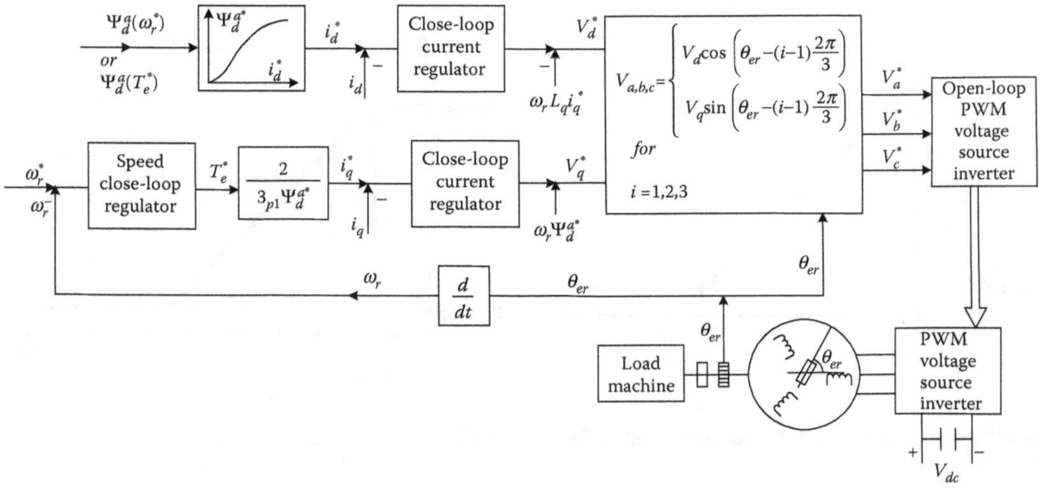

FIGURE 5.3 Generic active–flux–based combined voltage + current FOC of RSM.

Now

$$\Psi_d^{a*}(\omega_r) = (L_d - L_q) \cdot i_d^*, \tag{5.21}$$

with constant (invariable) parameters L_d, L_q or even with them variable: $L_d(i_m)$, $L_q(i_m)$, as described above.

Note: The main advantage of the active flux concept occurs in encoderless control when its position is identical to rotor position θ_{er}, irrespective of load, and the speed of the rotor $\omega_r = \omega_{\overline{\Psi}_d^a}$.

Alternatively, an early current-only FOC of the RSM implementation scheme is shown in Figure 5.4 [4,5] for an RSM with the data: two-pole ALA-rotor, $R_s = 0.95\ \Omega$, $L_{dm} = 136$ mH, $L_{qm} = 3.66$ mH, $L_{sl} = 4$ mH, $J = 2.5 \times 10^{-3}$ kgm^2 (1.5 kW at 60 Hz).

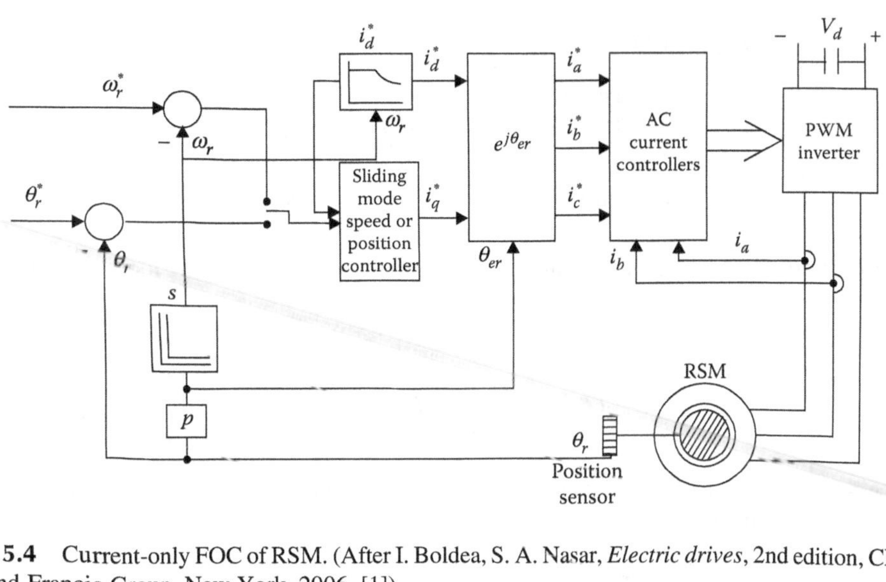

FIGURE 5.4 Current-only FOC of RSM. (After I. Boldea, S. A. Nasar, *Electric drives*, 2nd edition, CRC Press, Taylor and Francis Group, New York, 2006. [1])

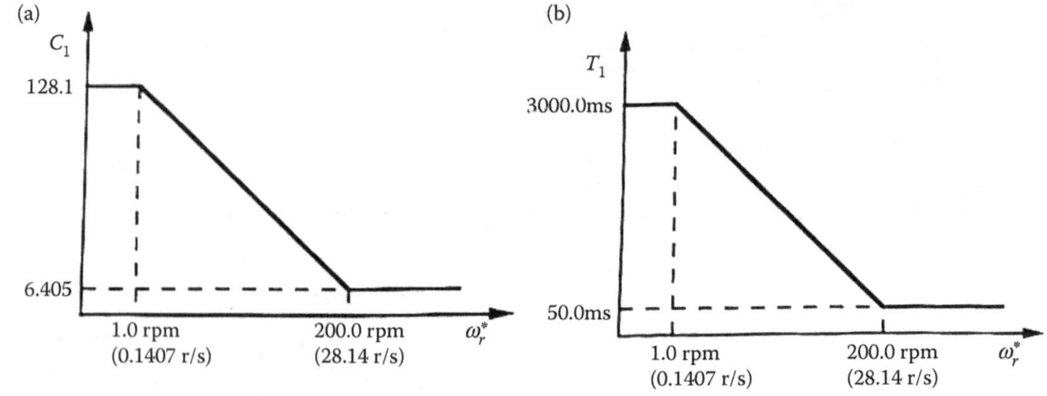

FIGURE 5.5 Gain (C_i), (a); and time constant (T_i), (b) of PI speed controller.

In this implementation, direct reference $i_d^*(\omega_r)$ and $i_{q\max}^*(\omega_r)$ limiters are provided and no emf compensation is included, as AC current controllers (instead of i_d, i_q controllers) are used.

An adaptive speed controller (Figure 5.5) was applied to provide for a wide speed range (1–2500 rpm) with flux weakening: i_d^* decreases with speed. A high-precision encoder (for position measurement and speedy online computation from the former) is also assumed available.

Reference speed and torque (Figure 5.6a), speed response (Figure 5.6b), and torque response (Figure 5.6c) are given as sample results from digital simulation.

Operation at 1.0 rpm with torque perturbation (Figure 5.7a and b) shows satisfactory performance, though during torque perturbation, the speed control shows a small steady-state error, proving that the speed controller, even adaptive, is not totally satisfactory.

5.4.1 Options for i_d^*, i_q^* Relationship

While $i_d^*(\omega_r)$or $i_d^*(T_e^*)$ or, instead, $\Psi_d^{a*}(\omega_r)$ or $\Psi_d^{a*}(T_e^*)$, functions may be derived, but a few methods have become generally accepted:

- Maximum torque per current

$$\frac{\partial T_e}{\partial i_d} = 0; \quad T_e = \frac{3}{2}p_1(L_d - L_q)i_d\sqrt{i_s^2 - i_d^2} \tag{5.22}$$

$$\text{Consequently, } i_d^* = i_q^* = i_s^*/\sqrt{2}; \quad tg\gamma_{ii} = i_q/i_d = 1 \tag{5.23}$$

This simple condition is no longer fulfilled if magnetic saturation is considered, when $\gamma_{ii} > 45^\circ$, (Figure 5.8).

The torque is:

$$(T_e)_{\text{MTPA}} = \frac{3}{2}p_1(L_d - L_q)\frac{i_s^{*2}}{2} \tag{5.24}$$

Note: $i_d^* = i_{sn}/\sqrt{2}$ in Equation 5.23 could be too large a current that might oversaturate the machine; thus, $i_s < i_{sn}$, and Equation 5.23 is applied for lower loads.

- Maximum power factor (MPF)

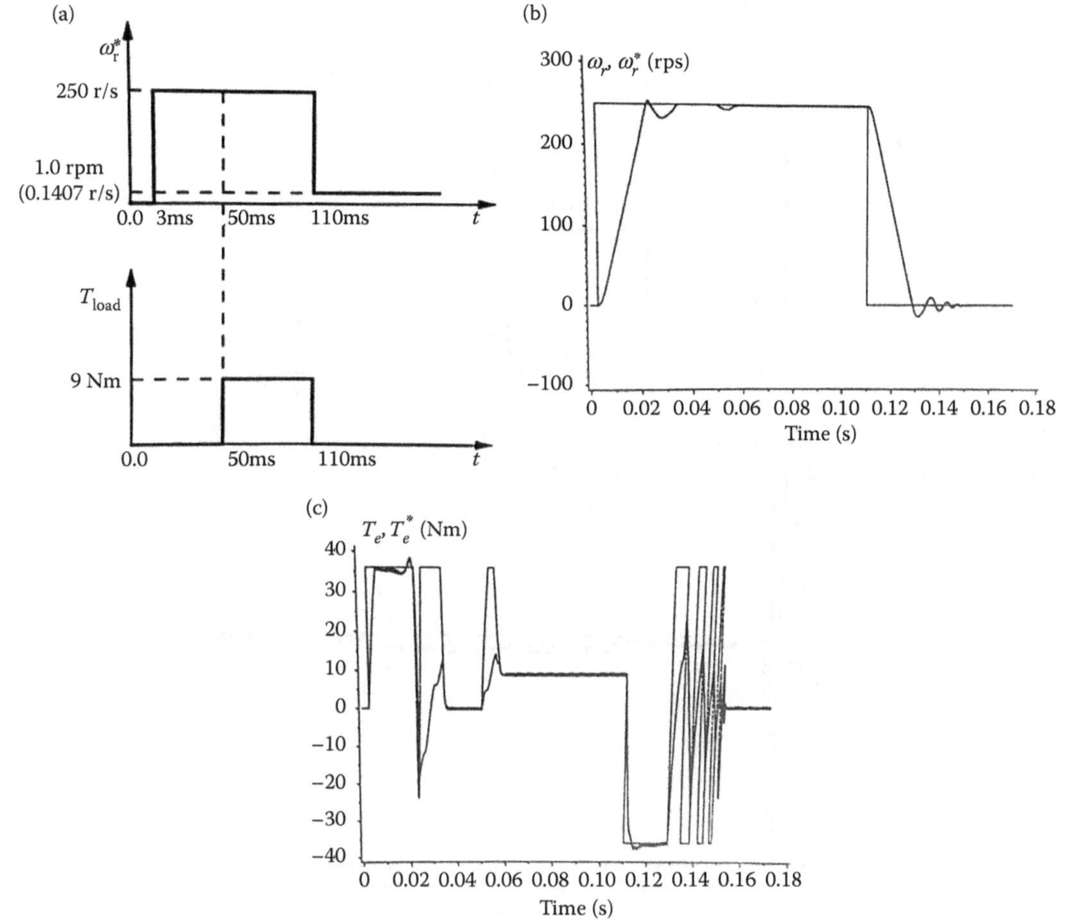

FIGURE 5.6 Speed transients with torque perturbation response of FOC-RSM: Reference speed and torque (a); (b) speed response; (c) torque response.

The tangent of the power factor tan φ with zero losses (from reactive/electromagnetic powers ratio) is:

$$\tan\varphi_1 = \frac{Q}{P} = \frac{L_d \frac{i_d^2}{2} + L_q \frac{i_d^2}{2}}{(L_d - L_q) i_d i_q} \tag{5.25}$$

The maximum of the power factor (minimum of tan φ_1) is obtained (for constant L_d and L_q):

$$\left(\frac{i_d}{i_q}\right)_{\varphi_{1\,\min}} = \sqrt{\frac{L_q}{L_d}} \tag{5.26}$$

$$T_{e_{\varphi 1\,\min}} = \frac{3}{2} p (L_d - L_q) i_d^2 \sqrt{\frac{L_d}{L_q}} > (T_e)_{MTPA} \tag{5.27}$$

$$(\cos\varphi)_{\max} = \frac{1 - L_q/L_d}{1 + L_q/L_d} \tag{5.28}$$

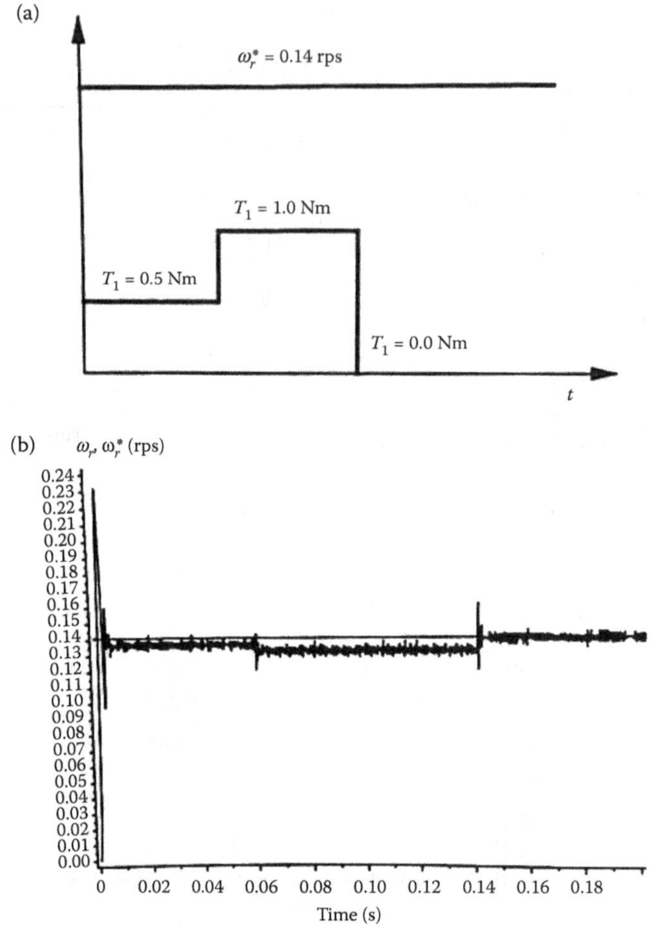

FIGURE 5.7 Speed buildup to 0.14 rps: (a) reference speed and load torque; (b) speed response in rad/s.

The RSM control at the maximum power factor thus secures a proper condition for rated (base) speed power design.

- Maximum torque per flux

This time, the stator flux Ψ_s^* is given:

$$\Psi_s^{*2} = (L_d i_d)^2 + (L_q i_q)^2 \tag{5.29}$$

$$\left(\frac{i_d}{i_q}\right)_{\Psi_s^*} = \frac{L_q}{L_d};\ \lambda_{d\Psi_S} = \lambda_{q\Psi_S} = \Psi_S/\sqrt{2} \tag{5.30}$$

with the corresponding torque:

$$T_{e\Psi_S} = \frac{3}{2} p(L_d - L_q) i_d^2 \frac{L_d}{L_q} \tag{5.31}$$

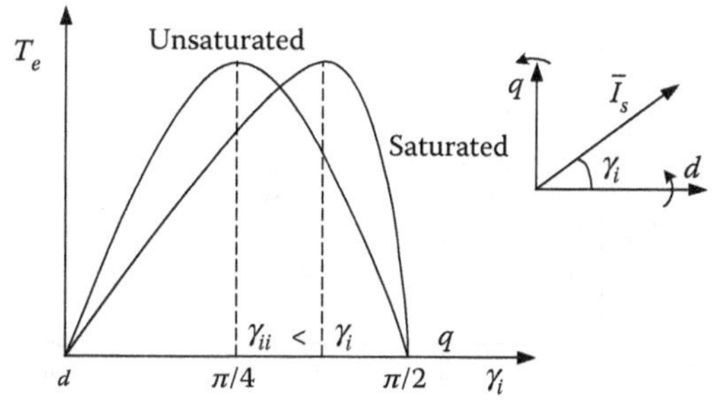

FIGURE 5.8 Qualitative T_e (γ_i) for MTPA for unsaturated and saturated RSM.

At base speed ω_b, where rated i_d^* (d axis flux) is still available,

$$\omega_b = \frac{V_{smax}}{L_d i_d^* \sqrt{2}} \tag{5.32}$$

It is evident that at base speed:

$$T_{eMTPA} < T_{e\varphi_{1\min}} < T_{e\Psi_s^*} \tag{5.33}$$

But the electromagnetic power P_e:

$$P_e = T_e \omega_r / p_1 \tag{5.34}$$

For the three main FOC strategies, P_e is:

$$(P_e)_{MTPA} = \frac{3}{2}(L_d - L_q)\frac{i_s^2}{2}\omega_r; \quad \sqrt{L_d^2 + L_q^2} \cdot \frac{i_s}{\sqrt{2}} \le \frac{V_{smax}}{\omega_r} \tag{5.35}$$

$$(P_e)_{\varphi_{i\min}} = \frac{3}{2}(L_d - L_q)i_s^2 \cdot \frac{\omega_r}{\sqrt{1 + \frac{L_q}{L_d}}}; \quad \sqrt{L_d L_q} \cdot i_s \le \frac{V_{smax}}{\omega_r} \tag{5.36}$$

$$(P_e)_{\Psi_S} = \frac{3}{2}(L_d - L_q)i_s^2 \cdot \frac{L_d L_q}{L_d^2 + L_q^2}\omega_r; \quad \frac{i_s L_d \sqrt{2}}{\sqrt{1 + \frac{L_d^2}{i_q^2}}} \le \frac{V_{smax}}{\omega_r} \tag{5.37}$$

It is now evident that for a given current and speed (base speed), the ideal maximum power is obtained for MTPA, but only if the magnetization current $(i_d)_{MTPA} = i_s/\sqrt{2}$. This is a rather high value, and this is how the maximum power factor condition comes into place for base power P_{eb} at base speed design.

Magnetic saturation makes the above formulae approximations, and more complex expressions are required for more precision. Getting wide CPSR is quite a challenge. The power envelope (up to maximum voltage V_{Smax}—above base speed) will never be flat and thus P_{emax} should be given in p.u. values with respect to base power P_{eb}:

$$P_{eb} = T_{eb} \cdot \frac{\omega_b}{p_1}; \quad P_{emax}(p.u.) = \frac{P_{emax}}{P_{eb}} \tag{5.38}$$

All depends on the saliency ratio L_d/L_q and the chosen base speed ω_b for a given maximum speed. Qualitatively, the situation is as in Figure 5.9.

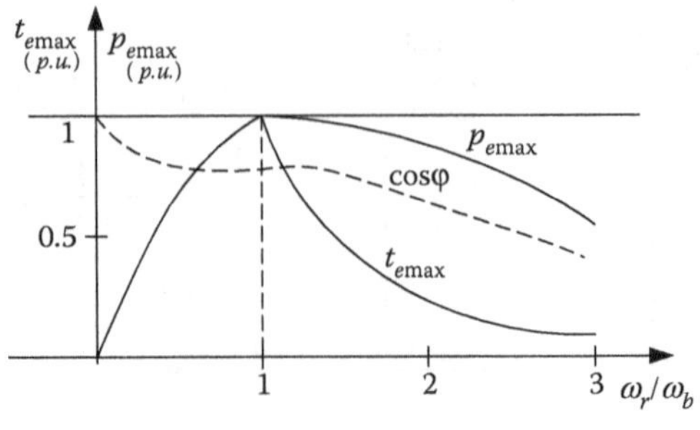

FIGURE 5.9 Maximum torque and power versus speed.

Also, as the speed increases, the dq current angle γ_i ($\tan\gamma_i = I_q/I_d$) increases from 45° (for MTPA) through minimum φ_i and minimum flux Ψ_s^* control conditions.

Also, the power factor above base speed decreases, with speed reaching:

$$\tan(\varphi_i)_{\Psi_s^*} = \frac{1 + \frac{L_q}{L_d}}{1 - \frac{L_q}{L_d}}; \quad (\varphi_i)_{\Psi_s^*} > 45^\circ \tag{5.39}$$

It is evident from the above that RSM, even with high magnetic saliency, is not directly suitable for wide CPSR, at least for the reason of lower power factors at high speeds, which means larger inverter currents, kVA, and losses.

As was understood early, magnetic saturation and core loss influence the transients, but the speed versus time (downward) shows nonlinearity, a sign that the problem was not fully solved. Such a solution is offered in [6], where the speed transients up and down are shown to be linear (Figure 5.10) [6], as referenced.

A robust (sliding mode) speed and position FOC (with encoder) of an RSM is illustrated in [1].

5.5 DIRECT TORQUE AND FLUX CONTROL

Direct torque and flux control (DTFC), introduced in 1985–1986 by Depenbrock and Takahashi as direct self-control and quick torque response control for IMs, was generalized as torque vector control (TVC) for synchronous motors [7,8] in 1988 and was coined by ABB in 1995 as DTC (direct torque control). DTFC (as TVC) was applied first to RSM in 1991 [9].

DTFC directly controls the stator flux linkage Ψ_S amplitude and the instantaneous torque by closed loops to yield the fastest torque response. Indirectly, DTFC, hunting for the fastest torque response, yields the performance of FOC, which is in principle decoupled i_d and i_q control.

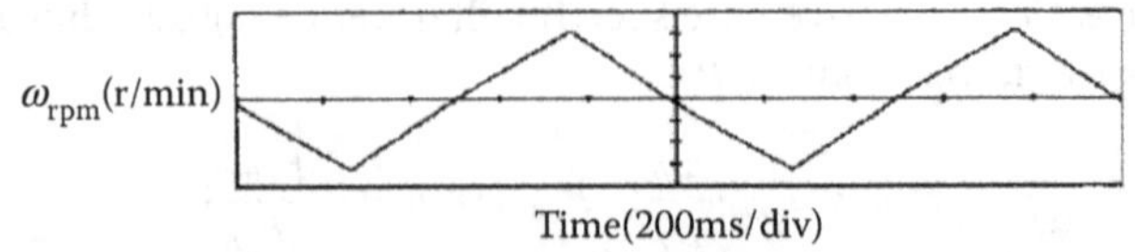

FIGURE 5.10 Speed versus time with core loss considered in RSM feedback linearization control. (After L. Xu et al., *IEEE Trans*, vol. IA–27, no. 5, 1991, pp. 977–985 [6]. 200 rpm/division, 200 ms/division.)

DTFC avoids the transformation of coordinates by selecting a certain voltage vector combination from the PWM inverter, based on stator flux amplitude and torque errors and the stator flux vector position with respect to phase a of RSM (PM-RSM).

But DTFC requires the estimation (observation) of stator flux $\overline{\Psi}_s$ (amplitude and position θ_{Ψ_s}) and torque.

Arguably the "troubles" with $\widehat{\Psi}_S$, $\widehat{T}_e$ online estimation are smaller than those with FOC, and the rotor position θ_{er} is used only in the estimators at low speeds but not in the control itself, where speed estimation is still required only if speed control mode is adopted.

Also, controlling stator flux (in flux weakening) DTFC is simpler to implement for wide CPSR. DTFC is based on same space vector motor model of RSM (Equation 5.6), but to explain the principle of DTFC, stator coordinates are used:

$$\overline{V}_s^s = R_s\bar{i}_s + \frac{d\overline{\Psi}_s}{dt}$$
$$\overline{V}_s^s = \frac{2}{3}\left(V_a + V_b e^{j\frac{2\pi}{3}} + V_c e^{-j\frac{2\pi}{3}}\right) \tag{5.40}$$

Neglecting stator resistance R_s (Equation 5.40) yields:

$$\Delta\overline{\Psi}_s = \overline{\Psi}_s^s(t) - \overline{\Psi}_s^s(0) \approx \int_0^t \overline{V}_s^s dt \tag{5.41}$$

A six-pulse voltage source inverter (Figure 5.11a) produces six nonzero voltage vectors (Figure 5.11b) along the machine phases axes; that is, with a definite special position. Equation 5.42 indicates that the stator flux variation falls along the applied voltage vector $\overline{V}_i$ direction for the turn-on time T_i:

$$\Delta\overline{\Psi}_s \approx \overline{V}_i T_i \tag{5.42}$$

Considering the stator flux vector initially in sector I—Figure 5.11c—if the flux should be increased, $\overline{V}_2$ is to be applied if the torque should increase (by flux acceleration). However, should the flux decrease, but torque still increase, $\overline{V}_3$ should be applied. To complete the picture, should the torque decrease, $\overline{V}_3$ is applied for flux increase and $\overline{V}_5$ for flux decrease.

The two zero-voltage vectors may also be applied either for reducing flux ripple (to reduce core loss) or for slower torque decrease in predictive control to reduce torque ripple [10].

A table of switchings (TOS) may be developed (Figure 5.11d) simply.

In subsequent implementations, the single voltage vector commands and the 2(3)-position hysteresis flux and torque regulators have been replaced with space vector modulation (SVM) methods that use a selected symmetric combination of voltage vectors within a 60°-wide sector or with predictive control methods mainly to reduce torque ripple for given maximum switching frequency in the converter.

The key factor is the state (stator flux observer) and torque estimator.

Among the many solutions so far for stator flux $\overline{\Psi}_s$ estimation (torque is calculated from $\widehat{\Psi}_s$ and measured i_a, i_b stator currents), the combined voltage and current model closed-loop observer has been proven to work well, down to speeds above 3–5 rpm, especially when an encoder (θ_{er}) is available (Figure 5.12).

The design of k_i, τ_i in the proportional integrator (PI) closed loop provides for a d-q current model ($\Psi_d = L_d i_d$, $\Psi_q = L_q i_q - \Psi_{PMq}$, $\Psi_{PMq} = 0$ for RSM) pre-eminence at low frequencies (a few Hz) and

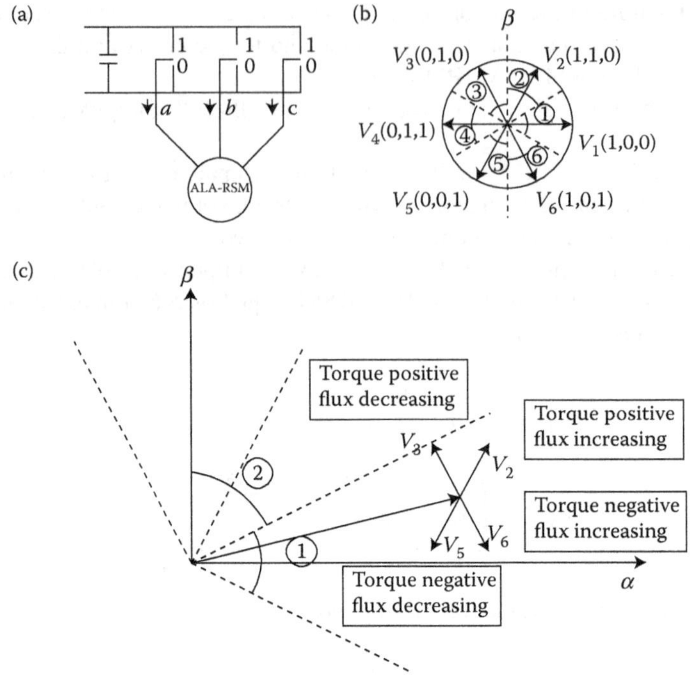

(d)

λ_s=1	τ=1	V_2	V_3	V_3	V_4	V_4	V_5	V_5	V_6	V_6	V_1	V_1	V_2
	τ=−1	V_6	V_1	V_1	V_2	V_2	V_3	V_3	V_4	V_4	V_5	V_5	V_6
λ_s=0	τ=1	V_3	V_4	V_4	V_5	V_5	V_6	V_6	V_1	V_1	V_2	V_2	V_3
	τ=−1	V_5	V_6	V_6	V_1	V_1	V_2	V_2	V_3	V_3	V_4	V_4	V_5
Stator flux position		1 2	3 4	5 6	7 8	9 10	11 12	13 14	15 16	17 18	19 20	21 22	23 24
τ=0		V_7	V_7	V_8	V_8	V_7	V_7	V_8	V_8	V_7	V_7	V_8	V_8

FIGURE 5.11 DTFC principles: (a) six-switch PWM voltage-source inverter; (b) the six nonzero voltage vectors; (c) suitable voltage vectors to apply with stator flux in sector 1 ($\pm 30°$ around phase a); (d) table of switchings.

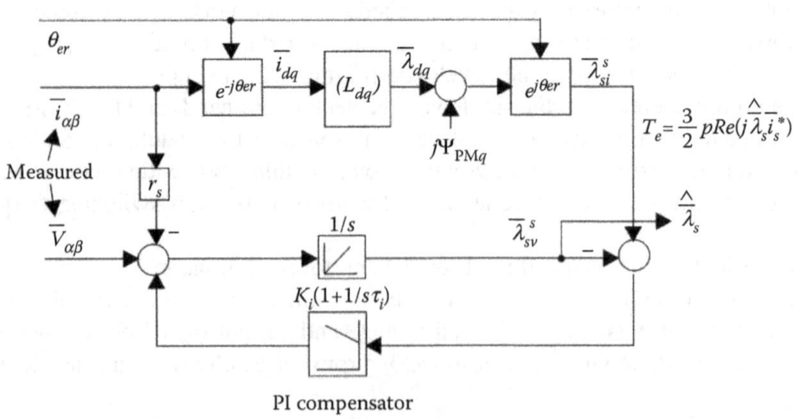

FIGURE 5.12 Voltage and current stator flux observer ($\Psi_{PMq} = 0$ for RSM).

voltage model dominance at higher frequency (speed). Typical vales are:

$$K_i = -(\omega_1 + \omega_2); \quad \tau_i \approx \frac{|k_i|}{\omega_1 \omega_2} \tag{5.43}$$

$$\text{with}\ \omega_1 = (3 - 10)\ \text{rad/s},\ \omega_2 = -(3 - 10)|\omega_1| \tag{5.44}$$

$$\text{The torque estimator is simply}\ T_e = \frac{3}{2} p_1 (\Psi_{s\alpha} i_\beta - \Psi_{s\beta} i_\alpha) \tag{5.45}$$

Even for encoderless drives, the v-i flux observer provides good performance if, instead of rotor, stator flux coordinates are used in an inherently positionless observer. Alternatively, active flux—$\overline{\Psi}_d^a$—coordinates may be used, when:

$$\overline{\Psi}_d^a = \overline{\Psi}_s - L_q \bar{i}_s \tag{5.46}$$

where $\overline{\Psi}_d^a$ is aligned to the rotor d axis in any load condition or speed, provided cross-coupling errors are reduced (cancelled), when rotor position $\widehat{\theta}_{er}$ and speed $\widehat{\omega}_r$ are equal to active flux vector position and speed. Later in the chapter, the active flux concept will be illustrated in a sensorless (encoderless) RSM case study.

A generic DTFC control scheme with dual hysteresis stator flux and torque controllers (as in the initial proposal of Takahashi) is shown in Figure 5.13 [9].

The sliding-mode speed controller, chosen for its robustness, (in Figure 5.13) has the functional:

$$S_{SM} = (\omega_1^* - \omega_r) - T_{SM} \frac{d\omega_r}{dt} \tag{5.47}$$

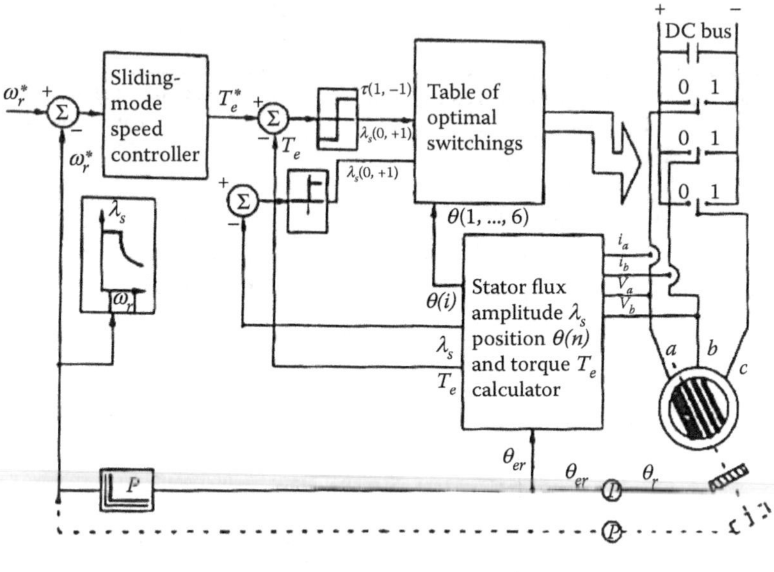

FIGURE 5.13 DTFC of RSM, with encoder. (After I. Boldea, Z. X. Fu, S. A. Nasar, *EMPS Journal*, vol. 19, 1991, pp. 381–398. [9])

with a digital filter to estimate rotor speed $\widehat{\omega}_r$ from the encoder feedback θ_{er}, the speed control law is:

$$T_e = +T_e^* \text{ for } S_{\text{SM}} > 0$$
$$T_e = -T_e^* \text{ for } S_{\text{SM}} < 0 \tag{5.48}$$

The sign of S_{SM} was read initially at an interval $T_s = 1.2$ ms for the case study in [9], but, at a constant inverter switching frequency of 15 kHz, the flux and torque errors will be read faster. The data of the RSM is: six poles, $L_q = 24$ mH, $L_d = 88$ μH for $i_d^* = 3A$ but strongly dependent on id, $R_s = 0.82\ \Omega$, $J = 0.0157$ kg·m^2, $V_{dc} = 80$ V$_{\text{dc}}$.

Despite neglecting magnetic saturation (strong along axis d), the speed build-up to 3000 rpm with torque perturbation is fully satisfactory (Figure 5.14a). Also, the speed response at 2 rpm (Figure 5.14b) is acceptable.

The base speed was chosen as 3000 rpm, while the motor could be accelerated on no load by flux weakening to 12,000 rpm.

As in any such DTFC that uses single voltage vector commands, the torque ripple is implicitly larger than for FOC at equal switching frequency. The high switching frequency reduces the torque ripple to a few percent, but SVM or predictive torque control [10] may still be used to reduce switching frequency (inverter losses) for the same (low) flux and torque ripple.

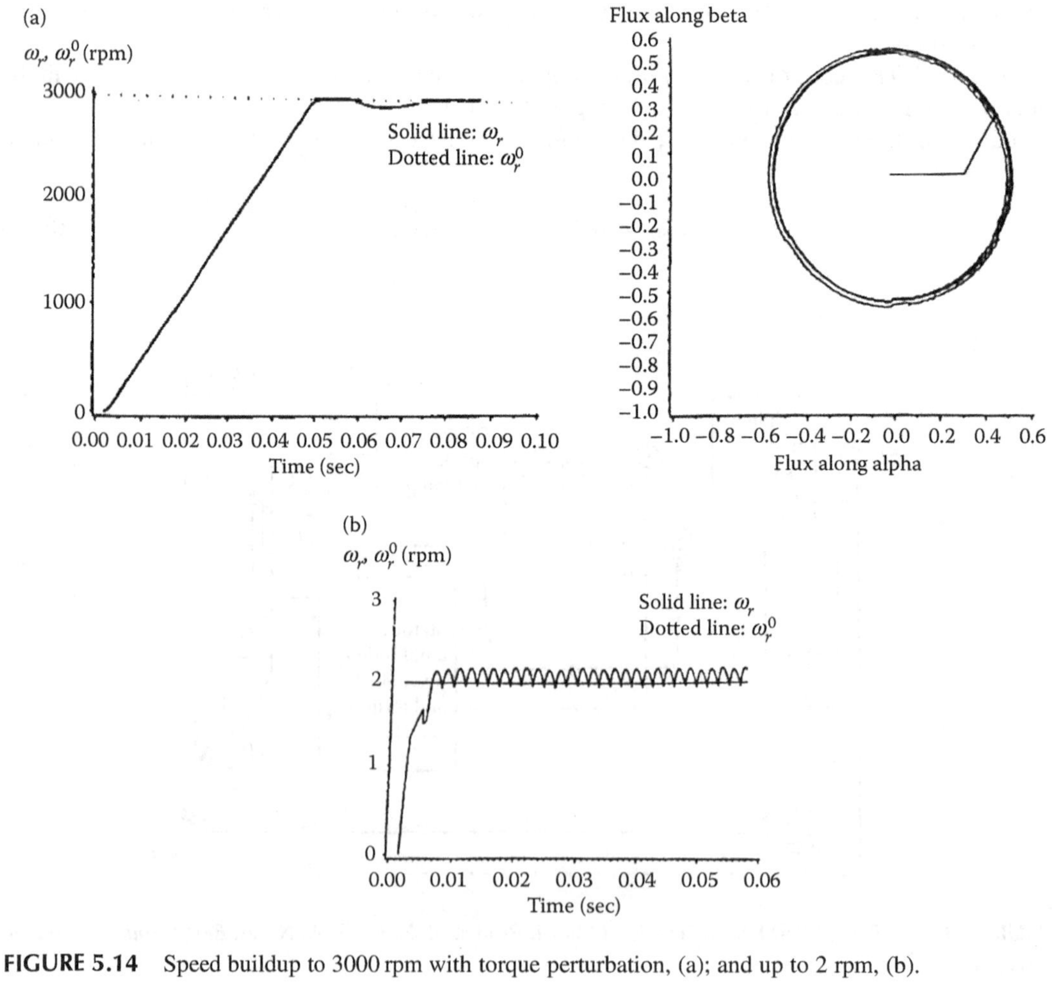

FIGURE 5.14 Speed buildup to 3000 rpm with torque perturbation, (a); and up to 2 rpm, (b).

5.6 FIELD-ORIENTED CONTROL AND DIRECT TORQUE AND FLUX CONTROL OF PERMANENT MAGNET–RELUCTANCE SYNCHRONOUS MACHINES FOR WIDE CONSTANT POWER SPEED RANGE

PM-RSMs involve high saliency ($L_d/L_q > 3.5$–5) and weak magnets (the PM emf at base speed is $e_{PMd} < 0.3\ V_{smax}$, in general, with magnets in axis q).

The FOC and DTFC principles remain similar to those for RSMs, but the "battle" for MTPA and maximum torque per flux (MTPF) is more complicated, as now:

$$\Psi_d = L_d i_d;\ \ \Psi_q = L_q i_q - \Psi_{PMq};\ \ L_d > L_q \tag{5.49}$$

The introduction of active flux [3] simplifies the math and thus will be used here:

$$\overline{\Psi}_q^a = \overline{\Psi}_s - L_d \bar{i}_s \tag{5.50}$$

The active flux is now aligned here to axis q (PM axis of the rotor), and the torque is:

$$T_e = \frac{3}{2} p_1 (\Psi_{PMq} + (L_d - L_q) \cdot i_q) \cdot i_d = \frac{3}{2} p_1 \Psi_q^a i_d \tag{5.51}$$

Now i_d is the torque current (reluctance torque is dominant) and:

$$\overline{\Psi}_q^a = \Psi_{PMq} + (L_d - L_q) \cdot i_q,\ \text{aligned to axis}\ q \tag{5.52}$$

The maximum torque/current standard relationship becomes:

$$2i_{qi}^2 + i_{qi}\frac{\lambda_{PM}q}{L_d - L_q} - i_s^2 = 0;\ \ i_{qi} > 0,\ \text{always} \tag{5.53}$$

which requires, even for constant λ_{PMq}, L_d, L_q, the online solution of a second-order equation.

By introducing the active flux (which may be estimated as discussed in section 5.4) Equation 5.53 becomes:

$$i_{qi} = i_s \sqrt{\frac{\left|\widehat{\Psi}_q^a\right| - \Psi_{PMq}}{2\left|\widehat{\Psi}_q^a\right| - \Psi_{PMq}}} \tag{5.54}$$

This equation is rather robust to magnetic saturation.

Equation 5.54, with $\widehat{\Psi}_q^a$ estimated and stator current measured, produces a notable simplification. For wide CPSR control, it is sufficient to operate at MTPA if the reference voltage V_s^* is $V_s^* \le V_{smax}$. When the reference voltage $V_s^* \ge V_{smax}$, an intervention Δi_{qi} is required on i_{qi} by reducing it to move the drive closer to maximum torque per flux conditions:

$$\Delta i_{qi} = -PI\left(\frac{V_s^* - V_{smax}}{V_{smax}}\right);\ \ \text{if}\ V_s^* > V_{smax} \tag{5.55}$$

A generic drive based on this methodology is portrayed in Figure 5.15.

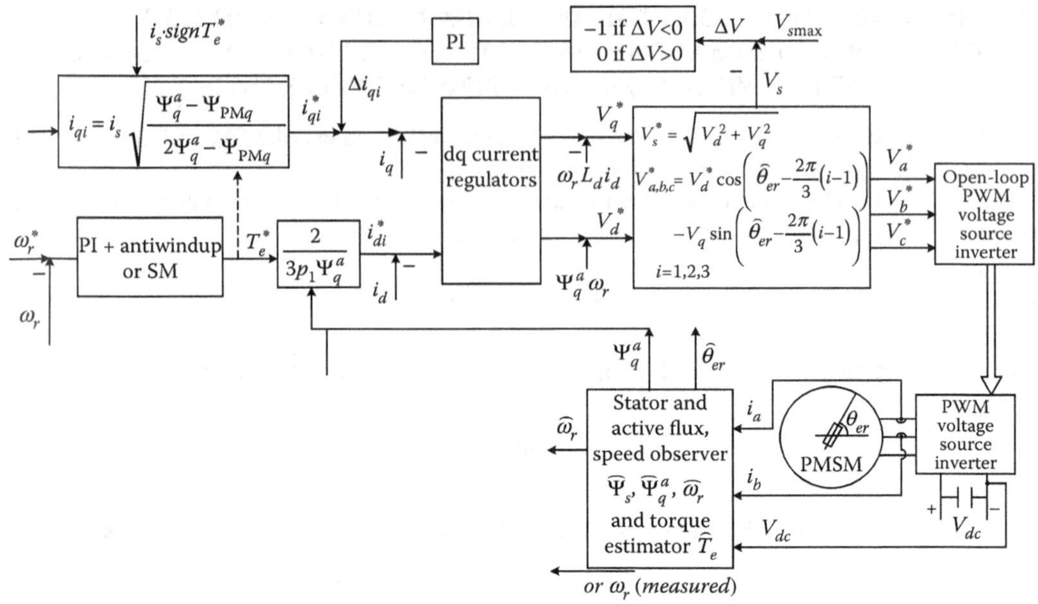

FIGURE 5.15 Generic wide CPSR-FOC of PM-RSM.

Alternatively, another approximation, after noticing that the reluctance torque component is dominant in PM-RSM, consists of defining a torque-speed envelope with $\Psi_q = 0$, which leads to:

$$L_q i_{qc} - \Psi_{\mathrm{PM}q} = 0 \tag{5.56}$$

This condition does not imply PM demagnetization, as part of L_q is the stator leakage inductance. The torque becomes:

$$T_{ek}^* = \frac{3}{2} p_1 \cdot L_d \cdot i_d \cdot i_{qc} = \frac{3}{2} p_1 \frac{L_d}{L_q} \cdot \Psi_{\mathrm{PM}q} \cdot i_d; \;\; i_{qc} = \Psi_{\mathrm{PM}q}/L_q \tag{5.57}$$

$$\Psi_s = L_d i_d; \;\; \Psi_q = 0; \;\; \Psi_s = \frac{V_s}{\omega_r}; \;\; \text{for} |\omega_r| > \omega_b \tag{5.58}$$

For $|\omega_r| < \omega_b$, $T_{ek}^* = (3/2) p_1 (L_d/L_q) \cdot \Psi_{\mathrm{PM}q} \cdot i_{db}^*$ represents the peak torque required by the drive. The current i_{db}^* produces the only flux in the machine and thus can be simply calculated for rated flux.

So i_d is limited to:

$$i_{d\,\mathrm{lim}} = \begin{cases} i_{db}^* & \text{for } |\omega_r| < \omega_b \\ i_{db}^* \cdot \dfrac{\omega_b}{|\omega_r|} & \text{for } |\omega_r| > \omega_b \end{cases} \tag{5.59}$$

The maximum torque-speed envelope is illustrated in Figure 5.16. As i_{dlim} is inversely proportional to speed, for $|\hat{\omega}_r| > \omega_b$, the envelope (maximum) electromagnetic power (P_{elmk}) is constant, so a wide CPSR is provided.

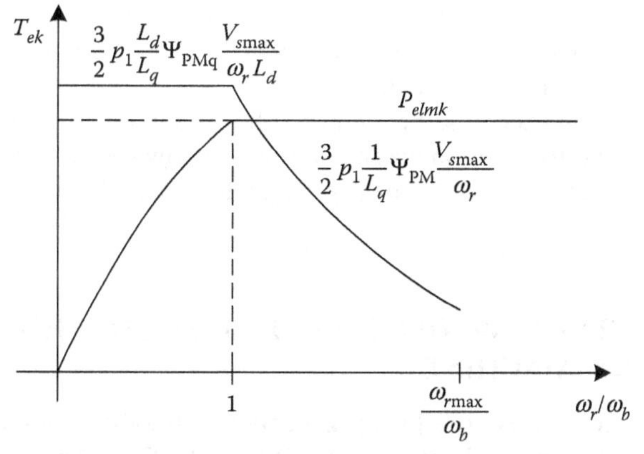

FIGURE 5.16 Envelope of torque T_{ek} versus speed for zero q axis flux; $i_{qc} = \Psi_{PMq}/L_q$.

Now, when the reference torque T_e^* is smaller than T_{ek} in Figure 5.16, we may relax (reduce) i_q^* from the i_{qc} value, by, say, a square-root function:

$$i_q^* = i_{qc}\sqrt{\frac{|T_e^*|}{T_{ek}}} \tag{5.60}$$

The copper losses will decrease for lower torque, and the efficiency will improve. However, in this case, the power factor, which otherwise will tend to unity at infinite speed (for $\Psi_q = 0$), will be smaller by the same 10–12% but still high, guaranteeing good system performance. Magnetic saturation may be considered simply by $L_d(i_d)$ found during drive self-commissioning or through FEM (or analytically) in the design stage.

The results obtained with such a simplified control methodology seem very close to those obtained by online computer-intensive methods for wide CPSR control.

The method will be illustrated for an encoderless PM-SRM drive later in this chapter.

Note: Methods to calculate online MTPA conditions keep being proposed, based on analytical online calculation or even on signal injection [11–15], but produce only incremental improvements.

Finally, deadbeat DTFC control of IPMSMs has been introduced to further define/improve the torque response in the millisecond range in the flux weakening region, with small but steady energy savings during such transients [16].

5.7 ENCODERLESS FIELD-ORIENTED CONTROL OF RELUCTANCE SYNCHRONOUS MACHINES

Encoderless control of RSMs (PM-RSMs) may be approached by FOC, DTFC, or scalar methods.

The core of such systems is the state observer, which now includes, besides stator flux or active flux, rotor position and speed, with the added torque calculator online. Numerous stator flux and rotor position observers have been proposed for RSM (PM-RSMs) but may be classified as:

- With fundamental model (emf and extended emf models [17,18])
- With signal injection (rotating voltage vector in stationary rotor coordinates [19–21], voltage pulse injection in stationary frame [21], inverter-produced stator current ripple processing [22–24], with a strong review analysis in [24]).

The signal injection state observers are used at very low speeds, and then the emf model–based state observers take over above a few Hz for a more than 500/1 speed range servomotor-like-performance encoderless drive with FOC or DTFC.

In general, signal injection state observers inject voltages and process the stator current negative second sequence to extract rotor position via saliency at very low speeds. The advantage of avoiding any integral (with its offset) is paid for dearly by at least two filters that produce delays and distortions inevitably.

5.8 ACTIVE FLUX–BASED MODEL ENCODERLESS CONTROL OF RELUCTANCE SYNCHRONOUS MACHINES

- The active flux concept was born [3] by generalizing (modifying) the virtual PM flux in IPMSMs [25] and the extended emf models [18] to all AC motors.
- The active flux $\overline{\Psi}_d^a$ for RSM here along axis d of the rotor ($L_d > L_q$) is obtained from the observed stator flux (by a closed-loop combined V-I model)—Figure 5.12 and Equation 5.50. This observer processes fluxes, not currents, but $\overline{\Psi}_d^a$ is aligned to axis d irrespective of load, speed, or voltage waveforms, where the integral offset is reduced by the PI loop in the Voltage and current (V&I) model state observer (Figure 5.12).
- Injecting, say, traveling voltage vectors in rotor coordinates at zero speed yields to a circular hodograph for the αβ stator flux vector components $\Psi_{s\alpha}$, $\Psi_{s\beta}$, to an elliptical current hodograph ($I_{s\alpha}$, $I_{s\beta}$), but to a fixed-axis AC signal for active flux at injecting frequency; the position of this vector is still the rotor position (Figure 5.17), even at standstill. The magnet axis is thus deciphered. PM polarity requires two $\pm$voltage vectors along the estimated rotor position at standstill and by measuring the larger current peak after a given time.
- So the same structure of state observer is used for the signal injection and emf model for the active flux–based state observers. The output is the active flux amplitude $\widehat{\Psi}_d^a$ and its position $\widehat{\theta}_{\overline{\Psi}_d^a}$.

A phase locked loop (PLL) additional observer that uses the motion equation is required to condition $\widehat{\theta}_{\overline{\Psi}_d^a}$ to obtain a cleaner rotor position estimation $\widehat{\theta}_{er}$ and, as a bonus, the rotor speed estimation $\widehat{\omega}_r$ (Figure 5.18) [26]. The motion equation gives the state observer [26] some robustness to torque perturbations (during speed transients).

Now, the signal injection observer is added at low fundamental frequency (speed) and fused with the active flux fundamental model to produce a smooth transition and thus cover speed control from zero to the maximum value (Figure 5.19).

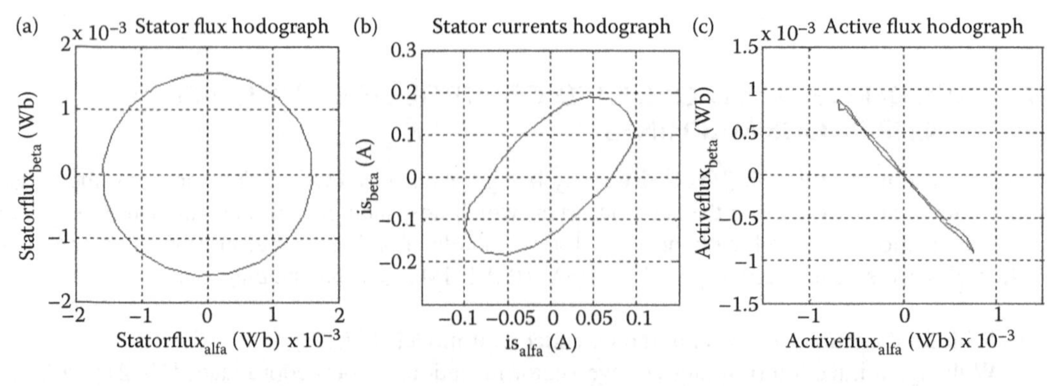

FIGURE 5.17 Hodographs for traveling voltage vector injection at standstill for an RSM: (a) stator flux hodograph; (b) stator current hodograph; (c) active flux hodograph.

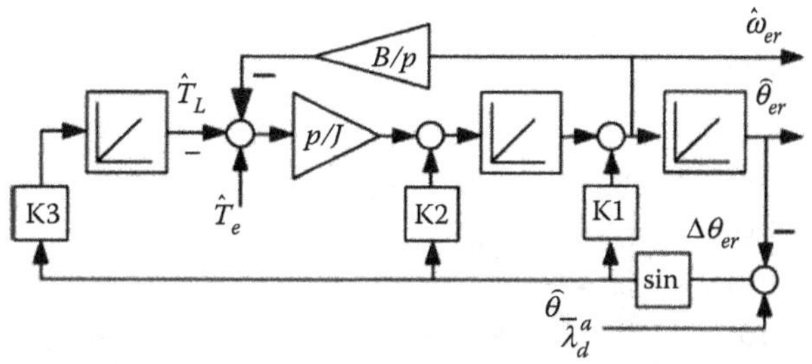

FIGURE 5.18 PLL-based position and speed observer based on $\hat{\theta}_{\Psi_d^a}$ input and motion equations. (After S. Agarlita, I. Boldea, F. Blaabjerg, *IEEE Trans. On*, vol. IA–48, no. 6, 2011, pp. 1931–1939. [26])

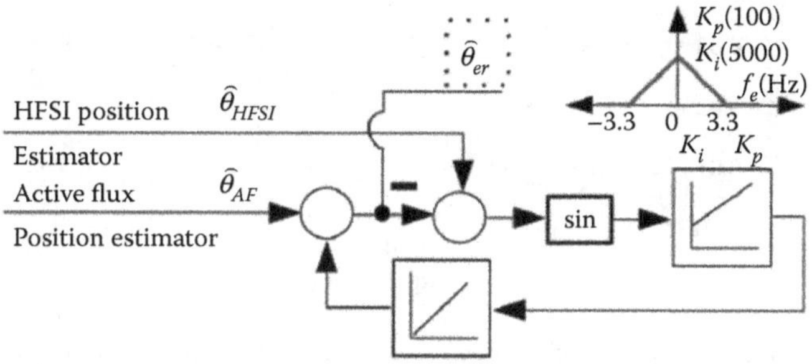

FIGURE 5.19 Rotor position observers' fusion.

The complete control scheme appears in (Figure 5.20).

All closed loops in the FOC and state observer are designed in [27] for an 157-W, 1500-rpm (base speed), four-pole RSM prototype with $R_s = 0.61\ \Omega$, $J = 10^{-3}\ \text{kg} \cdot \text{m}^2$, $V_{dc} = 42$ V and magnetization curves as in Figure 5.21, obtained from standstill flux decay tests.

Here, only some sample results are given [26]:

- Operation with step plus/minus full load (1 Nm) at zero speed (when signal injection provides for estimated $\hat{\theta}_{er}$ and $\hat{\omega}_r$) (Figure 5.22)
- Operation at 1 rpm with 50% plus/minus step load torque perturbation (Figure 5.23)
- ± 1500 rpm speed reversal with plus/minus 1 Nm step torque perturbation (Figure 5.24)
- ± 3000 rpm speed reversal with ± 0.2 Nm step load torque perturbation (Figure 5.25)

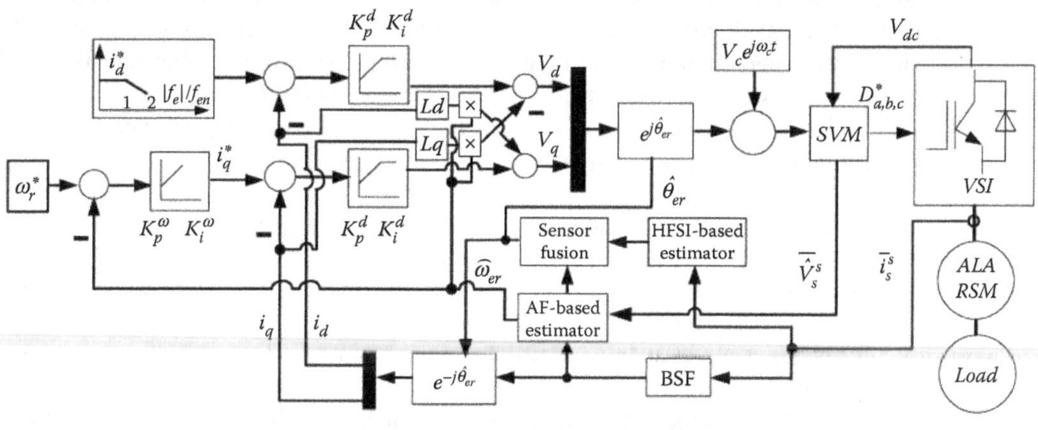

FIGURE 5.20 Complete active flux–based encoderless FOC of RSM. (After S. Agarlita, I. Boldea, F. Blaabjerg, *IEEE Trans. On*, vol. IA–48, no. 6, 2011, pp. 1931–1939. [26])

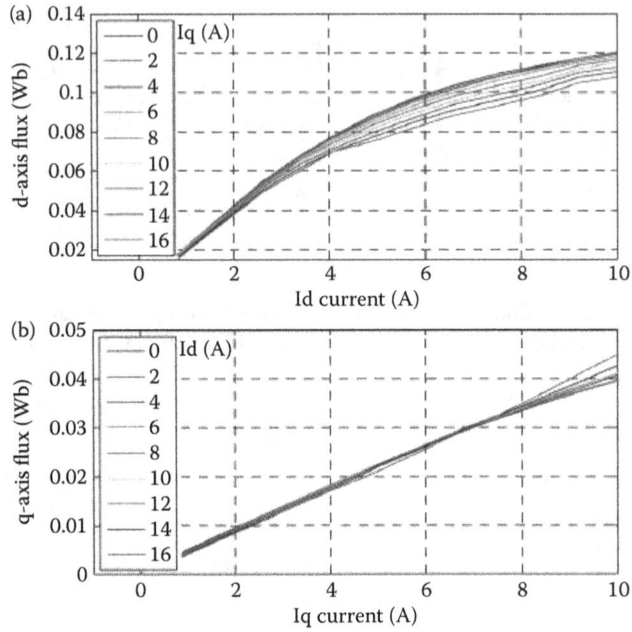

FIGURE 5.21 Magnetization curves of 157-W, four-pole, 1500-rpm RSM: (a) axis d; (b) axis q, by DC decay standstill tests. (After S. Agarlita, I. Boldea, F. Blaabjerg, *IEEE Trans. On*, vol. IA–48, no. 6, 2011, pp. 1931–1939. [26])

The load machine (Figure 5.26) is made of an indirect-current FOC encoder-IM drive capable of fast full load torque from zero speed.

A careful analysis of the experimental results in Figures 5.22 through 5.25 illustrates satisfactory performance with some flux weakening [by a $i_d^*(\omega_r)$ function] from 0 to 3000 rpm with step-torque perturbations and about 1 rpm speed observer steady-state error in an encoderless drive.

5.9 A WIDE SPEED RANGE ENCODERLESS CONTROL OF PERMANENT MAGNET–RELUCTANCE SYNCHRONOUS MACHINES

As already alluded to in section 5.6 of this chapter, a simplified wide speed range control of PM-RSM, which includes a $T_{ek}(\omega_r)$ envelope at $\Psi_q = L_q i_{qc} - \Psi_{PMq} = 0$, may be implemented to fuse good efficiency and wide speed range unique control by reducing $i_q^* = i_{qc}$ with torque, as in Equation 5.57. Reference 27 presents in detail a comparison between a torque-current referencer that provides MTPA when $V_s^* < V_{smax}$ and a reduction of i_q^* when the voltage limit is reached, and the above-mentioned zero Ψ_q with i_q^* reduction when $V_s^* > V_{smax}$, proportional to $\sqrt{\frac{|T_e^*|}{T_{ek}(\omega_r)}}$, for a given speed (Equation 5.60).

As the results of both methods are about the same, only sample results from the latter method are given here, after presenting the general control scheme (Figure 5.27).

Digital simulations have proven the control in Figure 5.27 to work well from 1 to 6000 rpm at 42 V_{dc}, while in experiments, 30 rpm was the minimum safe speed operation (Figure 5.28) on no load (the most difficult operation mode at low speed). The maximum speed was 3000 rpm, however, for 12 V_{dc} (Figure 5.29).

The running at low DC voltage ($V_{dc} = 12$ V) of the four-pole PM-RSM, with $R_s = 0.065\ \Omega$, $L_d(\text{sat}) = 2.5$ mH, $L_q = 0.5$ mH, $\Psi_{PMq} = 0.011$ Wb, $J = 10^{-3}$ kg · m^2, rated phase voltage $V_{PM\text{-}RMS} = 22$ V ($V_{dcn} = 42$ V), 750 W at 1500 rpm, explains part of the experimental difficulties in reducing

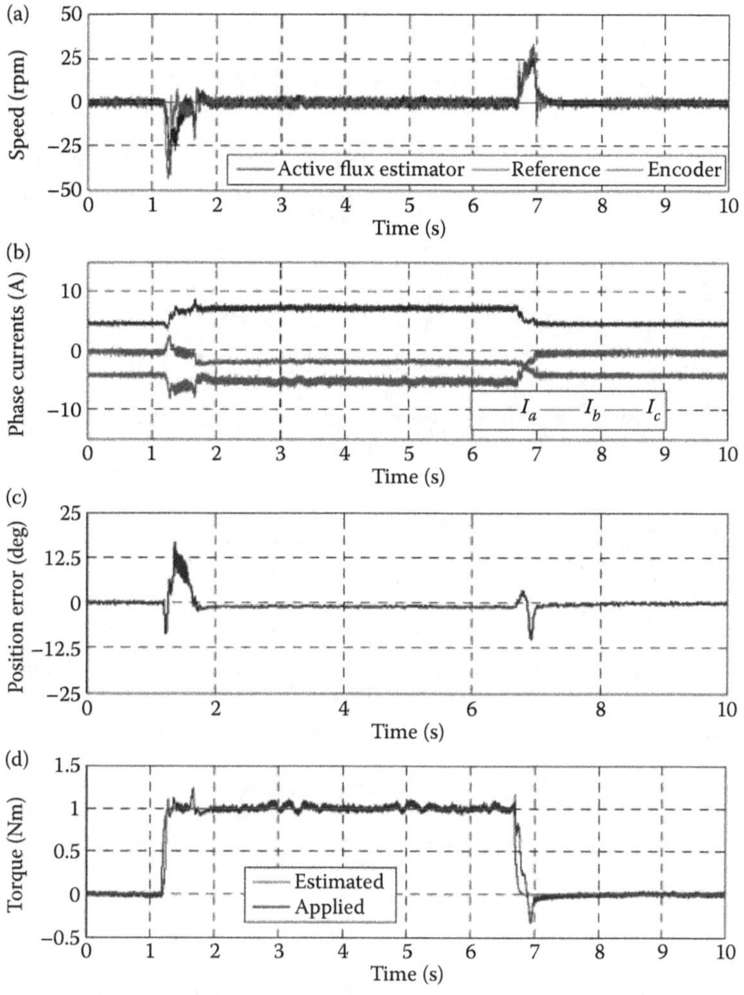

FIGURE 5.22 Zero speed operation at zero speed with ± 1 Nm torque perturbation. (a) speed; (b) phase currents; (c) position error; (d) torque. (After S. Agarlita, I. Boldea, F. Blaabjerg, *IEEE Trans. On*, vol. IA–48, no. 6, 2011, pp. 1931–1939. [26])

speed further down from 30 rpm and without signal injection; the rest is perhaps due to mechanical problems.

5.10 V/F WITH STABILIZING LOOP CONTROL OF PERMANENT MAGNET–RELUCTANCE SYNCHRONOUS MACHINE

For pump or compressorlike loads, an easy, though a bit hesitant, start without initial rotor position estimation is a functional asset. Simplified control with a large enough speed control range in encoderless implementation may suffice in such applications.

This is how the simple—open-loop—V/f control was improved, but for more stability and wider speed range by adding stabilizing loops. Still, V/f control does not have a speed closed loop or dq (or abc) stator current regulators, which is a plus in less demanding drives now served by IM advanced V/f drives. But such V/f drives should have the MTPA conditions embedded and provide wide constant power speed range, at least for redundancy.

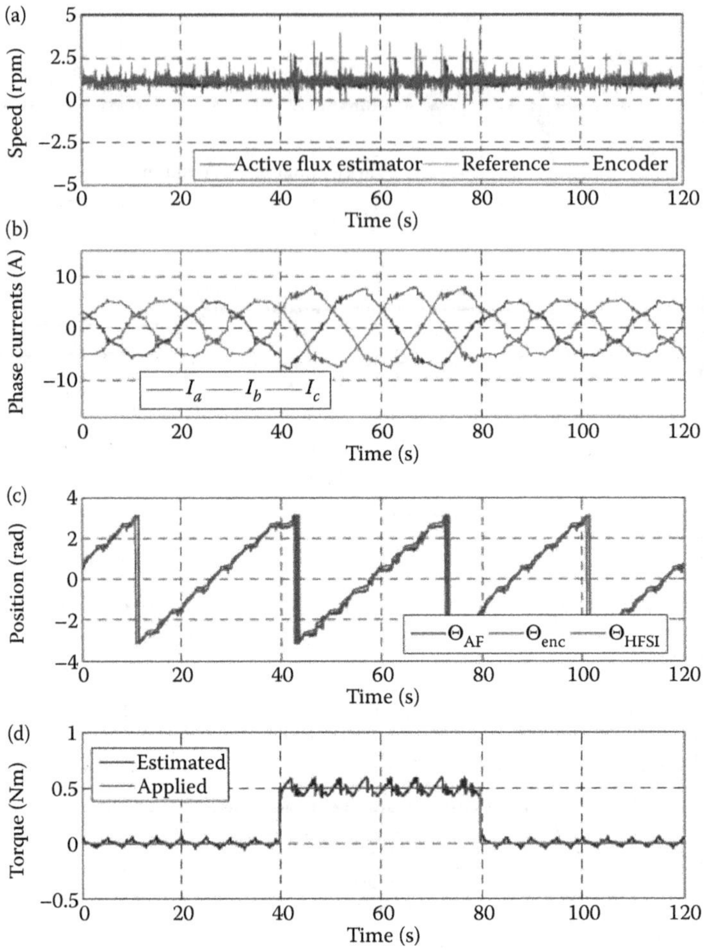

FIGURE 5.23 One-rpm speed operation with ± 0.5 Nm torque perturbation. (a) speed; (b) phase currents; (c) position error; (d) torque. (After S. Agarlita, I. Boldea, F. Blaabjerg, *IEEE Trans. On*, vol. IA–48, no. 6, 2011, pp. 1931–1939. [26])

Such a V/f modified system on the active flux model was developed in [28] and is characterized by:

- A stator flux combined V&I model closed-loop observer (Figure 5.12) with the active flux $\widehat{\Psi}_d^a = \widehat{\Psi}_s - L_q \bar{i}_s$, with PMs in axis d ($L_d < L_q$) and $\widehat{\Psi}_d^a = \Psi_{PMd} + (L_d - L_q) \cdot i_d$; $\Psi_d = L_d i_d + \Psi_{PMd}$.

Note: This change of PM position from axis $d \rightarrow q \rightarrow d$ axis in the active flux model state observers has been adopted to mitigate two similar machines: IPMSM with weak magnets (PM in axis d) and PM-RSM (PMs in axis q) are both present in the literature.

- The MTPA condition (Equation 5.54) is kept, but with $i_{di} < 0$ instead of $i_{qi} > 0$:

$$i_{di}^* = -i_s \cdot \sqrt{\frac{\left|\widehat{\Psi}_d^a\right| - \Psi_{PMd}}{2\left|\widehat{\Psi}_d^a\right| - \Psi_{PMd}}}; \quad \Psi_d^{a*} = \Psi_{PMd} + L_d i_{di}^* \tag{5.61}$$

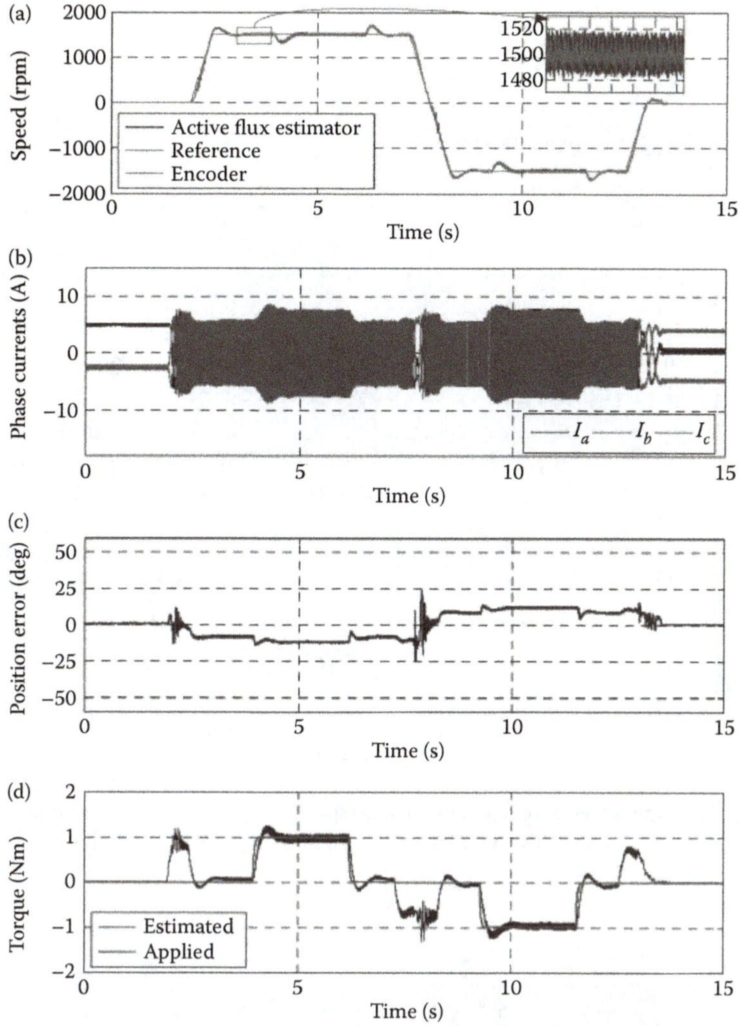

FIGURE 5.24 ±1500-rpm speed reversal transients with 1-Nm torque perturbation. (a) speed; (b) phase currents; (c) position error; (d) torque. (After S. Agarlita, I. Boldea, F. Blaabjerg, *IEEE Trans. On*, vol. IA–48, no. 6, 2011, pp. 1931–1939. [26])

- i_{di} is corrected by: $\Delta i_d^* < 0$ if the reference voltage $V_s^* > V_{smax}$:

$$i_d^* = i_{di}^* - \Delta i_d^*;\quad \Delta i_d^* = K_p\left(1 + \frac{1}{sT_i}\right)\Delta V_s;\quad \Delta V_s = V_s^* - V_{smax} \tag{5.62}$$

$$\text{For}\,\Delta V_s > 0,\ i_{di}^* < \frac{-\Psi_{PMd}}{L_d} \tag{5.63}$$

- The active flux error $\Delta\Psi_d^a = \Psi_d^{a*} - \widehat{\Psi}_d^a$ with Ψ_d^{a*} from Equation 5.61 and $\widehat{\Psi}_d$ from the active flux observer, as explained above, will be used to correct the voltage amplitude by ΔV^*:

$$\Delta V^* = -\Delta\Psi_d^a\left(k_{pV} + k_{iV}\frac{1}{s}\right) \tag{5.64}$$

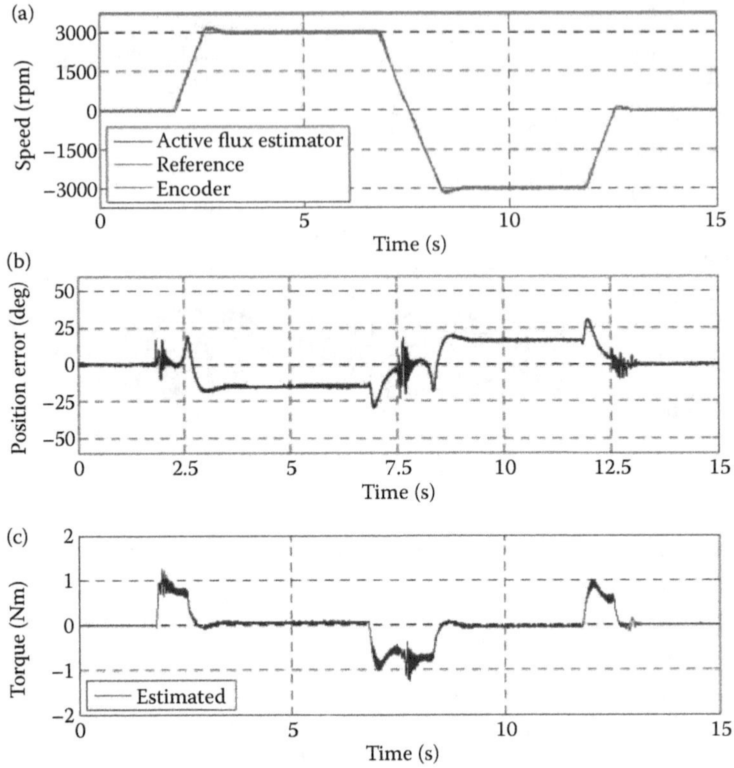

FIGURE 5.25 ±3000-rpm speed reversal transients: (a) speed; (b) position error; (c) torque. (After S. Agarlita, I. Boldea, F. Blaabjerg, *IEEE Trans. On*, vol. IA–48, no. 6, 2011, pp. 1931–1939. [26])

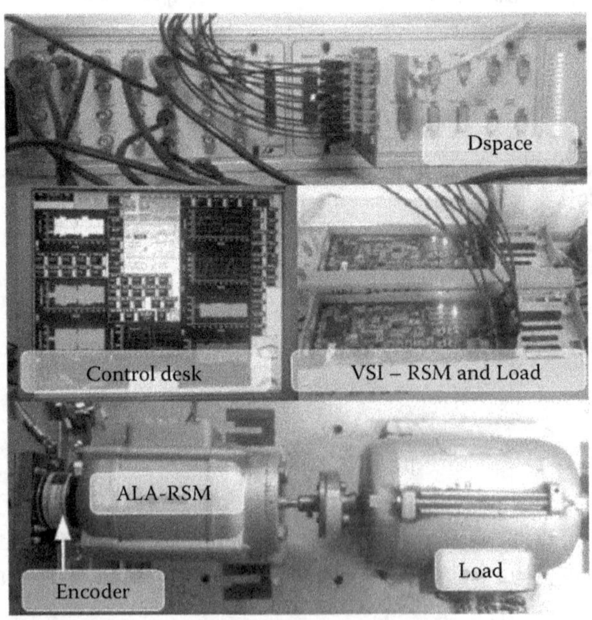

FIGURE 5.26 Experimental platform. (After S. Agarlita, I. Boldea, F. Blaabjerg, *IEEE Trans. On*, vol. IA–48, no. 6, 2011, pp. 1931–1939. [26])

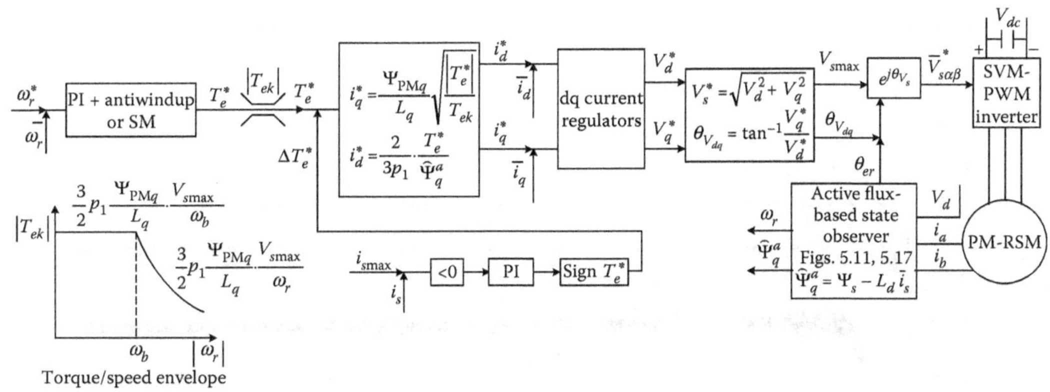

FIGURE 5.27 Simplified (zero Ψ_q $T_{ek}(\omega_r)$ envelope) wide-speed FOC encoderless PM-RSM drive.

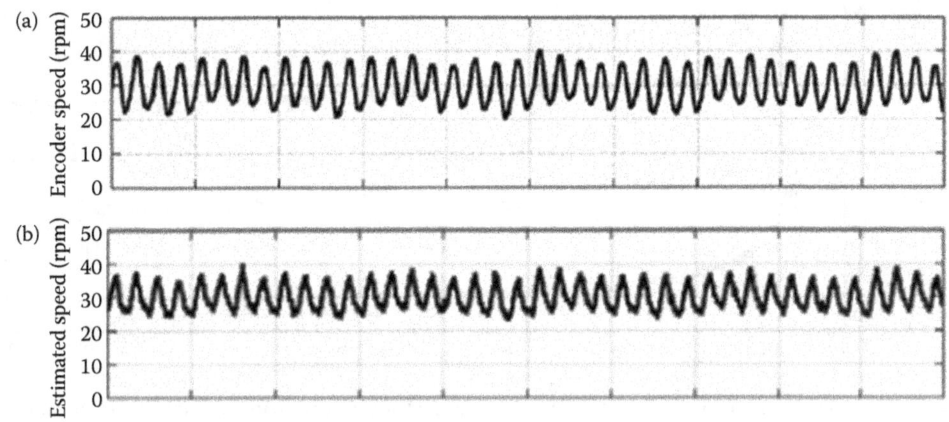

FIGURE 5.28 Steady-state encoderless operation of PM-RSM at 30 rpm: (a) speed; (b) estimated speed. (Test results after M. C. Paicu et al., *2009 IEEE Energy Conversion Congress and Exposition*, pp. 3822–3829. [27])

- The speed of the active flux is approximately (no PLL observer added):

$$\hat{\omega}_r = \omega\left(\hat{\Psi}_d^a\right) = \frac{\hat{\Psi}^a_{d\alpha(k-1)}\hat{\Psi}^a_{d\beta(k)} - \hat{\Psi}^a_{d\beta(k-1)}\hat{\Psi}^a_{d\alpha(k)}}{T_s\left|\hat{\Psi}^a_{d(k)}\right|^2} \tag{5.65}$$

with T_s—the sampling time;

- The speed error $\Delta\omega_r = \omega_r^* - \hat{\omega}_r$ will be used to correct the voltage vector assigned angle θ_{V_0} (Figure 5.29a) by $\Delta\theta_V$:

$$\Delta\theta_V^* = -\Delta\omega_r\left(k_{pi} + k_{ii}\frac{1}{s}\right) \tag{5.66}$$

Note: Alternatively, the DC link power may be measured and high-pass-filtered with its output integrated to give the voltage vector angle correction $\Delta\theta_V^*$.

The complete control system is shown in Figure 5.30b.

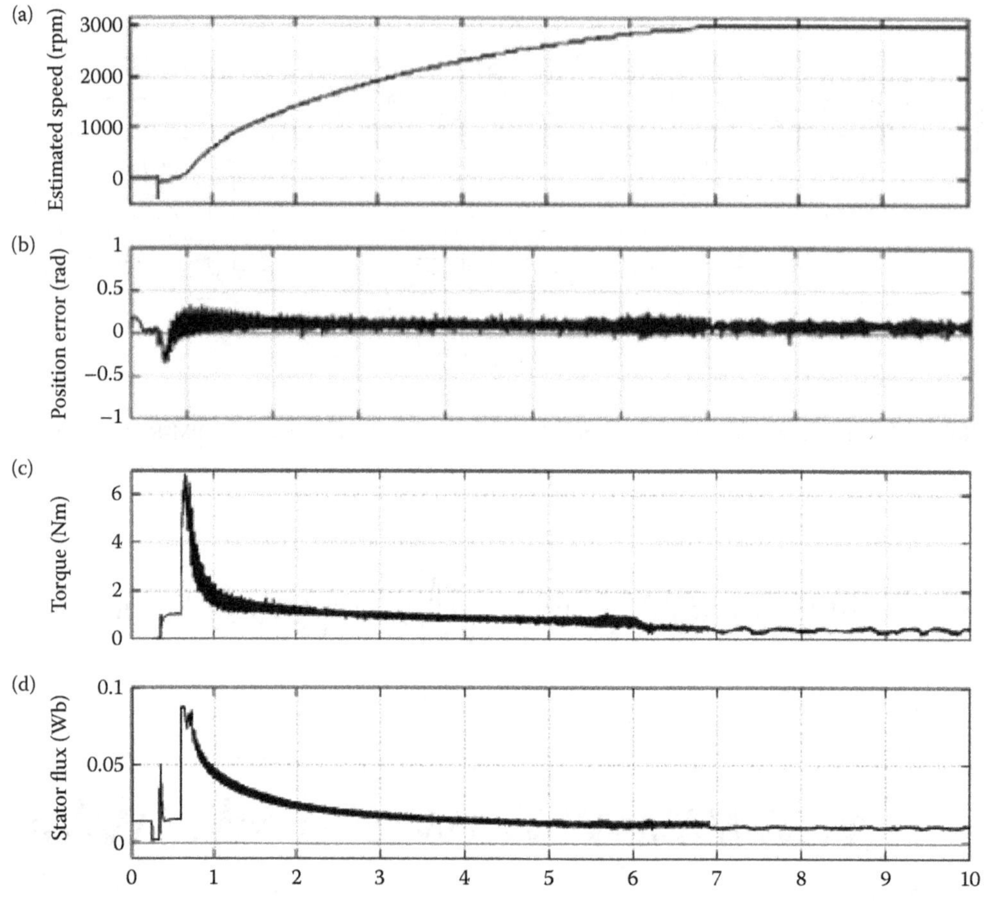

FIGURE 5.29 Acceleration to 3000 rpm of encoderless PM-RSM: (a) estimated speed; (b) position error; (c) torque; (d) stator flux. (Test results after M. C. Paicu et al., *2009 IEEE Energy Conversion Congress and Exposition*, pp. 3822–3829. [27])

For comparison, an encoderless active flux–based FOC system was used in experiments on the same motor with six poles, 12 Nm at 1750 rpm, line voltage 380 V (RMS), rated frequency 87.5 Hz, $R_p = 3.3\ \Omega$, $L_d = 41.6$ mH, $L_q = 57.1$ mH, $\Psi_{PMd} = 0.483$ Wb, $J = 10^{-3}$ kg·m^2, viscous friction coefficient $B_m = 2 \times 10^{-3}$ Nms/rad.

Both controls were capable of providing 0.3–100% speed range control. Comparative results at 5 rpm (the lowest ever with V/f control) are indicative of encoderless drive performance at low speeds without signal injection (Figure 5.31 and 5.32).

Acceleration to 1400 rpm of V/f control with a PM generator on resistive load (load torque is proportional to speed [50% full torque at 1400 rpm])—Figure 5.33—and the acceleration to the same 1400 rpm by encoderless FOC with 90% load torque applied only at 3.6 s—Figure 5.34—show similar performance. However, the encoderless FOC needed a dedicated strategy.

Note: It may be argued that the extension of the speed control range to 5 rpm in V/f control is obtained at the cost of controller-added complexity. But it has to be remembered that such a control provides implicitly for a direct, though a bit hesitant, start (without rotor initial position knowledge). Signal injection is not used, and speed and current control loops are absent.

An I-f start fused with active-flux-based encoder FOC has proved successful in driving RSMs [29] with speed reversal from 3000 rpm and zero speed operation [29].

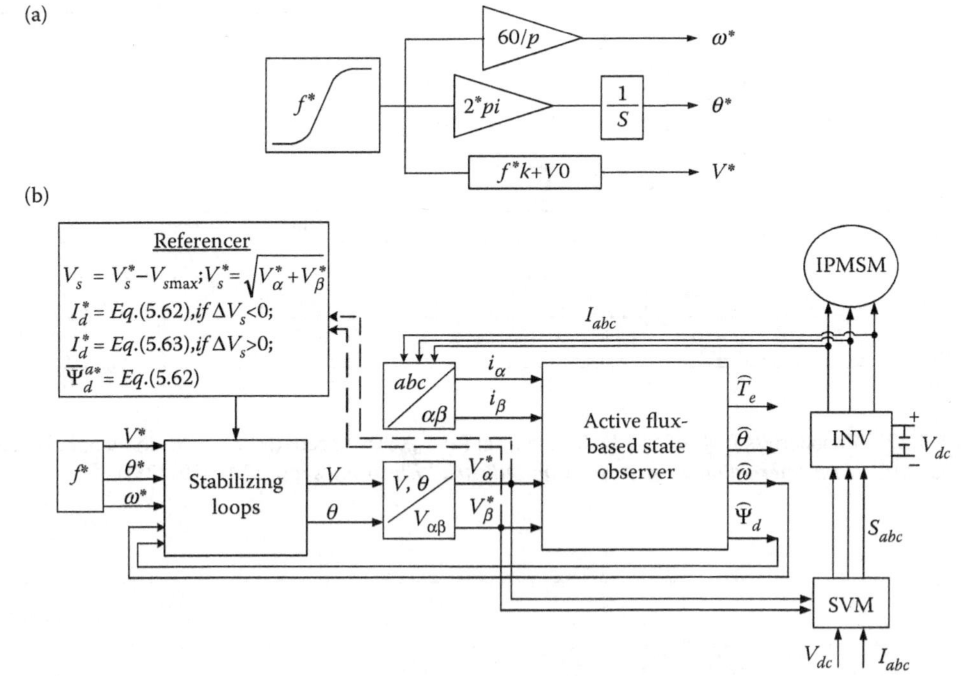

FIGURE 5.30 V/f control of PM-RSM with two stabilizing loops: (a) the voltage referencer; (b) the control system. (After A. Moldovan, F. Blaabjerg, I. Boldea, *2011 IEEE International Symposium on Industrial Electronics*, pp. 514–519. [28])

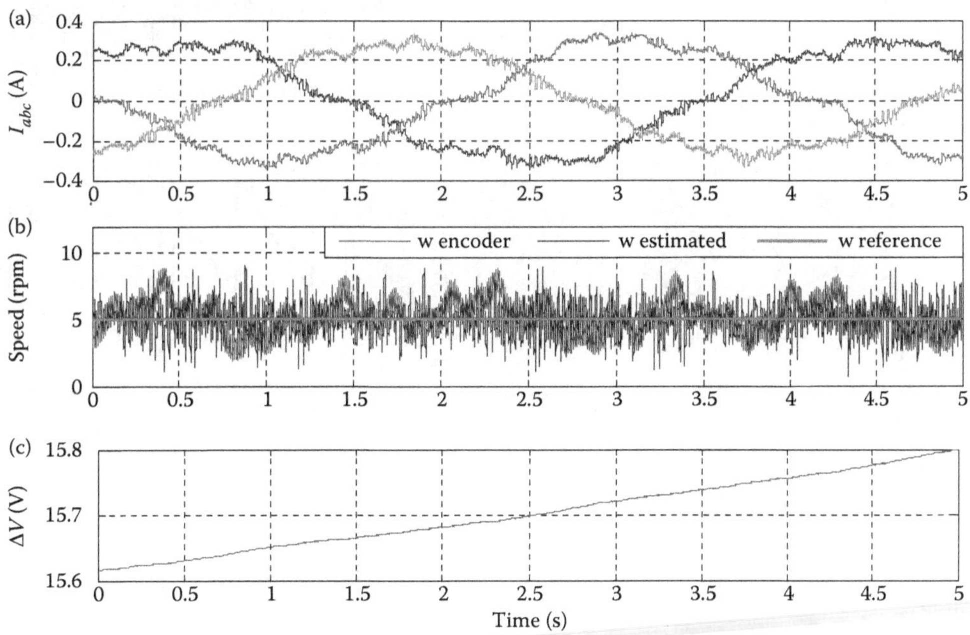

FIGURE 5.31 V/f control with two stabilizing loops at 5 rpm for PM-RSM: (a) $i_{a,b,c}$; (b) speed; (c) ΔV. (After A. Moldovan, F. Blaabjerg, I. Boldea, *2011 IEEE International Symposium on Industrial Electronics*, pp. 514–519. [28])

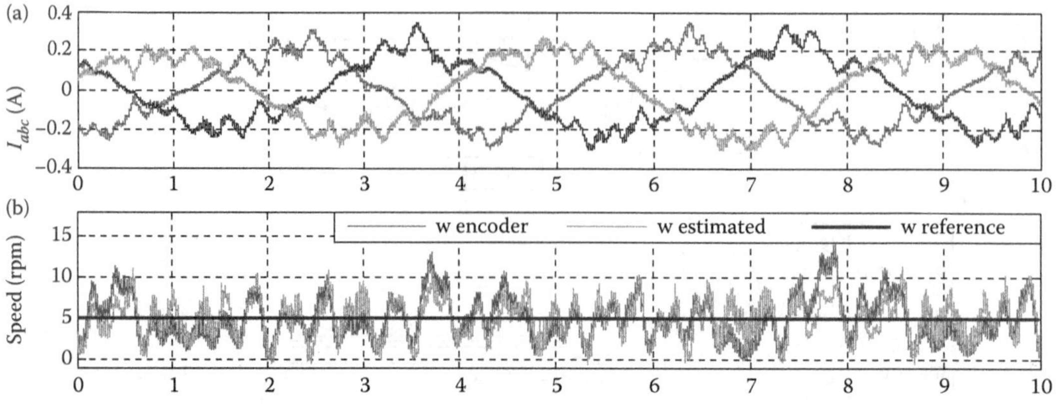

FIGURE 5.32 Encoderless FOC of PM-RSM at 5 rpm: (a) $i_{a,b,c}$; (b) speed. (After A. Moldovan, F. Blaabjerg, I. Boldea, *2011 IEEE International Symposium on Industrial Electronics*, pp. 514–519. [28])

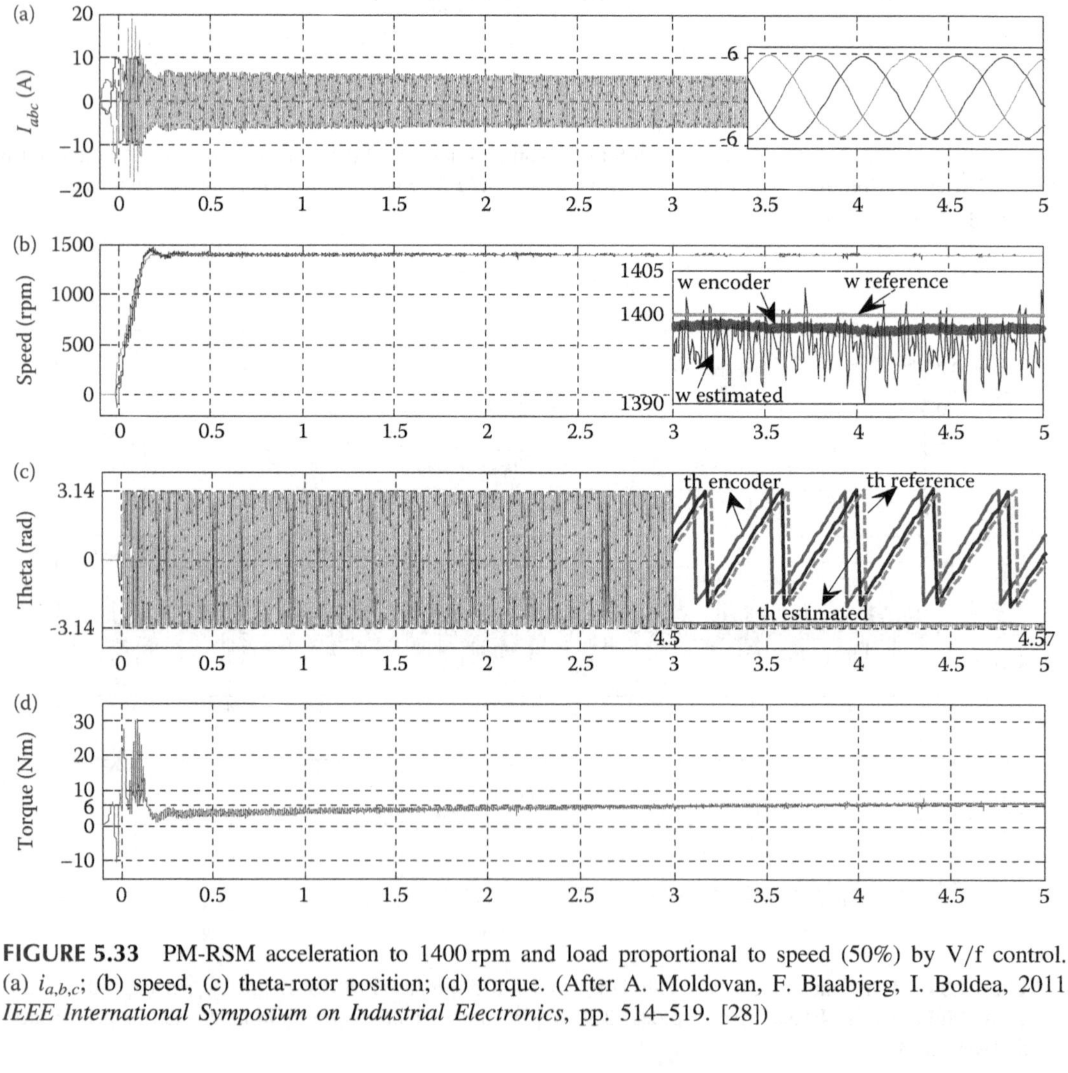

FIGURE 5.33 PM-RSM acceleration to 1400 rpm and load proportional to speed (50%) by V/f control. (a) $i_{a,b,c}$; (b) speed, (c) theta-rotor position; (d) torque. (After A. Moldovan, F. Blaabjerg, I. Boldea, 2011 *IEEE International Symposium on Industrial Electronics*, pp. 514–519. [28])

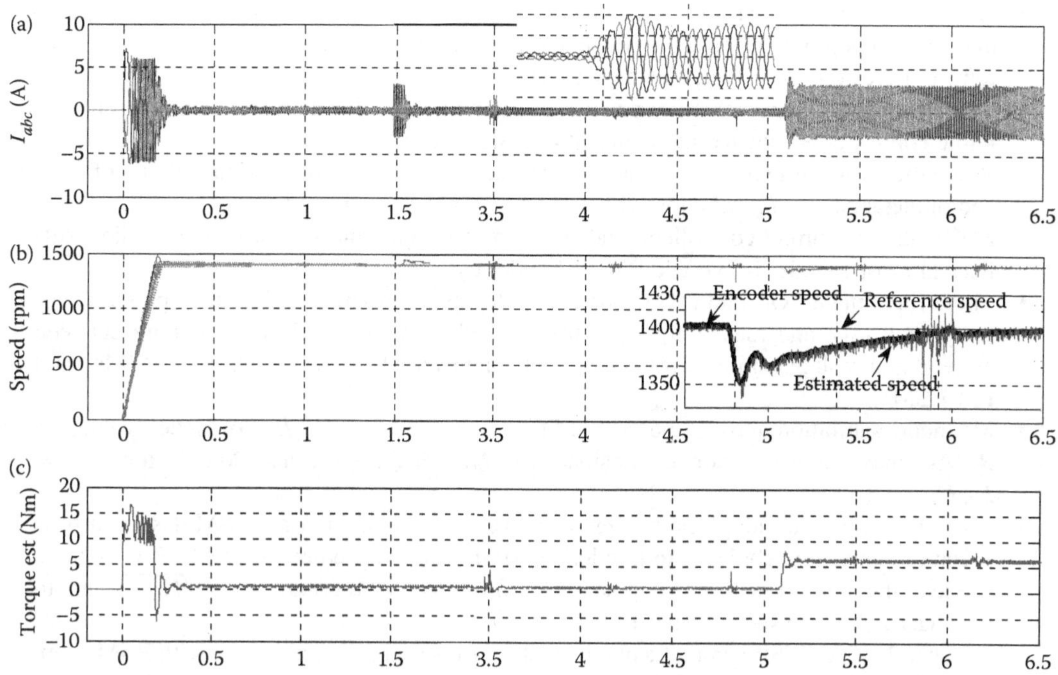

FIGURE 5.34 PM-RSM acceleration on no load to 1400 rpm with 50% load applied at 5.1 s with encoderless FOC. (a) $i_{a,b,c}$; (b) speed; (c) torque. (After A. Moldovan, F. Blaabjerg, I. Boldea, *2011 IEEE International Symposium on Industrial Electronics*, pp. 514–519. [28])

5.11 SUMMARY

- Control of VSDs is needed to vary speed and torque under closed loop position, speed, or torque according to the variation of the industrial process that is served in order to save energy and increase productivity.
- RSMs and PM-RSMs are synchronous machines, and their rich control knowledge heritage may be applied to it; so far, a myriad of control strategies have been proposed, and a few have reached the industrial stage.
- Energy conversion drive response, costs, and weight performance indexes are lined up to compare various solutions.
- To make some paths in these "woods," the control of RSMs and PM-RSMs is classified as with encoder and encoderless, with fast torque response (FOC and DTFC) and with slow torque response (scalar, V/f, or I-f), and control with stabilizing loops for reducing losses, increasing stability, and a wide speed control range in general (less-demanding) applications.
- The encoder-servo drive and traction drive divide is determined by control precision torque response quickness and CPSR > 2.5–3 for transportationlike applications. It seems that mostly PM-RSM (with Ferrite-PMs) qualify for wide CPSR, as the power factor is also enhanced notably though the assisting (weak) PMs producing an emf at base speed ω_b of less than 30% of maximum inverter produced voltage vector ($\overline{V}_{smax}$). No overvoltage inverter protection at maximum speed is thus needed.

- Core loss and magnetic saturation modify (reduce) steady-state performance but also introduce cross-coupling during transients that reduce torque response quickness, and in a very refined drive control system, they have to be accounted for, especially in encoderless drives where they produce notable rotor position estimation errors that have to be compensated for in the control or eliminated at the machine design stage.
- When the basics of FOC, DTFC, and scalar controls are lined up, the sinusoidality of emfs and inductances L_d, L_q variation with rotor position are considered.
- FOC with DC current controllers (and emf compensation) and AC current controllers suffices in most encoder RSM (and PM-RSM) drives.
- These three main FOC and DTFC control strategies, observing MTPA, max. power factor (only in RSM), and max. torque per flux (MTPF) conditions allow mitigating between current and voltage limits for wide speed range and low loss control of RSMs (and PM-RSMs).
- Magnetic saturation moves the max MTPA $\gamma_i = 45° = \tan^{-1} I_q/I_d$ to higher values in RSMs; max. power factor is obtained for $I_q/I_d = \sqrt{L_d/L_q}$, and MTPF for $I_q/I_d = L_d/L_q >> 1$.
- It is advised here to adopt the maximum power factor condition for RSM design at base speed as approximately leads to low losses for the entire speed range; if CPSR = 2–2.5 is needed, the control will move gradually from $\gamma_i = 45°$ through $\tan^{-1}\sqrt{L_d/L_q}$ and to $\tan^{-1}(L_d/L_q)$ as the speed increases.
- In general, wide CPSR cannot be obtained at a base speed full power level with RSM, even with high saliency ($L_d/L_q > 5$–8).
- Base speed ω_b corresponds to full flux for full T_{eb} torque and voltage $V_{s\max}$ for continuous or predominant driving duty cycle in the targeted application: base power $P_{eb} = T_{eb} \cdot \omega_b/p_1$; p_1-pole pairs.
- DTFC, introduced by Depenbrock in 1985 as "self-control" and in 1986 by Takahashi as "quick torque response" for IM drives, was generalized as TVC for all AC drives in 1988 [8] with application to RSM in 1991 [9]; the DTC term was introduced by ABB for IMs in 1995.
- DTFC means decoupled (fast) closed-loop control of stator flux amplitude and instantaneous torque by ordering a certain voltage vector (or a combination of them within a 60° sector) also dependent on the position of the stator flux vector in one of the 60° sectors around each phase, as given by a voltage source dual-level six-power-switch static converter (inverter).
- "Hunting" for the quickest response available, decoupled i_d, i_q control is eventually reached as in FOC, but only as a limit.
- As in DTFC, dq current controllers are replaced by stator flux amplitude and torque controllers, and the transformation of coordinates—back and forth—from abc to dq is eliminated. However, the troubles are moved to the state observer for stator flux and torque for encoder drives while FOC is lacking it. The situation swings to DTFC's favor in encoderless drives when the stator flux observer is needed anyway for rotor position and speed estimation, even for FOC.
- DTFC is more suitable for RSMs and PM-RSMs than for strong emf SPM-SMSs or IPMSMs when a good part of the stator flux is given by the fixed PM flux linkage, and definitely in encoderless drives. An early implementation of DTFC (as in TVC) [9] has shown remarkable control performance in an ALA-rotor RSM in 1991.
- Wide CPSR is typical of PM-RSMs with its wide speed range control observing current and voltage limits and loss reduction for a given maximum torque versus speed envelope. This time, the MTPA formula is different (Equation 5.53) and easier to express by adopting the active flux concept (Equation 5.54).

- Active flux here along axis q (where PMs are placed)—$\overline{\Psi}_q^a$—is the torque producing flux now with i_d as torque-current. $\overline{\Psi}_q^a$ is aligned to axis q (PM axis) irrespective of torque or speed at standstill (Figure 5.17). So it is very suitable for encoderless drives that combine signal injection with the active flux (emf) model in the active flux observer that finally delivers $|\overline{\Psi}_q^a|$, $\theta_{\overline{\Psi}_q^a}$.
- Based on these, a PLL with a motion equation refines $\theta_{\overline{\Psi}_q^a}$ to rotor position $\widehat{\theta}_{er}$ and speed $\widehat{\omega}_b$ estimation. Torque is estimated from stator flux and current $\alpha\beta$ components $T_e = \frac{3}{2}p_1(\Psi_{s\alpha}i_{s\beta} - \Psi_{s\beta}i_{s\alpha})$ in stator coordinates.
- Wide-speed (from zero to 3000 rpm) encoderless FOC of RSM has been demonstrated in a 1-Nm prototype for speed transients with step load torque perturbation [26], where a V&I stator (and active) flux observer with PLL produces complete state observance with signal injection assistance at very low speeds.
- For the PM-RSM, wide-speed DTFC encoderless control with a simplified approach ($\Psi_q = 0$) for the entire torque-speed envelope limit ($I_q^* = I_{qc}^* = \Psi_{PMq}/L_q$) is introduced; for lower torque demands, I_q^* is reduced by $\sqrt{|T_e^*/T_{ek}|}$ when, indirectly, the MTPA is approximately met if the voltage limit is not reached above base speed ($V_s^* < V_{smax}$).
- V/f with stabilizing loops was experimentally proven to produce safe control in PM-RSMs from 5 to 3000 rpm; the MTPA condition is implicitly—by control—met to provide energy savings at lower torque levels and a direct (though a bit hesitant) start (without the need of initial rotor position estimation with PM polarity) is provided. In comparison with a sensorless FOC of same PM-RSM prototype, similar dynamic performance was shown [28]. It has to be borne in mind that V/f control still does not contain speed or current regulators (with their tedious calibration).
- An I-f starting with active flux encoderless FOC of RSM was also demonstrated over a $\pm$3000 rpm range [29]. The I-f starting (with rated current) is able to start loads up to 75–80%, though with a bit of hesitance, which in most applications is allowed, but not so in robotics-like drives when an encoder drive is required at least for safety, with encoderless FOC, DTC, or V/f used for redundancy in such safety-critical applications.
- Design refinements of PM-RSM design for encoderless operation with lower rotor position estimation error are treated in [30] and so on.
- The PM-RSM design to allow absolute position encoderless control by asymmetrizing the stator winding and some rotor poles is investigated in [31], while the differences between stator tooth-wound and distributed windings for rotor position estimation are tackled in [32].
- Encoderless DTFC of RSM with SVM (to reduce torque pulsations) with experiments down to 15 rpm is given in [33].
- In another refinement, a signal injected-assisted full order observer, with parameter adaptation for RSM encoderless FOC, is treated in detail [34].
- The PM-RSM's MTPA encoderless control of PM-RSM for V/f control with two stabilizing loops (of the same source, however: Δi_d) benefits a solid stability investigation and regulators design [35].
- Though not in a definitive version, a self-commissioning sequence for encoderless IPMSM drives with signal injection is introduced in [36].
- An inspiring synthesis on position and speed observers of AC motors with application to RSMs is also available in [37].
- Efforts to increase robustness in PM-RSM drives are renewed in [38], via deadbeat DTFC, which is a model inverse digital control method that computes volt-second solutions in the inverter to drive both electromagnetic torque and stator flux linkage to their desired values in one inverter constant switching period.
- All of the above are many reasons for further investigations by the diligent reader.

REFERENCES

1. I. Boldea, S. A. Nasar, *Electric Drives*, book, 2nd edition, CRC Press, Taylor and Francis Group, New York, 2006.
2. I. Boldea, S. A. Nasar, Unified treatment of core loss and saturation in the orthogonal—axis model of electric machines, *Proc. IEE*, vol. 134–B, no. 6, 1987, pp. 355–363.
3. I. Boldea, M. C. Paicu, G. D. Andreescu, Active flux concept for motion—sensorless unified a.c. drives, *IEEE Trans.*, vol. PE–23, no. 5, 2008, pp. 2612–2618.
4. I. Boldea, Z. X. Fu, S. A. Nasar, Digital simulation of a vector current controlled axially-laminated anisotropic (ALA)-rotor synchronous motor servo-drive, *EMPS Journal*, vol. 19, no. 4, 1991, pp. 415–424.
5. M. Ferrari, N. Bianchi, E. Fornasiero, Rotor saturation impact in synchronous reluctance and PM assisted reluctance motors, Record of IEEE—ECCE, 2013, pp. 1235–1242.
6. L. Xu, X. Xu, T. A. Lipo, D. W. Novotny, Vector control of synchronous reluctance motor including saturation and iron losses, *IEEE Trans*, vol. IA–27, no. 5, 1991, pp. 977–985.
7. H. D. Lee, S. J. Kang, S. Ki Sul, Efficiency optimized direct torque control of synchronous reluctance motor using feedback linearization control, *IEEE Trans*, vol. IE–46, no. 1, 1999, pp. 192–198.
8. I. Boldea, S. A. Nasar, Torque vector control (TVC) a class of fast and robust torque, speed and position digital controllers for electric drives, *EMPS Journal*, vol. 15, 1988, pp. 135–148.
9. I. Boldea, Z. X. Fu, S. A. Nasar, Torque vector control (TVC) of axially-laminated anisotropic ALA rotor RSMs, *EMPS Journal*, vol. 19, 1991, pp. 381–398.
10. M. Pacas, R. Morales, A predictive torque control for the synchronous reluctance machine taking into account the magnetic saturation, *IEEE Trans*, vol. IE–54, no. 2, 2007, pp. 1161–1167.
11. Y. Inoue, Sh. Morimoto, M. Sanada, Comparative study of PMSM drive systems based on current control and DTC in flux weakening control region, *IEEE Trans*, vol. IA–48, no. 6, 2012, pp. 2382–2389.
12. D. Stojan, D. Drevensek, Z. Plantic, B. Grcar, G. Stumberger, Novel field weakening control scheme for PM–SMs based on voltage angle control, *IEEE Trans.*, vol. IA–48, no. 6, 2012, pp. 2390–2401.
13. J. M. Kim, S. K. Sul, Speed control of IPMSM drive for the flux weakening operation, *IEEE Trans.* vol. IA–33, no. 1, 1997, pp. 43–48.
14. S. Bolognani, S. Calligaro, R. Petrella, F. Pogni, Flux weakening in IPM motor drives: comparison of state-of-art algorithms and a novel proposal for controller design, Power Electronics and Applications (EPE 2011), Proceedings of the 2011—14th European Conference on, pp. 1–11.
15. X. Zhang, G. Hock Beng Foo, D. M. Vilathgamuwa, D. L. Maskell, An improved robust field-weakening algorithm for direct-torque-controlled synchronous-reluctance-motor drives, *IEEE Trans*, vol. IE–62, no. 5, 2015, pp. 3255–3264.
16. J. S. Lee, R. D. Lorenz, M. A. Valenzuela, Time-optimal and loss-minimizing deadbeat—direct torque and flux control for interior permanent magnet synchronous machines, *IEEE Trans*, vol. IA–50, no. 3, 2014, pp. 1880–1890.
17. P. Guglielmi, M. Pastorelli, A. Vagati, Impact of cross-saturation in sensorless control of transverse—laminated synchronous reluctance motors, *IEEE Trans*, vol. IE–53, no. 2, 2006, pp. 429–439.
18. S. Ichikawa, M. Tomita, S. Doki, S. Okuma, Sensorless control of synchronous reluctance motors based on extended emf models considering magnetic saturation with online parameter identification, *IEEE Trans*, vol. IA–42, no. 5, 2006, pp. 1264–1274.
19. P. L. Jansen, R. D. Lorenz, Transducerless field orientation concepts employing saturation-induced saliencies in induction machines, *IEEE Trans*, vol. IA–32, no. 6, 1996, pp. 1380–1393.
20. C. Caruana, G. M. Asher, M. Sumner, Performance of HF signal injection techniques for zero-low-frequency vector control of induction machines under sensorless conditions, *IEEE Trans*, vol. IE–53, no. 1, 2006, pp. 225–238.
21. M. Schroedel, Sensorless control of a.c. machines at low speed and standstill based on INFORM method, *Record of IEEE—IAS*, 1996, vol. 1, pp. 270–277.
22. T. Matsuo, T. A. Lipo, Rotor position detection scheme for synchronous reluctance motor based on current measurements, *IEEE Trans*, vol. IA-31, no. 4, 1995, pp. 860–868.
23. L. Kreindler, A. Testa, T. A. Lipo, Position sensorless synchronous reluctance motor drive using the phase voltage 3rd harmonic, *Record of IEEE—IAS*, 1993, pp. 679–686.
24. F. Briz, M. W. Degner, Rotor position estimation, *IEEE Trans*, vol. IE Magazine–5, no. 2, 2011, pp. 24–36.
25. S. Koonlaboon, S. Sangwongwanich, Sensorless control of interior permanent-magnet synchronous motors based on a fictitious permanent-magnet flux model, Fortieth IAS Annual Meeting. Conference Record of the 2005 Industry Applications Conference, 2005, vol. 1, pp. 311–318.

26. S. Agarlita, I. Boldea, F. Blaabjerg, High-frequency-injection-assisted "active flux"- based sensorless vector control of reluctance synchronous motors with experiments from zero speed, *IEEE Trans On*, vol. IA-48, no. 6, 2011, pp. 1931–1939.
27. M. C. Paicu, L. Tutelea, Gh.-D. Andreescu, F. Blaabjerg, C. Lascu, I. Boldea, Wide speed range sensorless control of PM-RSM via 'active flux model', 2009 IEEE Energy Conversion Congress and Exposition, pp. 3822–3829.
28. A. Moldovan, F. Blaabjerg, I. Boldea, Active-flux based V/f with stabilizing loops versus sensorless vector control of IPMSM drives, 2011 IEEE International Symposium on Industrial Electronics, pp. 514–519.
29. S. Agarlita, M. Fătu, L. N. Tutelea, F. Blaabjerg, I. Boldea, I-f starting and active flux based sensorless vector control of reluctance synchronous motors, with experiments, 2010 12th International Conference on Optimization of Electrical and Electronic Equipment, pp. 337–342.
30. N. Bianchi, S. Bolognani, Sensorless oriented design of PM motors, *IEEE Trans. On*, vol. IA–45, no. 4, 2009, pp. 1249–1257.
31. Y. C. Kwon, S-Ki Sul, N. A. Baloch, S. Murakami, S. Morimoto, Improved design of IPMSM for sensorless drive, with absolute rotor position estimation capability, *IEEE Trans, On*, vol. IA–52, no. 2, 2016, pp. 1441–1451.
32. I. P. Brown, G. Y. Sizov, L. E. Brown, Impact of rotor design on IPMSM with concentrated and distributed windings for signal injection–based sensorless control and power conversion, *IEEE Trans, On*, vol. IA–52, no. 1, 2016, pp. 136–144.
33. I. Boldea, L. Janosi, F. Blaabjerg, A modified direct torque control (DTC) of reluctance synchronous motor sensorless drive, *EMPS Journal*, vol. 28, no. 2, 2000, pp. 115–128.
34. T. Tuovinen, M. Hinkkane, Signal-injection-assisted full-order observer with parameter adaptation for synchronous reluctance motor drives, *IEEE Trans, On*, vol. IA–50, no. 5, 2014, pp. 3392–3402.
35. Z. Tang, X. Li, S. Dusmez, B. Akin, A new V/f based sensorless MTPA control for IPMSMs, *IEEE Trans On*, vol. PE–31, no. 6, 2016, pp. 4400–4415.
36. S. A. Odhano, P. Giangrande, R. I. Bojoi, C. Gerada, Self-commissioning of interior permanent-magnet synchronous motor drives with high-frequency current injection, *IEEE Trans, On*, vol. IA–50, no. 5, 2014, pp. 3295–3303.
37. L. Harnefors, H. P. Nee, A general algorithm for speed and position estimation of ac motors, *IEEE Trans. On*, vol. IE–47, no. 1, 2000, pp. 77–83.
38. J. S. Lee, R. D. Lorenz, Robustness analysis of dead beat-direct torque and flux control for IPMSM drives, 2013 15th European Conference on Power Electronics and Applications (EPE), pp. 1–10.

6 Claw Pole and Homopolar Synchronous Motors
Modeling, Design, and Control

6.1 INTRODUCTION

In an effort to reduce DC excitation rotor power in synchronous machines, the claw-pole rotor was proposed by Lundell and implemented as an alternator with diode-rectified output on all cars and trucks in 1970–1980.

The claw-pole rotor is at its best when a large number of rotor poles is considered ($2p > 8$), to benefit from the "blessing" of a single circular DC excitation coil, which explains the drastic DC excitation power reduction.

It is also possible to add a solid core on the stator and place two circular DC coils on the stator to excite the claw-pole rotor and thus avoid the slip ring brush hardware; however, in this case, the machine volume and weight increase notably. On the other hand, a contactless power transformer to transfer power to the rotor circular DC coil, which requires volume, weight, and costs (high-frequency transformer plus inverter and a fast diode rectifier on rotor) has recently been proven feasible (at 3–5 kW) for a regular DC-excited multipole rotor synchronous motor proposed for HEVs and EVs at 100 kW power/unit.

Also, other methods to transfer power to DC excitation rotor windings, such as frequency injection in the stator-connected inverter and an auxiliary winding on the rotor, have been proposed recently. All these methods would be more beneficial on claw-pole SMs because the DC-transferred power would be a few times smaller (more than two times, anyway, for 3% down to 1%−1.5% of rated power!).

Moreover, CP-SMs may benefit by the replacement of the existing conductor DC excitation coil by a single disk-shaped axially magnetized Ferrite-PM. The "blessing" of the disk shape intuitively leads to the possibility of reduced PM weight as the number of poles produced by the claw-pole rotor increases; as there is enough volume between the rotor claws, the PM placement should not impose any additional volume. Also, increasing the number of poles for a pole pitch reduction to 20–30 mm leads to the use of tooth-wound windings (with 0.6–0.65 slot fill factor) and thus reduces further copper losses in the stator for better efficiency.

The main drawback, even with a disk-shaped Ferrite-PM claw-pole rotor, is considered the limited stator stack length "dictated" by the limitation of rotor claw radial thickness, which leads to large-diameter (pancake-shaped) designs. But there are applications such as small hydrogenerators or traction motors where a large-diameter rotor with many poles and a big hole inside it may use this space for the integrated hydroturbine rotor or for a mechanical transmission.

Yes, the claw-pole structure (Figure 6.1) typically holds a variable reluctance structure, which qualifies the machine as the reluctance type, though the machine saliency $L_d/L_q < 1.5$. For DC excitation, $L_d/L_q > 1$ unless the machine is heavily saturated, when $L_d/L_q < 1$, for the PM claw-pole rotor.

The claw-pole concept leads to the production of a multipolar magnetic field in the airgap, but some 20%–30% of the rotor-produced magnetic flux is "lost" between the adjacent claws; to reduce this effect, interclaw Ferrite-PMs (Figure 6.1) may be placed; they not only increase emf by

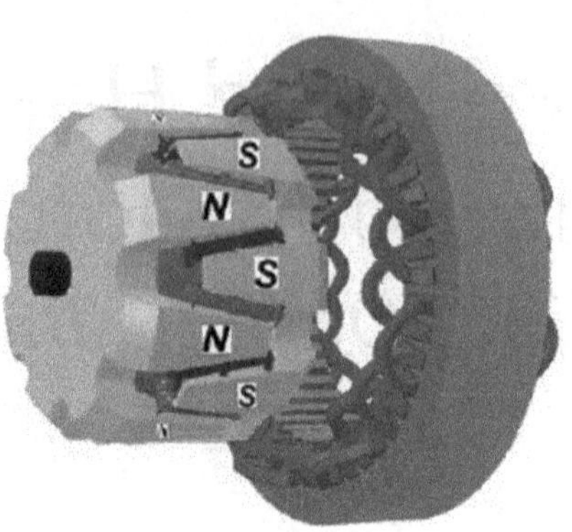

FIGURE 6.1 Typical automotive claw pole (machine/alternator) with interpole Ferrite PMs and DC excitation on the rotor.

20%–30% but also desaturate the rotor. All the above discussion leads to the idea that there are strong reasons to further investigate CP-SMs with DC excitation or with PMs on the rotor. The first mild HEV (Crown Royal by Toyota) used this technology for electric motor/generator traction/braking assistance, as it is not only technically competitive but already a technology accepted in industry.

The homopolar SM has a variable-reluctance passive solid iron rotor, but the stator is provided with a circular DC excitation coil between the two twin stators that contains a three-phase AC winding.

Due to its rotor ruggedness, despite smaller efficiency and higher weight, the homopolar SM-PM-less was applied for high-speed drives in high-temperature environments and for some "inertia batteries."

In what follows, the principles, topologies of practical interest, modeling, performance, and design, with a few case studies and control of CP-SMs and homopolar synchronous machines (H-SMs), are presented.

Note: Claw-pole stator core made by of Somaloy and regular SPM or IPM rotor machine have also being introduced [2] but they are still far from industrial stage even for small torque ratings.

6.2 CLAW POLE–SYNCHRONOUS MOTORS: PRINCIPLES AND TOPOLOGIES

The existing (Lundell) CP-SM with $2p_1 > 8$ pole, claw-pole rotor with DC-excitation singular coil and distributed ($q_1 = 1, 2$) stator winding is depicted in Figure 6.1, where, in order to increase output (and efficiency), interclaw pole Ferrite-PMs are placed. The interclaw PMs produce a small emf at maximum speed, as for no load (zero stator currents), their magnetic field closes mainly in the rotor; this is why cogging torque (at zero stator and DC excitation currents) is small, too. But the PMs could be placed in three main locations in the rotor (Figure 6.2).

Interclaw PMs have been shown to influence the output of CP alternators on cars by 25%, while SPMs increase output about 50% [3].

The main reason this solution did not reach production so far is PM (machine) additional cost, at least for automotive applications. But as more electric energy is needed in vehicles, its production efficiency (cost) becomes imperative, and this is why increased-efficiency CP alternator/motors are required ([4] shows an efficiency of 80% for 11 kW at 42 Vdc (generator operation by the stator of a 3.2-kW, 14-Vdc, up-to-date CP alternator with an efficiency of 70%) after small design alterations such as DC voltage, turn/coil, airgap, use of interclaw PMs). As low-speed applications in wind and hydro variable-speed (dual inverter–controlled) generator drives are looming, the PM-only rotor multipole CP machine is to be reconsidered. This is mainly due to the reduction in PM weight (cost) when the number of poles increases and eventually a tooth-wound-coil winding is used in the stator to reduce the pole pitch to 20–30 mm (or even less) when such a claw-pole pitch

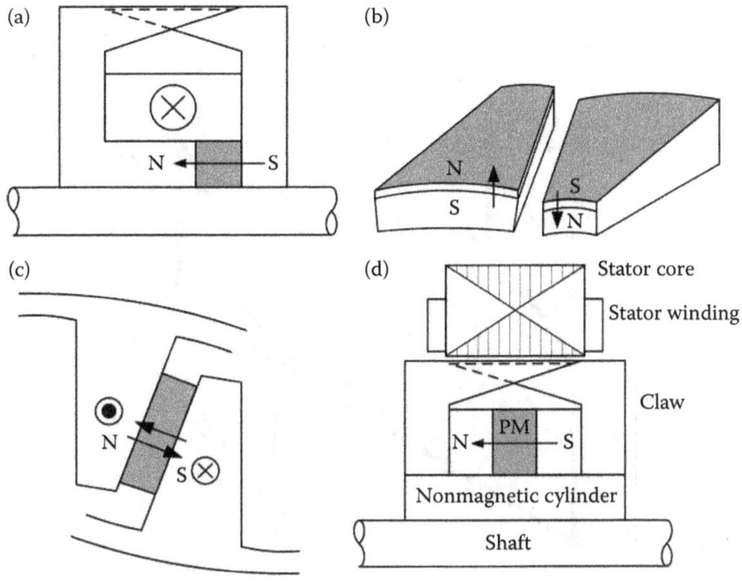

FIGURE 6.2 Locations to place PMs in the claw-pole rotor: (a) cylinder shape on the nonmagnetic shaft; (b) on the surface of claw pole (SPMs); (c) interclaw PMs; (d) disk-shaped PM.

is not feasible with regular distributed AC windings in machines with 20 Nm because of lack of space in slots with a normal slot aspect ratio and reasonable slot leakage inductance. As the analysis of CP-alternators for car systems is thoroughly treated in [4], here we will present in synthesis the modeling and performance assessment approaches to CP-SMs as motors/generators in variable-speed drives fed through PWM voltage converters.

6.3 CLAW POLE–SYNCHRONOUS MOTORS MODELING

6.3.1 Three-Dimensional Magnetic Equivalent Circuit Modeling

As the CP-SM inductances L_d and L_q and R_s stator resistance are defined as for regular synchronous machines [1], the main interest here is emf expressions. Let us consider the case of PM excitation; since the excitation magnetic flux lines are three-dimensional (Figure 6.3), a 3D-equivalent magnetic circuit or 3D-FEM modeling is required.

In Figure 6.3b (valid also for DC excitation), the excitation mmf $w_f i_f$ is replaced simply by $H_{cPM}h_{PM}$, while R_{esc} will now be the magnetic reluctance of the PM itself (it was the leakage-path DC coil magnetic reluctance); H_{cPM} and h_{PM} are the coercive field and the thickness of the disk-shaped PM in the rotor (Figure 6.2d).

The other magnetic reluctances in Figure 6.3b are: R_g—for airgap

$$R_g \approx \frac{g \cdot k_c}{\mu_0 \cdot \alpha_p \cdot \tau \cdot l_{\text{stack}}} \tag{6.1}$$

g—airgap, α_p—claw-pole span angle (average), k_C—Carter coefficient for stator slotting, τ—pole pitch of claw rotor, l_{stack}—stator length, R_{st}—stator tooth, R_{sy}—stator yoke, R_{ca}—claw axial, R_{cr}—claw radial, R_{cy}—rotor yoke, R_{ctl}, R_{cal}—claw tangential and axial leakages. Except R_g, R_{ctl}, R_{cal}, R_{csl}, which refer to the air zone, the other magnetic reluctances depend on the magnetic load (magnetic saturation level). So, the 3D MEC (or 3D FEM) should allow for local magnetic saturation levels. The MEC yields the magnetic flux in different regions and, finally, the distortion of flux density in the airgap B_α, B_z along the tangential (α) and axial (z) directions. With an average fundamental

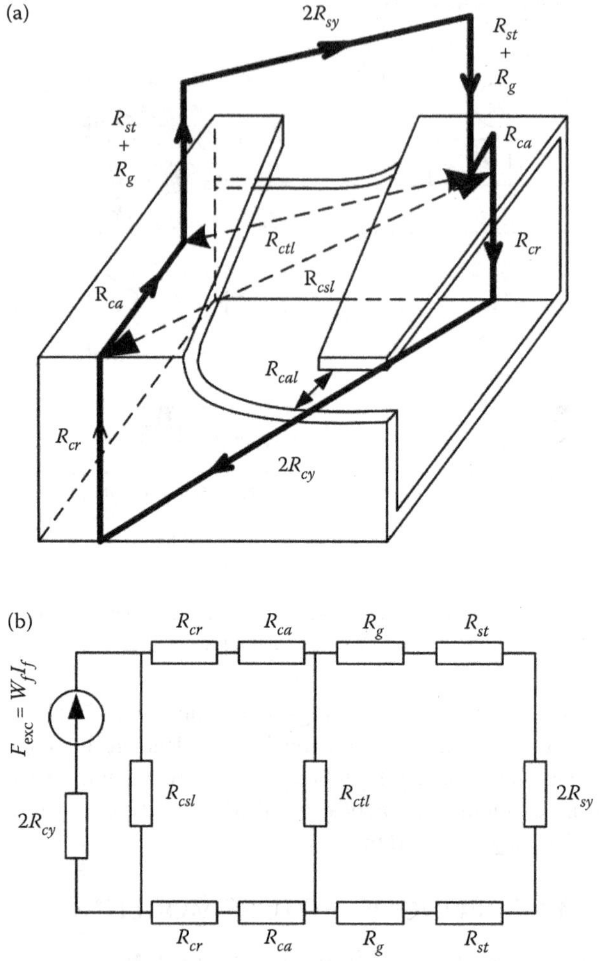

FIGURE 6.3 Typical emf 3D field lines, (a); and a simplified equivalent magnetic circuit, (b) $w_f i_f$, h_{PM} for PM-CP-rotors.

flux density at no load B_{g10}, the emf E_1 (the fundamental component) is obtained:

$$E_1 = \pi\sqrt{2}np_1 w_1 k_{w1} \frac{2}{\pi} B_{g10} \tau l_{\text{stack}} \qquad \text{(rms)} \tag{6.2}$$

A new, equivalent, and again nonlinear MEC has to be developed for load conditions (I_d and/or $I_q \neq 0$); see Figure 6.4.

Here the two PMs refer to magnets when the excitation mmf has to be replaced by $H_{CPM}\, h_{PM}$ and refers to the main disk-shaped magnet (Figure 6.2b). This multiple MEC would allow calculating both E_1 and L_{dm}, and L_{qm}, the magnetization inductances of the CP-SM, with the magnetic saturation included.

6.3.2 Three-Dimensional Finite Element Method Modeling

3D-FEM modeling requires full geometry, materials data, and stator currents i_d, i_q for a steady-state (static) analysis. This data may be obtained from an optimal design code based on MEC, given the complexity of the problem. 3D FEM should be used wisely to save computation time/effort. For a

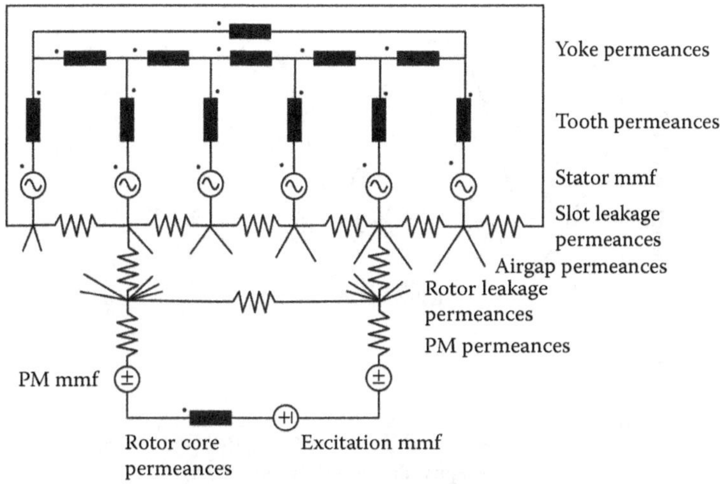

FIGURE 6.4 Multiple MEC of CP-SM for one pole pair, under load.

16-pole (48 slots) CP-SM with interclaw NdFeB magnets ($B_r = 1.13\ T$, $H_c = 860$ kA/m at 20°C) comparable with a 2.5-kW 14-V DC CP-alternator but redesigned for 42 Vdc, such a 3D-FEM investigation has shown a no-load airgap flux density rich in harmonics, even in the axially middle zone of the machine ($z = 0$)—Figure 6.5.

The torque pulsations (with and without interpole magnets) and 1500 Aturns of mmf of the field DC coil are shown in Figure 6.6. This mmf could be obtained alternatively with a Ferrite-PM ($H_{cPM} = 300$ kA/m) with a thickness h_{PM}:

$$h_{PM} = \frac{w_F i_F}{H_{cPM}} = \frac{1500}{300} = 5 \text{ mm} \tag{6.3}$$

This is a reasonable value for $g = 1$ mm (airgap). For a larger airgap, larger $w_F i_F$ values are required, but, again, the ratio of h_{PM}/g should be the same in principle. Now, for safety, PM

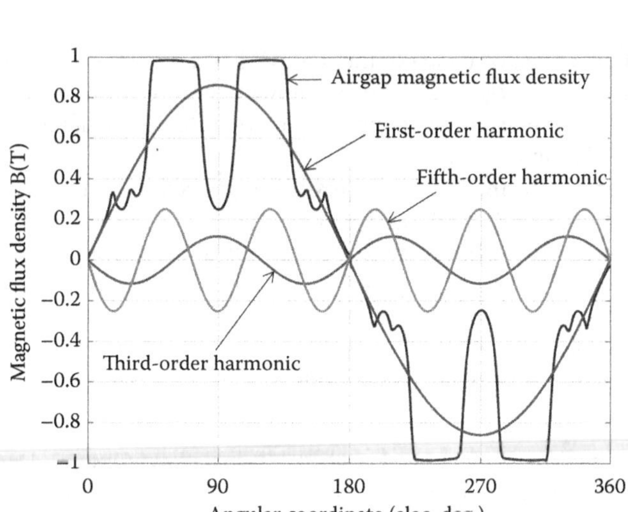

FIGURE 6.5 No-load airgap flux density in am IPM-CP-SM in the middle axial position ($z = 0$) by 3D FEM.

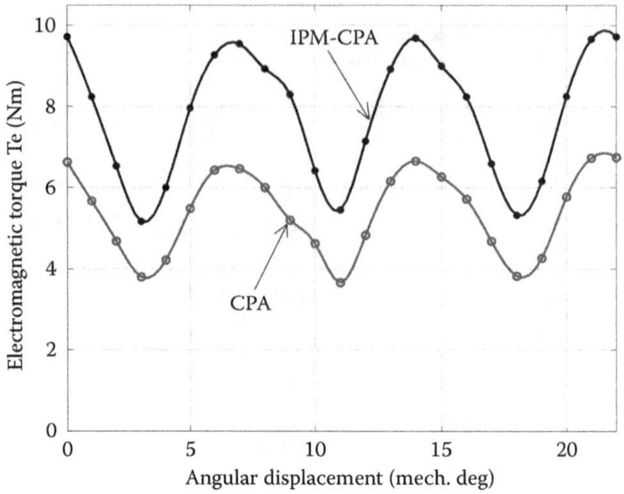

FIGURE 6.6 3D-FEM torque with given i_d, i_q, $w_F i_F = H_{cPM} h_{PM} = 1500$ Aturns versus rotor position (sinusoidal stator currents).

demagnetization, and as Ferrite-PMs are not expensive or very heavy, we may double it to 10 mm while also perhaps doubling the airgap to reduce L_{dm}, L_{qm} to meet the voltage ceiling in the inverter. The problem is that the PM has to produce the magnetic flux of all, say, north poles in the airgap:

$$A_{PM} B_m = \frac{p_1 \cdot \tau \cdot B_{g10} \cdot \frac{2}{\pi} \cdot l_{stack}}{1 + k_{f\ ring\ e}} \tag{6.4}$$

Also, the rotor claws on one side have to be able to handle this magnetic flux without entering saturation too deeply. This is how the larger-diameter but short-length rotor and short core stack stator should be adopted. If the outer and inner diameters of the disk-drive PM are about 0.7 D_{is} and 0.4 D_{is}, the PM area is:

$$A_{PM} = \frac{\pi}{4}(0.7^2 - 0.4^2)D_{is}^2; \qquad \pi D_{is} = 2p_1\tau \tag{6.5}$$

Equations 6.3–6.4 give a realistic idea of the rotor PM size required for a given stack length l_{stack} and no-load airgap flux density B_{g10} value. After showing how a given DC excitation mmf may be translated into a disk-shaped Ferrite-PM to produce $B_{g10} = 0.65$–0.8 T in the airgap, we may resume the analysis and results in Figure 6.5.

The interclaw PM emf is so small that no overvoltage protection for the uncontrolled generator is needed, if and only if the CP-SM is DC-excited with $i_F = 0$.

The interclaw magnets increase the no-load airgap flux by as much as around 30%. Similar improvements are expected in torque for given i_d, i_q, $w_F i_F$ ($H_{cPM} h_{PM}$).

As the CP-SM still has DC excitation, the saturation level in the rotor yoke is high and thus torque pulsations are high; the three slots/pole situation also produces torque pulsations despite rotor effective double skewing (by the very shape of the claw poles).

Note: The PM-only claw-pole rotor case will lead to less saturation, but the armature reaction field will remain within pole tops (axis q) and go between adjacent claws (axis d). Demagnetization avoidance of both main and interclaw magnets should be checked for most critical operation modes.

The good thing is that the Ferrite-PM performance increases at higher temperatures, and then electric resistivity is high (low eddy currents in PMs by various magnetic field harmonics).

6.4 CLAW POLE–SYNCHRONOUS MOTORS: THE *dq* CIRCUIT MODEL FOR STEADY STATE AND TRANSIENTS

The dq model of a synchronous machine in rotor coordinates may be used if sinusoidal emf, and distributed windings are considered in the stator of the CP-SM.

We consider here both cases (DC or PM excitation). The interclaw magnet contribution in fact increases the emf of the main (disk-shaped) magnet by reducing the fringing coefficient in Equation 6.3, and the latter (main magnet) is replaced by its mmf ($w_F i_F = H_{cPM} h_{PM}$), as discussed earlier. Finally, the field current influence is replaced by the constant ψ_{PM0} and the field circuit equation is removed:

$$\begin{aligned}
&\bar{i}_s R_s - \overline{V}_s = -\frac{\partial \overline{\psi}_s}{\partial t} + j\omega_r \overline{\psi}_s; \quad \overline{\psi}_s = \psi_d + j\psi_q; \quad \bar{i}_s = i_d + ji_q; \quad \overline{V}_s = V_d + jV_q \\
&\psi_d = L_{sl} i_d + \psi_{dm}; \quad \psi_{dm} = L_{dm} i_d + L_{dm} i_F; \quad L_{dm} i_F = \psi_{PM} \quad \text{if PM is used} \\
&\psi_q = L_{sl} i_q + \psi_{qm}; \quad \psi_{qm} = L_{qm} i_q
\end{aligned} \tag{6.6}$$

If there exists a DC excitation system:

$$\begin{aligned}
&i'_F R'_F - V'_F = -\frac{\partial \psi'_F}{\partial t}; \quad \psi'_F = L'_{Fl} \cdot i'_F + \psi_{dm} \\
&T_e = \frac{3}{2} p_1 \mathrm{Re}(j\overline{\psi}_s i_s^*) = \frac{3}{2} p_1 (\psi_d i_q - \psi_q i_d)
\end{aligned} \tag{6.7}$$

$$\begin{aligned}
&\frac{J}{p_1}\frac{d\omega_r}{dt} = T_e - T_{\text{load}} - B\omega_r; \quad \frac{d\theta_r}{dt} = \omega_r \\
&\overline{V}_{d,q} = \frac{2}{3}\left(V_a + V_b e^{j\frac{2\pi}{3}} + V_{bc} e^{-j\frac{2\pi}{3}}\right) e^{-j\theta_r} = V_d + jV_q
\end{aligned} \tag{6.8}$$

Though there are Joule losses in the rotor claws, as there are in any solid iron rotor, their value is limited by design (larger airgap, etc.) and may be neglected in the dq model for simplicity and added in the end as claw losses when efficiency is calculated:

$$P_{\text{claws}} \approx \frac{3}{2} R_{\text{claw}} i_s^2 \tag{6.9}$$

Core losses are also neglected in Equations 6.5 through 6.7, but they also should be added when efficiency is determined:

$$P_{\text{iron}} \approx \frac{3}{2} \frac{V_{s1}^2}{R_{\text{core}}(\omega_r)} \tag{6.10}$$

R_{claw} and $R_{\text{core}}(\omega_r)$ may be determined by separate FEM calculations (for example, in AC at standstill) and, respectively, in running on load ($\omega_r \neq 0$).

For steady state (and PM-only claw-pole rotor), Equations 6.5 through 6.7 simplify to:

$$\begin{aligned}
&\bar{i}_{s0}R_s - \overline{V}_{s0} = -j\omega_r\overline{\psi}_{s0} \qquad \left(\frac{d}{dt}=0\right)\\
&\psi_d = L_d i_d + \psi_{PM}; \quad \psi_q = L_q i_q \qquad (6.11)\\
&T_e = \frac{3}{2}p_1(\psi_{PM} + (L_d - L_q)\cdot i_d)\cdot i_q = \frac{3}{2}p_1\cdot\psi_d^a\cdot i_q
\end{aligned}$$

So, in fact, the CP-PMSM is an IPMSM ($E_1/V_{s\,max} > 0.75$!) and all the analysis, design, and control heritage of the latter may be used here also. This is why we do not insist here on this case. However, when DC excitation is used, the machine has one more degree of freedom in control, since

$$\psi_d = L_d i_d + L_{dm} i_F \qquad (6.12)$$

and the torque T_e is

$$T_e = \frac{3}{2}p_1(L_{dm}i_F + (L_d - L_q)\cdot i_d)\cdot i_q \qquad (6.13)$$

There are two favored operation modes for the DC-excited CP-SM:

- Pure i_q operation ($i_d = 0$) for full torque below base speed
- Unity power factor operation ($\varphi_1 = 0$) for flux weakening (CPSR)—above base speed

The pure i_q control instead of maximum torque per winding losses (stator + rotor [with DC excitation]) is performed for simplicity, since the magnetic saliency is small $L_d/L_q < 1.5$ (in deep saturation conditions: $0.8 < L_d/L_q < 1$) and thus the use of reluctance torque is not safe and requires too many Joule losses in the stator (through $V_d < 0$), though $i_d < 0$ means less flux (saturation) in axis d.

The vector diagrams in Figure 6.7 illustrate the $\varphi_1 = 0$ and $i_d^* = 0$ motor operation in steady state: $\psi_{s0\varphi1} < \psi_{s0q}$ for the same i_{q0}, i_{F0}

As the stator flux ψ_{s0} for the same i_{q0} and $i_{F0}(\psi_{PM})$ is smaller for the unity power factor ($\varphi_1 = 0$; $i_d < 0$) it means that, for the same speed, it requires less voltage, and this is why it is used for flux weakening as long as i_F is variable. For PM-CP-SMs, this is not the case, and the controls of IPMSM apply.

The field current reduced to the stator i_F' required for unity power factor operation and a given voltage V_{smax} and torque is obtained as below:

$$V_{s\,max} = R_s i_{s0} + \omega_{r0}\cdot\psi_{s0}^*; \quad \psi_{s0}^*, T_e^* \text{ given}; \quad i_{s0}^* = T_e^*\frac{2}{3\cdot p_1\cdot\psi_{s0}^*} \qquad (6.14)$$

So, for a given torque T_e^*, motor flux ψ_{s0}, and voltage $V_{s\,max}$, the speed ω_{r0} may be calculated. Also from Figure 6.7a:

$$\tan(\delta_\psi^*) = \frac{L_q i_{s0}^*}{\psi_{s0}^*}; \quad i_{q0}^* = i_{s0}^*\cos(\delta_\psi^*); \quad i_{d0}^* = -\sqrt{i_{s0}^{*2} - i_{q0}^{*2}} < 0 \qquad (6.15)$$

Finally, the required field current flux $L_{dm}\, i_{F0}$ is:

$$L_{dm}i_{F0}^{*\prime} = \psi_{s0}^*\cos(\delta_\psi^*) + L_{dm}i_{s0}^*\sin(\delta_\psi^*) \qquad (6.16)$$

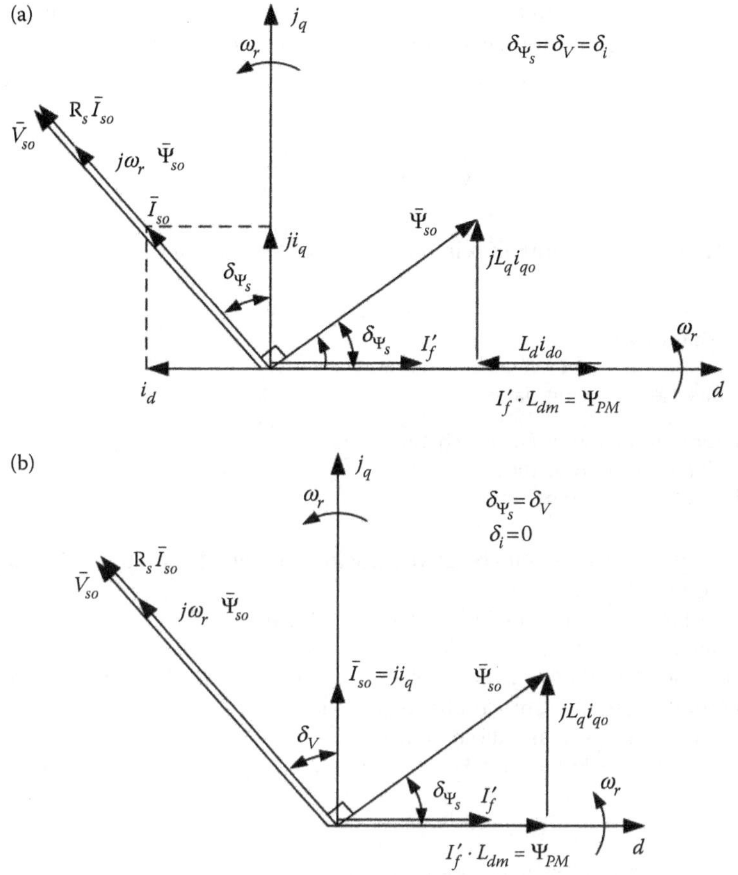

FIGURE 6.7 Vector diagrams of CP-MS at steady state: (a) at unity power factor ($\varphi_1 = 0$); (b) $i_d^* = 0$.

As expected, field control i_{F0} increases with torque to maintain the unity power factor. The ideal no-load speed ω_{r0} is:

$$(\omega_{r0})_{i_{s0}=0} = \frac{V_{s\max}}{\psi_{s0}^*}, \tag{6.17}$$

as in the separately excited DC brush motor. Equation 6.12 with torque T_e^* instead of current becomes:

$$V_{s\max} = \frac{2R_s T_e}{3p_1\psi_{s0}^*} + \omega_r\psi_{s0}^* \tag{6.18}$$

A linear mechanical characteristic is obtained at the unity power factor, which is known to be ideal for control; in addition, it is available during flux weakening (wide CPSR). On top of this, i_F decreases in wide CPSR operation and thus the efficiency remains good, despite $i_d < 0$ (and its associated Joule losses) still existing. For zero i_d (this regime is also feasible for PM excitation of CP-SM):

$$\psi_d = L_{dm}i_F = \psi_{\text{PM}};\quad \psi_q = L_q i_q;\quad T_e = \frac{3}{2}p_1 \cdot \psi_{\text{PM}} \cdot i_q = \frac{3}{2}p_1(L_{dm}i_F')i_q \tag{6.19}$$

As $\omega_r < \omega_b$, in general, the voltage is not limited: $V_s \leq V_{s\ max}$. The control will have to consider how to drive i_q, i_F in producing torque T_e (Equation 6.19) for similar current density values in the stator and excitation circuit (if any); then:

$$i_q^* \approx i_{F0}^* = \sqrt{\frac{2 \cdot T_e^*}{3p_1 L_{dm}}} \tag{6.20}$$

may be adopted for close to minimum winding losses in the machine.

NUMERICAL EXAMPLE 6.1

A PM-CP-SM has the following data:

Interior stator core diameter $D_{is} = 178$ mm
Stator core length $l_{stack} = 40$ mm
Pole pitch $\tau = l_{stack} = 40$ mm

Rotor PM (Ferrite) excitation airgap flux density fundamental $B_{g10} = 0.7$ T; $B_r = 0.45$ T, $H_{cPM} =$ 320 kA/m for Ferrite-PM excitation.

Outer/inner diameter of disk (cylinder)-shaped PM is $0.75D_{is}/0.4D_{is}$; magnet thickness $h_{PM} =$ 10 mm (initial airgap $g = 1$ mm).

The main magnet flux fringing coefficient $K_{f\ ring\ e} = 0.5$ is reduced to 0.2 (by interclaw Ferrite-PMs whose properties are the same as above).

Rotor pole pitch average span ratio $a_p = 0.7$.

Emf at base speed $n_b = 500$ rpm is $E_1 = 0.75\ V_{s\ max}$, $V_{s\ max} = 220{*}1.41$ (peak value/phase).

Calculate:

a. The number of poles $2p_1$ and base frequency f_b

$$p_1 = \frac{\pi D_{is}}{2\tau} = \frac{\pi \cdot 178}{2 \cdot 40} \approx 7 \text{ pole pairs} \tag{6.21}$$

$$f_{10} = n_p p_1 = \frac{500}{60} \cdot 7 = 58.33 \text{ Hz} \tag{6.22}$$

The peak airgap PM flux

$$\psi_{PM1} = \frac{2}{\pi} \cdot B_{g10} \cdot \tau \cdot l_{stack} \cdot w_1 \cdot k_{w_1} = \frac{2}{\pi} \cdot 0.7 \cdot 0.04 \cdot 0.04 \cdot w_1 \cdot k_{w_1}$$
$$= 1.783 \cdot 10^{-3} \cdot w_1 \cdot k_{w_1} \tag{6.23}$$

where w_1 and k_{w1} are the number of turns per phase (one current path only) and the fundamental winding factor, respectively.

Is the disk-shaped Ferrite-PM capable of producing the required airgap flux linkage in no load?

According to Equation 6.3:

$$A_{PM} \cdot B_m = \frac{p_1 \cdot \tau \cdot B_{g10} \cdot l_{stack}}{1 + k_{f\,ring\,e}} \tag{6.24}$$

and as above and Equation 6.4

$$A_{PM} = \frac{\pi}{4}(0.75^2 - 0.4^2) \cdot 0.18^2 = 0.0102 \text{ m}^2 \tag{6.25}$$

So, the required $B_m < 0.45$ T is determined from Equations 6.23 and 6.24:

$$B_m = \frac{7 \cdot 0.7 \cdot (2/\pi) \cdot 0.04 \cdot 0.04}{(1 + 0.2) \cdot 0.0102} = 0.408 \text{ T} < 0.45 \text{ T} \quad (6.26)$$

Now, with a 20-mm-thick magnet that "serves" even a 2 × 2 mm airgap and $B_r = 0.45$ T (more above 25°C) is capable of producing the required PM flux density.

b. The number of turns per phase w_1, for a $q_1 = 1$ stator-distributed (42 slots for 14 poles), one-layer winding ($k_{w1} = 1$), is:

$$w_1 = \frac{0.7 \cdot V_s \cdot \sqrt{2}}{1.783 \cdot 10^{-3} \cdot 1 \cdot 2\pi \cdot f_{10}} \quad (6.27)$$

Consequently:

$$w_1 = \frac{0.7 \cdot 220 \cdot \sqrt{2}}{1.783 \cdot 10^{-3} \cdot 1 \cdot 2\pi \cdot 58.33} \approx 332 \frac{\text{turns}}{\text{phase}} \quad (6.28)$$

c. With a slot height/width ≈4/1, equal slot/tooth width, and current density $j_{co} = 10$ A/mm^2, we may calculate the rated current I_{sn}. First, there are seven coils per phase and thus the number of turns/coil (slot) n_s is:

$$n_s = \frac{w_1}{p_1 q_1} = \frac{332}{7} \approx 47 \text{ turns/slot} \quad (6.29)$$

The slot pitch τ_s comes from:

$$\tau_s = \frac{\tau}{m \cdot q} = \frac{\tau}{3 \cdot 1} = \frac{40}{3} = 13.33 \text{ mm} \quad (6.30)$$

with an average tooth width of $b_{ts} = 6$ mm, the average slot width (along slot height) is about $w_s = 9$ mm and the slot depth (active part) $h_{as} = 28$ mm; the slot useful area A_{slot} is:

$$A_{\text{slot}} = h_{as} \cdot w_{sav} = 28 \cdot 9 = 252 \text{ mm}^2 \quad (6.31)$$

Now the ampere-turns per slot $n_s I_{sn}/\sqrt{2}$ is:

$$\frac{n_s \cdot I_{sn}}{\sqrt{2}} = A_{\text{slot}} \cdot k_{\text{fill}} \cdot j_{\text{con}} = 252 \cdot 0.45 \cdot 10 = 1134 \frac{A\,\text{turns}}{\text{slot}}; \text{ (rms)} \quad (6.32)$$

So

$$I_{sn} = \frac{1134\sqrt{2}}{47} = 34\,A \text{ (peak value per phase)} \quad (6.33)$$

d. Supposing that pure I_q control is performed, calculate the base-speed-torque:

$$T_{er} = \frac{3}{2} p_1 \cdot i_{qn} \cdot \Psi_{\text{PM1}} = \frac{3}{2} \cdot 7 \cdot 1.783 \cdot 10^{-3} \cdot 332 \cdot 34 = 211.328 \text{ Nm} \quad (6.34)$$

e. Calculate the base power at 500 rpm:

$$P_{eb} = T_{em} \cdot 2\pi \cdot n_b = 211.328 \cdot 2\pi \cdot \frac{500}{60} \approx 11 \text{ kW} \tag{6.35}$$

f. Calculate the outer rotor diameter D_{as}, the torque/volume, and the shear stress on the stator inner core:

$$D_{os} = D_{is} + 2(h_{as} + 0.003) + 2\frac{B_{g1}}{B_{ys}} \cdot \frac{\tau}{\pi} = 0.254 \approx 0.26 \text{ m (for safety)} \tag{6.36}$$

The stator core cylinder volume is $V_{os} = \frac{\pi}{4} D_{os}^2 l_{stack}$, and the torque per volume is:

$$\frac{T_{en}}{V_{os}} = \frac{211.32}{(\pi/4) \cdot 0.26^2 \cdot 0.04} \approx \frac{100 \text{ Nm}}{\text{litre}} \tag{6.37}$$

The tangential shear stress f_t is:

$$f_t = \frac{T_{en}}{(\pi/2) \cdot D_{is}^2 l_{stack}} = \frac{211.32}{(\pi/2)0.18^2 \cdot 0.04} = 10.5 \frac{\text{N}}{\text{cm}^2} \tag{6.38}$$

The above are mighty high values in term of the last IPMS designs for racecars, and in all probability, the airgap should be increased from 1 mm to perhaps 2 mm to reduce armature reaction flux density to a reasonable value at base (peak) torque, T_{en}: $B_{agq1} < 0.8B_{ag1}$ to leave a reasonable 0.8 power factor. The "segmented appearance of claw poles to stator field" also helps in reducing the armature airgap field (I_q control, $I_d = 0$). The already-adopted $h_{pm} = 20$ mm thickness for the disk-shaped Ferrite-PM also might hold, but the above preliminary performance assessment should continue with the magnetic design of the rotor and computation of L_{dm}, L_{qm}, L_d, L_q, R_s to prepare all data to verify the above performance claims by more elaborate (MEC and FEM) modeling.

As $E_{1PM} = 0.75\ V_{s\,max}$, the condition $B_{agq1} < 0.8B_{ag10}$ should suffice to make the design practical: $\omega_1 L_q i_q / E_1 \approx B_{agq1}/B_{ag10}$ and $V_{s\,max} = \sqrt{E_{1PM}^2 + (\omega_r L_q i_q)^2}$ for pure I_q control.

g. Calculate approximately the B_{agq1} for the base (peak) torque T_{en}.

First, we notice the kind of segmented structure of the rotor in axis q; thus, the airgap may be considered doubled in axis q, while only $\tau_p/\tau = 0.7$ of the maximum q axis mmf is active. So, approximately:

$$\frac{3 w_1 k_{w1} I_q (\tau_p/\tau)}{\pi p_1} = \frac{B_{agq1} 2g}{\mu_0}; \quad g = 2 \text{ mm} \quad (k_c \text{ is included})$$

$$B_{agq1} = \frac{3 \cdot 332.1 \cdot 34 \cdot 0.7 \cdot 1.286 \cdot 10^{-6}}{2 \cdot 2 \cdot 10^{-3} \cdot \pi \cdot 7} = 0.338 \text{ T} \tag{6.39}$$

As magnetic saturation was not considered, some reserve must be maintained. Also, even for double (rated) current, $B_{agq1\,max} = 0.676\text{ T} < 0.7\text{ T}$, which prevents demagnetization.

Final Note: The preliminary design numerical example here should only be indicative of the performance range; consequently, an optimal design methodology is needed and will follow.

6.5 OPTIMAL DESIGN OF CLAW POLE–SYNCHRONOUS MOTORS

An optimal design methodology and code based on the nonlinear 3D-MEC model, as presented earlier in this chapter, has been developed and put in place with promising results (after key FEM validation) for mild-like HEVs and then extended to 100 kW at 6000 rpm for the case of DC-excited interclaw PMs with and without NdFeB or Ferrite. Sample results are given in Table 6.1 [6,7].

To prepare for the case of the Ferrite disk-shaped PM claw-pole rotor CP-SM, we describe here only the essentials of the optimal design's MATLAB code.

6.5.1 The Objective Function

The objective function contains three terms: initial machine cost c_i, energy loss cost c_e over its life, and the penalty function cost c_p. In the initial cost, in addition to the active materials cost, the passive materials and fabrication costs may be added:

$$c_i = c_{\text{copper}} + c_{\text{stator core}} + c_{\text{rotor iron}} + c_{\text{PM}} + c_{\text{passive}} + c_{\text{fabrication}} \tag{6.40}$$

$$c_e = p_{\text{energy}}\left(P_{n1}\frac{t_{r1}}{\eta_1} + P_{n2}\frac{t_{r2}}{\eta_2}\right)h_{py}n_y \tag{6.41}$$

with: p_{energy}—energy price; P_{n1}, t_{r1}, η_1—electromagnetic power (in kW), relative time operation, efficiency at 3000 rpm; P_{n2}, t_{r2}, η_2—electromagnetic power, relative time operation, efficiency at 6000 rpm; h_{py}—hours in use per year, n_y—number of operation years.

The main penalty function related to power unrealization is c_p:

$$c_p = k_{pp}\left((P_n > P_{n1\ \max})\cdot\frac{p_n}{p_{n1\ \max}} + (P_n > P_{n2\ \max})\cdot\frac{p_n}{p_{n2\ \max}} + (P_n > P_b)\cdot\frac{p_n}{p_b}\right)\cdot c_i \tag{6.42}$$

where k_{pp} is a coefficient such that it always renders the optimal design capable of power realization at all critical speed speeds, base speed (P_b), within the maximum voltage of the inverter $V_{s\ \max}$. The $(P_n > P_{n1\ \max})$ expression (logical comparisons) is "1" if the maximum generator power at the speed is smaller than required power, and it is "0" if the machine is able to realize the required power.

Other penalty function components such as the PM demagnetization avoidance condition, say, for (1.5–2) I_{sn} (pure I_q control) may be added. The condition in numerical example 6.1, $(B_{agq1})_{Iq\ \max}/B_{ag10} < 0.8$, could be such a penalty function component.

Stator over temperature may also be a penalty function component to be expressed by a simple thermal model referring to stack radial outer area A_{sout} and to an equivalent heat-transmission

TABLE 6.1
100-kW CP-SM Optimal Design Results

	CPA without IPM	IPM-CPA with NdFeB PMs	IPM-CPA with Ferrite PMs
Efficiency at 3000 (rpm)(−)	0.934	0.9468	0.9425
Power produced at 3000 rpm (W)	105280	108500	159260
Efficiency at 6000 (rpm) (−)	0.9262	0.942	0.9396
Power produced at 6000 rpm (W)	120000	130720	207000
Active material cost (EUR)	132.71	213.98	148.15
Airgap length (mm)	1.15	1	1.45
Total cost (EUR)	296710	307110	296250
Stator outside diameter (mm)	395.5	389	388
Stator core length (mm)	36.5	51	47.5

coefficient $\alpha_{Temp} = (14–100)\ W/m^2/°C$, dependent on the cooling system for the application:

$$\Delta T = \frac{P_{\text{loss in critical operation mode}}}{\alpha_{Temp} A_{sout}} < \Delta T^* \quad (6.43)$$

The rotor temperature penalty may also be considered, especially due to eddy currents in the solid iron claw poles (by field harmonics from all sources), even if the Ferrites are better (higher B_r) with higher temperatures. The ferrite PMs have to be protected against low temperatures, even by preheating them at standstill via AC stator currents in cold mornings before operation.

As the nonlinear multiple MEC model, already introduced, we mention here the set of variables for optimization (the variable vector $\overline{X}$).

$$\overline{X} = (\text{sDo, sDci, rDi, lc, shy, stw, sMs, hag, poles, sb, ap, rpl, wrd, lrc, rpk, lextmax, kippm}) \quad (6.44)$$

where

sDo—stator outer diameter (mm)
sDi—stator inner diameter (mm)
rDci—rotor coil inner diameter (mm)
lc—stator core length (mm)
shy—stator yoke thickness (mm)
stw—stator tooth width (mm)
sMs—slot opening (mm)
hag—airgap (mm)
ap—polar span (p.u.)
rpl—rotor pole length (mm)
wrd—width of rotor end disk (mm)
lrc—rotor length variation due to disk (mm)
rpk—rotor pole height (mm)
lextmax—maximum ampere-turns of field coil (or $H_{cPM} h_{PM}$ for PM excitation only) in Aturns
kippm—interclaw magnet fill factor

Note: For the disk-shaped Ferrite PM-excited claw-pole rotor, the changes in the MEC are minimal (as shown earlier in this chapter). The disk PM radius R_{PMo}, R_{PMi} and its thickness h_{PM} have to enter the variable vector; other than that, the optimal design code may also be used for the disk-PM excited claw-pole rotor. The optimization algorithm is based on the Hooke–Jeeves method with 10–20 random starting variable vector runs to avoid (most probably) local optimums.

6.6 OPTIMAL DESIGN OF A PERMANENT MAGNET–EXCITED CLAW POLE–SYNCHRONOUS MOTOR: A CASE STUDY

Based on Numerical Example 6.1, the optimal design code was put in place with FEM key validation. (Am NdFeB or SM investigation by preliminary analytical design with 3D-FEM heavy usage is available in [8].)

6.7 CLAW POLE–SYNCHRONOUS MOTOR LARGE POWER DESIGN EXAMPLE 6.2 (3 MW, 75 RPM)

In an attempt to investigate the suitability of DC-excited CP-SM for hydro or wind generators, a few specifications are put in place to start:

- Rated power: $P_n = 3\ MW$
- Rated speed: $n_n = 75\ rpm$
- Tentative efficiency: $\eta_n > 0.9$

- Rated power factor: $\cos \rho_1 = 1$
- Airgap: $g = 8$ mm
- Number of poles: $2p_1 = 40$
- Number of phases: $m = 3$
- Number of current paths: $a_c = 10$
- Tangential specific force: $f_t = 6\,\text{N/cm}^2$
- Tentative inner stator diameter: sDi = 4.5 m; no load flux density $B_{gd0} = 0.8$ T
- Phase voltage F_{fn}: 2900 V (rms)
- DC link voltage: $\text{Vdc} = (\pi/\sqrt{2})V_{fn} = 6346$ V at 75 rpm
- DC excitation

A diode rectifier is used to interface with a DC voltage bus and thus $\cos \varphi_1 = 1$; sinusoidal current in the machine is assumed.

The design is based on same nonlinear MEC model as was depicted in this chapter [9] and thus here only sample results are given to acquire a sense of magnitudes:

- The excitation mmf $w_F i_F = 32.789$ kAturns
- Stack length $l_{\text{stack}} = 0.22$ m
- Excitation losses $P_{\text{exc}} = 45.801$ kW
- Current density $j_{\text{con}} = 6\,\text{A/mm}^2$
- Stator copper losses $p_{\text{copper}} = 97.127$ kW
- Iron losses $p_{\text{iron}} = 3.14$ W/kg $G_{\text{iron}} = 6.109$ kW
- Claw pole losses 0.3% $P_n = 9$ kW
- Active weight 10.4 tons
- Efficiency $= \dfrac{p_n}{p_n + p_{\text{exc}} + p_{\text{copper}} + p_{\text{iron}} + p_{\text{claw}} + p_{\text{mec}}} = 0.941$
- Weights: stator core (1945 kg), stator cooper weight (1412 kg), rotor iron + excitation coil (mover) weight (7038 kg!).

A few design sensitivity investigations have shown results as in Figure 6.8 [1].

By increasing the stator inner diameter from 3 m up reduces stator and active material weight but increases the total copper losses. The sDi = 4.5 m and 8-mm airgap yield the above performance, which is considered acceptable for a preliminary investigation. As the excitation coil losses have not been considered, the efficiency of 0.94 is an upper limit for the disk-shaped Ferrite-PM CP-SM design.

Similar designs for 1 MW, 150 rpm and 100 kW, 600 rpm [9] have yielded efficiencies in excess of 0.93 (excitation losses not considered) for active weights of 3431 kg and 1063 kg, respectively. It is inferred here that the optimal design code described earlier in this chapter may be exercised on disk-shaped Ferrite-PM claw pole–excited version of the above design for hydro and wind generators or other low-speed high-torque applications. But this is beyond the scope/space here.

6.8 CONTROL OF CLAW POLE–SYNCHRONOUS MOTORS FOR VARIABLE SPEED DRIVES

The control strategies for CP-SMs are:

- With encoder
- Encoderless

and, on principle:

- Field-oriented control
- DTFC
- Scalar (V/f or I-f) with stabilizing loops

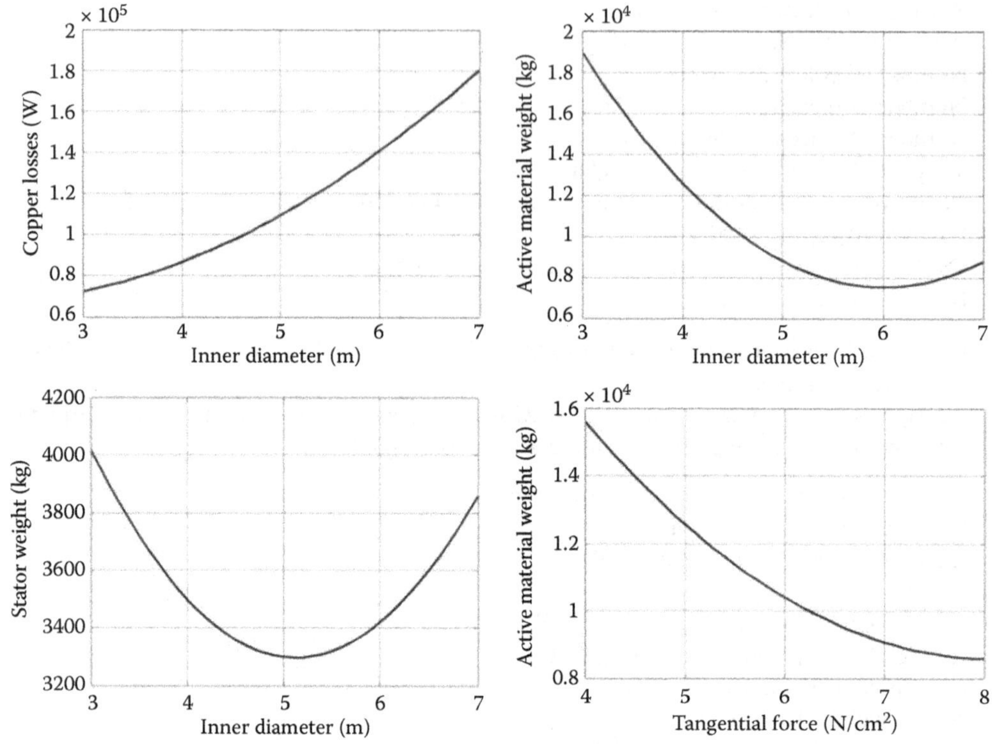

FIGURE 6.8 Design sensitivity to inner rotor diameter and tangential specific force (N/cm^2).

Disk-shaped PM-excited claw-pole rotor SMs are similar to IPMSMs with strong PMs ($E_1/V_{s\,max} > 0.75$) and with variable magnetic saturation ($0.7 < L_d/L_q < 1.2$). Their control is treated richly in the literature, and this is why we will leave this part out here.

The circular coil DC-excited claw-pole-rotor SM, on the other hand, has not enjoyed such attention. In fact, only encoder FOC strategy has been implemented for this machine so far. Here is such a FOC for CP-SM for automobile tractionlike applications that is run in the torque mode control. A calculator (Figure 6.9) gathers information on vehicle speed, M/G speed, V_{dc} of battery on load, and SOC of same battery. For start, let us consider the case of pure I_q control, though unity power factor control is needed, too, to extend the CPSR. Such a control scheme is illustrated in Figure 6.9.

The output of the reference calculator prescribes the field current (I_F) versus speed and the $\psi_F(I'_F)$ function. It also yields the reference torque T_e^*. A coefficient transforms I'_F (reduced to stator) to I_F in the field coil. Subsequently, with measured field current I_F, the field current error enters a PI+ sliding mode robust controller that controls the voltage in the DC–DC converter that feeds the excitation circuit on the rotor. As the saliency is small, from the reference T_e^*, the q axis reference current I_q^* is:

$$I_q^* \approx \frac{2T_e^*}{3p_1\psi_F^*(I'_F)} \tag{6.45}$$

Now, $I_d^* = 0$ unless the voltage limit $V_{s\,max}^*$ in the inverter is reached. Otherwise, a nonzero (negative) I_d^* is considered through a robust PI+SM controller, as is required from the inverter (through the regulators of I_d, I_q closed-loop PI+SM regulators V_d^*, V_q^*; $V_s = \sqrt{V_d^{*2} + V_q^{*2}}$).

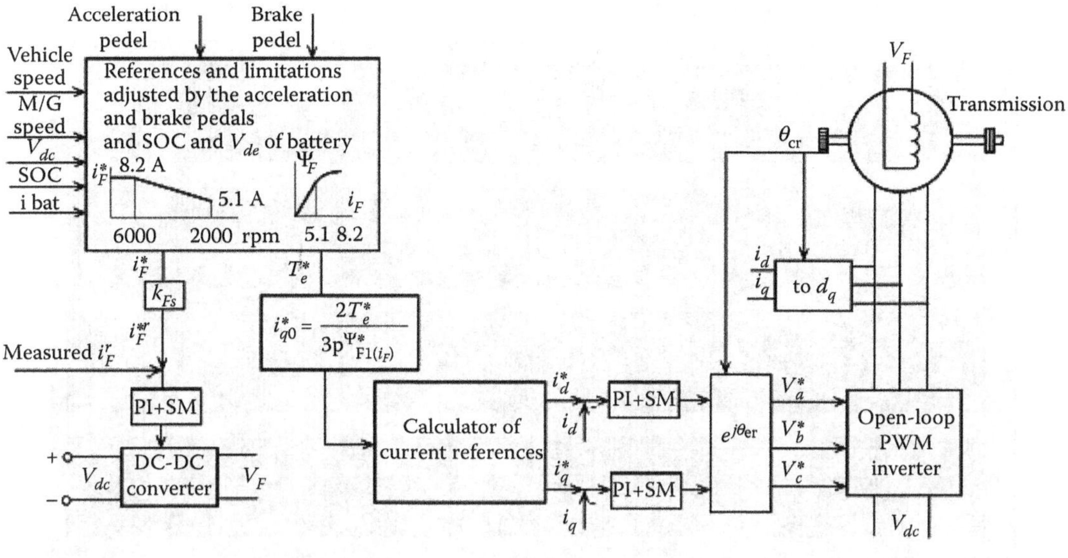

FIGURE 6.9 Encoder FOC of CP-SM.

If $V_s^* < \frac{V_{dc}}{\sqrt{2}}$ for motoring mode, a negative I_d^* is prescribed:

$$I_d^* = -\left(V_s^* - \frac{V_{dc}}{\sqrt{2}}\right) \cdot \left(k_p + \frac{k_i}{s}\right) + k_m \,\text{sign}\left(V_s^* - \frac{V_{dc}}{\sqrt{2}}\right) \tag{6.46}$$

Now, if $V_s^* < (V_{dc}/\sqrt{2})$ for generating mode, a positive I_d^* is ordered through the loop in Equation 6.46.

The slight increase for motoring and decrease in generating may be required to allow for the 10%–15% reluctance torque contributions.

For the 100-kW, 6000-rpm, CP-SM design already mentioned earlier in the chapter and with $V_{dc} = 500$ V, $R_s = 0.213\ \Omega$, $R_{exc} = 142.1\ \Omega$, $L_d = 1.515$ mH, $L_q = 1.6$ mH, stator leakage inductance $L_{sl} = 0.303$ mH, excitation leakage inductance $L_{el} = 2.09$ H, the above-mentioned FOC control (Figure 6.9) was implemented with results as in Figures 6.10 through 6.12. Figure 6.10 shows results at 2000 rpm in motoring at 100 kW in terms of V_α, V_β, I_α, I_β, I_F, P_{mech}, P_{input}.

Similar results at 100 kW for motoring and generating at 6000 rpm are given in Figures 6.11 and 6.12.

The above results show the control capable of handling the 2000–6000 rpm speed range at 100 kW. Full voltage is needed at 6000 rpm for full power, as expected, and a negative I_d^* is implemented; otherwise, $I_d^* = 0$ FOC. The response in field current and powers are stable.

Note: DTFC and scalar controls (V-f, I-f with stabilizing loop [as developed for PM-RSM]) may also be applied to CP-SMs with DC excitation or PM excitation.

6.9 THE HOMOPOLAR–SYNCHRONOUS MOTOR

Two typical topologies of H-SM are presented in Figure 6.13a and b.

The two configurations are characterized as follows:

In configuration a) the DC field does not use a special magnetic circuit but overlaps the AC field in the laminated core of the stator, which is now made of two identical parts with an AC winding that goes through both of them or two AC windings connected in series or in parallel.

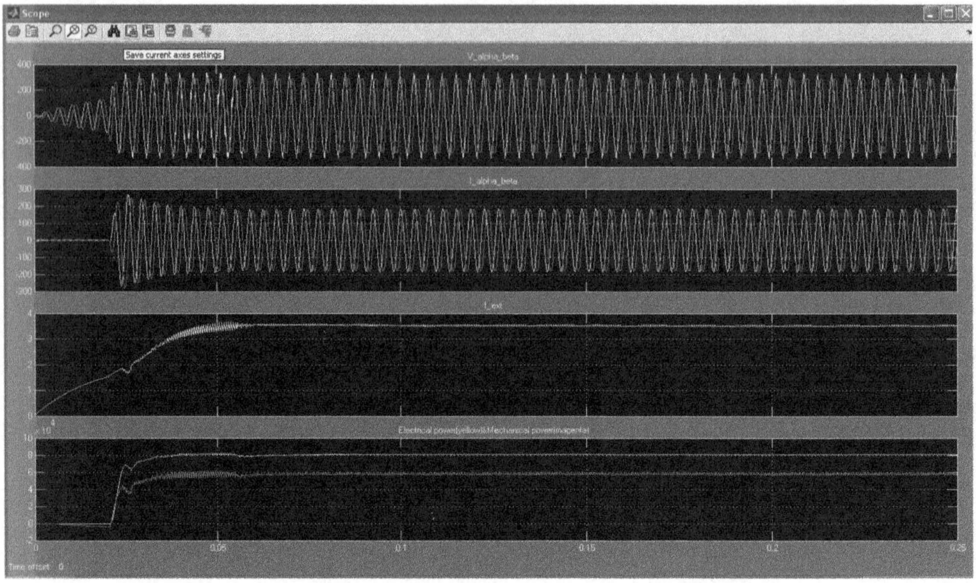

FIGURE 6.10 Motoring at 2000 rpm, 100 kW, 500 V_{dc} with the excited-PM-CP-SM.

Configuration a) looks simpler, but the AC coils have longer end connectors and thus AC copper losses are larger; also, DC coil excitation copper losses end up larger, as the diameter of the DC coil is larger.

Configuration b), by overlapping DC and AC fields, is exposed to larger magnetic saturation levels unless the stator core is larger.

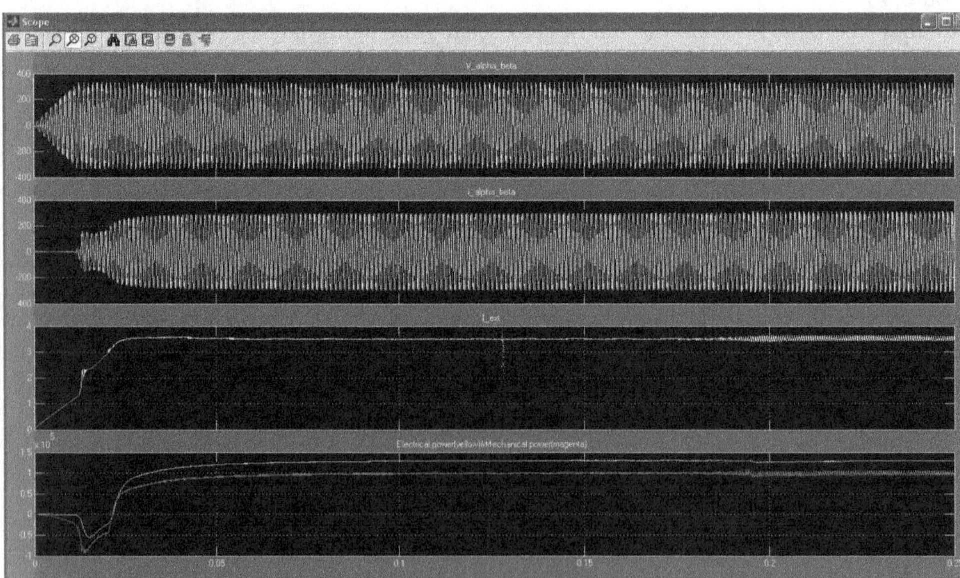

FIGURE 6.11 Motoring at 6000 rpm, 100 kW, 500 V_{dc} with the excited-PM-CP-SM.

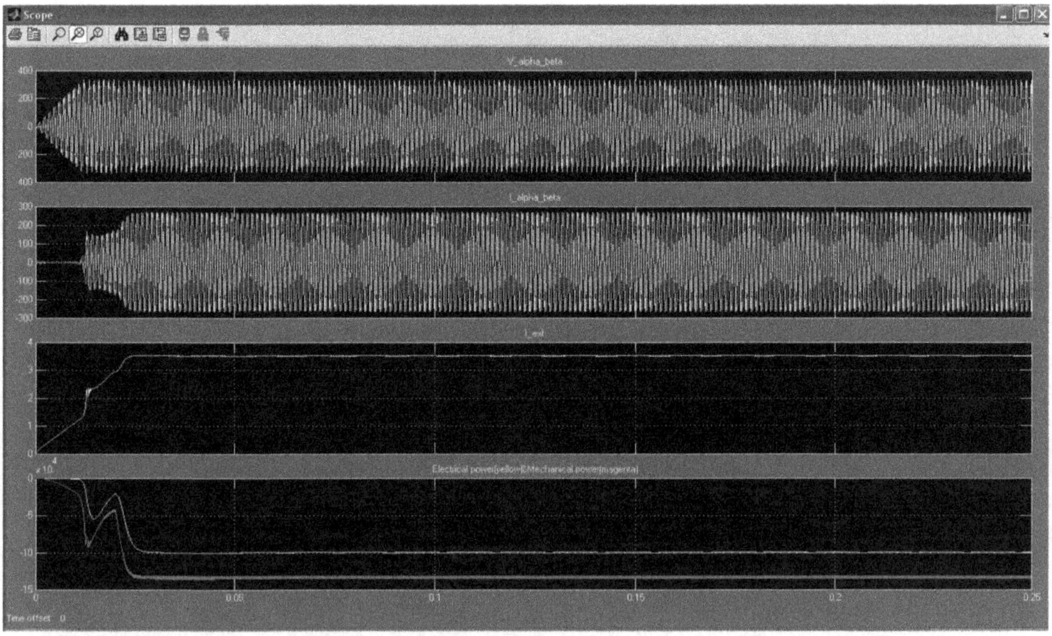

FIGURE 6.12 Generating at 6000 rpm, 100 kW, 500 V_{dc} with the excited-PM-CP-SM.

Configuration b) needs a special heavy/voluminous solid iron additional core that "embraces" the laminated AC core provided with an AC winding (even a tooth-wound three-phase winding may be used to reduce AC copper losses and the total length of the stator).

Configuration b) does not transmit any DC homopolar magnetic field through the single AC stator core and thus magnetic saturation is smaller and so are the AC coil ends; the rotor structure is simple (segmented).

If the DC coils are placed at a smaller diameter (closer to the rotor diameter), the DC excitation losses will be smaller than for configuration a).

Both configurations may be built with external rotors, but only configuration a) in Figure 6.13 seems competitive/practical.

Which of the two configurations is finally better depends on the application specifications.

Note: Concerning the feasibility of replacing the DC excitation stator coils with stator PMs, no configuration offers enough space in the solid iron stator core for PMs to use Ferrite-PM flux concentration (Figure 6.14). Other EM configurations (such as flux switching or flux reversal machines) offer such opportunities for convenient PM location in the stator.

H-SMs have not come close to market status on a large scale, and they show small torque/volume (weight). They may be used for special high-speed (energy-storage) applications or in small hydrogenerators where the rugged solid iron rotor may be placed above the integrated turbine rotor in one piece. An H-SM is basically an SM that can be considered as such. Consequently, we end here this short treatment of H-SMs, referring the reader to some representative literature [10–13]. However, starting from the application of linear H-SMs for integrated propulsion and levitation on magnetic levitations (MAGLEVs), we may imagine the multiple large-diameter rotary H-SM made of sectors—used, say, with a hydroturbine-integrated rotor for self-suspending the entire rotor by controlling the DC excitation accordingly.

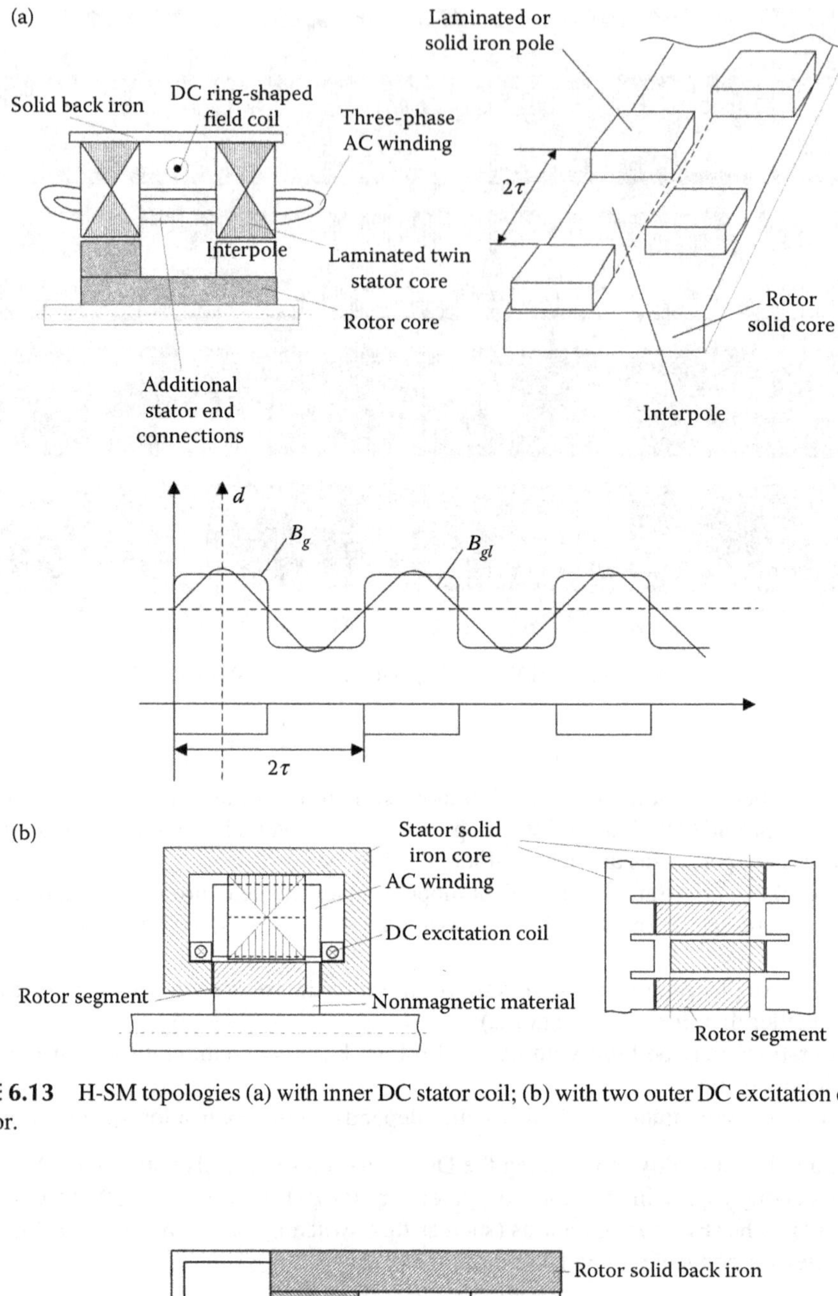

FIGURE 6.13 H-SM topologies (a) with inner DC stator coil; (b) with two outer DC excitation coils and single AC stator.

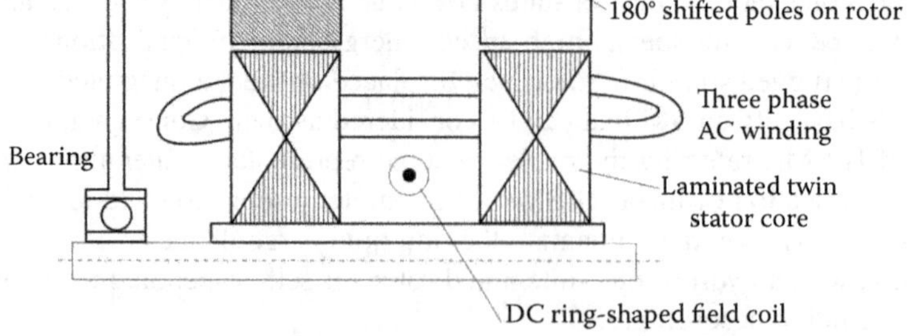

FIGURE 6.14 Outer rotor H-SM for inertia battery.

6.10 SUMMARY

- Claw-pole rotor synchronous machines with more than eight poles are characterized by at least two times the reduction of DC excitation losses in comparison with pole-by-pole N S N S DC excitation on regular SMs with same number of poles.
- This reduction of DC excitation power from 3% of SM rated power by more than two times increases efficiency and reduces the power transfer to rotor system size and cost and the rotor temperature, which has been shown recently to be the bottleneck in the thermal design of a regular SM with contactless power transfer to the rotor for 100-kW HEV applications [14].
- A high number of poles and a large diameter in a pancake-shaped machine are required for CP-SM to be competitive in high-torque low-speed applications (hundreds to tens of rpm, typical of hydro and wind generators and direct drives).
- A disk-shaped Ferrite-PM claw-pole rotor in SMs is shown in the chapter to produce in such large diameter/length machines an airgap multipolar flux density on the order of 0.7 T, which may prove sufficient for high torque (rotor shear force) density. It is clear that, in this way, PM weight may be saved due to the PM disk-shaped "blessing," similar to the blessing of the DC coil circularity in the Lundell machine (CP-SM). The DC coil may thus be eliminated with the CP-SM, turning it into a strong PM-emf ($E_1 > (0.7–0.75)V_{s\,max}$ IPMSM.
- The 3D shape of DC excitation (or PM) field lines through the CP-SM leads to the use of 3D-equivalent MEC or 3D-FEM approaches, to calculating airgap flux density, emf E_1, inductances L_{dm}, L_{qm}, and torque T_e.
- The nonlinear 3D-MEC method determines the performance rather satisfactorily in less than 10% of the computation time of 3D-FEM and thus the optimal design code for CP-SM preferably uses the 3D-MEC method for modeling the machine. 3D-FEM may be used to validate the 3D-MEC optimal design.
- Such an optimal design code with multiple-term objective function was presented, with representative results and 3D-FEM validation for an automotive-like CP-SM with and without interclaw PMs (to boost torque) in the 10 kW range; the results are encouraging.
- Based on the circuit dq model, suitable for steady state and transients, through Numerical Example 6.1, an 11-kW, 600-rpm, 211-Nm CP-SM with disk-shaped Ferrite PM claw-pole rotor is evaluated for steady and preliminary design with a potential 10 Nm/kg and 10 N/cm^2 rotor shear stress for a current density $j_{cu} = 10$ A/mm^2, at 58.44 Hz/14 poles, with efficiency well above 90%.
- Numerical Example 6.1 shows the potential of a Ferrite-PM CP-SM for high torque density ([5] claims a recent record 80 Nm/liter for 60 kW at 6000 rpm via IPMS with Sm_xCo_y magnets).
- The optimal code developed for the Ferrite-PM CP-SM at 11 kW and 600 rpm has confirmed Numerical Example 6.1's performance, leading to the conclusion that this machine should be investigated further, at least for low-speed, high-torque drive.
- Another numerical example (6.2) gives the results for DC-excited 3-MW, 15-rpm 1-MW, 150-kW, 100-kW, 600-rpm CP-synchronous generators for wind or hydro turbines, which show promising performance.
- Recently, another hybrid excited claw-pole rotor SM has been proved capable of doubling the emf by adding axial Ferrite-PMs [15].
- Though, basically, the control of CP-RSM resembles that of SM (if they have DC excitation) or of strong emf ($E_1 > 0.75V_{s\,max}$) IPMSMs, there are very few solutions that have been experimented with for CP-SM. An example related to an encoder FOC of a DC-excited CP-SM, designed for 100 kW from 2000–6000 rpm at 500 Vdc, gives good performance in terms of power, current, voltage, and field current responses over the entire speed-power range.

- DTFC and scalar controls (V/f with stabilizing loops) are still due in encoder or encoderless implementation for CP-SM variable-speed drives.
- The homopolar SM (DC and AC on stator and variable reluctance rotor) in two main implementations is introduced and discussed qualitatively. A detailed optimal design code and advanced control methods for H-SM are still due, as the literature on the subject is still sparse, though application to inertia batteries seems promising.
- The extension of homopolar SMs with stator PMs and variable reluctance rotor for high-speed bearingless motor drives [16] should also be considered, as in such a case, the low torque density of the machine does not lead to a bulky motor/kW, since forced cooling already increases the machine volume.

REFERENCES

1. I. Boldea, *Synchronous Generators*, 2nd edition chapter 2, book, CRC Press, Taylor & Francis Group, New York, 2016.
2. A. Reinap, M. Alaküla, Study of three-phase claw-to-claw pole machine, *IEMDC'03, 2003*, pp. 325–329.
3. R. Block, G. Hennenberger, Numerical calculation and simulation of claw pole alternator, *Record of ICEM–1992*, vol. 1, 1992, pp. 127–131.
4. I. Boldea, *Variable Speed Generators*, 2nd edition chapter 6, book, CRC Press, Taylor & Francis Group, New York, 2016.
5. M. Popescu, I. Folley, D. A. Staton, J. E. Gross, Multi-physics analysis of a high torque density motor for electric racing cars, *Record of IEEE-ECCE 2015*, pp. 6537–6544.
6. B. Simo, D. Ursu, L. Tutelea, I. Boldea, Automotive generator/motor with claw pole PM-less, NdFeB or ferrite IPM rotors for 10 and 100 kW a 6krpm: optimal design performance and vector control dynamics, Internal Report-UPT, 2014.
7. L. Tutelea, D. Ursu, I. Boldea, S. Agarlita, IPM claw-pole alternator system for more vehicle breaking energy recuperation, *JEE, UPT*, www.jee.ro (online only). vol. 13, no. 3, 2012, pp. 211–220.
8. F. Jurca, I. A. Viorel, K. Biro, Steady state behavior of a permanent magnet claw pole generator, *Record of OPTIM*, 2010, pp. 291–296 (IEEE xplore).
9. C. R. Bratiloveanu, DTC Anghelus, I. Boldea, A comparative investigation of three PM-less MW power range wind generator topologies, *Record of OPTIM 2012*, pp. 535–543 (IEEE xplore).
10. R. Rebhi, A. Ibala, A. Masmoudi, MEC-based sizing of a hybrid-excited claw pole alternator, *IEEE-Trans*, vol. IA–51, no. 1, 2015, pp. 211–223.
11. Z. Lou, K. Yu, L. Wang, Z. Ren, C. Ye, Two-reaction theory of homopolar inductor alternator, *IEEE Trans*, vol. EC–25, no. 3, 2010, pp. 677–679.
12. P. Tsao, An integrated flywheel energy storage system with homopolar inductor motor/generator and high-frequency drive, PhD thesis, University of California, Berkeley, 2003.
13. E. A. Erdelyi, Influence of inverter loads on the airgap flux of aerospace homopolar alternators, *IEEE Transactions on Aerospace*, vol. IA–3, no. 2, 1965, pp. 7–11.
14. C. Stancu, T. Ward, K. Rahman, R. Dawsey, P. Savagian, Separately excited synchronous motor with rotary transformer for hybrid vehicle application, *Record of ECCE 2014*, pp. 5844–5851.
15. G. Dajaku, B. Lehner, Xh. Dajaku, A. Pretzer, D. Gerling, Hybrid excited claw pole rotor for high power density automotive alternators, *ICEM-2016*, pp. 2536–2543 (IEEE xplore).
16. W. Gruber, M. Rothböck, R. T. Schöb, Design of a novel homopolar bearingless slice motor with reluctance rotor, *IEEE Trans*, vol. IA–51, no. 2, 2015, pp. 1456–1464.

7 Brushless Direct Current–Multiple Phase Reluctance Motor Modeling, Control, and Design

7.1 INTRODUCTION

Quite a few electric machine technologies without PMs, such as reluctance synchronous and DC-excited synchronous machine drives, are currently used, both for line-start (constant-speed) and variable-speed applications, where, however, for the latter case, a PWM converter to produce variable voltage amplitude and frequency is required.

The absence of PMs in RSMs and DC-excited SMs leads to lower cost but a lower power factor (for RSMs) and, respectively, to larger-volume machines with DC joule losses (temperature-limited) rotors fed either through a brush slip-ring device or a contactless (rotary transformer) power transformer to the rotor (less than 3% power rating, though).

A passive (windingless and PM-less) rotor as in RSMs but with a larger airgap and better PWM converter voltage utilization (better equivalent power factor) would eventually lead to a better drive at reasonable costs and performance. Such a machine should also be capable of working at low and high speeds due to a larger airgap that allows enough space for a reinforcing carbon fiber rotor ring. A departure from sinusoidal to trapezoidal current control is also expected to be beneficial in a multiphase machine ($m > 3$ phases). A higher torque density is expected in a given geometry by using m_F phases as excitation phases, and the remaining $m_T = m - m_F$ phases are torque phases. Each phase plays both roles in relation to the distributed magnetic anisotropy (MFBA or ALA) rotor, similar to the case of RSMs, but with a higher airgap.

From such targets, the hereby called BLDC-MRM was born in 1986 [1]. It stems, in our interpretation, from a DC brush motor stripped of its DC exitation, but with the brushes moved to the corner of the stator poles; the latter is provided with multiple flux barriers to reduce armature reaction (Figure 7.1a). Finally, the stator and rotor are inversed, and the stator, now with $m > 3$ phases ($2p$ poles) and diametrical ($y = \tau$) coils, is fed from a multiphase PWM inverter triggered by rotor position (Figure 7.1b).

The multiphase inverter may be built as a null point configuration (Figure 7.2b) or as multiple one-phase inverter bridges in parallel (Figure 7.3).

The number of phases $m \geq 3$ may be an odd number (5, 7, 9, 11, 13, 15…) or an even number (6, 8, 10, 12, 14, 16…).

For trapezoidal phase current control, the summation of phase currents (the null current in null current inverter-fed motors is the DC link current in $m \times$ 1-phase inverters) is no longer zero (as it is with symmetrical windings with $m \geq 3$ phases and sinusoidal current).

It has been shown that for m, an odd number of phases, the null current (or phase current summation) in amplitude is one time the phase current but two times the phase current for m as an even number.

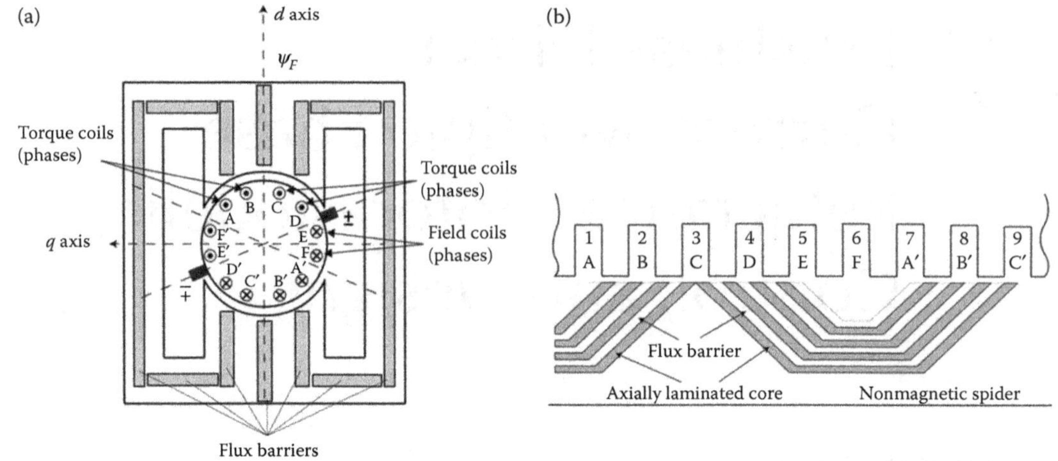

FIGURE 7.1 Derivation of BLDC-MRM from a DC brush machine ($m = 6$ phase, $2p = 2$poles): (a) original DC brush machine; (b) BLDC-MRM counterpart.

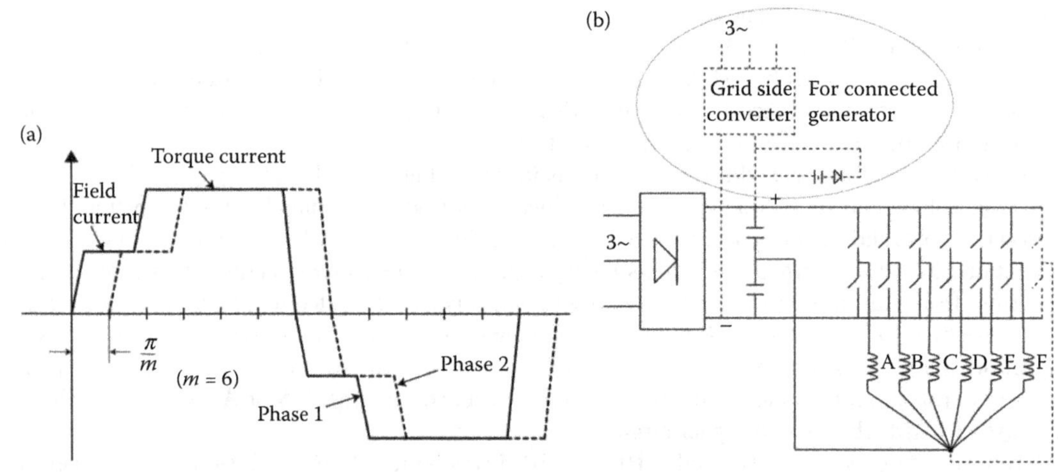

FIGURE 7.2 Phase 1 and 2 current waveform (a); and a null-point six-phase inverter (b).

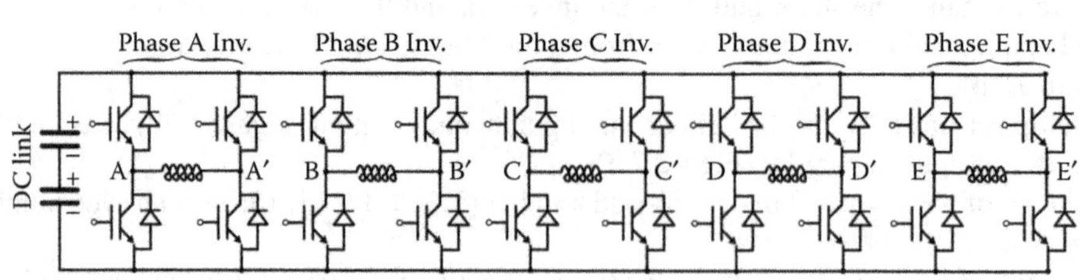

FIGURE 7.3 Five one-phase inverter topology.

Irrespective of the number of phases, the ones counted 1, 3, … have their winding ends swapped, and the voltage applied is (−V) so that power on that phase remains positive, but the null current (or phase current summation) is reduced to the values mentioned above.

So, apparently, an odd-phase current machine ($m = 2k + 1$) is better than an even-phase current machine ($m = 2k$). However, for simple redundancy improvements, an $m = 3k$ machine allows the segregation of $m/3$ subinverter units and thus at least one of them can continue the operation in case of one phase fault. Such an operation at lower power requires small changes in control.

The airgap g of the machine needs not to be very small (as in RSMs) in order to quicken the current commutation when one phase after the other switches roles (from a field phase, when its coils are placed between rotor poles, to a torque phase, when the coils "fall" under rotor poles [Figure 7.1b]). But the airgap g should not be too large, either, because then too much of an mmf (current I_F) is required to produce full flux density in the airgap (0.7–0.85 T).

As the stator slotting has to be uniform, because all phases switch between field and torque roles, an optimum airgap should be found. This airgap mitigates between sufficiently quick current commutation at maximum speed and reasonable copper losses in the field phase stages.

For motoring, when the field current top i_F and the torque current top i_T levels are equal to each other ($i_F = i_T$), the operation at the MTPA condition is met implicitly without any online decision making in the control so typical in sinusoidal current-controlled AC machines. In general, the torque expression with dual flat-top trapezoidal phase current (i_F, i_T) is simply:

$$T_e = p\psi_F i_T (m - m_F) \tag{7.1}$$

The flux linkage by the m_F field phases in a torque phase is Ψ_F (per phase):

$$\psi_F = \frac{\mu_0 \cdot n_s \cdot I_F \cdot m_F}{2 \cdot g \cdot K_c \cdot (1 + K_{sd})} \cdot \left(1 - \frac{m_F}{m}\right) \cdot \tau \cdot l_{\text{stack}} \cdot n_s \cdot p_1 \tag{7.2}$$

where: n_s—turns /coil (slot), p_1—pole pairs, g—airgap, K_c—carter factor, K_{sd}—magnetic saturation in axis d (for the excitation field flux lines: phases E and F in Figure 7.1b).

There is no apparent need to use the dq model, as the stator mmf is no longer a traveling wave and the currents are not sinusoidal.

Phase coordinates may be adopted for simplicity, despite the fact that the self and mutual inductance all vary with rotor position.

7.2 TORQUE DENSITY AND LOSS COMPARISONS WITH INDUCTION MOTORS

Torque density and loss comparison with the induction motor (IM) may be performed if the actual (trapezoidal) current shapes are replaced with rectangular ones (Figure 7.4) [3].

The ideal excitation flux ends up trapezoidal (Figure 7.4b), while the emf E becomes rectangular.

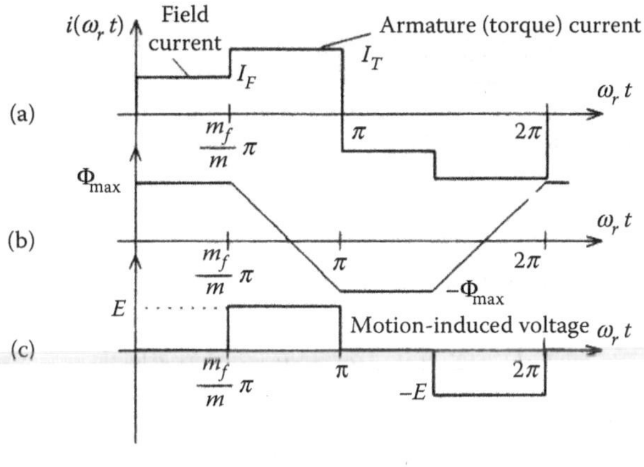

FIGURE 7.4 Ideal phase current (a); phase flux (b); and emf (c).

The ideal electromagnetic power P_e is thus simply:

$$P_e = m_T \cdot E \cdot i_T;\ E = 2 \cdot B_{\text{gav}\,0} \cdot w_1 \cdot l_{\text{stack}} \cdot 2 \cdot p \cdot \tau \cdot n \tag{7.3}$$

where $B_{\text{gav}\,0}$ is the average airgap flux density in a torque phase, $w_1 = n_s \cdot p$ turns per phase, τ—pole pitch, n—speed in rps. With two-level rectangular flat bipolar ideal currents, the RMS value of stator current I_{RMS} is [3]:

$$I_{\text{RMS}} = i_T\sqrt{\frac{m_T + m_F\left(I_F/I_T\right)}{m_T + m_F}} \tag{7.4}$$

But the current sheet A_s is:

$$A_s = \frac{2(m_T + m_F) \cdot w_{12 \cdot p_1}}{2 \cdot p_1 \cdot \tau} \cdot I_{\text{RMS}} \tag{7.5}$$

The tangential specific force (rotor shear stress), f_T is:

$$f_T = \frac{P_e}{2 \cdot \pi \cdot n \cdot p \cdot \left(2 \cdot p \cdot l_{\text{stack}}\right)\tau} = K_I \cdot B_{g\,\text{avg}} \cdot A_s \cdot \left(\text{N/m}^2\right) \tag{7.6}$$

with K_I from Equations 7.4 through 7.6 as:

$$K_I = \frac{m_T}{\sqrt{(m_T + m_F)\left(m_T + \left(I_F/I_T\right)^2 \cdot m_F\right)}} \tag{7.7}$$

Let us denote L_{dm}, L_{qm} as the magnetization inductances of a phase when the latter is placed with its field in axis d (field phase), and, respectively, in axis q (torque phase). The expression of average airgap flux density $B_{g\,\text{avg}}$ (also used in Equation 7.2) and produced by the field phases B_{gk} that also account for armature reaction is [3]:

$$B_{g\,\text{avg}} = K_a \cdot K_f \cdot B_{gk} \tag{7.8}$$

with

$$K_a = \frac{1}{1 + \dfrac{L_{qm}}{L_{dm}} \cdot \dfrac{m_T}{2 \cdot m_F} \cdot \dfrac{I_T}{I_F}};\ K_f = \frac{4 \cdot m_F \cdot I_F/I_T - 1}{4 \cdot m_F \cdot I_F/I_T} \tag{7.9}$$

Consequently, the tangential specific force f_T becomes:

$$f_T = K_I \cdot K_a \cdot K_f \cdot B_{gk} \cdot A_s\,[\text{N/m}^2] \tag{7.10}$$

For an induction motor, this specific force f'_T is:

$$f'_T = \frac{1}{\sqrt{2}} k_{w1} \cdot B_{gk1} \cdot A'_{s1} \cdot \frac{\cos\varphi_1}{1 + \cos\varphi_1} \tag{7.11}$$

with k_{w1}—stator winding factor, φ_i—internal power factor angle, and A'_{s1}—the total current sheet. Adopting the same current sheet A_s, A'_{s1} and resultant airgap flux density B_{gk}, the ration of specific forces f_T/f_T' for the BLDC-MRM and IM is [3]:

$$\frac{f_T}{f_{T'}} = \sqrt{2}\frac{K_a \cdot K_T \cdot K_i}{K_{w1}}\frac{(1+\cos\varphi_i)}{\cos\varphi_i} \tag{7.12}$$

As an example, for $m = 6$ ($m_F = 2$, $m_T = 4$), $I_F/I_T = 1$, $L_{qm}/L_{dm} = 0.2$, and $K_{w1} = 0.933$, cos φ_i, we obtain $f_T/f_{T'} = 1.558$.

Alternatively, if the same power (torque) density in Equation 7.3 is targeted, the ratio of current sheets $A_s/A'_{s1} = f_{T'}/f_T = 0.642$.

So the ratio of copper losses is: $P_{co}/P_{co'} = \left(A_s/A'_{s1}\right)^2 = 0.412$. Consequently, the BLDC-MRM with six phases, in our case, turns to be, ideally, superior to three-phase IM either in torque density, copper losses (efficiency), or both (at lower levels). The passive (windingless and PM-less) rotor in BLDC-MRM leads to a more efficient cooling (exclusively in the stator) and thus a higher peak torque density.

7.3 CONTROL PRINCIPLES

The consideration of stator slotting, rotor MFB or ALA structure, magnetic saturation, current trapezoidal two-level bipolar phase currents may lead to a reduction in torque density or efficiency improvements calculated above. However, they are still worth revisiting by more practical (technical) theories/models, but not before illustrating here the simplicity of the control system (Figure 7.5).

In Figure 7.5, i_F^* is made dependent on speed, allowing for flux weakening (by reducing i_F) with speed. The torque reference current value i_T^* is the output of the speed controller; a limiter may be added, especially for low speeds.

The current reference waveforms, whose two top values are i_F^* and i_T^*, are given as a function of a rotor position whose zero values are related to phase a^* in rotor axis d.

Let us note that this simple control may be augmented to correct, say $i_F^*(w_r)$ or $i_F^*(T_e^*, w_r)$, to operate at $i_F^* = i_T^*$ or, when the voltage limit is reached, i_F^* is further reduced and i_T^* is limited.

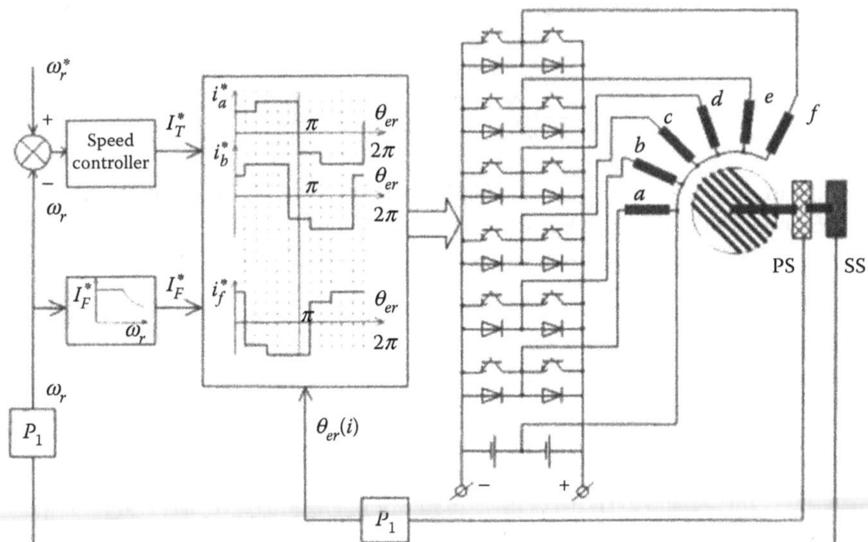

FIGURE 7.5 Generic control system of encoder BLDC-MRM ($m = 6$ phases, $m_T = 4$ phases, $m_F = 2$ field phases).

Finally, the machine may switch from motoring to generating or back by simply changing the sign of i_T^* in Figure 7.5. In principle, we could change, instead, the sign of i_F^*, but the transients will be slower and the torque response sluggish, as it is in the separately excited DC brush machine.

7.3.1 A Technical Theory of Brushless Direct Current–Multiple Phase Reluctance Motor by Example

Let us consider the ideal specifications in Figure 7.6.

To illustrate the technical (very preliminary) theory here, let us consider a seven-phase machine ($m_F=2$, $m_T=5$), $V_{DC}=400$ V, $2p=6$ poles ($f_{1\max}=300$ Hz, at 6000 rpm), outer stator diameter $D_{os}=0.3$ m, airgap $g=2$ mm, and torque density $t_v=45$ Nm/liter at T_{eb}. A typical application would be a traction drive, but, in a similar way, dedicated specifications could be assigned for, say, an autonomous DC output generator or a variable-speed hydro/wind generator, and so on.

7.3.1.1 Solution

Let us consider $i_T=i_F$ up to base speed. The stator stack length l_{stack} is straightforward:

$$l_{\text{stack}}=\frac{P_{eb}/(2\cdot\pi\cdot n_b)}{t_v\cdot\pi\cdot\dfrac{D_{os}^2}{4}}=\frac{50/(2\cdot\pi\cdot 25)}{45\cdot\pi\cdot\dfrac{0.3^2}{4}}=0.1\text{ m} \tag{7.13}$$

Considering an airgap flux density produced by the two field phases ($m_F=2$) of 0.75 T, we obtain:

$$B_{gF}=\frac{\mu_0\cdot m_F\cdot i_{Fb}\cdot n_s}{2g\cdot K_C(1+K_{SF})} \tag{7.14}$$

with $K_C=1.12$, $K_{SF}=0.25$:

$$n_b i_{Fb}=\frac{0.75\cdot 2\cdot 2\cdot 10^{-3}\cdot 1.12\cdot 1.25}{1.256\cdot 10^{-6}\cdot 2}=1.672\cdot 10^3\text{ Aturns/slot} \tag{7.15}$$

with $2p_1=6$ poles and a stator core interior diameter $D_{is}=0.6\cdot D_{os}=0.6\cdot 300=180$ mm, the pole pitch of coils τ is:

$$\tau=\frac{\pi\cdot D_{is}}{2p_1}=\frac{\pi\cdot 180}{2\cdot 3}=94.2\text{ mm} \tag{7.16}$$

The aspect ratio of stator core $l_{\text{stack}}/\tau=100/94.2=1.06$ is close to an intuitive low copper loss case (due to reasonably long coil ends).

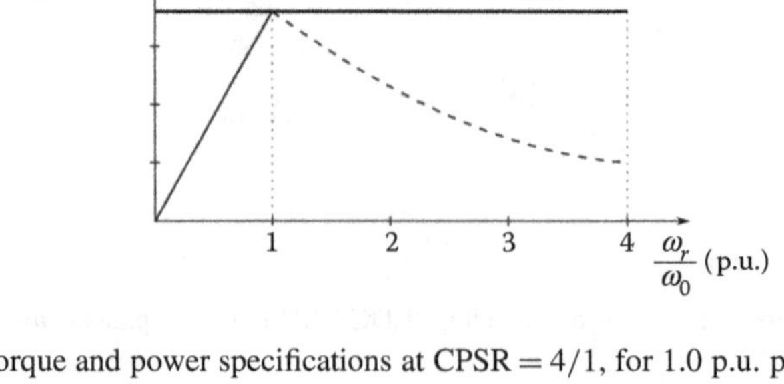

FIGURE 7.6 Typical torque and power specifications at CPSR = 4/1, for 1.0 p.u. power.

The stator yoke depth h_{ys} is:

$$h_{ys} = \frac{B_{g\,load}}{B_{ys}}\frac{\tau}{\pi} = \frac{1.1 \cdot 0.75}{1.5} \cdot \frac{94.2}{\pi} = 16.5 \text{ mm} \tag{7.17}$$

So the height left for the stator slot h_{st} becomes:

$$h_{st} = \frac{D_{os} - D_{is}}{2} - h_{is} = \frac{300 - 180}{2} - 16.5 = 43.5 \text{ mm} \tag{7.18}$$

Considering the wedge and slot neck height $h_{s0} + h_w = 2.5$ mm, the useful height of the trapezoidal-shape stator slot h_{su}:

$$h_{su} = h_{st} - (h_{s0} + h_w) = 43.5 - 2.5 = 41 \text{ mm} \tag{7.19}$$

Here comes the first (of three) verifications, which is related to torque-producing capability for $i_F^* = i_T^*$. First, the base torque T_{eb} is calculated:

$$T_{eb} = \frac{P_{eb}}{2 \cdot \pi \cdot n_b} = \frac{50 \cdot 10^3}{2 \cdot \pi \cdot 25} = 318 \text{ Nm} \tag{7.20}$$

But

$$\begin{aligned} T_{ebc} &\approx B_{gF} \cdot l_{\text{stack}} \cdot n_s \cdot i_{Fb} \cdot m_T \cdot 2 \cdot p_1 \cdot \frac{D_{is}}{2} \\ &= 0.75 \cdot 0.1 \cdot 1672 \cdot 5 \cdot 6 \cdot \frac{0.18}{2} \\ &= 338 \text{ Nm} > 318 \text{ Nm} \end{aligned} \tag{7.21}$$

As $T_{ebc} > T_{eb}$, there is some guarantee that the machine may deliver the required base (peak) torque at the calculated stator mmf (1672 Aturns/slot).

Note: A more realistic modeling, based on MEC or 2D-FEM, has been applied to verify in detail this torque claim in detail; see Equation 7.21.

The second verification is related to the slot area sufficiency for the base (peak) torque in terms of current density (which then defines the copper losses).

The slot pitch at the airgap τ_s is:

$$\tau_s = \tau/m = 94.2/6 = 15.7 \text{ mm} \tag{7.22}$$

By adopting a rectangular tooth with a width $b_{ts} = 8.7$ mm, the lower width of the slot is: $b_{s1} = \tau_s - b_{ts} = 15.7 - 8.7 = 7$ mm (Figure 7.7). The slot width b_{s2} is:

$$b_{s2} = \frac{\pi \cdot (D_{os} - 2 \cdot h_{ys})}{2 \cdot p_1 \cdot m} - b_{ts} = \frac{\pi \cdot (320 - 38)}{6 \cdot 6} - 7 = 18 \text{ mm} \tag{7.23}$$

The slot useful area A_{slot} emerges as:

$$A_{\text{slot}} = h_{su} \cdot \frac{b_{s1} + b_{s2}}{2} = 41 \cdot \frac{7 + 18}{2} = 512.5 \text{ mm} \tag{7.24}$$

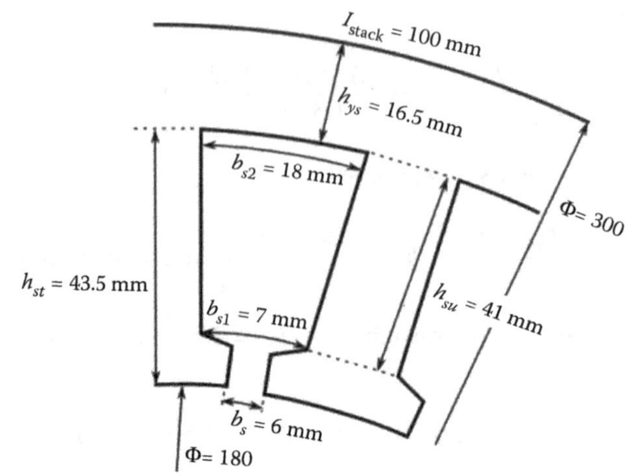

FIGURE 7.7 Stator slotting.

Consequently, the current density j_{con} (with perfectly rectangular bipolar current $i_F^* = i_T^*$) is:

$$j_{con} = \frac{n_s \cdot i_{Fb}^*}{A_{slot} \cdot k_{fill}} = \frac{1672}{512.5 \cdot 0.45} = 7.25\ \text{A/mm}^2 \tag{7.25}$$

The length of a turn of the stator coils l_c calculates:

$$\begin{aligned} l_c &\approx 2 \cdot l_{stack} + 2 \cdot 1.33 \cdot \tau \\ &= 2 \cdot 0.100 + 2 \cdot 1.33 \cdot 0.0492 \\ &= 0.4505\ \text{mm} \end{aligned} \tag{7.26}$$

So the copper losses at base (peak) torque are:

$$\begin{aligned} p_{cob} &\approx m \cdot \rho_{cob} \cdot n_s \cdot l_c \cdot p_1 \cdot i_{Fb} \cdot j_{cob} \\ &= 7 \cdot 2.3 \cdot 10^{-8} \cdot 0.4505 \cdot 1672 \cdot 3 \cdot 7.25 \cdot 10^{-6} \\ &= 2637.6\ \text{W} \end{aligned} \tag{7.27}$$

This represents 5.27% of base power P_{eb}. Assuming 1500 rpm, a reasonable efficiency (after core and mechanical losses are still to be added) is expected.

The peak flux density produced by armature reaction at the corner of the rotor poles in the stator $B_{agq}^b(i_{Tb}^* = i_{Fb}^*)$:

$$B_{agq}^b = \frac{\mu_0 \cdot m_F \cdot (n_s \cdot i_{Fb})}{2 \cdot g + m_{FB\,p} \cdot b_b} = \frac{1.256 \cdot 10^{-6} \cdot 5 \cdot 1672}{2 \cdot 2 \cdot 10^{-3} + 6 \cdot 4.5 \cdot 10^{-3}} = 0.338\ \text{T} \tag{7.28}$$

The armature reaction flux density is reasonably small in comparison with the 0.75 T of flux phases. Also, the ratio L_{qm}/L_{dm}:

$$\frac{L_{qm}}{L_{dm}} \approx \frac{2 \cdot g}{2 \cdot g + m_{FB\,p} \cdot b_b} = \frac{2 \cdot 2}{2 \cdot 2 + 6 \cdot 4.5} = 0.129 \tag{7.29}$$

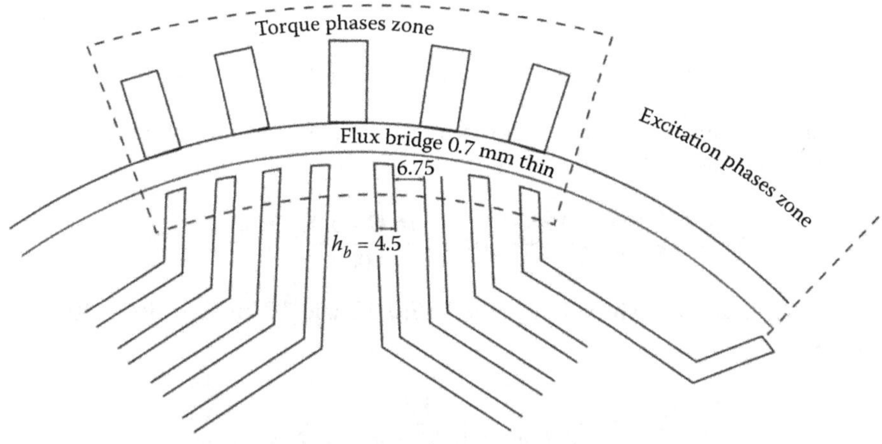

FIGURE 7.8 The rotor geometry.

The flux bridges on the rotor (Figure 7.8) exist to maintain the integrity of rotor laminations and might increase this ratio; thus, methods used for the RSM may be used here to calculate the contribution of flux bridges to the increase of L_{qm}.

Also, we may infer from above that even doubling the torque current (now $i^*_{\text{Toverlaod}} = 2 \cdot i^*_{Fb}$) will double the torque and the machine will still be able to cope with the situation, albeit being more saturated.

The technical theory so far may be extended to calculate the number of turns per coil n_s.

A very approximate way to do it is to use the equation of a torque phase between commutations but with current chopping (still) at base speed and maximum speed. Thus (as for regular BLDC PM motors!):

$$V_{DC} = R_s \cdot i_{Tb} + E + \Delta E_{\text{chopping}} \tag{7.30}$$

$$\begin{aligned} E(\omega_b) &= 2 \cdot p \cdot n_s \cdot l_{\text{stack}} \cdot B_{gF} \cdot \pi \cdot D_{is} \cdot \underline{n_b} \\ &= 2 \cdot 3 \cdot 0.1 \cdot 0.75 \cdot \pi \cdot 0.18 \cdot 25 \cdot n_s \\ &= 2.437 \cdot n_s \end{aligned} \tag{7.31}$$

But

$$R_s i_{Tb} \approx \frac{p_{cob} \cdot n_s}{m(i^*_{Tb} \cdot n_s)} = \frac{2637.6}{7 \cdot 1672} \cdot n_s = 0.225 \cdot n_s \tag{7.32}$$

To calculate the chopping voltage ΔE, we need the torque phase equivalent inductance L_q:

$$\begin{aligned} (L_q)_{\text{phase}} &\approx 1.2 \cdot (L_{qm})_{\text{phase}} \\ &= 1.2 \cdot L_{dm} \cdot \frac{L_{qm}}{L_{dm}} = 1.2 \cdot L_{dm} \cdot 0.129 \end{aligned}$$

$$\begin{aligned} L_{dm} &\approx \frac{6 \cdot \mu_0 \cdot (n_s \cdot p)^2 \tau_p \cdot l_{\text{stack}}}{\pi^2 \cdot g \cdot K_C \cdot (1 + K_{SF}) \cdot p} \\ &= \frac{6 \cdot 1.256 \cdot 10^{-6} \cdot (3 \cdot n_s)^2 \cdot (5/7) \cdot 94.2 \cdot 10^{-3} \cdot 0.1}{\pi^2 \cdot 2 \cdot 3 \cdot 1.12 \cdot 1.25 \cdot 10^{-3}} \\ &= 5.51 \cdot 10^{-6} \cdot n_s^2 \end{aligned} \tag{7.33}$$

So

$$L_q \approx 1.2 \cdot 0.129 \cdot 5.51 \cdot 10^{-6} \cdot n_s^2 = 0.853 \cdot 10^{-6} \cdot n_s^2 \tag{7.34}$$

But if the chopping frequency of the current is $f_{ch} = 5$ kHz and the hysteresis band of current chopping is 1% of the rated (base) current, then:

$$\Delta E = L_q \cdot \frac{\Delta i}{T} = \frac{0.853}{\pi^2} \cdot 10^{-6} \cdot \frac{1672 \cdot 0.01}{5000^{-1}} \cdot n_s = 0.713 n_s \tag{7.35}$$

So, using ΔE, $R_s i_{TB}$, and $E(\omega_s)$ from Equations 7.35, 7.32, and 7.31 in Equation 7.30, the number of turns per coil n_s is:

$$n_s = \frac{400}{2.437 \cdot n_s + 0.225 \cdot n_s + 0.713 \cdot n_s} \approx 118 \text{ turns/coil (slot)} \tag{7.36a}$$

Note: We introduced for Vdc 400 V, which means that we assumed implicitly that the 6 × 1 phase inverter is used. Now we may verify if, at max speed and its torque for CPSR, the voltage is sufficient. Let us suppose that we reduce the field current $i_F^* \; n_{max}/n_s = 4$ times, but keep the torque current to provide a ¼ torque (CPSR = 4/1).

By applying the same Equation 7.30, the resistive voltage drop stays the same and so does the emf E (speed increases by four times, but the flux (field current) decreases by four times also.

The chopping voltage drop also stays the same, as i_T^* is the same and so are f_{ch} and $\Delta i_T = 1\%$ of the base torque current.

So, in this case, the operation of torque phases is not posing any problem. For the same (base) power at max. speed, even the copper losses will be somewhat reduced, as i_F^* is reduced by four times, while i_T^* stays the same. A similar approach may be applied for the phases when they commutate, which will find, probably, that the number of turns/coil n_s has to be somewhat diminished.

However, still, during current commutation from torque to excitation mode at maximum speed, another voltage equation stands:

$$V_{dc} > \left| L_d \frac{di_{\text{phase}}}{dt} \right| \approx \left| \frac{L_d\left(i_{T\max} - \left(-i_{T\max}/4\right)\right)}{n_{\max}^{-1} \cdot \left(2p \cdot m \cdot 2\right)^{-1}} \right| = \frac{1.2 \cdot 5.51 \cdot 10^{-6} \cdot 118}{100^{-1} \cdot (42 \cdot 2)^{-1}} \cdot \frac{5}{4} \cdot 1672 = 137 \text{ V} \tag{7.36b}$$

Apparently, from Equation 7.36b, the current commutation at maximum speed is guaranteed, but the motion-induced voltage in the commutated phase is still neglected. As there is a generous voltage reserve in Equation 7.36b and as commutation duration was considered half the stator slot pitch angle time, the result still holds by being on the conservative side.

It is hoped that the technical theory so far has opened a window in understanding how this machine works. It is thus now time for a thorough FEM investigation to bring the necessary precision in results and further clues to various phenomena in BLDC-MRMs.

7.4 FINITE-ELEMENT MODEL–BASED CHARACTERIZATION VERSUS TESTS VIA A CASE STUDY

As the machine has a passive rotor and both stator and rotor cores are laminated, only static 2D-FEM is used to characterize the machine, emphasizing a 35-Nm laboratory prototype of 5 phases, 30 slots, and 6 poles (Figure 7.9):

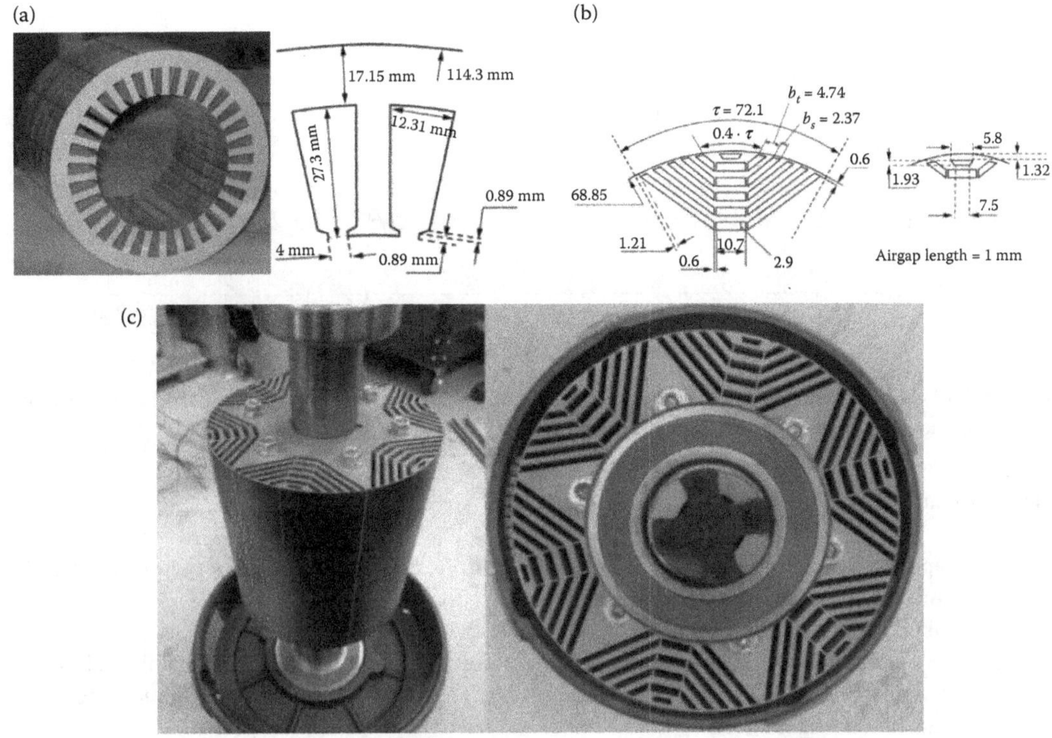

FIGURE 7.9 Five-phase six-pole BLDC MRM, (a) stator; (b) rotor; (c) experimental rotor.

Magnetic field distribution at load (Figure 7.10) [6].
Magnetic field distribution at full load (Figure 7.11) [9].
Self and mutual phase a inductances versus rotor position at rated current $I^*_{Fb} = I^*_{Tb}$ (Figure 7.12) [9].
Average torque and torque density versus torque phase current density i_T for a few field current densities i_F [9] (Figure 7.13).
Torque pulsations have been explored on a six-phase ALA-rotor prototype [9] at standstill with all phases in series (Figure 7.14).
Total torque ripple with current actual trapezoidal waveforms are shown in Figure 7.15 [9].

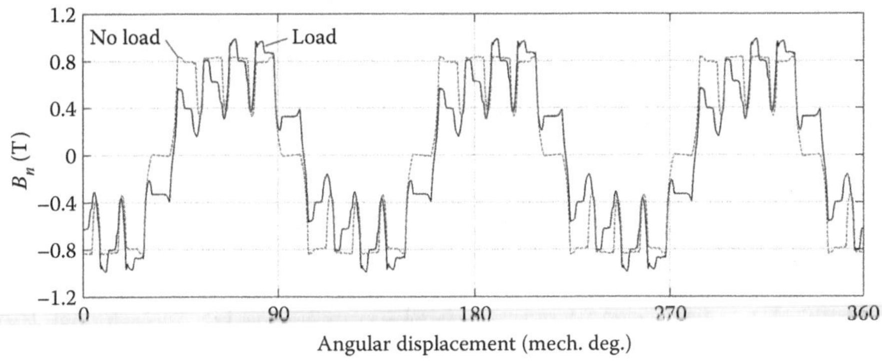

FIGURE 7.10 FEM-calculated radial component of airgap flux density at no load ($i_F = 4$A, $i_T = 0$) and on full load ($i_F = i_T = 4$A, $T_{em} = 33$ Nm).

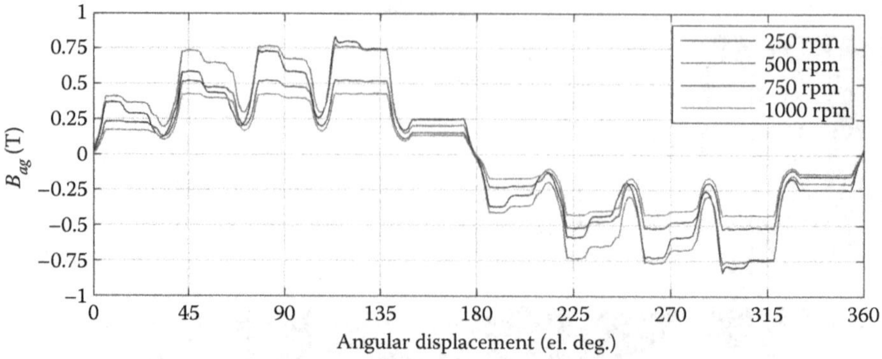

FIGURE 7.11 Radial component of airgap flux density (on load) over two poles at low speeds: flux weakening is viable and is due to i_F^* decreasing with speed and $i_T^* =$ ct.

The above sample results warrant comments such as:

The airgap flux density is within limits and has pulsations mostly "dictated" by stator slot openings, and the armature reaction (on load) is mild (Figure 7.10).

Flux weakening by reducing i_F is effective (Figure 7.11).

The phase inductances (self and mutual) depend on rotor position (and current), but they should be curve fitted by an average value plus a $\cos(2\theta_{ev})$ term, as done with salient-pole synchronous machines (Figure 7.12).

The average torque versus torque current density shows the machine capable of delivering rated (base) torque 33 Nm for $j_F = j_T = 5\ \text{A/mm}^2$, but it is capable of producing up to 160 Nm ($j_{Co} = 12\ \text{A/mm}^2$) for a short duration of time (Figure 7.13).

The static torque pulsations are notable, but FEM results fit experimental results. They may be cut in half by making the rotor two parts, shifted by half a slot pitch (Figure 7.15).

FEM-extracted flux lines prove a gain in the armature reaction at rated (base) load (Figure 7.16) in the 6-phase, 12-pole, 190-Nm, 200-rpm prototype [2].

FEM investigation may be used further to reduce torque pulsations and assess magnetic saturation influences, especially if heavy overloading is needed. But also due to the complexity of the geometry, FEM may be mandatory to be used to calculate iron losses, since the rotor core losses are not negligible.

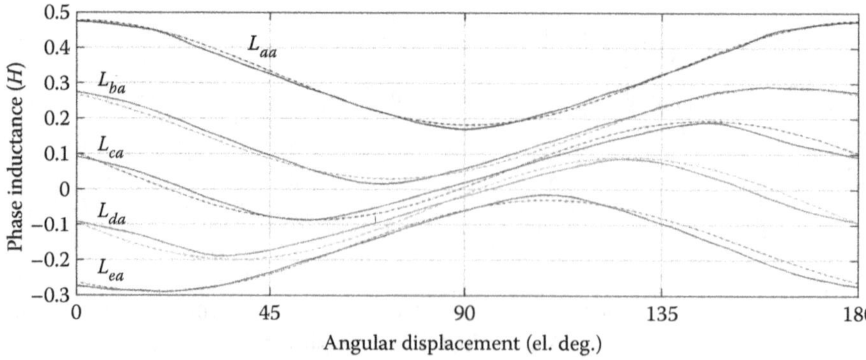

FIGURE 7.12 FEM (continuous lines) and curve-fitted phase a inductances (self and neutral) in the five-phase machine at rated current versus rotor position at standstill.

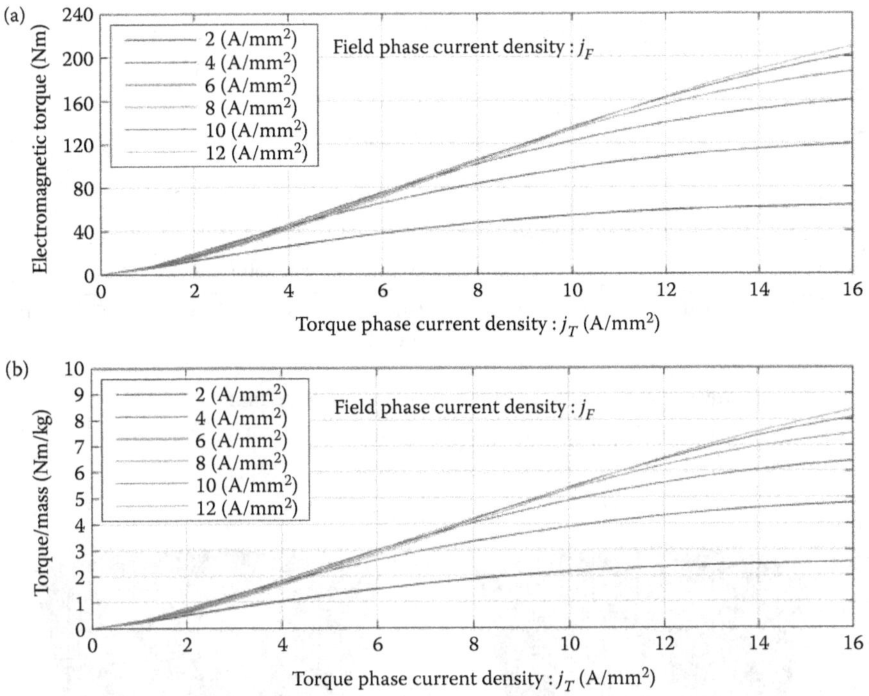

FIGURE 7.13 FEM-calculated average torque (a); and torque density (b) versus torque current density for the five-phase machine.

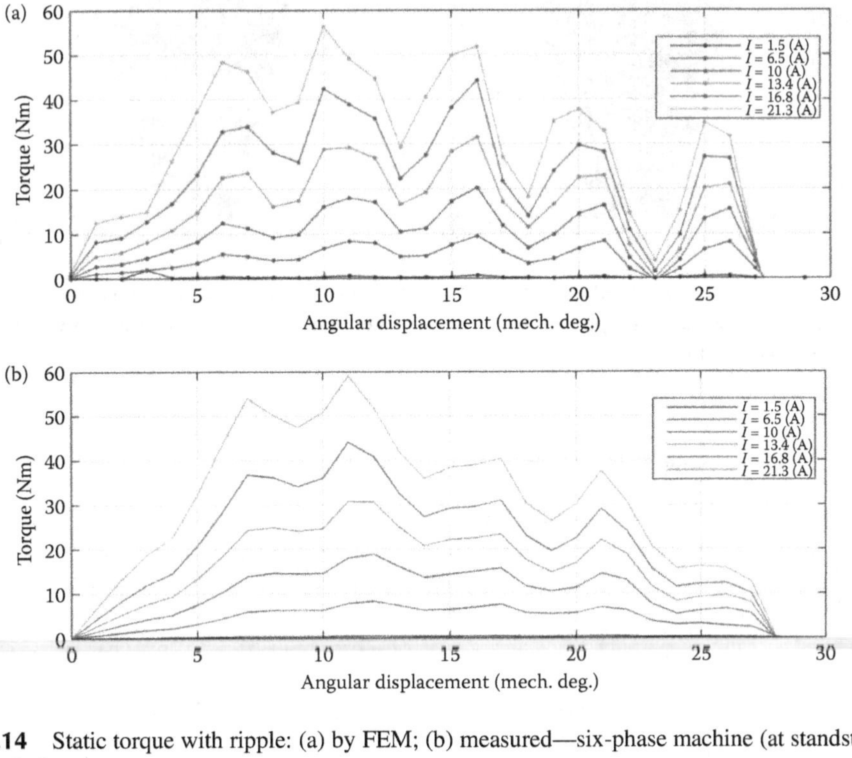

FIGURE 7.14 Static torque with ripple: (a) by FEM; (b) measured—six-phase machine (at standstill with all phases in series).

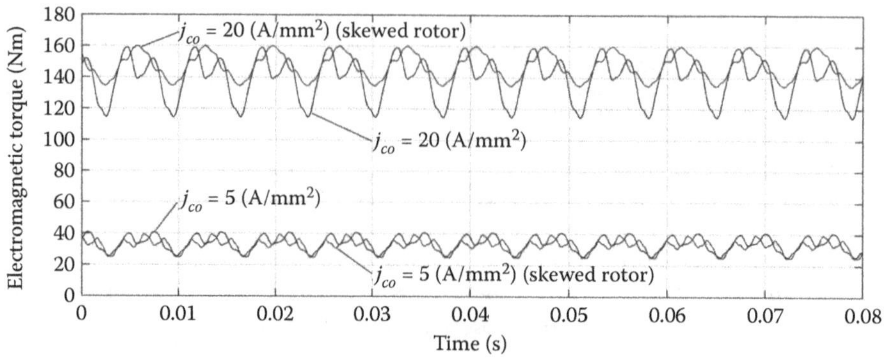

FIGURE 7.15 Total torque of a six-phase machine for $j_F = j_T = 20$ A/mm^2.

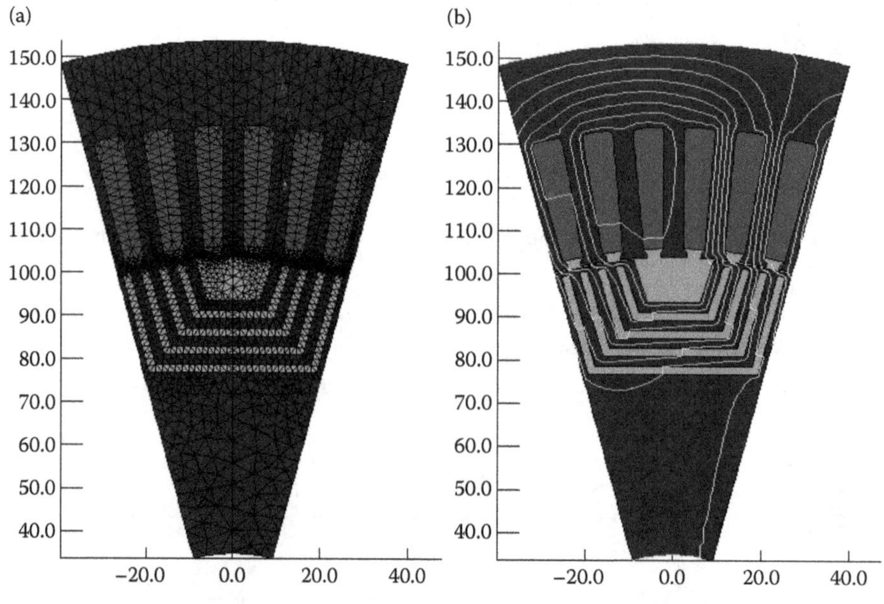

FIGURE 7.16 Six-phase/12 pole BLDC-MRM: 2D FEM model (a); and flux lines at rated load (190 Nm), (b). (After D. Ursu, "Brushless DC multiphase reluctance machines and drives", PhD Thesis, University Politehnica Timisoara, Romania, 2014 (in English). [9])

7.4.1 Iron Loss Computation by Finite-Element Model

To reduce the computation effort, we use here 2D-FEM only to calculate the flux density variation with time in all FEM elements (in fact with rotor position for given phase current waveforms at given speed), and these flux density variations in all FE elements are added up to get the core losses by an analytical formula traced back to Steinmetz-Jordan-Bertotti [10]:

$$\begin{aligned} p_{Fe} &= K_h \cdot f \cdot B^{\alpha} + K_e \cdot f^2 \cdot B^2 + K_a \cdot f^{1.5} \cdot B^{1.5}\ [\mathrm{W/m^3}] \\ K_e &= \pi \cdot \sigma \cdot d^2/6 \end{aligned} \tag{7.37}$$

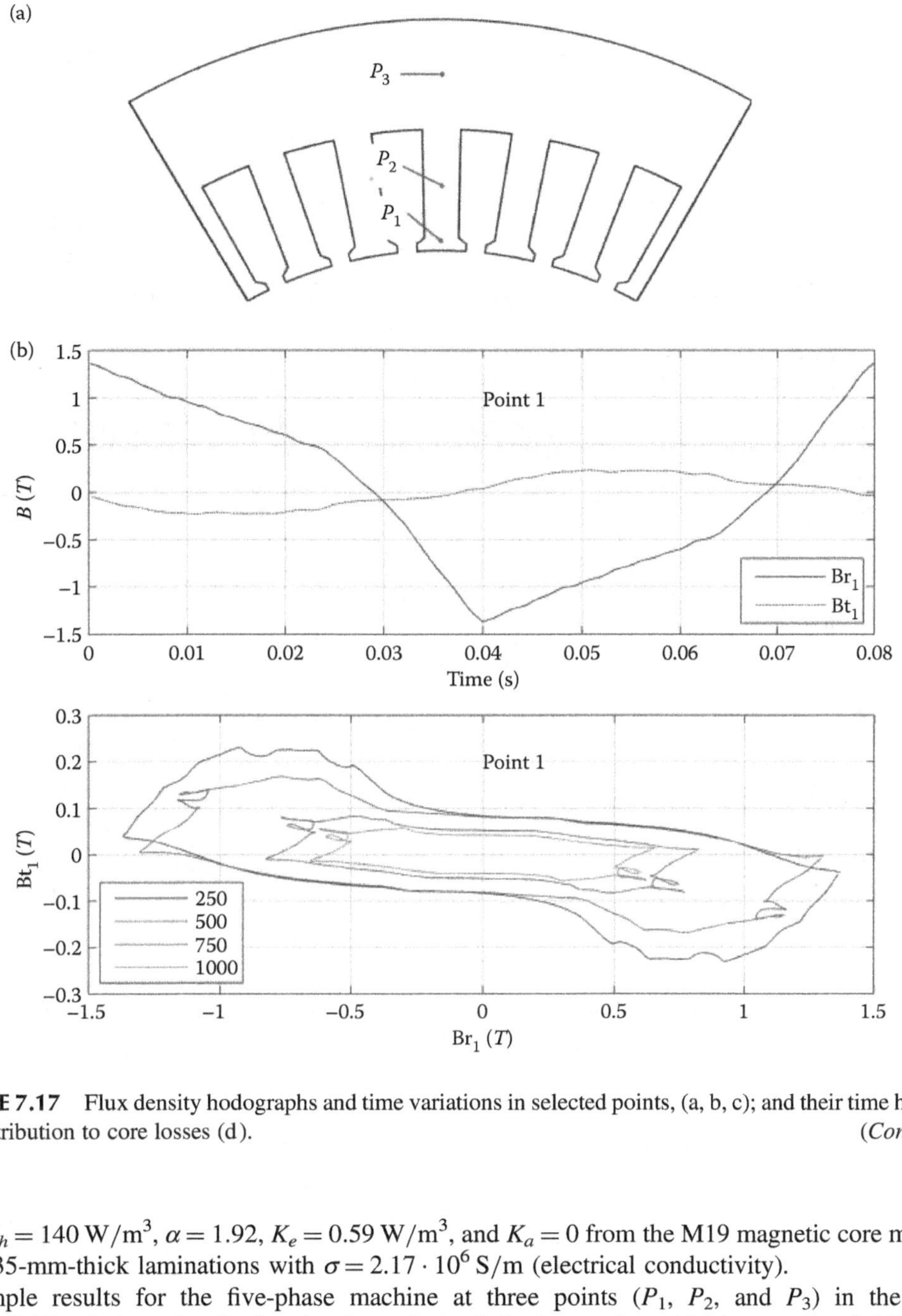

FIGURE 7.17 Flux density hodographs and time variations in selected points, (a, b, c); and their time harmonics contribution to core losses (d). *(Continued)*

with $K_h = 140\ \mathrm{W/m^3}$, $\alpha = 1.92$, $K_e = 0.59\ \mathrm{W/m^3}$, and $K_a = 0$ from the M19 magnetic core made of $d = 0.35$-mm-thick laminations with $\sigma = 2.17 \cdot 10^6$ S/m (electrical conductivity).

Sample results for the five-phase machine at three points (P_1, P_2, and P_3) in the stator (Figure 7.17a–d) show that the flux density variation is far from a sinusoidal (or traveling field) waveform and thus decomposition in harmonics in both stator and rotor elements is required and their contribution added up. Calculations have been made for four speeds (250, 500, 750, 1000 rpm).

It seems that in our rather small speed (frequency) machine (200 rpm), the core losses are not large (10%–20% of rated copper losses) but the rotor core loss contribution is up to 50%, despite its reduced volume with respect to stator core volume.

This is due to the fact that higher-order time flux density harmonics ($r > 7$) are larger in the rotor core, and we should remember that this is not a traveling field AC machine but rather closer to the operation mode of a DC brush machine.

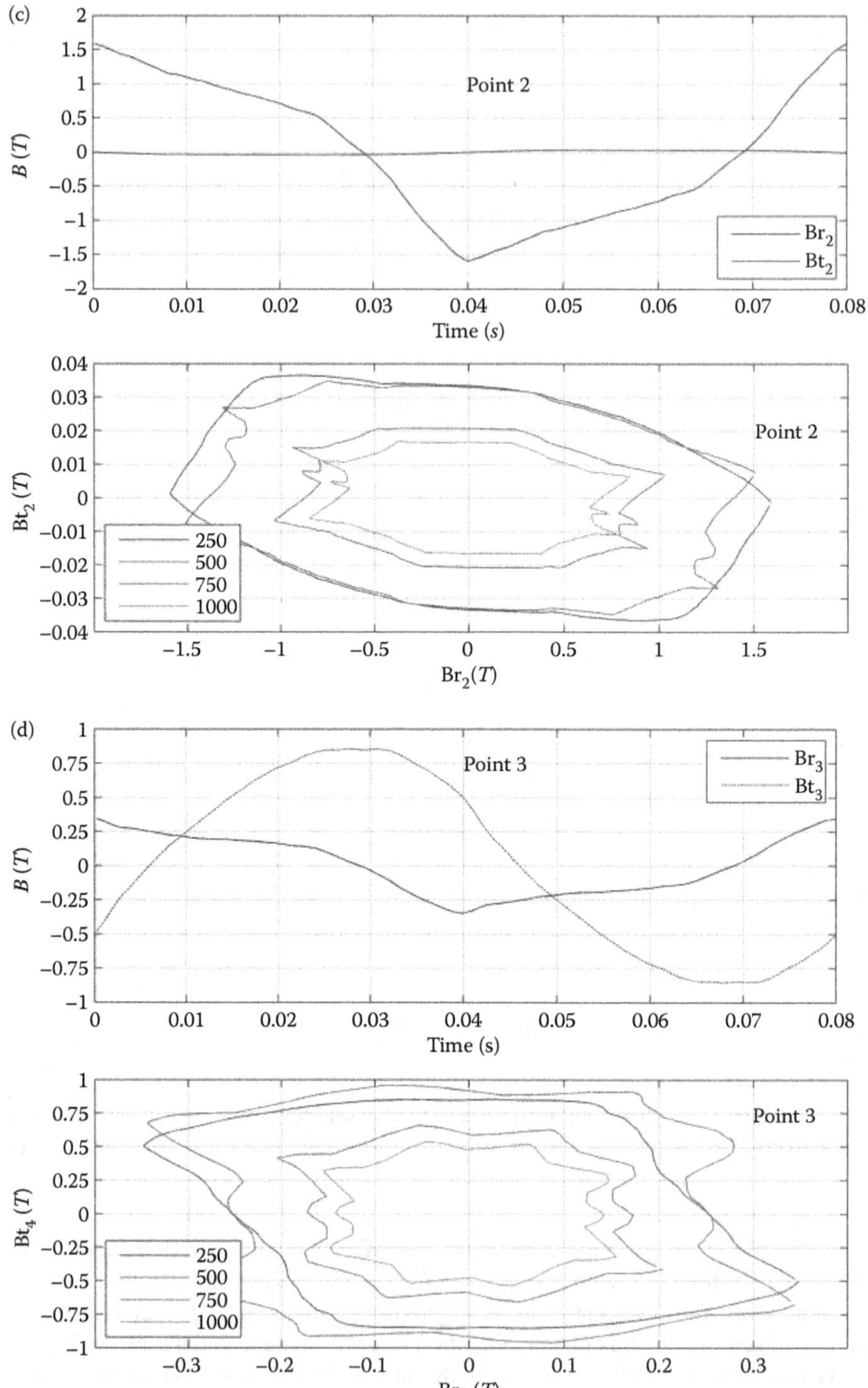

FIGURE 7.17 (Continued) Flux density hodographs and time variations in selected points, (a, b, c); and their time harmonics contribution to core losses (d).

7.5 NONLINEAR MAGNETIC EQUIVALENT CIRCUIT MODELING BY A CASE STUDY

As known, FEM provides precision, but requires large computation time and resources. In view of optimal design of BLDC-MRMs, a theory that is more complete than the technical one developed earlier in this chapter (which, however, may offer a good starting point geometry) is needed, without too large of a computation time.

The nonlinear MEC method [11] may be used successfully for the scope (as done for the claw-pole SM in Chapter 6).

MEC is based on defining flux tubes all over the machine. The flux tubes have two equipotential planes with magnetic scalar potentials u_1 and u_2 (which are virtual variables in magnetics, in contrast to electrostatics).

The magnetic reluctance R_m and its relations to magnetic flux Φ and $u_2 - u_1 = \Theta$ (mmf) are:

$$R_m = \int_0^x \frac{dx}{\mu(x) \cdot A(x)}$$
$$\text{for } \mu(x) = ct,\ A(x) = ct\text{: } R_m = \frac{L}{\mu A} = \frac{1}{G} \tag{7.38}$$

$$R_m = \frac{u_2 - u_1}{\varphi};\ u_2 - u_1 = \int_0^x \bar{H} \cdot d\bar{x};\ \varphi = \iint_A \bar{B} \cdot d\bar{a} \tag{7.39}$$

where G—permeance.

The solution of MEC is similar to that of a multiple resistive electric circuit with $u_2 - u_1 = \theta_i$, as voltages, Φ_i as currents, and R_{mi} as resistances.

To simplify the modeling in the stator and rotor yokes, the geometric permeances G_{sy}, G_{ry} relate only to flux density tangential components, while only radial components are considered for the stator and rotor teeth permeances G_{st} and G_{rt}.

The more elements (and nodes), the higher the precision, but the larger the computation time.

Permeances may be divided according to their magnetic permeability in constant (airgap) permeability and variable permeability permeances.

Finally, we consider the permeances that vary with rotor position in relation to the airgap where the energy conversion takes place.

For example, the airgap permeance G_{ij}, between i stator tooth and the jth rotor tooth (between barriers), may be defined as (Figure 7.18) [9]:

$$G_{ij} = \begin{cases} C_f G_{\max}, & 0 \le \gamma \le \gamma_t \\ C_f \dfrac{G_{\max}}{2}\left(1 + \cos\left(\pi \dfrac{\gamma - \gamma_t'}{\gamma_t - \gamma_t'}\right)\right), & \gamma_t' < \gamma < \gamma_t \\ C_f \dfrac{G_{\max}}{2}\left(1 + \cos\left(\pi \dfrac{\gamma - 2 \cdot \pi + \gamma_t'}{\gamma_t - \gamma_t'}\right)\right), & 2\pi - \gamma_t' \le \gamma < 2 \cdot \pi \\ 0 & \end{cases} \tag{7.40}$$

with

$$\gamma_t' = 2\frac{(w_{st} - w_{rt})}{sDi - g};\ \gamma_t = 2\frac{w_{st} + w_{rt} + sMs + rMs}{sDi - g} \tag{7.41}$$

$$G_{\max} = \mu_0 \cdot l_{\text{stack}} \cdot w_t / g;\ w_t = \min(w_{st}, w_{rt}) \tag{7.42}$$

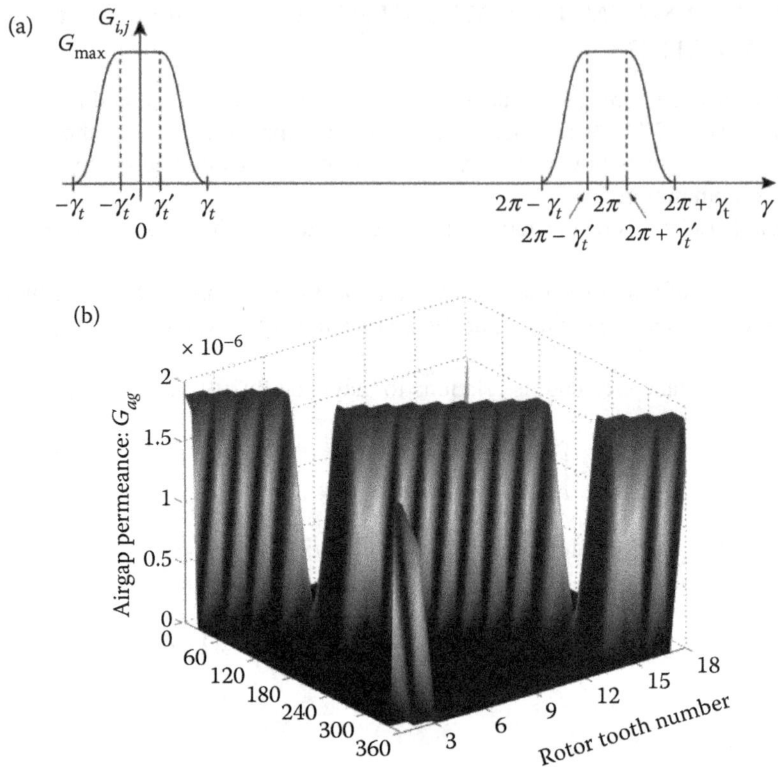

FIGURE 7.18 Airgap permeance variation: (a) with rotor position; (b) between a stator tooth and all rotor teeth (between flux barriers).

There is some fringing in the airgap that may be accounted for by a coefficient C_f (Equations 7.40 through 7.42) that multiplies G_{max} [11]:

$$C_f \approx \frac{sMs}{w_{st} \cdot \ln\left(1 + \dfrac{sMs}{w_{st}}\right)} \tag{7.43}$$

As the airgap permeance of different airgap-related flux tubes varies with rotor position, new connections between them have to be considered, with motion.

The sources in the MEC are based on the slot mmfs (Figure 7.19). The MEC equation in matrix form is simply:

$$[G][V] = [\varphi_t] \tag{7.44}$$

$[G]$—permeance matrix: (53 × 53) for the six-phase, six-pole (36 slots), nine rotor slots/pole
$[V]$—node potentials (53 × 1)
$[\Phi_t]$—tooth fluxes (52 × 1)

Observing the machine symmetry, only two poles are simulated, using, though, adequate boundary conditions [11].

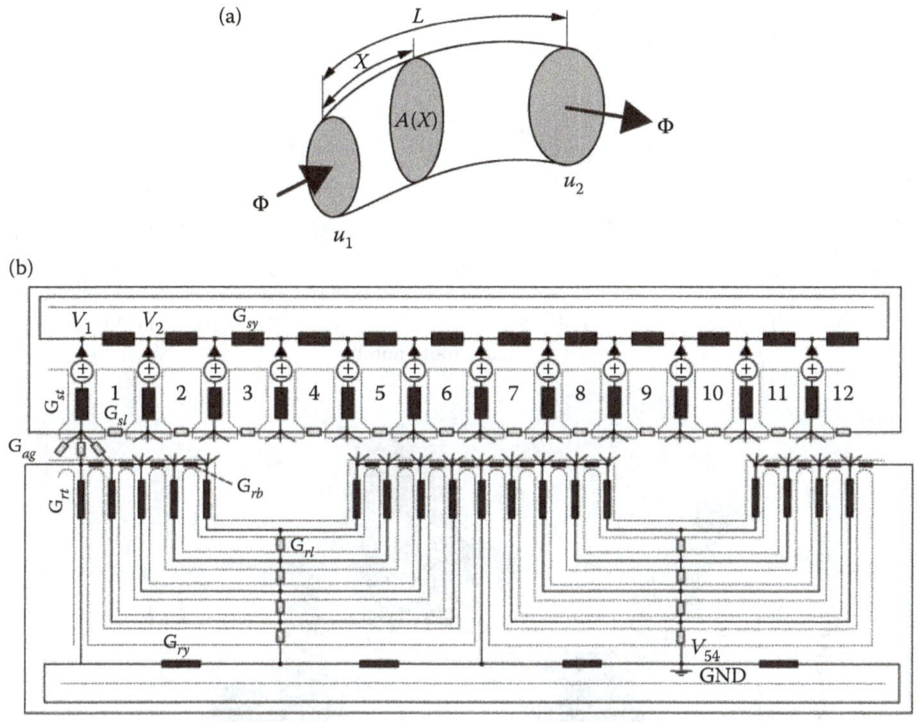

FIGURE 7.19 The magnetic flux tube, (a); and the generic MEC of BLDC-MRM, (b). (After D. Ursu, "Brushless DC multiphase reluctance machines and drives", PhD Thesis, University Politehnica Timisoara, Romania, 2014 (in English). [9])

The electromagnetic torque T_e is:

$$T_e = p_1 \sum_{i=1}^{N_s} \sum_{j=1}^{N_r} (V_i - V_j) \frac{dG_{i,j}}{d\gamma} \tag{7.45}$$

with N_s—stator slots/two poles ($N_s = 12$) and N_r—rotor teeth/two poles ($N_r = 18$).

For given sources (mmfs) with known fluxes, the program solves iteratively for node potentials. Iterations are needed to account for magnetic saturation.

A typical tooth flux density calculated by the MEC method for one rotor position, its 2D variation with stator tooth current (Figure 7.20a), and its variation with rotor position (Figure 7.20b) show the capability of the MEC model to fairly portray the geometrical slotting complexity and nonlinearity due to magnetic saturation. The torque was subsequently calculated (Equation 7.45) and compared with FEM results (Figure 7.21).

As is evident in Figure 7.21, the MEC-calculated torque has its pulsations much too large in comparison with the 2D-FEM results. So, at best, the MEC peak torque may be used as average torque. However, the computation time was 4 seconds versus 3700 seconds for 2D-FEM.

FEM may be used for one electrical degree time-step after MEC-based model optimal design code to check the optimal solution and calculate the core losses, as mentioned in section 7.3.1. But in this case, if FEM is to be embedded in the MEC optimal design, a time-step of mechanical degree should perhaps be used for 130 s FEM computation time. This way, the total optimal computation time will be kept within a few hours on a single contemporary desktop computer.

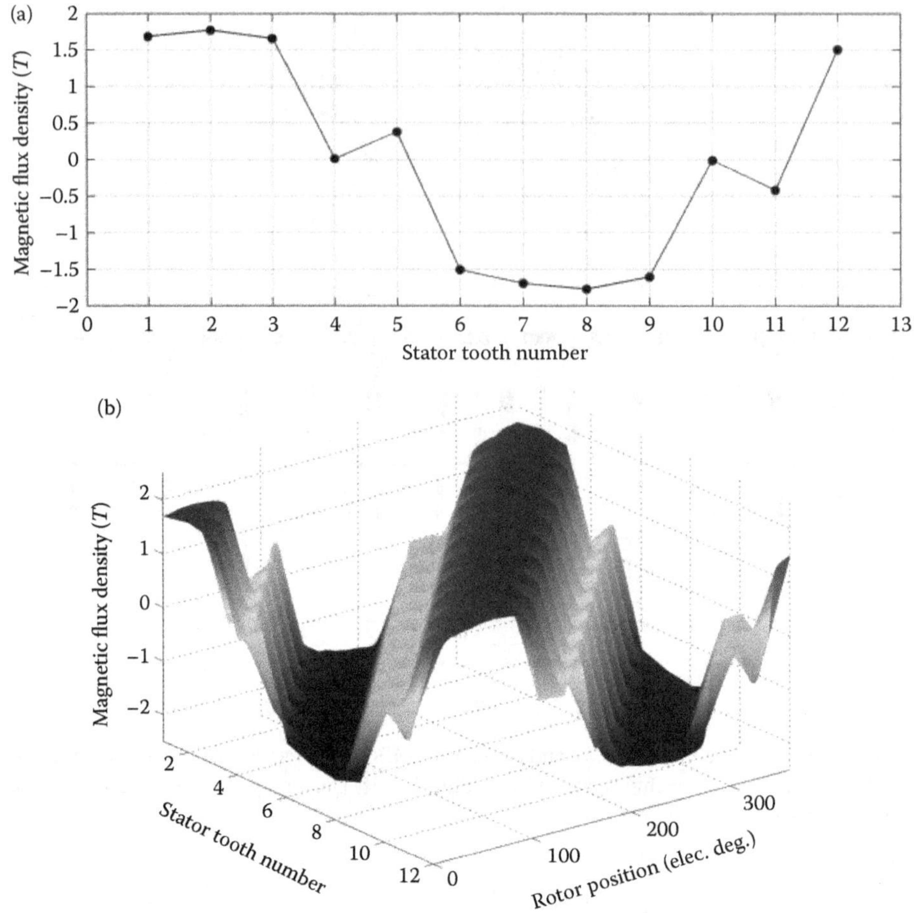

FIGURE 7.20 Stator tooth magnetic flux density: (a) versus stator tooth number; (b) versus rotor position.

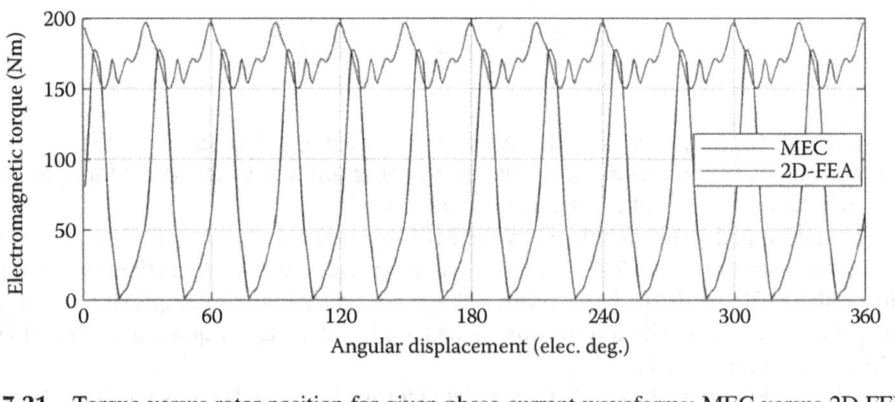

FIGURE 7.21 Torque versus rotor position for given phase current waveforms: MEC versus 2D FEM.

7.6 CIRCUIT MODEL AND CONTROL

Technical theory, MEC, and FEM modeling are all useful to investigate performance on an existing device and for designing it, mainly for steady-state operation. However, to quickly investigate steady results, transients, and control design, with an inductance matrix [L] imported after curve fitting from the above methods, a circuit model is needed.

As already implied, a phase coordinate model is practical despite the fact that its inductance matrix depends heavily on rotor position and currents (or magnetic saturation), because phase current control is operated with flat-top two-level bipolar currents (rather than with sinusoidal currents) for a five-phase machine [6]. The voltage equation in matrix form is:

$$[V_s] = R_s[i_s] + \frac{d}{dt}\left([L(\theta_{er}, i_s)][i_s]\right) \tag{7.46}$$

$$\begin{aligned} &[V_s] = [V_a, V_b, V_c, V_d, V_e]^T \\ &[i_s] = [i_a, i_b, i_c, i_d, i_e]^T; \\ &[\psi_s] = [L(\theta_{er}, i_s)][i_s] \\ &[\psi_s] = [\psi_a, \psi_b, \psi_c, \psi_d, \psi_e, \psi_f]^T \end{aligned} \tag{7.47}$$

$$T_e = \frac{1}{2}[i_s]^T \frac{\partial(L(\theta_{er}, i))}{\partial\theta_{er}} \cdot [i_s] \tag{7.48}$$

$$\frac{J}{p}\frac{d\omega_r}{dt} = T_e - T_{\text{load}} - B\cdot\omega_r; \quad \frac{d\theta_{er}}{dt} = \omega_r; \ \omega_r = 2\cdot\pi\cdot n\cdot p \tag{7.49}$$

where n is the rotor speed in rps.

The inductance matrix (L) for a five-phase machine is "full":

$$[L] = \begin{bmatrix} L & L_{ab} & L_{ac} & L_{ad} & L_{ae} \\ L_{ba} & L_{bb} & L_{bc} & L_{bd} & L_{be} \\ L_{ca} & L_{cb} & L_{cc} & L_{cd} & L_{ce} \\ L_{da} & L_{db} & L_{dc} & L_{dd} & L_{de} \\ L_{ea} & L_{eb} & L_{ec} & L_{ed} & L_{ee} \end{bmatrix} \tag{7.50}$$

The phase inductances may be approximated as for salient-pole synchronous machines by introducing a single term in $2\theta_{er}$. For phase a:

$$\begin{aligned} L_{aa} &= L_{aa0} + L_{aa2}\cdot\cos(2\theta_{er}) \\ L_{ab} &= L_{ab0} + L_{ab2}\cdot\cos(2\theta_{er} - \pi/5) \\ L_{ac} &= L_{ac0} + L_{ac2}\cdot\cos(2\theta_{er} - 2\pi/5) \\ L_{ad} &= L_{ad0} + L_{ad2}\cdot\cos(2\theta_{er} - 3\pi/5) \\ L_{ae} &= L_{ae0} + L_{ae2}\cdot\cos(2\theta_{er} - 4\pi/5) \end{aligned} \tag{7.51}$$

The values of $L_{aa0}, \ldots, L_{ae0}, L_{aa2}, \ldots, L_{ae2}$ for phase a, but similarly for the other phases, may be FEM (or MEC) model-calculated or measured (from, say, steady-still flux decay tests) by regression methods.

Typical such comparisons are shown in Figure 7.12. For given inputs $[V]$, T_{load}, and initial values of variables $[i_s]$, the circuit model (Equations 7.46 through 7.51) can be solved through a numerical method. True, for each computation step, the matrix $[L(\theta_{er}, i)]$ has to be inverted. It may be more feasible to solve the same model in $[\psi_s]$ as the variable vector:

$$[V_s] = R_s[i_s] + \frac{d[\psi_s]}{dt} \tag{7.52}$$

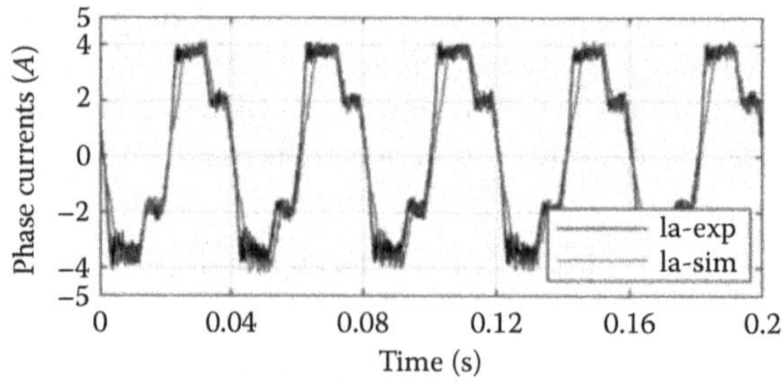

FIGURE 7.22 Phase a rated current waveform at 500 rpm/five-phase machine. (After D. Ursu et al., *Record of OPTIM-2014*, pp. 354–161 (IEEEXplore). [6])

with $[\psi_s]$ from Equation 7.47 for each integration step, $[i_s(k)]$ has to be calculated after inverting $[L]$, but, in this case, the derivatives $\partial[L]/\partial\theta_{er}$ are not modeled, which may turn out to be an important advantage.

If enough time is allowed when starting with zero (or given) speed ω_r and zero phase currents, steady-state operation is obtained with the same circuit model (Equations 7.46 through 7.51).

The calculated waveform of phase a current versus time at 500 rpm (for the five-phase prototype already introduced in the chapter) shows good agreement with test results in Figure 7.22 [6].

Also, measured and circuit model–calculated torque for given (experimental) currents compare acceptably (Figure 7.23).

The higher-frequency torque ripples in the calculations have perhaps been introduced mainly by the Runge-Kutta numerical solving algorithm, while the lower-frequency torque ripple differences between theory and experiments are due to machine manufacturing imperfections and the neglecting of magnetic saturation variations in inductances. FEM-calculated torques for same-rated currents $i_F^* = i_T^*$ are notably closer to measurements (Figure 7.23c).

The control system for the five phase BLDC-MRM, fed from a 5×1 phase H-bridge inverter, implies independent control of all phases. The speed and current control scheme are shown in Figure 7.24.

The speed PI controller with a limiter has $K_p = 0.005$, $K_i = 30$ while hysteresis current controllers are applied (band = 0.05A) with $V_{DC} = \pm 300$ V.

The simulation model is illustrated in Figure 7.25.

A 4/1 speed ratio was demonstrated (Figure 7.26) though not at rated power in tests, as i_F^* was not reduced inversely proportional to speed, in order to keep system efficiency high: (90% motor + inverter maximum efficiency was measured).

Note: A comprehensive FEM characterization and four-quadrant control of a six-phase BLDC-MRM is offered in [5]. From there, we gather here only a few sample results, including:

Speed reversal transients (Figure 7.27).

No-load motor to generator transients (Figure 7.28).

For this machine, a null point six-phase inverter was used (Figure 7.29). The three main control system parts are shown in Figure 7.30a–c.

Making use of a lower power switch count in the null point inverter implies controlling not the null current but the null point voltage pulsations (Figure 7.30c) in order to stabilize phase currents that each have PI controllers.

7.7 OPTIMAL DESIGN METHODOLOGY AND CODE WITH A CASE STUDY

In an optimal design methodology, a few main steps are to be followed:

- Choose a proper (rapidly enough convergent) optimization algorithm.
- Define the optimization variable vector and its domain for the case in point.

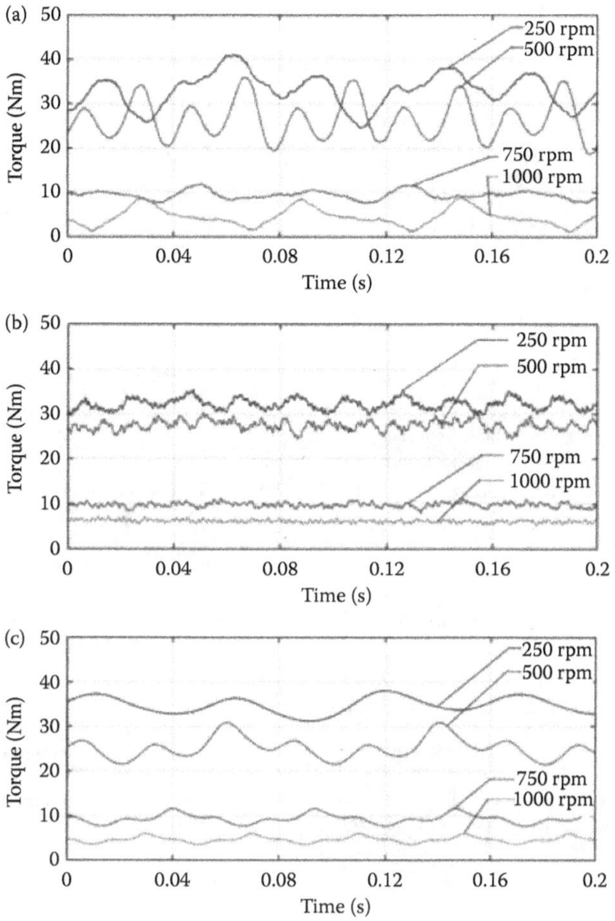

FIGURE 7.23 Measured (a) and circuit model-calculated torque versus time for measured phase currents waveforms at four different speeds a); from circuit model, (b); and FEM, (c). (After D. Ursu et al., *Record of OPTIM-2014*, pp. 354–161 (IEEEXplore). [6])

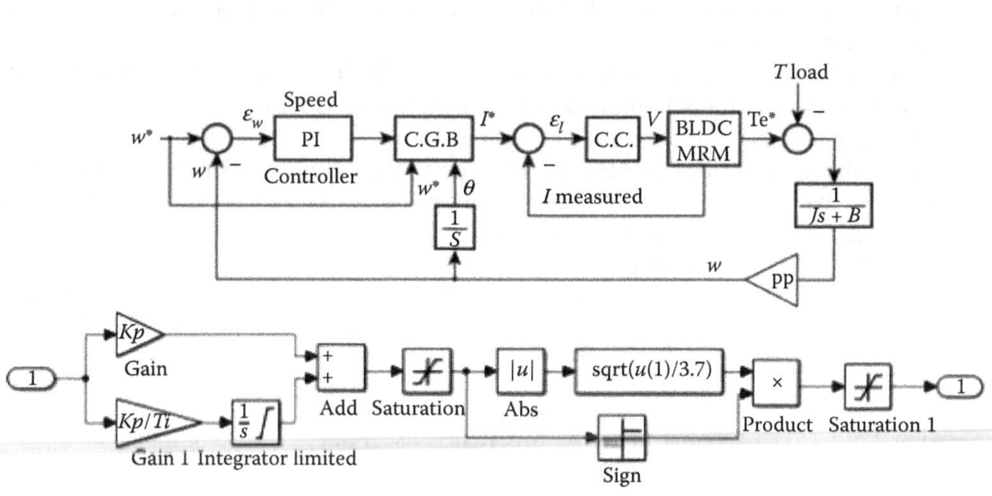

FIGURE 7.24 Speed and phase current control scheme (a); and speed PI controller with limiter (b), for the five-phase BLDC-MRM fed from 5 × 1 phase inverter.

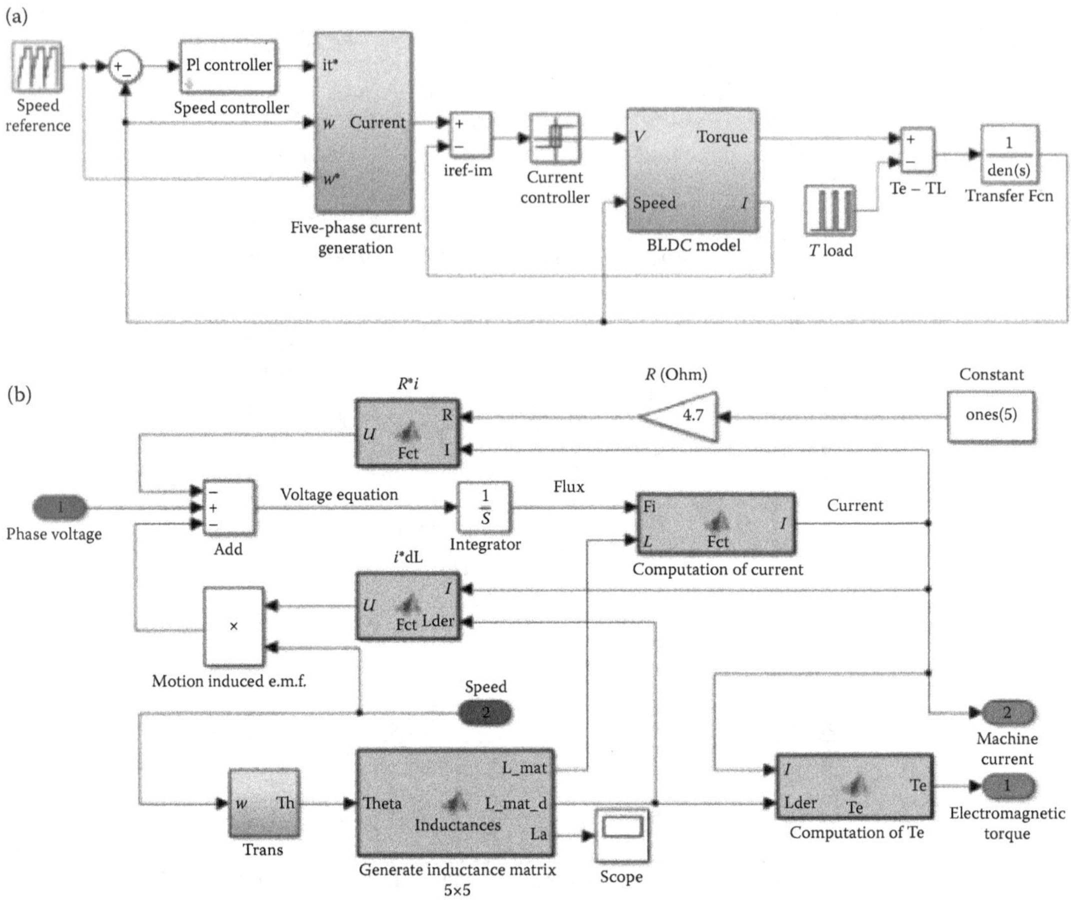

FIGURE 7.25 System simulation model of the five-phase BLDC-MRM (a); and the BLDC-MRM model (b). (After D. Ursu et al., *Record of OPTIM-2014*, pp. 354–161 (IEEEXplore). [6])

- Define the objective (cost-fitting) function components: cost functions and penalty functions, in an unitary way.
- Choose a machine (drive) nonlinear model: MEC- or 2D-FEM–based, whose imported data will feed the circuit plus control model of the motor/generator drive.
- Define input and output files as text and graphs.
- Verify offline or embedded by FEM the main performance of the optimal design machine and correct the machine model (if requested) by fudge factors until the FEM and non-FEM models are close enough to each other.
- If thermal and mechanical design follows, then that part of optimal design has to be added, but these steps are beyond our scope here. A simple thermal model here, though, excludes all designs with a high average or critical-point over-temperature.

After comparing four different metaheuristic optimal algorithms [9], the modified particle swarm optimization (PSO_M) was applied to BLDC-MRM, because it can be seen as artificial intelligence in the sense that it follows five principles:

- Proximity: simple space and time calculation made by population
- Quality: simple space and time calculation carried out by population

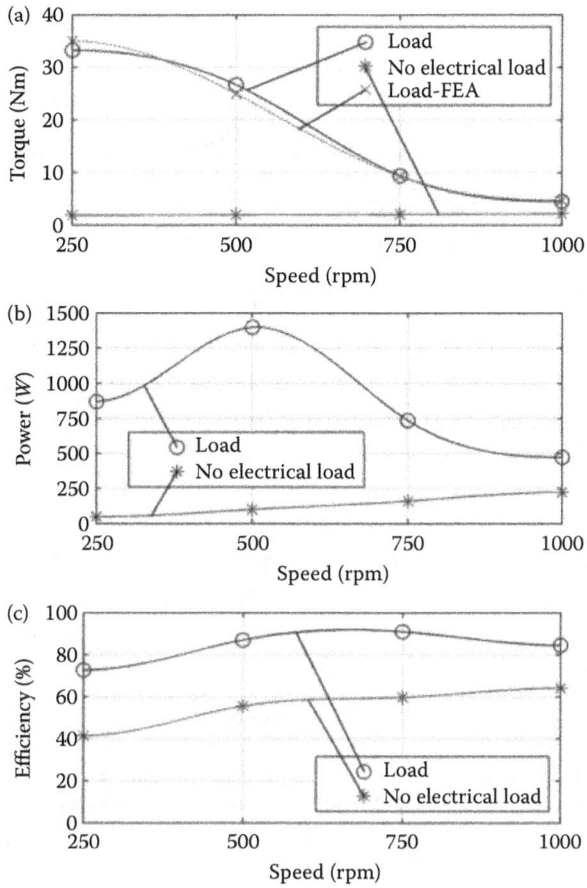

FIGURE 7.26 Torque, (a); power, (b); and drive efficiency, (c) versus speed. (After D. Ursu et al., *Record of OPTIM-2014*, pp. 354–161 (IEEEXplore). [6])

- Diverse response: no excess resources are allocated in too narrow a space
- Stability: population should not change significantly with environmental change
- Adaptability: population should change behavior if the comparative price is worth the trouble

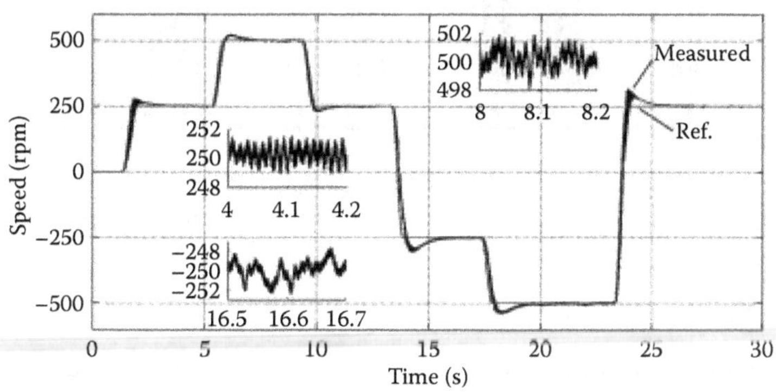

FIGURE 7.27 Speed reversal transients of a six-phase BLDC-MRM. (Measured, Adapted from D. Ursu et al., *IEEE-Trans.*, vol. IA–51, no. 3, 2015, pp. 2105–2115. [5])

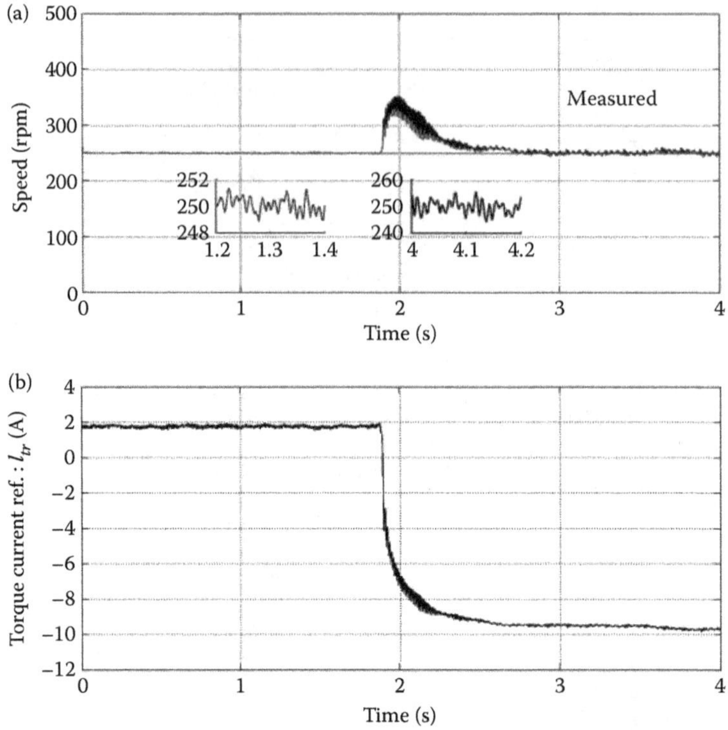

FIGURE 7.28 rom no-load motoring to generating at 2500 rpm with the six-phase BLDC-MRM (a) speed transients; (b) torque current reference. (Adapted from D. Ursu et al., *IEEE-Trans.*, vol. IA–51, no. 3, 2015, pp. 2105–2115. [5])

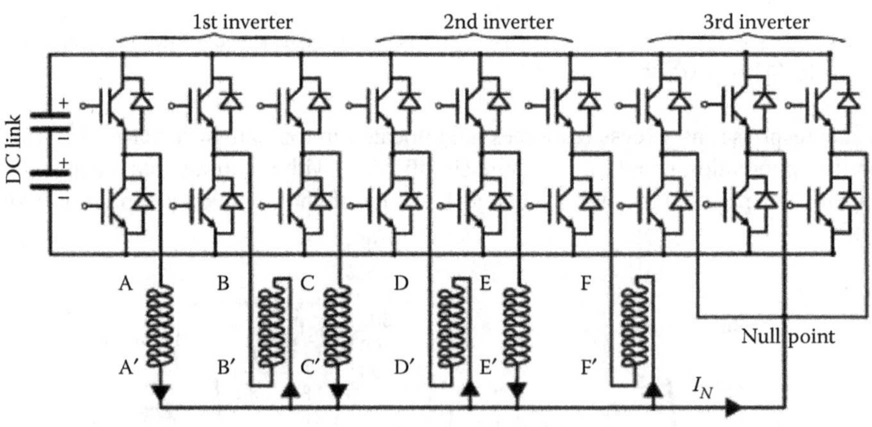

FIGURE 7.29 Six-phase null-point inverter with inversed connections of phases 2, 4, 6 to reduce the null point current. (Adapted from D. Ursu et al., *IEEE-Trans.*, vol. IA–51, no. 3, 2015, pp. 2105–2115. [5])

7.7.1 Particle Swarm Optimization

PSO emulates bird flocking or fish schooling [12–13]. Its foundation is based on the fact that cooperative populations (that share info) are eventually superior to the noncooperative ones, especially when food resources have an unpredictable distribution, when cooperation surpasses competition effects. Each particle has not only a position x_i but also a speed v_i.

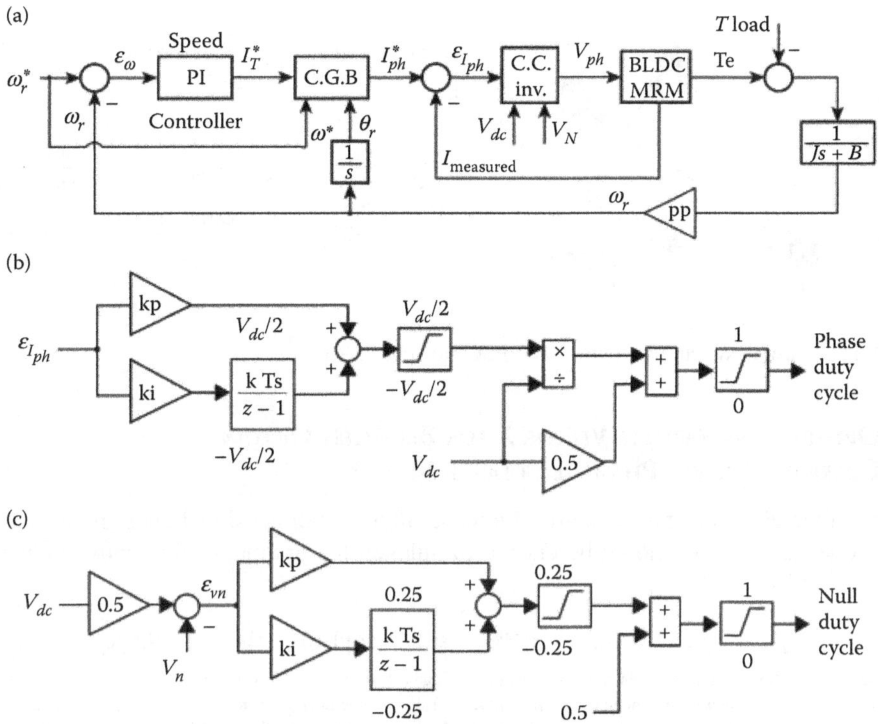

FIGURE 7.30 Speed control scheme (a), ($K_p = 3$, $K_i = 300$) phase current controller ($V_{DC} = 250$ V, $K_{pi} = 5$, $K_{ii} = 0.002$) (b), and null point voltage regulator (c). (Adapted from D. Ursu et al., *IEEE-Trans*, vol. IA–51, no. 3, 2015, pp. 2105–2115. [5])

After a random initialization of particles (variable vector), their fitting (objective) function values are evaluated, and the best position in space (q_{best}) and the best position of each particle (p_{best}) are memorized.

A new velocity $v(t+1)$ at $t+1$ time moment is computed as:

$$v(t+1) = w(t)\cdot v(t) + c_1\cdot \text{rand}(0,1)\cdot\left(p_{\text{best}} - x(t)\right) + c_2\cdot \text{rand}(0,1)\left(q_{\text{best}} - x(t)\right) \quad (7.53)$$

$$x(t+1) = x(t) + v(t+1) \quad (7.54)$$

where $c_1 < 2$ is the self-confidence factor, $c_2 < 2$ is the swarm factor, and w is the weight factor (it decreases from 0.9 to 4 during optimization).

To speed up convergence, the weight factor $w(t)$ is replaced in Equation 7.53 by coefficient k

$$k = \frac{2}{2 - \varphi - \sqrt{\varphi^2 - 4\varphi}}; \quad \varphi = c_1 + c_2 > 4 \quad (7.55)$$

Example: With $\varphi = 4.1$, $k = 0.729$, and $c_1 + c_2 = 1.5$ in Equation 7.53, the situations in Equations 7.53 and 7.54 are equivalent.

Synthetic particle swarm movement and the PSO flowchart are illustrated in Figure 7.31.

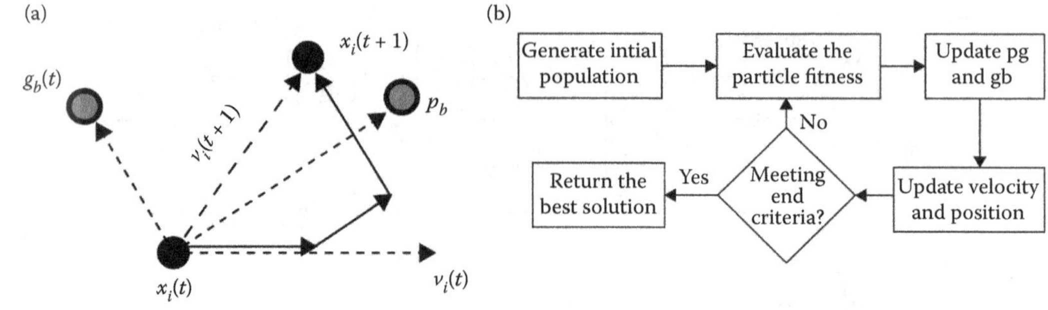

FIGURE 7.31 Particle swarm motion (a); and PSO flowchart, (b).

7.7.2 Optimization Variable Vector $\overline{X}$ for Brushless Direct Current–Multiple Phase Reluctance Motor

The choice of variable vector $\overline{X}$ is an art. The most influential dimensional ones and as many nondimensional ones as possible should be chosen to quicken the optimal design convergence process. In our case, $\overline{X}$ is [9]:

$$\overline{X} = [f_{\tan},\ B_{ag},\ j_s,\ \lambda,\ b_z P\tau s,\ sMs,\ shyP\tau p,\ shyP\tau r\ \ hagq,\ hag,\ rhyP\tau p] \tag{7.56}$$

$f_{\tan}$	B_{ag}	j_s	λ	$b_zP\tau s$	sMs	shyPτp	shyPτr	hagq	hag	rhyPτp
tangential force in (N/cm^2)	airgap flux density in (T)	current density in (A/mm^2)	lstack/pole pitch ratio	stator yoke pole pitch ratio (−)	stator slot width in (mm)	stator yoke pole pitch ratio (−)	rotor tooth/rotor slot (flux barrier) pitch ratio (−)	equivalent airgap in axis q in (mm)	machine airgap in (mm)	rotor yoke width/pole pitch

A population of 100 particles (variable vectors), followed for 50 generations in a hypercube, at the designer choice, suffices:

$$X_{\min} < X < X_{\max} \tag{7.57}$$

In our case study, the specifications refer to a six-phase BLDC-MRM of $T_{eb} = 160$ Nm, six poles, at 3000 rpm (50 kW). For this case, $X_{\min}$ and $X_{\max}$ are:

$$\begin{aligned} X_{\min} &= [0.5,\ 0.4,\ 2,\ 0.6,\ 0.3,\ 1,\ 0.25,\ 0.35,\ 5,\ 0.35,\ 3] \\ X_{\max} &= [3,\ 1,\ 10,\ 2.4,\ 0.7,\ 3,\ 0.7,\ 0.65,\ 20,\ 1.2,\ 0.7] \end{aligned} \tag{7.58}$$

The random initial population (of 100 particles) is:

$$\begin{aligned} &\overline{X}_i = \text{rand}\left(X_{\min} + \text{rand}(0,1)\left(\overline{X}_{\max} - \overline{X}_{\min}\right),\ X_r\right) \\ &\text{for } i = 1 : P = 100 \end{aligned} \tag{7.59}$$

After each particle speed computation and position updating, the new position is rounded according to X_r, again chosen by the designer, in our case:

$$\overline{X}_r = [0.05,\ 0.01,\ 0.1,\ 0.05,\ 0.1,\ 0.1,\ 0.05,\ 1,\ 0.05,\ 0.1] \tag{7.60}$$

If an initial design is available from a preliminary method, as the technical theory in section 7.2, it may be introduced as $\overline{X}_{\text{init}}$.

Now, let us define the cost or fitting or objective function $t_{\cos t}$.

7.7.3 Cost Function Components

As we are concerned with both initial cost (or volume or weight) and lower losses, a complex cost function $t_{\cos t}$ is defined here:

$$t_{\cos t} = m_{\cos t} + e_{\cos t} + \text{pen}_{\cos t} \tag{7.61}$$

with $m_{\cos t}$—material costs (copper, laminations, and shaft) and passive (framing, etc.) materials

$$m_{\cos t} = \text{copper}_{\cos t} + \text{Lam}_{\cos t} + \text{shaft}_{\cos t} + \text{PasM}_{\cos t} \tag{7.62}$$

The energy loss cost $e_{\cos t}$ is:

$$e_{\cos t}\begin{cases} e_{\text{price}} \cdot P_n \cdot \left(\dfrac{1}{\eta_n} - \dfrac{1}{\eta_{\min}}\right) \cdot hpy \cdot ny,\ \eta < \eta_{\min} \\ e_{\text{price}} \cdot P_n \cdot \left(\dfrac{1}{\eta} - 1\right) \cdot hpy \cdot ny,\ \eta \geq \eta_{\min} \end{cases} \tag{7.63}$$

where *hpy* is the number of hours at full power per year and *ny* is the number of years in operation. It is also possible to define a "most frequent load" level for, say, a standard driving cycle, for which to calculate $e_{\cos t}$.

Penalty cost $\text{pen}_{\cos t}$ is here the sum of two terms:

- Overtemperature penalty $\text{Temp}_{\cos t}$
- Power nonrealization penalty: $\text{Pow}_{\cos t}$

$$\text{Temp}_{\cos t} = \max[0, (T_w - T_m)m_{\cos t}] \tag{7.64}$$

with

$$T_w = \frac{p_{\text{copper}} + p_{\text{iron}}}{\alpha_T \cdot A_{rea}} + T_{\text{ambient}};\ T_{\text{ambient}} = 40°\text{C} \tag{7.65}$$

α_T—equivalent heat transmission coefficient calculated at the outer stator core area. It also depends on the cooling system: $\alpha_T = 14\text{–}100\ \text{W/m}^{2}{}^\circ\text{C}$.

$\text{Pow}_{\cos t}$ is:

$$\text{Pow}_{\cos t} = \begin{cases} 2(P_n - P_{\min}) \cdot m_{\cos t}, & P_n < P_{\min} \\ 0, & P_n \geq P_{\min} \end{cases} \tag{7.66}$$

All components in $t_{\cos t}$ are expressed in terms of money, and the penalty functions are also proportional to the materials cost, $m_{\cos t}$.

7.7.4 Optimal Design Sample Results

Sample results from the best particles of the swarm are presented in Figures 7.32 through 7.40 [9].

The initial and optimal cross-section are illustrated in Figure 7.40.

The optimal design code investigated 5000 machines (100 × 50) within 20,000 seconds on an 2.26-GHz Intel i5 processor.

The sample results in the case-study in point (Figures 7.32 through 7.40) for 160 Nm, 3000 rpm (50 kW) may be characterized as follows:

- Keeping the efficiency above 94%, the active weight dropped from 67 kg to 32 kg; also, the $t_{\cos t}$ dropped from 670 to 250 USD.

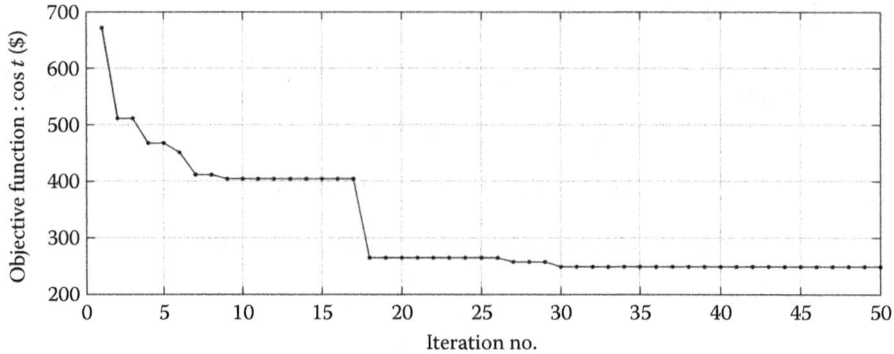

FIGURE 7.32 $t_{\cos t}$ function evolution.

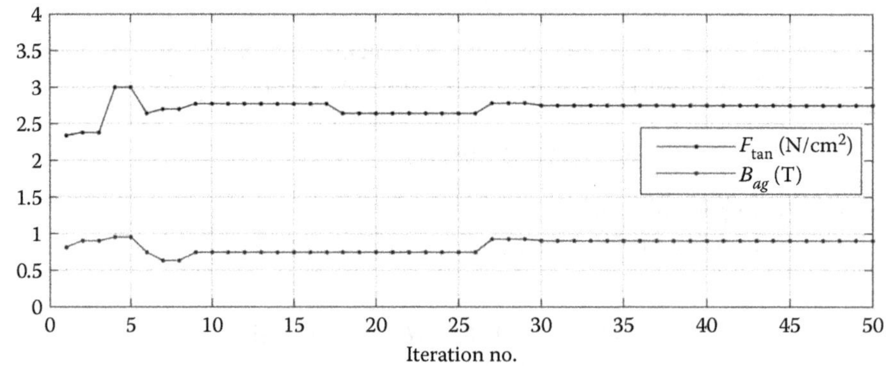

FIGURE 7.33 Tangential force and airgap flux density.

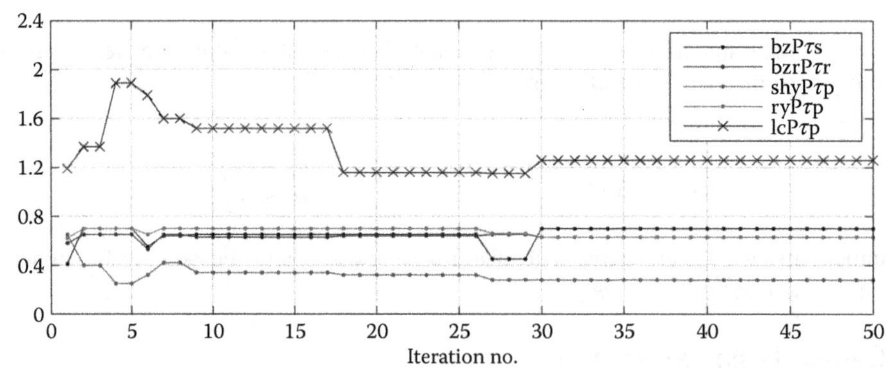

FIGURE 7.34 Main machine cross-section ratios.

- The outer stator diameter was reduced from 342 to 271 mm, with the stack length about the same: $l_{\text{stack}} = 123$ mm.
- The core and copper losses are not far away from each other in the optimal solution.
- The current density went from 1.5 to 9.4 A/mm^2.
- The stator slot opening and airgap were reduced to 1.4 mm and 0.4 mm, respectively. As the airgap is small, the commutation of the current (trapezoidal with $i^*_{Fb} = i^*_{Tb}$) at 3000 rpm and $V_{DC} = 500$ V within one slot pitch time has to be checked via circuit model.

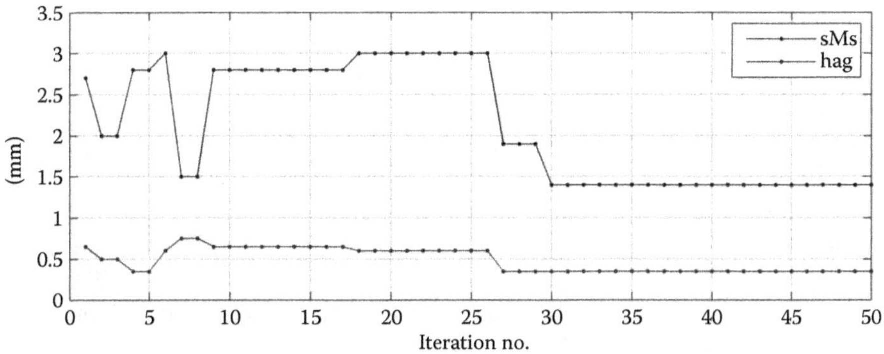

FIGURE 7.35 Stator slot opening and airgap evolution.

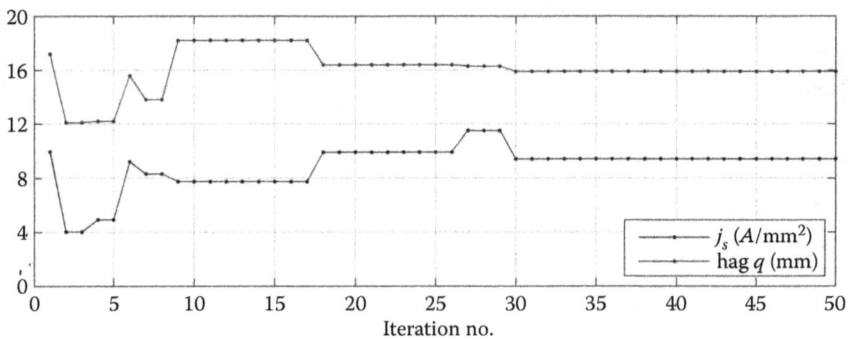

FIGURE 7.36 Current density and q-axis air-depth (gap) evolution.

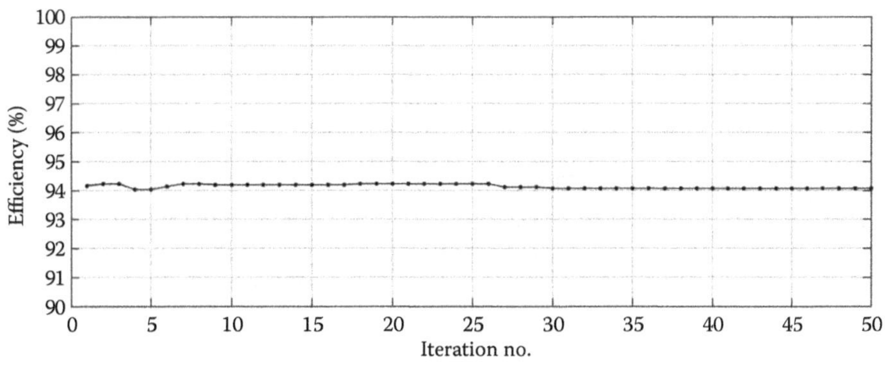

FIGURE 7.37 Efficiency evolution.

But, increasing the airgap to improve current commutation will also increase copper losses in field phases and thus should reduce efficiency somewhat.

If the machine were intended for a CPRS = 4, $n_b = 3000$ rpm, and $n_{\max} = 12{,}000$ rpm, a penalty function for power unrealization at maximum speed should be added, while, implicitly, the airgap would increase. The efficiency at maximum speed might hold if the field current $i^{*}_{Fn\max} = i^{*}_{Fn}/\text{CPRS}$.

The torque density is only 25 Nm/liter, which is to be expected because the efficiency was kept above 94%.

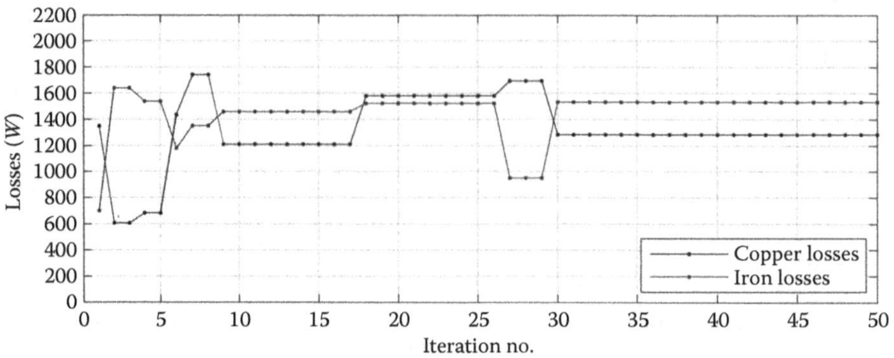

FIGURE 7.38 Copper and iron loss evolution.

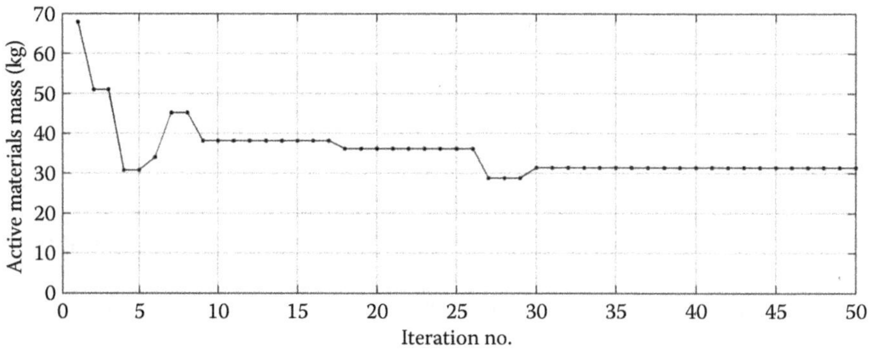

FIGURE 7.39 Active materials cost evolution.

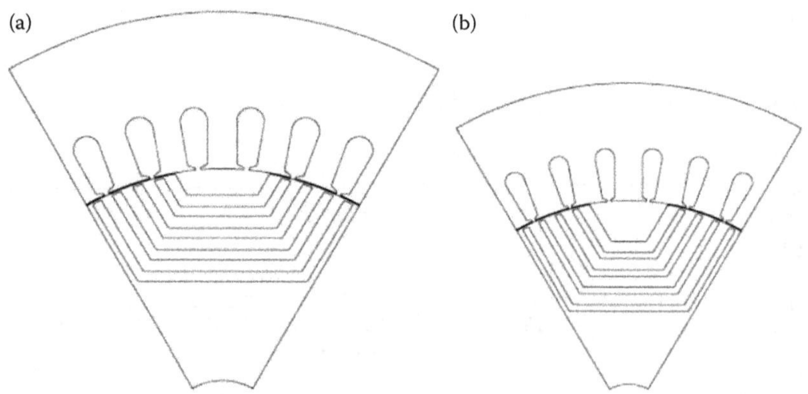

FIGURE 7.40 Initial and optimal machine cross-section.

7.8 SUMMARY

- BLDC-MRM stems from an upside-down DC brush motor with stripped-off excitation and brushes shifted from a neutral axis to the corners of former inductor poles. Starting with a two–stator pole, six–rotor slot (coils) DC brush machine, we thus obtain a two-pole six-phase BLDC-MRM where the stator holds the six-phase AC coils, and the multiple flux barriers or axially laminated anisotropic passive rotor provide(s) for low armature reaction, thus securing high torque density. Out of six phases, two (m_F) of them are situated at some moment between the rotor poles and play the role of excitation (i_F—current), while the other four (m_T), under the rotor poles momentarily, play the role of torque phases (i_T current) [1–7]. The principle may be used in multiphase IM drives as well [8].

- Mechanical commutation is replaced by electronic commutation, and the machine becomes brushless.
- The natural situation corresponds to $i_F = i_T$, when the torque-speed curve is similar to that of a series DC brush machine, known to be ideal for traction (wide CPSR):
- Torque is: $T_e = k_T \cdot m_T \cdot i_F \cdot i_T$
- A multiphase reluctance machine was obtained, which, however, operates as a DC brush machine with a power angle oscillating a little around the value of 90°.
- The current waveforms, intact with rotor position, are trapezoidal bipolar with two flat tips (i_F^* and i_T^*).
- The commutation transition of currents between + and – polarities and the i_F^* and i_T^* tops, almost linear, depends on machine inductances. This is why the machine airgap should not be too small (as is in reluctance synchronous machines), but it may not be too high, in order to keep the excitation losses ($m_F R_s i_F^2$) reasonably small for good efficiency.
- A mild magnetic saliency in the rotor, with $L_{qm}/L_{dm} \approx 4.5$, sometimes suffices to secure low enough armature reaction for high torque density before the stator core teeth oversaturate.
- The BLDC-MRM may be run:
  - At constant i_F^* but variable i_T^* (with reference torque)
  - At $i_F^* = i_T^*$ as long as the voltage ceiling in the inverter permits; this is the implicit maximum torque per current condition (MTPA: so simple to achieve here)
  - for wide CPSR

$$i_F^* = i_{Fb}^* \frac{\omega_b}{\omega_r}, \quad \text{for } |\omega_r| > |\omega_b|$$

  and torque current i_T^* decreasing to reference torque
- Stator coordinates are used for the circuit model, with rotor position–dependent machine inductance matrix.
- The torque density (Nm/Kg or Nm/liter) has been proven 40% more than in an IM with same stator and total copper losses.
- The larger airgap allows a retaining carbon fiber on the rotor to permit operation at high speeds.
- The efficiency remains high during flux weakening for wide CPRS as i_F^* decreases with speed.
- A technical theory was developed first to calculate a preliminary (initial) geometry for given specifications.
- 2D-FEM characterization of the BLDC-MRM; one of five phases and one of six phases were presented to shed light on the complexity of machine geometry, magnetic saturation, and stator and rotor core loss computation.
- Alternatively, a nonlinear magnetic equivalent circuit method was developed for BLDC-MRM and proved useful for machine modeling in optimal design. The computation time is hundreds of times smaller than for 2D-FEM, but the precision is lower, too.
- The circuit model in phase coordinates was developed with the inductance "full" matrix dependent on the rotor position by $\cos(2\theta_{er})$ as in salient pole synchronous machines; curve fitting from 2D FEM or from a flux decay standstill test is required to get simple analytical formulae for the inductance matrix.
- The circuit model may then be used to portray steady-state operation and especially transient operation without or with closed-loop speed and current control.

- With m one-phase inverters, the DC line voltage ($\pm$) is applied to each phase and phase currents are all controlled individually. The same is valid for a null-point m-phase inverter, but now the voltage applied to each phase is smaller than V_{DC}, and, in addition, the null-point voltage pulsations have to be controlled to bring symmetric current in all phases. The null current is equal to the phase current top value in odd-phase count machines and twice as much in even-count phase motors. However, the latter uses fewer power switches and allows for simpler fault-tolerant operation by grouping the phases in groups of three.
- A 4/1 or more CPSR was demonstrated at high efficiency in a five-phase BLDC-MRM.
- Machine overloading is permissible, as no PMs are present.
- A PSO-based optimal design code (ODC) with MEC for machine modeling was put in place and applied with encouraging results for high efficiency (94%) for an 160-Nm, 3000-rpm drive (for 6 hours of computation time on a single laptop). Wide CPRS designs are also feasible with ODC.
- 2D-FEM may be used parsimoniously as embedded in the ODC after the optimization cycle is done to check the key performance (torque and inductance matrix) and correct the MEC model online by fudge factors to speed up convergence. Still, tens of hours of computation on a single laptop are required.
- In conclusion, BLDC-MRM is worth trying both in drives with and without a wide CPSR, at small and high speeds and torques; the not-so-small airgap leads to room for a rotor carbon fiber retaining rig, but the remaining airgap also leads to small enough mechanical losses in high speed applications.
- BLDC-MRM may be considered a strong competitor to the brushless excited SM in most variable-speed motoring/generating drive applications.

REFERENCES

1. R. Mayer, H. Mosebach, U. Schroder, H. Weh, Inverter fed multiphase reluctance machine with reduced armature reaction and improved power density, *Record of ICEM-1986*, pp. 1138–1141.
2. I. Boldea, L. Tutelea, D. Ursu, BLDC multiphase reluctance machines for wide range applications a revival attempt, *Record of EPE-PEMC-ECCE Europe*, 2012 (IEEEXplore).
3. J. D. Law, A Chertok, T. A. Lipo, Design and performance of the field regulated reluctance machine, *Record of IEE-IAS*, 1992, vol. 1, pp. 234–241.
4. I. Boldea, G. Papusoiu, S. A. Nasar, Z. Fu, A novel series connected switched reluctance motor, *Record of ICEM-1990*, Part 3, pp. 1217–1217.
5. D. Ursu, V. Gradinaru, B. Fahimi, I. Boldea, Six phase BLDC reluctance machines: FEM based characterization and four-quadrant control, *IEEE-Trans*, vol. IA–51, no. 3, 2015, pp. 2105–2115.
6. D. Ursu, P. Shamsi, B. Fahimi, I. Boldea, 5 phase BLDC-MRM: Design, control, FEA and steady state experiments, *Record of OPTIM-2014*, pp. 354–161 (IEEEXplore).
7. E. T. Ragati, M. J. Kamper, A. D. Le Roux, Torque performance optimally designed six-phase reluctance DC machine, *Record of IEEE-IAS-2006*, vol. 3, pp. 1186–1192.
8. Y. L. Ai., M. J. Kamper, A. D. Le Roux, Move/direct flux and torque control of 6 phase IM with newly square airgap flux density, *IEEE Trans*, vol. IA–43, no. 6, 2007, pp. 1534–1541.
9. D. Ursu, Brushless DC multiphase reluctance machines and drives, PhD thesis, University Politehnica Timisoara, Romania, 2014 (in English).
10. G. Bertotti, A. Boglietti, M. Chiampi, D. Chiarabaglio, F. Fiorillo, M. Lazzari, An improved estimation of iron losses in rotating electrical machines, *IEEE Trans*, vol. MAG–27, no. 6, 1991, 5007–5009.
11. V. Ostovic, *Dynamics of Saturated Electric Machines*, book, Springer Verlag, New York, 1989.
12. J. Kennedy and R. C. Eberhart, Particle swarm optimization. *Record of IEEE International Conference on Neural Networks*, 1995, vol. 4, pp. 1942–1948.
13. M. Clerc, J. Kennedy, The PSO-explosion, stability and converge in a multicultural complex space, *IEEE Trans on Evolutionary Computation*, 2002, vol. 6, pp. 58–73.

8 Brushless Doubly-Fed Reluctance Machine Drives

8.1 INTRODUCTION

As implied in Chapter 1, the commercially named BDFRM may be considered a flux modulation reluctance machine that has two windings in the stator (both AC three-phase, for example, or one AC three-phase and one DC-fed) placed in semiclosed uniform slots and a salient pole rotor with P_r poles. The two stator AC-distributed windings, primary and secondary (controlled), have different numbers of pole pairs p_p, p_s and are called primary (power) secondary windings. The latter are fed at ω_s = variable through an AC-DC-AC (or AC-AC) PWM converter such that a speed ω_r is:

$$\omega_r = \frac{\omega_p + \omega_s}{p_p + p_s};\ p_p + p_s = p_r;\ \omega_s >/< 0 \tag{8.1}$$

where p_r is the number of rotor equivalent salient poles.

In principle, it may also be feasible to have $p_p - p_s = p_r$, but then performance is lower. The so-called primary (power) and secondary (control) windings are both on the stator, while the flux modulator is represented by the salient-pole rotor, which may be built in four main configurations (Figure 8.1).

Though the number of rotor poles p_r may be odd or even, even numbers lead to fewer radial force peaks and thus to lower noise and vibrations.

It has also been demonstrated that the higher the magnetic saliency in the rotor, the better the magnetic coupling of the two windings and thus the higher the torque for given primary (power) and secondary (control) winding mmfs. Also, it has been proven that the numbers of stator windings pole pairs p_s and p_r should not be too far apart, though this would mean that they should be rather large numbers and, consequently, BDFRM is more suitable for low-speed applications for primary fundamental frequency $f_p = 50$ (60) Hz.

The coupling between the two windings (or the flux modulation) is performed by the first harmonic (with p_r periods per revolution) of airgap magnetic permeance [1]. Alternatively, the same coupling may be performed by a nested cage rotor with p_r poles, when the brushless doubly fed induction motor (BDFIM)—proposed a century ago—is obtained [1–3]. The rotor nested cage on the rotor and its losses are notable demerits of BDFIM.

Let us consider, as an example, a two-pole primary (power) winding and a six-pole secondary (control) winding, with a four-pole rotor. The fundamentals of the mmfs of the two windings (Figure 8.2a) interact through the first (p_r periods) harmonic of the airgap magnetic permeance $(1/g)$—Figure 8.2b and c.

For simplicity, the stator slot opening effect is neglected in Figure 8.2.

The BDFRM may be built with two separate rotor distributed AC windings placed in the same slots or as a single winding (divided into three or two sections) connected to handle both roles (or one of p_p and one of p_s pole pair windings); see Figure 8.3a c.

When the secondary (control) winding is DC-fed ($i^*_c = i_{d0}$), the machine behaves as a synchronous machine with p_r pole pairs: $\omega_r = \omega_p/p_r$ ($p_r = p_p + p_s$) [4], when the power factor is improved as in any synchronous motor.

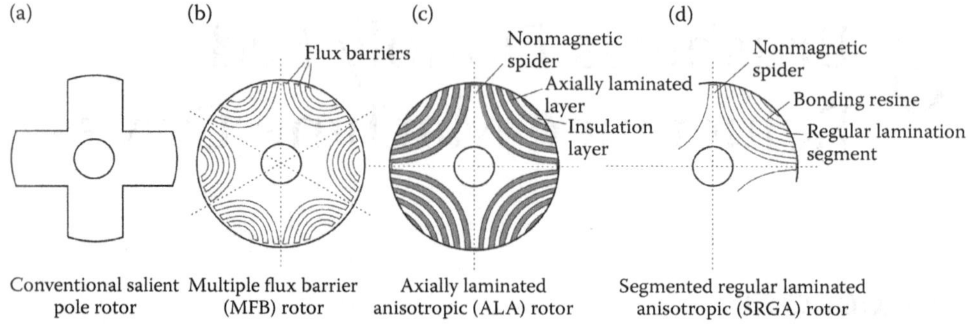

FIGURE 8.1 Typical magnetically anisotropic rotors: (a) conventional; (b) MFB-rotor; (c) ALA-rotor; (d) segmented regular laminated anisotropic (SRGA)-rotor.

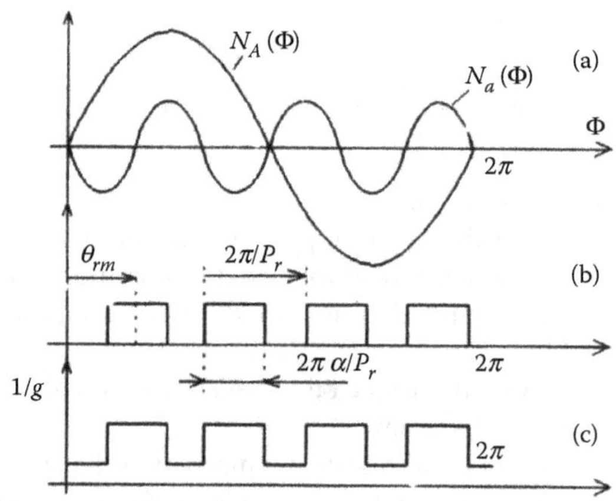

FIGURE 8.2 Winding function fundamentals $N_p(\Phi)$, $N_s(\Phi)$, (a); ideal inverse airgap, (b); actual inverse airgap $1/g$, (c).

The fabrication simplification brought in by the simple winding solution (Figure 8.2a,b,c) leads to a higher slot filling factor but also to additional copper losses in the winding with a lower number of poles. So, in general, two separate windings, eventually phase shifted by 90° mechanical degrees, may be the practical solution in many applications. It is to be shown in subsequent sections that for a limited speed range $|\omega_s| < \omega_p/3$, the power rating of the secondary (control) stator winding P_s is:

$$P_s \geq P_p \frac{\omega_{s\max}}{\omega_p} \tag{8.2}$$

With slip defined as $S = -\omega_s/\omega_p$, the BDFRM becomes somewhat similar to BDFIM, but it has better efficiency, as only core losses occur in the winding-less (passive) rotor.

This is how the main merit of BDFRM, besides being brushless, comes into play: partial rating of secondary (control) winding power converter. If the power winding unity power factor is targeted, the KVA rating of the converter has to be raised notably. To grasp the BDFRM operation, its phase coordinate and dual dq model are to be treated in what follows.

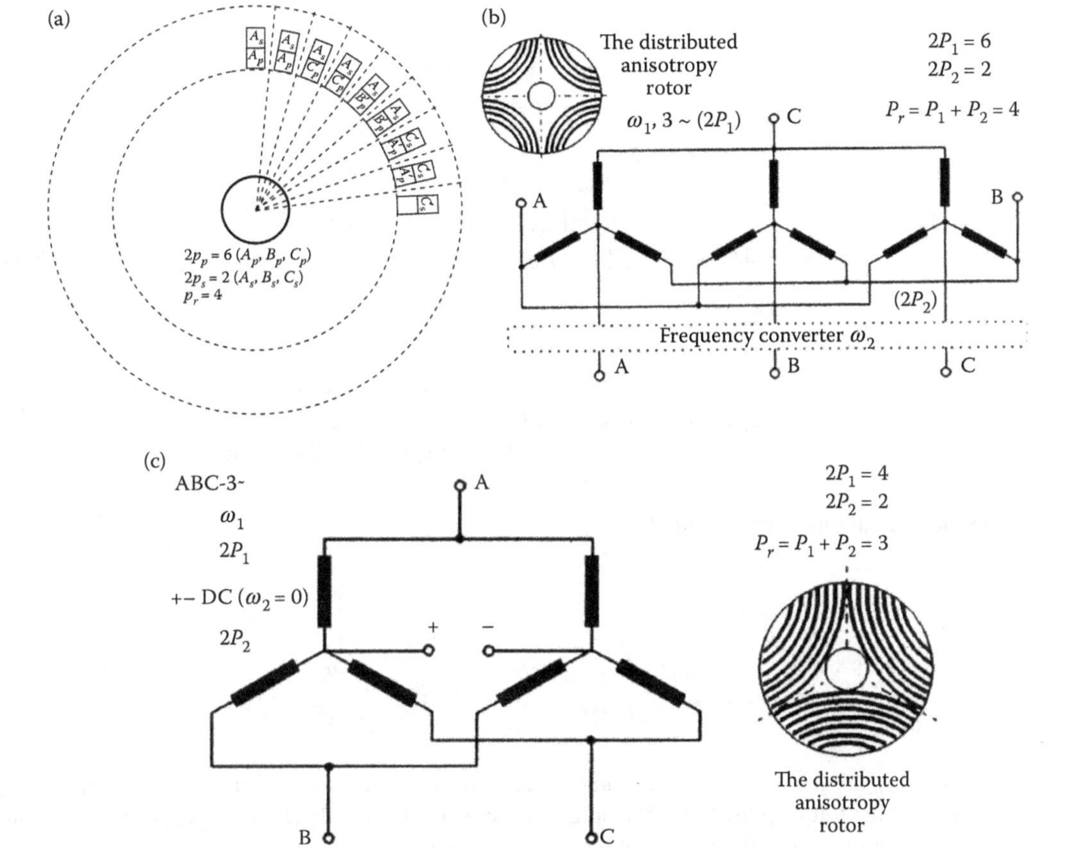

FIGURE 8.3 BDFRM stator windings: (a) separate AC windings ($2p_p = 6, 2p_s = 2, p_r = 4$); (b) single AC + AC windings ($2p_p = 6$, $2p_s = 2$, $p_r = 4$); (c) single AC + DC winding ($2p_p = 4$ AC, $2p_s = 2$, DC $p_r = 3$).

8.2 PHASE COORDINATE AND dq MODEL

The voltage equations in phase (stator) coordinates are simply:

$$[V_{A_pB_pC_p}] = [R_p][i_{A_pB_pC_p}] \tag{8.3}$$

$$[V_{As}V_{Bs}V_{Cs}] = [R_s][i_{A_sB_sC_s}] + \frac{d}{dt}[\psi_{A_sB_sC_s}] \tag{8.4}$$

where $[R_p] = \text{Diag}\ R_p$, $[R_s] = \text{Diag}\ R_s$ are three component diagonal matrixes and $[V_{A_pB_pC_p}]$, $[i_{A_sB_sC_s}]$, $[\psi_{A_pB_pC_p}]$ are three term column matrixes (the first two) and a 3×3 matrix (the third-order variable: flux linkage matrix).

The relationship between the phase flux linkage and currents are standard:

$$\underset{3\times1}{[\psi_{A_pB_pC_p}]} = \underset{3\times3}{[L_{A_pB_pC_pA_pB_pC_p}]}\underset{3\times1}{[i_{A_pB_pC_p}]} + \underset{3\times3}{[L_{A_pB_pC_pA_sB_sC_s}]}\underset{3\times1}{[i_{A_pB_pC_p}]} \tag{8.5}$$

$$\underset{3\times1}{[\psi_{A_sB_sC_s}]} = \underset{3\times3}{[L_{A_sB_sC_sA_sB_sC_s}]}\underset{3\times1}{[i_{A_sB_sC_s}]} + \underset{3\times3}{[L_{A_pB_pC_pA_sB_sC_s}]}\underset{3\times1}{[i_{A_pB_pC_p}]} \tag{8.6}$$

with

$$[L_{A_pB_pC_pA_pB_pC_p}] = \begin{vmatrix} L_{A_pA_p} & L_{A_pB_p} & L_{A_pC_p} \\ L_{B_pA_p} & L_{B_pB_p} & L_{B_pC_p} \\ L_{C_pA_p} & L_{C_pB_p} & L_{C_pC_p} \end{vmatrix} \quad (8.7)$$

$$[L_{A_sB_sC_sA_sB_sC_s}] = \begin{vmatrix} L_{A_sA_s} & L_{A_sB_s} & L_{A_sC_s} \\ L_{B_sA_s} & L_{B_sB_s} & L_{B_sC_s} \\ L_{C_sA_s} & L_{C_sB_s} & L_{C_sC_s} \end{vmatrix} \quad (8.8)$$

$$[L_{A_pB_pC_pA_sB_sC_s}(\theta_{rm})] = \begin{vmatrix} L_{A_pA_s} & L_{A_pB_s} & L_{A_pC_s} \\ L_{B_pA_s} & L_{B_pB_s} & L_{B_pC_s} \\ L_{C_pA_s} & L_{C_pB_s} & L_{C_pC_s} \end{vmatrix} \quad (8.9)$$

The motion equations may be added:

$$\begin{aligned} & J\frac{d\omega_{rm}}{dt} = T_e - T_{\text{load}} \\ & T_e = \frac{1}{2}[i_{A_pB_pC_p}]\frac{\partial[L_{A_pB_pC_pA_sB_sC_s}(\theta_{rm})]}{\partial\theta_{rm}}[i_{A_pB_pC_p}]^T; \quad \frac{d\theta_{rm}}{dt} = \omega_{rm} \end{aligned} \quad (8.10)$$

where ω_{rm} and θ_{rm} are the mechanical angular speed and rotor axis angle with the axis of phase A_p. In order to reduce the torque pulsations, it would be practical to have only $[L_{A_pB_pC_pA_sB_sC_s}]$ dependent on rotor position, like in a nonsalient-pole rotor synchronous machine.

The stator slotting will be considered here only by the Carter coefficient (Kc), and the rotor slotting (due to flux barriers) is neglected in terms of airgap permeance pulsations. Thus, only the simplified (equivalent) 1/(gKc) stepwise function in Figure 8.2b is considered.

Pursuing the method of winding functions and considering only their fundamentals [5] per pole as $M_p(\varphi)$ and $M_s(\varphi)$, Figure 8.2a, the neutral (airgap) inductance between any two phases of the two windings $L_{ij}(\theta_{rm})$ is:

$$L_{ij}(\theta_{rm}) = \mu_0 \cdot l_{\text{stack}}\frac{p_p\tau_p}{\pi} \cdot \int_0^{2\pi} g_e^{-1}(\theta_{rm}, \varphi) \cdot N_i \cdot \sin(p_p, \varphi) \cdot N_j \cdot \sin(p_s, \varphi)d\varphi \quad (8.11)$$

where τ_p is the pole pitch of the primary (power winding) and l_{stack} is the stator core axial length. The equivalent inverse airgap function $g_e^{-1}(\theta_{rm}, \varphi)$—Figure 8.2b—is:

$$\begin{aligned} & g_e^{-1}(\theta_{rm}, \varphi) = (g \cdot K_c)^{-1} \\ & \theta_{rm} + \frac{\pi}{p_r}(2n - \alpha) \le \varphi \le \theta_{rm} + \frac{\pi}{p_r}(2n + \alpha) \\ & g_e^{-1}(\theta_{rm}, \varphi) = 0 \\ & \theta_{rm} + \frac{\pi}{p_r}(2n + \alpha) < \varphi < \theta_{rm} + \frac{\pi}{p_r}(2n + 2 - \alpha) \end{aligned} \quad (8.12)$$

Here $n = 0, 1, 2, p_r - 1$.

The ideal inverse airgap function here was adopted for simplicity, while a more involved one may be adopted for more precision. In such conditions, $L_{A_pA_s}$ is:

$$L_{A_pA_s} = \mu_0 \frac{p_p \tau_p}{\pi} \frac{l_{\text{stack}}}{gK_c} N_p N_c \cdot \begin{pmatrix} -\frac{\sin(p_p - p_s)}{p_p - p_s} \cdot \frac{\pi\alpha}{p_r} \cdot \\ \sum_{n=0}^{p_r-1} \left(\cos\left[(p_p - p_s)\theta_{rm} + \frac{2\pi n(p_p - p_s)}{p_r} \right] \right) \\ + \sin\frac{p_p + p_s}{p_r} \cdot \frac{\pi\alpha}{p_r} \cdot \sum_{n=0}^{p_r-1} \left(\cos\left[(p_p + p_s)\theta_{rm} + 2\pi n \right] \right) \end{pmatrix} \tag{8.13}$$

If $(p_p - p_s)/p_r$ is not an integer, the first term in Equation 8.13 becomes zero and $L_{A_pA_s}$ is $(p_p + p_s = p_r)$:

$$L_{A_pA_s} = L_{mps} \cos(p_r \cdot \theta_{rm}) \tag{8.14}$$

$$L_{mps} = \mu_0 \cdot p_p \cdot \tau_p \cdot l_{\text{stack}} \frac{N_p N_s \alpha}{g \cdot K_c} \frac{\sin(\pi\alpha)}{\pi\alpha} \tag{8.15}$$

Similarly:

$$\begin{aligned} L_{B_pC_s} &= L_{C_pB_s} = L_{A_sA_p} = L_{A_pA_s} = L_{mps} \cdot \cos(p_r \cdot \theta_{rm}) \\ L_{A_pB_s} &= L_{B_pA_s} = L_{C_sC_p} = L_{C_pC_s} = L_{mps} \cdot \cos\left(p_r \cdot \theta_{rm} - \frac{2\pi}{3} \right) \\ L_{A_pC_s} &= L_{C_pA_s} = L_{C_sC_p} = L_{C_pC_s} = L_{mps} \cdot \cos\left(p_r \cdot \theta_{rm} + \frac{2\pi}{3} \right) \end{aligned} \tag{8.16}$$

Now if $2p_p/p_r$ and $2p_s/p_r$ are also not integers, the self and mutual inductances within the two windings end up independent of rotor position θ_{rm}:

$$\begin{aligned} L_{A_pA_p} &= L_{B_pB_p} = L_{C_pC_p} = L_{p\sigma} + L_{mp} \\ L_{A_pB_p} &= L_{A_pC_p} = L_{B_pC_p} = -\frac{1}{2} L_{mp} \\ L_{A_sA_s} &= L_{B_sB_s} = L_{C_sC_s} = L_{s\sigma} + L_{ms} \\ L_{A_sB_s} &= L_{A_sC_s} = L_{B_sC_s} = -\frac{1}{2} L_{ms} \end{aligned} \tag{8.17}$$

with:

$$L_{mp} = \frac{\mu_0 \cdot p_p \cdot \tau_p \cdot l_{\text{stack}} \cdot N_p^2 \cdot \alpha}{g \cdot K_c}; \; L_{ms} = \frac{\mu_0 \cdot p_s \cdot \tau_s \cdot l_{\text{stack}} \cdot N_s^2 \cdot \alpha}{g \cdot K_c} \tag{8.18}$$

and:

$$N_p = \frac{2}{\pi} \frac{N_1 \cdot K_{w1}}{p_p}; \; N_s = \frac{2}{\pi} \frac{N_2 \cdot K_{w2}}{p_s} \tag{8.19}$$

with N_1, N_2, K_{w1}, K_{w2} the number of turns per phase and the winding factor of the primary (power) and secondary (control) windings.

So:

$$L_{mp} = \frac{4}{\pi^2}\frac{\tau_p \cdot l_{\text{stack}}(N_1 K_{w1})^2 \cdot \alpha}{p_p \cdot g \cdot K_c}; \; L_{ms} = \frac{4}{\pi^2}\frac{\tau_s \cdot l_{\text{stack}}(N_2 K_{w2})^2 \cdot \alpha}{p_s \cdot g \cdot K_c} \quad (8.20)$$

The space phase vectors of the flux linkages of the two windings are in their phase coordinates:

$$\begin{aligned}\bar{\psi}^*_{A_pB_pC_p} &= \frac{2}{3}(\psi_{A_p} + a \cdot \psi_{B_p} + a^2 \cdot \psi_{C_p}) = L_p \bar{i}_{A_pB_pC_p} + L_m \bar{i}^*_{A_sB_sC_s} \cdot e^{j\theta_{er}} \\ \bar{\psi}^*_{A_sB_sC_s} &= \frac{2}{3}(\psi_{A_s} + a^2 \cdot \psi_{B_s} + a \cdot \psi_{C_s}) = L_s \bar{i}_{A_sB_sC_s} + L_m \bar{i}^*_{ApBpCp} \cdot e^{-j\theta_{er}}\end{aligned} \quad (8.21)$$

with

$$\begin{aligned}&\theta_{er} = p_r\theta_{rm} \quad \text{and} \quad L_p = L_{p\sigma} + \frac{3}{2}L_{mp} \\ &L_s = L_{s\sigma} + \frac{3}{2}L_{ms},\, L_m = \frac{3}{2}L_{mps}\end{aligned} \quad (8.22)$$

$L_{p\sigma}$ and $L_{s\sigma}$ are the leakage inductances per phase.

The *dq* (space phase) model is needed to eliminate the rotor position angle in Equation 8.21. A random *dq* reference system may be used, as defined by its angle θ_b, because only the coupling inductances between two windings depend on rotor position.

For the secondary (control) winding, the transfer matrix is applied to conjugates:

$$\bar{f}_{dqp} = \frac{2}{3} f_{A_pB_pC_p} e^{-j\theta_b}, \bar{f}_{dqs} = \frac{2}{3}\bar{f}^*_{A_sB_sC_p} e^{-j(\theta_b - \theta_{er})} \quad (8.23)$$

As visible in Equation 8.23, the Park transformations are valid in two different systems of coordinates, defined by θ_b and $\theta_b - \theta_{er}$ for the two windings.

The Park transformation matrixes are used to get the *dq* variables directly:

$$\begin{aligned}f_{d_pq_pn_b} &= [P(\theta_b)] f_{A_pB_pC_p} \\ f_{d_sq_sn_s} &= [P(\theta_b - \theta_{er})] f^*_{A_sB_sC_s}\end{aligned} \quad (8.24)$$

with

$$[P(\theta_{er})] = \frac{2}{3}\begin{bmatrix} \cos(-\theta_{er}) & \cos\left(-\theta_{er} + \frac{2}{3}\pi\right) & \cos\left(-\theta_{er} - \frac{2}{3}\pi\right) \\ \sin(-\theta_{er}) & \sin\left(-\theta_{er} + \frac{2}{3}\pi\right) & \sin\left(-\theta - \frac{2}{3}\pi\right) \\ \frac{1}{2} & \frac{1}{2} & \frac{1}{2} \end{bmatrix} \quad (8.25)$$

Finally, Equations 8.21 through 8.25 lead to the *dq* model of BDFRM:

$$\begin{aligned}\bar{V}_{d_pq_p} &= R_p \bar{i}_{d_pq_p} + \frac{d}{dt}\bar{\psi}_{d_pq_p} + j\omega_b \bar{\psi}_{d_pq_p};\; \omega_b = \frac{d\theta_b}{dt} \\ \bar{V}_{d_sq_s} &= R_s \bar{i}_{d_sq_s} + \frac{d}{dt}\bar{\psi}_{d_sq_s} + j(\omega_b - \omega_r)\bar{\psi}_{d_sq_s};\; \omega_r = p_r \cdot \omega_{rm}\end{aligned} \quad (8.26)$$

and

$$\bar{\psi}_{d_pq_p} = L_p \cdot \bar{i}_{d_pq_p} + L_m \cdot \bar{i}_{d_sq_s}$$
$$\bar{\psi}_{d_sq_s} = L_s \cdot \bar{i}_{d_sq_s} + L_m \cdot \bar{i}_{d_pq_p} \tag{8.27}$$

The model (Equations 8.26 and 8.27) is similar to that of the doubly fed induction machine if we only define slip as: $S = -\omega_s/\omega_p$.

Synchronous coordinates ($\omega_b = \omega_p$) may be profitable for field-oriented control, while stator coordinates ($\omega_b = 0$) are adequate for the case of $\omega_s = 0$: DC excitation of secondary (control) winding, when the case of a stator "DC-excited cageless rotor nonsalient pole" synchronous machine is met.

The electromagnetic torque in the *dq* model is simply:

$$\begin{aligned} T_e &= \frac{3}{2}p_r(\psi_{d_s}i_{q_s} - \psi_{q_s}i_{d_s}) \\ &= \frac{3}{2}p_rL_m(i_{d_p}i_{q_s} - i_{q_p}i_{d_s}) \\ &= \frac{3}{2}pL_m(i_{d_s}i_{q_p} - i_{q_p}i_{d_s}) \end{aligned} \tag{8.28}$$

We may now reduce the secondary (control) winding to the primary (power) winding to obtain:

$$\begin{aligned} L'_{ms} &= L_{ms} \cdot \left(\frac{N_p}{N_s}\right)^2 = L_{mp};\ L'_m = \frac{3}{2}L_{mps} \cdot \frac{N_p}{N_s} = \frac{3}{2}L_{mp} \cdot K_p \\ K_p &= \frac{\sin(\pi\alpha)}{\pi\alpha} < 1;\ \text{with } \alpha = \frac{1}{2} \Rightarrow K_p = \frac{2}{\pi}! \end{aligned} \tag{8.29}$$

Now $L'_m < (3/2) \cdot L_{mp}$, which is the strong difference with respect to the doubly fed inductance machine (DFIG) counterpart as:

$$\begin{aligned} L_p - L'_m &= L_{p\sigma} + \frac{3}{2}L_{mp}(1 - K_p) > L_{p\sigma} \\ L'_s - L'_m &= L'_{s\sigma} + \frac{3}{2}L_{mp}(1 - K_p) > L'_{s\sigma} \end{aligned} \tag{8.30}$$

So there is an additional large leakage inductance $L_{m\sigma}$ that corresponds to the average airgap of the machine: $L_{m\sigma} = (3/2)L_{mp}(1 - K_p)$. In the *dq* model (with the same reference systems for both windings ω_b), all variables have the same frequency (in both windings), but in reality, it is ω_p in the primary (power) winding that refers to stator coordinates ($\omega_b = 0$) and ω_s for the secondary (control) winding in the same coordinates.

So, for steady state $d/dt \rightarrow j\omega_p$ and $j\omega_s$ for the dq voltages of the two windings. The *dq* equivalent circuit for steady state in stator coordinates is shown in Figure 8.4.

The *dq* model of BDFRM leads to remarks such as the following:

- The additional inductance ($L_{m\sigma}$) in both sides of Figure 8.4 dependent on ($1 - K_p$) should be made as small as possible to secure an acceptable power factor and thus lower the kilo volt ampere (KVA) rating in the PWM converter that feeds the secondary (control) windings. To do so, p_p and p_q should be close to each other, but the shaping of rotor saliency will also help.
- To do so, the MFBA-rotor or the ALA-rotor (Figure 8.1b and c) should be adopted, as explained later in the chapter.
- The slip S expression ($S = -\omega_s/\omega_p$) contains the sign minus because $\omega_r = \omega_{rm}(p_p + p_s) = \omega_p + \omega_s$, where ω_s is positive (same phase sequence with primary windings) in

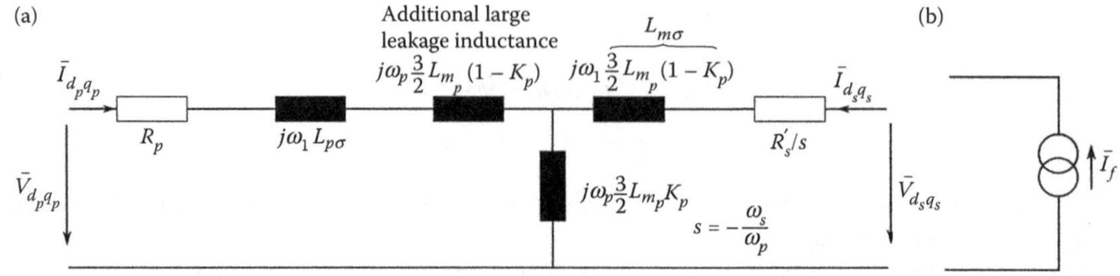

FIGURE 8.4 *dq* equivalent circuit of BDFRM in steady state: both windings are AC fed (a); secondary (control) winding is DC fed from a three phase winding (b).

oversynchronous operation and negative (opposite sequence of phase in the secondary windings) in undersynchronous operation. Synchronous operation is considered for $\omega_s = 0$ and thus $\omega_r = \omega_p/p_r$.

- When the secondary (control) winding is DC-fed $S = 0$, synchronous machine-like operation is obtained.
- The right part of the equivalent circuit in Figure 8.4 becomes in this case a current DC source (Figure 8.4b):

$$\bar{i}_F = i_{dc};\ V_s = \frac{3}{2}R_s \cdot i_{dc} \tag{8.31}$$

- As the secondary (control) winding is three phase, its DC resistance is $(3/2)R_s$, which explains the DC control voltage V_s expression in Equation 8.31.
- Finite-element analysis of ALA-rotor BDFRM has shown that, despite nonsinusoidal airgap flux density, the mutual inductances between the two windings vary rather sinusoidally with rotor position [6], as in Equations 8.14 through 8.16.
- The additional inductance $L_{m\sigma}$ (Figure 8.4) both in the primary (power) and secondary (control) windings will have a beneficial effect in the sense of reducing the overcurrents during large symmetric and asymmetric voltage sags or swells in the power grid, in contrast to the regular DFIG (with slip-rings and brushes and the secondary winding on the rotor), again, at the price of a lower power factor.
- It should be mentioned that a BDFRM may also operate as an induction motor if the secondary (control) winding is short-circuited; in this case, ω_s is given by ω_p and the speed ω_r: $\omega_s = \omega_r - \omega_p$.
- Also, it should be noted that the key torque factor (Equation 8.29) is

$$L_m = \frac{3}{2}L'_{mps} = \frac{3}{2}L_{mp} \cdot \alpha \cdot \frac{\sin(\pi\alpha)}{\pi\alpha} = \frac{3}{2}L_{mp} \cdot K_{ps} \tag{8.32}$$

both α (ratio of rotor pole span/pole pitch) and $\sin(\pi\alpha)/\pi\alpha$ are smaller than unity; consequently, the torque density is inevitably smaller than for DFIG, for similar mmfs in the windings. Minimizing this effect leads to the optimal design of the MFBA or ALA-rotors (Figure 8.1) for BDFRM. Using a sinusoidal analytical model has led to values of K_{ps} from 0.11 to 0.5 for various pole pair combinations (p_p, p_s) and for given airgap. The higher coupling factor is obtained with $p_p > p_s$ [7].

- As expected, the maximum torque is obtained when the self-induced and induced voltages in each of the two windings are at 90°, or when the corresponding phase currents in the *dq* model, say, in stator coordinates, are at 90° with each other. With zero stator resistance, say, for motoring, in the primary (power) winding, the phasor diagram may be simplified

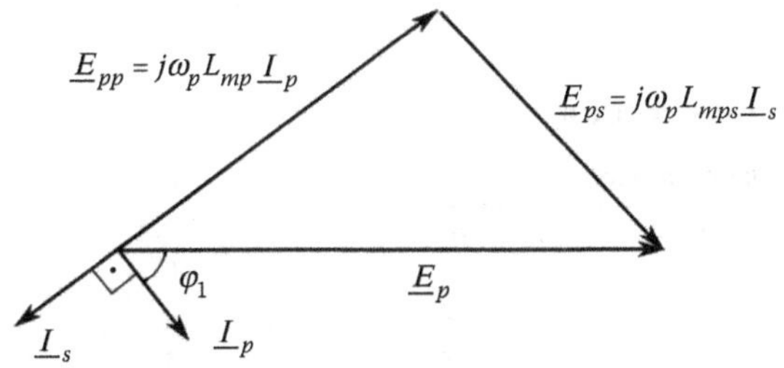

FIGURE 8.5 Power winding factor for maximum torque (motoring).

to that of Figure 8.5 where φ_s is the ideal power factor angle. It may be demonstrated [7] that the ratio of current loadings (A_p, A_s) in Aturns/m is, in this case:

$$\frac{A_p}{A_s} = \tan\varphi_s \cdot G.F. \tag{8.33}$$

$$G.F. = \frac{p_p C_{ps}}{p_s C_{pp}} \tag{8.34}$$

- $G.F.$ is the goodness factor, with C_{ps} and C_{pp} as coupling and self-coupling coefficients (C_{ii}, C_{jj}, C_{ij}), as in Tables 8.1 and 8.2 [7].

  It should be noticed that $G.F.$ depends on the combination of pole pairs p_p and p_s and that for $p_p > p_s$, the grid winding electrical loading A_p may be increased (by increasing $G.F.$) without affecting the power factor (in Equation 8.33) too much; fortunately, this would mean higher ideal coupling coefficients (Cij).

TABLE 8.1
Ideal Coupling Coefficient

Pole Numbers			Coupling Coefficients		
p_r	p_i	p_j	C_{ii}	$C_{ij} = C_{ji}$	C_{jj}
2	4	8	0.00	0.00	0.00
4	2	6	0.1817	0.3183	0.661
5	4	6	0.3831	0.4677	0.5780
6	2	10	0.0865	0.2067	0.5827
6	4	8	0.2933	0.4135	0.6034
8	2	14	0.0498	0.1501	0.5643
8	4	12	0.1817	0.3183	0.6061
8	6	10	0.3499	0.4502	0.5900
10	2	18	0.0323	0.1169	0.5520
10	4	16	0.1216	0.2523	0.5946
10	6	14	0.2477	0.3784	0.6081
10	8	12	0.3831	0.4677	0.5780
20	10	22	0.4454	0.4918	0.5447

Source: A. M. Knight et al., *IEEE Trans.*, vol. IA–49, no. 1, 2013, pp. 50–58. [7]

TABLE 8.2
Ideal Goodness Factor

Pole Numbers				
p_r	p_g	p_s	$\frac{p_s C_{pp}}{p_p C_{ps}}$	$\frac{C_{pp} C_{ss}}{C_{ps}^2}$
6	2	10	0.4780	1.1793
6	4	8	0.7050	1.0349
6	8	4	1.3706	1.0349
6	10	2	1.7741	1.1793
8	2	14	0.4300	1.2493
8	4	12	0.5840	1.0869
8	6	10	0.7718	1.0189
8	10	6	1.2716	1.0189
8	12	4	1.5755	1.0869
8	14	2	1.8614	1.2493

Source: A. M. Knight et al., *IEEE Trans.*, vol. IA–49, no. 1, 2013, pp. 50–58. [7]

- If a similar procedure is followed for the secondary (control) winding, one obtains [7]:

$$\frac{A_s}{A_p} = \tan\varphi_s \left(\frac{p_s C_{sp}}{p_p C_{ss}}\right) \tag{8.35}$$

Now multiplying Equation 8.33 with Equation 8.35 yields:

$$G_{PF} = \frac{C_{pp} C_{ss}}{C_{ps} C_{sp}} = \tan\varphi_p \cdot \tan\varphi_s;\ C_{ps} = C_{sp} \tag{8.36}$$

G_{PF} in Equation 8.36 was calculated in Reference 7 and is also given in Table 8.2. As expected, the lower the coupling ($C_{ps}C_{sp}$), the larger the power factor angles φ_p, φ_s and thus the lower the power factor. Also, for maximum torque conditions, the two power factors may not be maximized simultaneously. This comes as no surprise, as the machine magnetization may be provided either from one winding, from the other, or by both. If the primary (power) winding is run at the unity power factor ($\cos\varphi_p = 1$), the above conditions fall and thus the machine may not operate for maximum torque conditions in this situation.

- Again, for maximum torque/current (maximum torque as discussed above) in space vectors, one obtains:

$$T_e = \frac{3}{2} p_r |\bar{I}_p| \cdot |\bar{I}_s| \cdot L_m;\ \frac{\bar{I}_s}{|\bar{I}_s|} = j\frac{\bar{I}_p}{|\bar{I}_p|} \tag{8.37}$$

For steady state, the voltage equations become:

$$\begin{aligned} \bar{V}_p &= \underline{R}_s \cdot \bar{I}_p + j\omega_p(L_p\bar{I}_p + L_m\bar{I}_s) \\ \bar{V}_s &= R_s \cdot \bar{I}_s + j\omega_s(L_m\bar{I}_p + L_s\bar{I}_s) \end{aligned} \tag{8.38}$$

These expressions, with L_m, L_p, L_s calculated preliminarily, as discussed above, allow one to calculate approximately the maximum torque capability of the machine and then the active and reactive powers for each winding.

Such a calculus done in Reference 7 for an outside stator diameter of 430 mm and stack length $l_{stack} = 251$ mm BDFRM showed it capable of producing 300 Nm over a wide speed range with a power factor of 0.8 lagging in the power winding and 0.5–0.6 in the control winding. The torque density proved to be small (8.6 Nm/liter) for $A_p/A_s = 16220$ A/m/15780 A/m and grid frequency $f_p = 50$ Hz, and synchronous (normal) speed ($\omega_s = 0$) of 500 rpm. The speed range around synchronism defines the KVA of the secondary (control) winding and its PWM converter rating by the same principle as for DFIG: $P_s \geq (|\omega_{s\ max}|/\omega_p) \cdot P_p$.

- To conclude, after this preliminary investigation by electric circuit (phase coordinates and *dq* model) theories, we have to notice that the machine torque density in Nm/liter is smaller than that of the competitor's solutions, but there are no brushes. However, the power factor is impaired by an additional notable airgap inductance present in both windings ($L_{m\sigma}$ in Figure 8.4); this will lead to higher kVA in the secondary control winding PWM converter if high or even unity power factor is required in the primary (power) winding.

8.3 MAGNETIC EQUIVALENT CIRCUIT MODELING WITH FINITE-ELEMENT MODEL VALIDATION

The phase coordinate and *dq* model with winding-function-method-based inductances may be used for preliminary performance investigation and control. But the consideration of magnetic saturation and stator and rotor slotting effects on circuit model parameters on average torque, torque ripples, and radial forces is imperative for practical design purposes.

2D-FEM is the way to go, but, for optimal design codes, it requires prohibitive computation time. So FEM is to be used with care, rather to validate optimal design codes that use nonlinear analytical (realistic machine) models: MEC, for example.

Given the complexity of stator and rotor slotting, using FEM-only for optimal design codes for BDFRM within a few hours of computation time on a single desktop computer are still due.

The MEC case study here [8] refers to a four-pole power and eight-pole control three-phase winding 540-Watt, 1380-rpm BDFRM with its main specifications in Table 8.3 and main given geometry data in Table 8.4.

A rather complex equivalent magnetic circuit including the airgap, stator and rotor slots, flux barriers, and teeth is required. There are 48 slots on the stator and 11 equivalent slots/pole (6 poles) on the rotor (Figure 8.6).

Considering this structure, the MEC for three stator slots and four flux rotor barriers is as in Figure 8.7.

TABLE 8.3
Specifications

Item	Description	Value
4-pole winding	Rotor speed	1380
	Phase voltage [V]	100
	Phase current [A]	5.3
	Frequency [Hz]	60
8-pole winding	Phase voltage [V]	179
	Phase current [A]	12.1
	Frequency [Hz]	78

Source: M. F. Hsieh et al., *IEEE Trans.*, vol. Mag-50, no. 11, 2014, pp. 1–5. [8]

TABLE 8.4
Main Dimensions

Description	Value	Description	Value
Length of air gap	0.5 mm	Height of the stator shoe	1.1
Stator slot number	48	Width of the rotor tooth (induction rotor)	4.2 mm
Stack length	81 mm	Width of the rotor tooth	4.2 mm
Coil turns (4- and 8-pole)	15	Width of the rotor rim	2 mm
Stator tooth width	3.8 mm	Height of the rotor shoe (reluctance rotor)	1 mm
Width of the stator yoke	10 mm	Outer slot radius	48.4 mm
Rotor radius	70 mm	Height of the stator slot	21.1 mm

Source: M. F. Hsieh et al., *IEEE Trans.*, vol. Mag–50, no. 11, 2014, pp. 1–5. [8]

As the MEC method was described in previous chapters (see Chapter 7), we will present it only briefly here.

The main element is represented by the magnetic reluctances. R_g for airgap; R_{ts}, R_{tr} for stator and rotor teeth; R_{sn}, R_{rn} for stator and rotor poles; and the leakage reluctances R_{ls}, R_{lr} for stator and rotor shoe gaps.

There is a direct relationship between the magnetic flux in different magnetic tubes and various field sources (mmfs: F_{s1}, F_{s2}, … in Figure 8.7) via a nodal approach:

$$[G_m][V] = [\varphi_t];\ [G_m] = [R_m]^{-1} \tag{8.39}$$

It is implicit that the magnetic reluctances (R_{gi}) and their magnetic permeances G_{gi} depend on rotor position angle; thus, torque T_e is produced:

$$T_e = p_r \sum\nolimits_{i=1}^{N_s} \sum\nolimits_{j=1}^{N_r} (V_i - V_j)^2 \frac{dG_{ij}}{d\gamma} \tag{8.40}$$

The fluxes in different teeth are summed to get the flux linkage of all coils of all phases at different moments in time (with simultaneous values of all currents as assigned). Magnetic saturation is considered iteratively in the MEC code (Figure 8.8) [8].

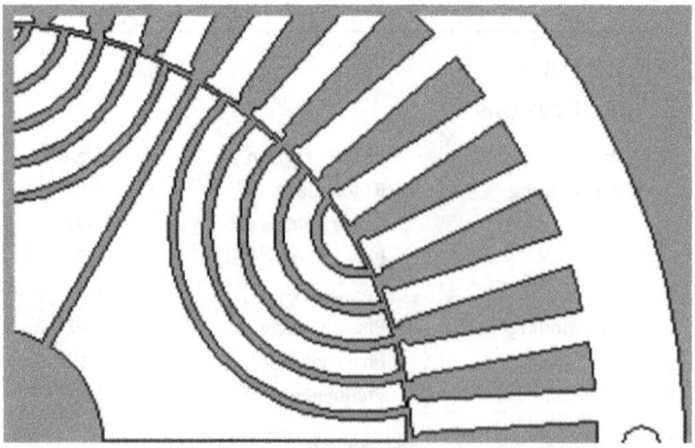

FIGURE 8.6 One-quarter cross-section: the white part represents steel and the black part refers to slots (or flux barriers). (After M. F. Hsieh et al., *IEEE Trans.*, vol. Mag–50, no. 11, 2014, pp. 1–5. [8])

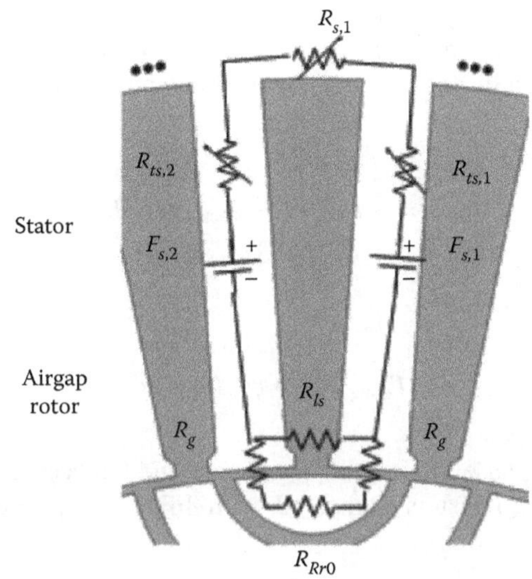

FIGURE 8.7 MEC of stator airgap and rotor section. (After M. F. Hsieh et al., *IEEE Trans.*, vol. Mag–50, no. 11, 2014, pp. 1–5. [8])

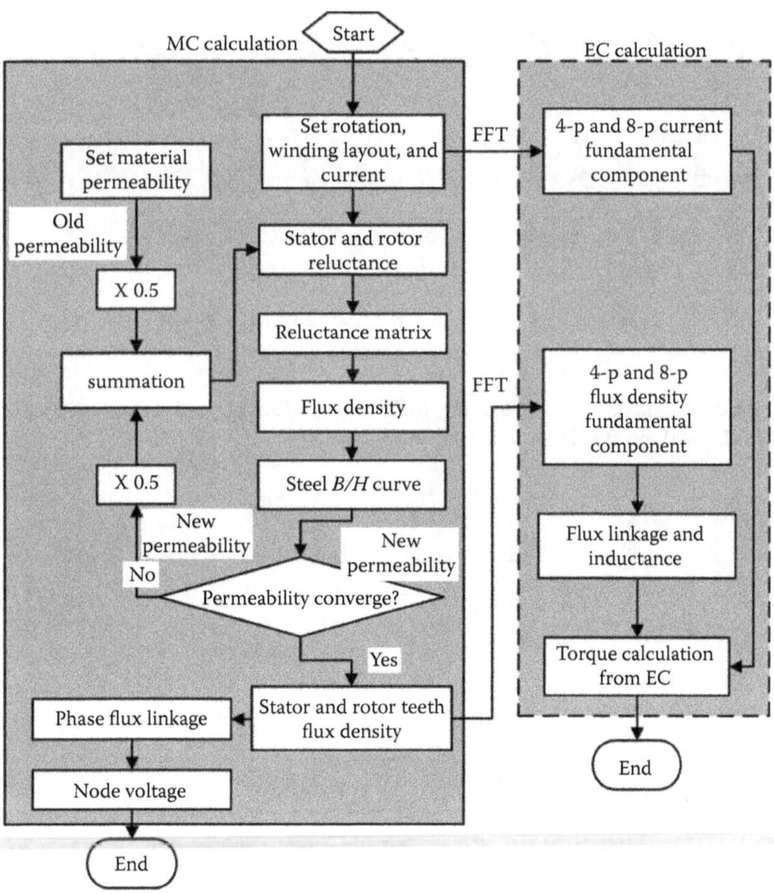

FIGURE 8.8 MEC code flowchart.. (After M. F. Hsieh et al., *IEEE Trans.*, vol. Mag–50, no. 11, 2014, pp. 1–5. [8])

The airgap flux density linking the two windings (coupling torque producing) flux density is separated by using, say, only currents in the primary (power) winding and calculating the flux linkage in the secondary (control) winding after extracting the fundamental for the secondary (control) winding period ($2\tau_s$):

$$\psi_{ps} = \frac{2}{\pi} B_{g1s} \cdot \tau_s \cdot l_{\text{stack}} \cdot N_2 \cdot K_{w2} \tag{8.41}$$

Then the average torque $T_{e\ \text{avg}}$ is:

$$T_{e\,\text{avg}} = \frac{3}{2} p_r \psi_{ps} I_p \sin \varphi_i \tag{8.42}$$

where ψ_{ps} and I_p are peak phase values and φ_i is the angle between the induced voltage and the current. Sample MEC (V_{mec}) results compared with FEM (V_{sim}) show acceptable agreements (Figures 8.9 and 8.10) [8].

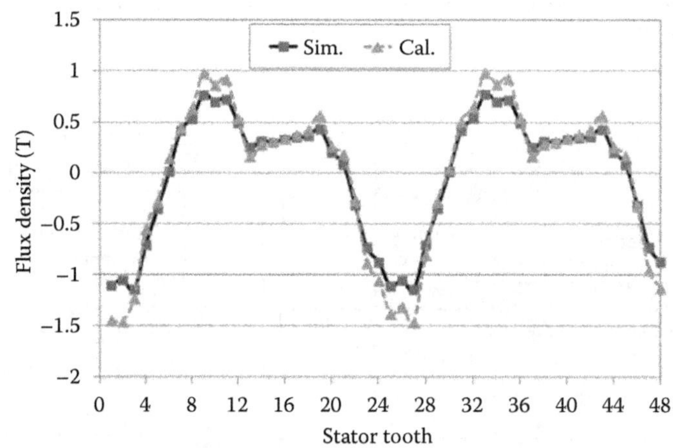

FIGURE 8.9 Flux density in the 48 stator teeth by MEC and by FEM, at rated currents. (After M. F. Hsieh et al., *IEEE Trans.*, vol. Mag–50, no. 11, 2014, pp. 1–5. [8])

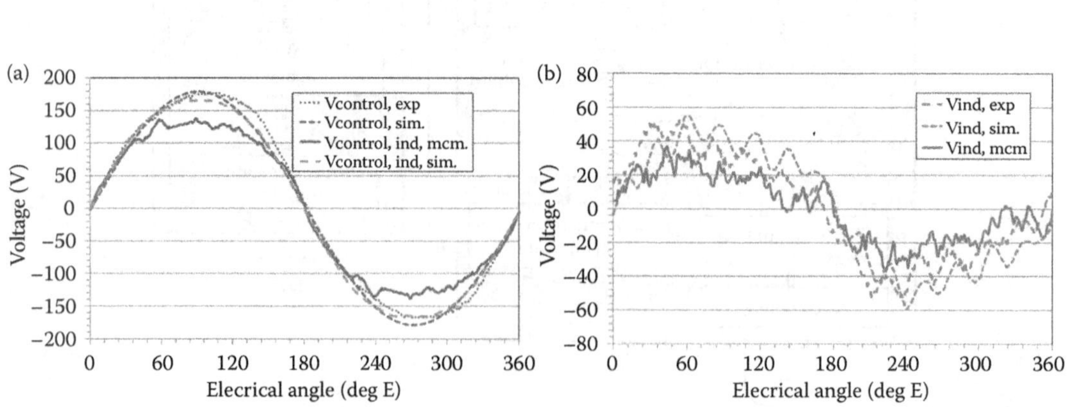

FIGURE 8.10 MEC versus FEM phase voltage waveforms (a) eight-pole winding; (b) four-pole winding. (After M. F. Hsieh et al., *IEEE Trans.*, vol. Mag–50, no. 11, 2014, pp. 1–5. [8])

The stator teeth (airgap) flux density in Figure 8.9 is calculated rather acceptably by MEC, but, as the voltage waveforms show (under steady state), there are still notable differences with respect to both experiments and FEM on the small-power prototype under investigation here (Figure 8.10).

The reduction, up to two orders of magnitude, of computation time by MEC with respect to FEM may justify the lower precision of the former method and its use in optimal design codes.

8.4 CONTROL OF BRUSHLESS DOUBLY-FED RELUCTANCE MACHINES

Both BDFIMs (with nested-cage rotor) and BDFRMs (with reluctance rotor) are credited with safer operation (lower overcurrents for low grid voltage ridethrough) due to the additional (leakage) inductance present both in the power and control windings. This way, the crowbar protection of secondary (control) winding is not necessary, in contrast to regular DFIMs. This is a solid reason to pursue BDFRM control and its implementation for both generator and motor modes in undersynchronous ($\omega_s < 0$, $S > 0$) and oversynchronous ($\omega_s > 0$, S < 0) operation modes. The rotor speed ω_{rm} is:

$$\omega_{rm} = \frac{\omega_p}{p_r}\left(1 + \frac{\omega_s}{\omega_p}\right) = \frac{\omega_p}{p_r}(1 - S);\ S = -\omega_s/\omega_p \tag{8.43}$$

while the mechanical power (at zero losses) P_m is:

$$P_m = T_e \cdot \frac{\omega_p}{p_r} + T_e\frac{\omega_r}{p_r} = P_p + P_s = P_p(1 - S) \tag{8.44}$$

There is rich literature on BDFRM control:

- Scalar control [9]
- Voltage vector oriented (VC) [10]
- Power (primary) winding flux oriented control [10]
- DTFC [11]

Other controlled strategies such as torque and reactive power control (TQR) or direct power control (DPC) are derivatives of DTFC.

All the above control strategies may be implemented with encoder feedback or encoderless, but DTFC (and DPC) seem particularly adequate for encoderless control. Here, we investigate in some detail VC and FOC as they reveal the essentials of active and reactive power control in the primary (power) winding:

$$P_p = \frac{3}{2}(V_\alpha i_{\alpha p} + V_\beta i_{\beta p});\ i_{\alpha p} = i_{Ap},\ i_{\beta p} = \frac{i_{Ap} + 2i_{Bp}}{\sqrt{3}} \tag{8.45}$$

$$\begin{aligned} Q_p &= \frac{3}{2}(V_{\beta p} i_{\alpha p} - V_{\alpha p} i_{\beta p}) \\ V_{\alpha p} &= \frac{2V_{Ap} - V_{Bp} - V_{Cp}}{3} \\ V_{\beta p} &= \frac{V_{Bp} - V_{Cp}}{\sqrt{3}} \end{aligned} \tag{8.46}$$

VC and FOC are both vector controls, but for VC, the voltage vector $\bar{V}_p$ of primary (power) winding is oriented along axis q_p, while for FOC, the same power winding flux vector is oriented along axis d_p. This difference, which reflects only the primary (power) winding resistive voltage drop ($R_p i_p$) influence leads to some coupling of active and reactive power control for VC while decoupling

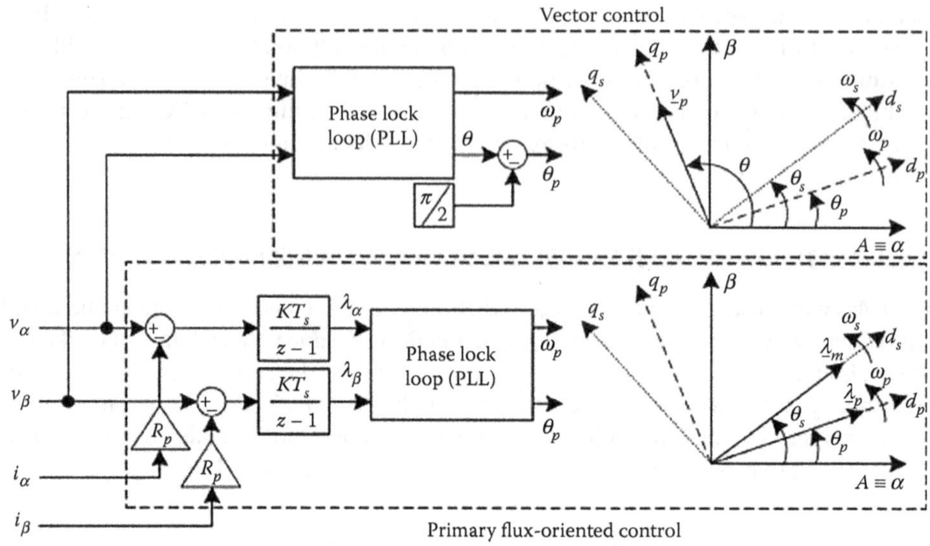

FIGURE 8.11 Primary (power) winding voltage vector angle and speed (θ_p, ω_p), its flux angle and speed estimators. (After S. Ademi et al., *IEEE Trans.*, vol. EC–30, no. 2, 2015, pp. 596–604. [10])

control is secured for FOC, at the price of an additional primary (power) winding flux estimator (Figure 8.11).

Making use of the already developed dq model, but this time at speed ω_p for the primary (power) winding and at speed ω_s for the secondary (control) winding (to secure DC variables during steady state), the torque T_e and active and reactive power P_p, Q_p may be expressed by:

$$(T_e)_{VC} = \frac{3p_r}{2}(\psi_{md}i_{sq} - \psi_{mq}i_{sd}) \tag{8.47}$$

$\psi_{md} = \psi_{mp}$ for FOC (Figure 8.11):

$$(P_p)_{VC} = (T_e)_{vc} \cdot \frac{\omega_p}{p_r};\ (Q_p)_{VC} = \frac{3}{2}\omega_p\left[\left(\frac{\psi_p^2}{L_p} - \psi_{md}i_{sd}\right) - \psi_{mq}i_{sq}\right] \tag{8.48}$$

$$(P_p)_{FOC} = \frac{3}{2}\omega_p\frac{L_m}{L_p}\psi_p i_{sq};\ (Q_p)_{FOC} = \frac{3}{2}\omega_p\psi_p i_{pd} \tag{8.49}$$

Equations 8.47 through 8.49 evince the uncoupling and, respectively, coupling character of P_p and Q_p for FOC and, respectively, for VC.

A generic VC or FOC system is portrayed in Figure 8.12 [10].

Only the secondary (control) winding-side PWM converter is shown in Figure 8.12. The grid-side converter (as part of an AC-DC-AC system) is not shown, as its control (in general V_{dc}^*, Q_c^* control by i_{dg}, i_{qg} control) is standard as for DFIG and independent of the other part, though, during voltage sags and the like, its behavior should be mandatorily considered.

Typical experimental results on a lab prototype (its data is shown in Table 8.5) are given in Figures 8.13 and 8.14 but for FOC and VC.

The results in Figures 8.13 and 8.14 warrant remarks such as:

- The two control methods (FOC and VC) show very similar speed, torque, and active power responses.

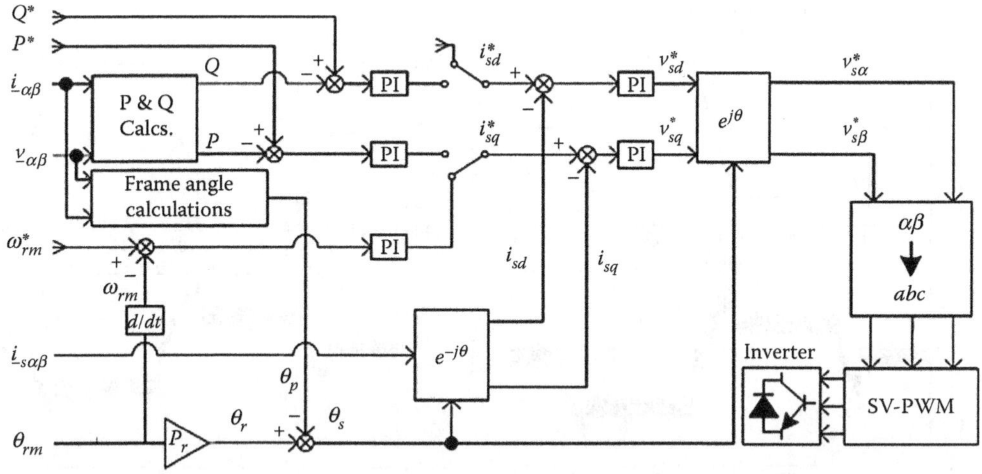

FIGURE 8.12 Generic VC or FOC system for BDFRM. (After S. Ademi et al., *IEEE Trans*, vol. EC–30, no. 2, 2015, pp. 596–604. [10])

- The results differ in terms of power winding current $i_{\varphi d}$ and reactive power, probably due to the small coupling of d-q axes in VC.
- The q-axis currents i_{pq}, i_{sq} are similar to active power and torque responses, as they are directly related to each other.
- The speed transition through synchronism (Figure 8.14)—750 rpm—shows clearly that the sequence of secondary (control) winding currents changes at synchronism ($\omega_S = 0$) when the latter become, for a while, DC currents.
- The lagging reactive power of 1200 VAR is a direct result of $i_{sd}^* = 0$ (pure active power handled by the secondary [control] windings) to secure minimum converter current (KVA in the inverter).

TABLE 8.5
BDFRG Lab Prototype

Rotor inertia $[J]$	$0.1\ \text{kg}\cdot\text{m}^2$
Primary resistance $[R_p]$	$11.1\ \Omega$
Secondary resistance $[R_s]$	$13.5\ \Omega$
Primary inductance $[L_p]$	0.41 H
Secondary inductance $[L_s]$	0.57 H
Mutual inductance $[L_m]$	0.34 H
Rotor poles $[p_r]$	4
Primary power $[P_r]$	1.5 kW
Rated speed $[n_r]$	1000 rev/min
Stator currents $[I_{p,s}]$	2.5 A/rms
Stator voltage $[V_p]$	400 V/rms
Stator frequency $[f_p]$	50 Hz
Winding connections	Y/Y
Stator poles $[p/q]$	6/2

Source: S. Ademi et al., *IEEE Trans.*, vol. EC–30, no. 2, 2015, pp. 596–604. [10]

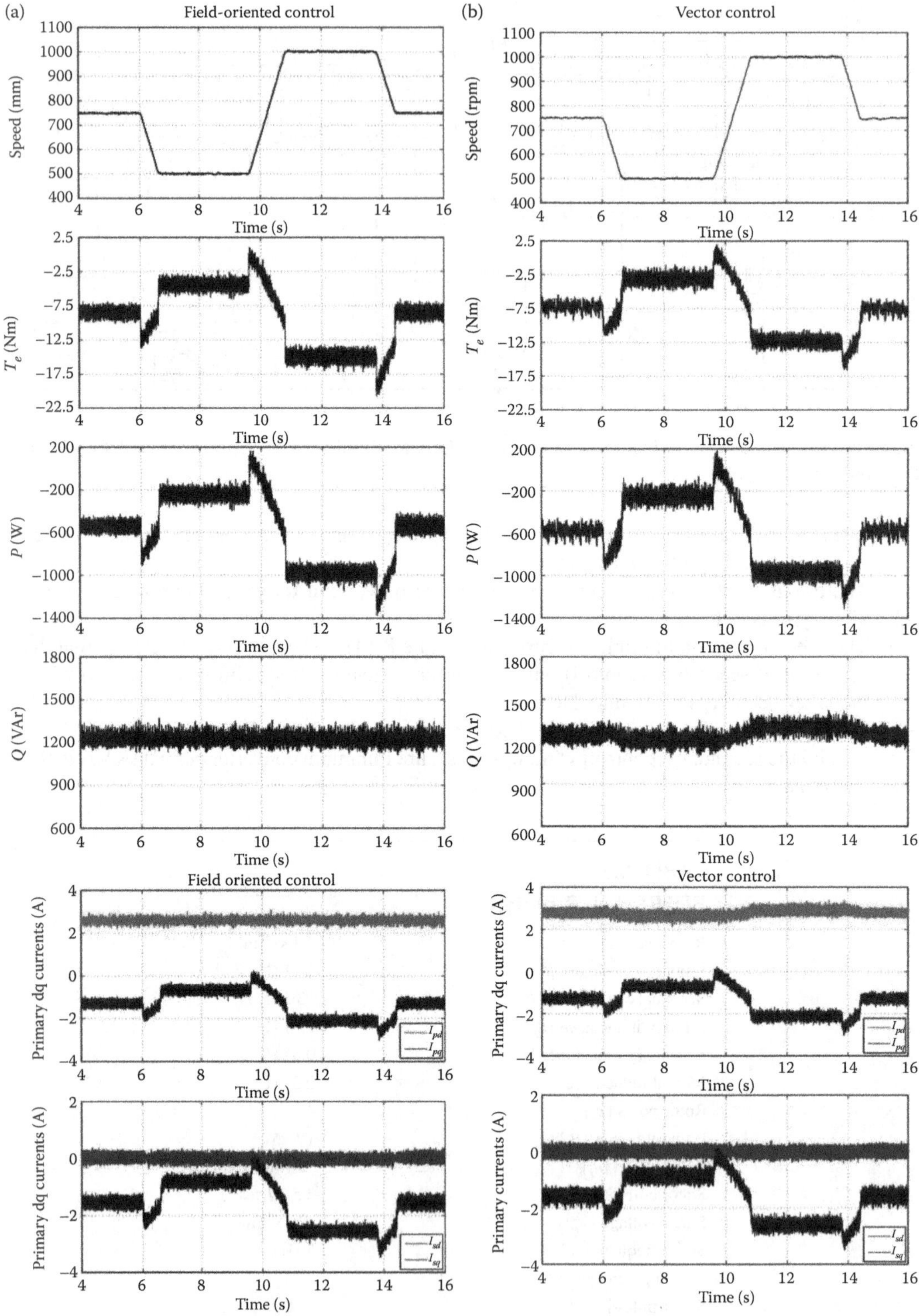

FIGURE 8.13 Experimental speed, torque, primary (power) and secondary (control) winding dq current transients: (a) FOC; (b) FC. (After S. Ademi et al., *IEEE Trans.*, vol. EC–30, no. 2, 2015, pp. 596–604. [10])

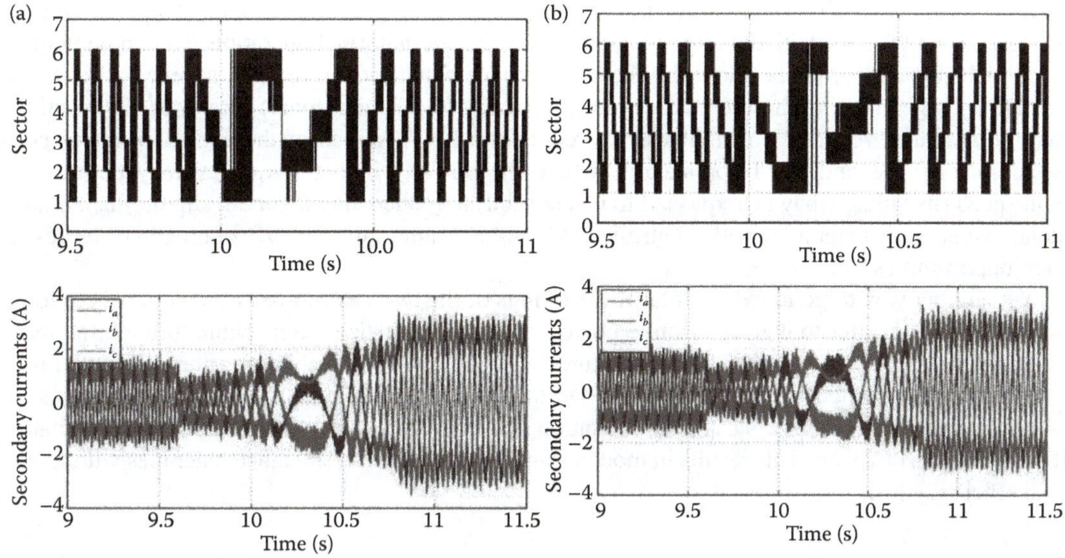

FIGURE 8.14 Experimental secondary (control) winding voltage and currents for a speed transition through synchronism ($\omega_s = 0$): (a) FOC; (b) FC. (After S. Ademi et al., *IEEE Trans.*, vol. EC–30, no. 2, 2015, pp. 596–604. [10])

- If unity power factor in the stator is needed, the control should have $i^*_{sd} \neq 0$ and the inverter KVA ratings (and cost) increase notably (even by 100% from $|S_{max}|/P_p$).
- Quantitative analysis of BDFRM behavior in the presence of asymmetric deep power grid voltage sags, smooth transitions from grid to autonomous (islanding), back operation, and so on all still need to be investigated. Notable refinements in encoderless control are also expected.

8.5 PRACTICAL DESIGN ISSUES

The design of BDFRM was treated far less than its control [12–14]. It is suggested that this machine may be advantageously designed even for a 2-MW wind generator for $\omega_p = 50$ Hz, $f_s \leq 50$ Hz, and speed range from 500 rpm to 1000 rpm when the power and control windings each deliver 50% of full power 2 MW at 1000 rpm. At 750 rpm, $f_s = 25$ Hz, the control winding may produce 0.66 MW with the power winding delivering 1.14 MW [12]. In both cases, the power factors are low (0.87 lag and 0.12 lagging for 1000 rpm and 0.58 lag and 0.33 lag for 750 rpm, both at about 1.8–1.9 MW delivered). It is evident that such a small power factor in the control winding (0.2, then 0.3) would lead to a hardly acceptable KVA rating of the converters, which have to deliver 0.66 to 0.9 MW of active power in the design.

The additional (leakage) inductance—due to average airgap permeance—that occurs both in the power and control winding by principle leads to this situation. Yes, this additional inductance limits the currents in both windings in case of power grid deep voltage sags or swells, but the price in the control winding inverter oversizing hardly seems practical. Moreover, the 7–10 Nm/liter torque density, in a 0.74 m outer stator diameter, 200 kW (at 1200 rpm, with minimum speed of 600 rpm), though offering 94% efficiency for 70% of operation points in the speed-torque envelope, is far less than with competitive solutions. The PWM dual AC-DC-AC converter KVA sizing at the unity power factor in the power winding should be close to 100%—as, for example, in cage rotor induction generators with full ratings (100%) and similar PWM converters. A pertinent analysis of these aspects is available in [15], but for a BDFIM (with nested cage rotor).

It seems that disruptive improvements, eventually adding capacitors in series with power winding to compensate for the additional airgap inductance is needed in normal operation. To handle voltage sags, such capacitors may be short-circuited. Resonance conditions have to be avoided all over the operation range. Trading these capacitors for drastically lower (additional) converter KVA ratings may be practical. Another potential way to reduce the KVA ratings of the control winding converter may be the finding of applications where the power winding frequency may be variable with speed (its voltage may be requested to be constant) to reduce the maximum slip (ω_s/ω_p). Standalone (vehicular) generators with controlled AC voltage amplitude or DC output may represent such opportunities.

Yet another way to go to reduce total KVA ratings of the two back-to-back converters is to force the grid side converter to deliver reactive power to partly magnetize the machine through the main winding. Still, that additional inductance (Figure 8.4) raises the KVA to higher values than in DFIGs.

Finally, when the control winding is DC fed ($\omega_s = 0$) a synchronous machine with a large synchronous inductance and a slow DC excitation response (all due to the additional inductance [Figure 8.4]) is obtained. This results in moderate-high power factor reluctance machines with favorable efficiency [4].

8.6 SUMMARY

- BDFRMs, similar to BDFIMs, have a stator with two AC windings of p_p and p_s pole pairs placed in uniform semiclosed slots of a laminated iron core. Their rotor has $p_r = p_p + p_s$ poles, either salient (reluctance) or of nested cages (BDFIM).
- One power winding of p_p pole pairs is connected to the grid or kept in general at constant V_r, f_p while the other is fed from a power converter at variable V_s, f_s.
- The two windings interact through the first harmonic of airgap permeance (for BDRM) to produce constant torque (power) if the speed $\omega_{rm} = (\omega_p + \omega_s)/p_r$.
- But the constant/average component of airgap permeance is reflected in an additional large constant inductance $L_{m\sigma}$ (Figure 8.4) in both windings (same value if the secondary [control] winding is reduced to the primary [power] winding). This additional inductance does not produce torque, but consumes reactive power, thus leading to inherent low power factor.
- The torque for given geometry and mmfs in the two windings is maximum when the two windings currents (in a single-speed *dq* reference system) are phase shifted 90° (in time).
- Though BDFRM has a salient pole rotor, it behaves like a nonsalient pole machine in the sense that only the coupling inductances between two windings vary with rotor position.
- Coupling inductance is characterized by a coupling factor $K_{ij} \leq 0.5$, indicating limited torque density in comparison with other AC machines with the same geometry and airgap. But it is brushless, and the ideal power rating of secondary control winding $P_s = P_p(|\omega_{s\,\max}|/\omega_p)$; $\omega_s < 0$ for undersynchronous and $\omega_s > 0$ for over synchronous speed ω_{syn} operation ($\omega_{\text{syn}} = (\omega_{rm}) = \omega_r/p_r$). Unfortunately, the sizing of the secondary (control) winding PWM converter, even for $P_s < 0.3P_p$, is not possible unless all reactive power in the machine is extracted through the primary (power) winding from the power grid when the power factor is rather low.
- At the other extreme of operation, at the unity power factor in the power winding, a notable oversizing (+100%) of the KVA in the control winding converter is required.
- To alleviate the situation, the grid-side PWM converter (part of the AC-DC-AC system that feeds the secondary winding) may be controlled not only at given DC link voltage, but also to deliver reactive power ($Q_q > 0$) from the DC link capacitor and thus counteract the reactive power required in the stator with the unity power factor in the control winding.

- Apart from simplified parameter formulae in phase and dq coordinates, useful for control, MEC and FEM are used to account for magnetic saturation and stator and rotor slotting.
- Quite a few design hints and examples are available to provide a feeling of magnitudes.
- Control of active and reactive power delivery from the power winding may be scalar, voltage vector–oriented control, power winding flux–oriented vector control (FOC), or DTFC or TQC or DPQC, with or without encoder feedback over a 2/1 speed range with application for motoring and generating (for wind energy conversion) at small speeds (as $p_r > 4$ in general), but they may also be used in speeds of thousands of rpm in autonomous generators, where power winding frequency may also vary with speed for AC or DC constant (increase) voltage output (avionics, ship, train generators) with a 2/1 speed range. This way, $S = -\omega_s/\omega_p$ can be kept small over the entire speed range and thus the KVA rating of the AC-DC-AC converter may be kept reasonably small, making the situation economically feasible.
- A series AC capacitor in the power winding, provided with a fast short-circuit hardware, will further reduce the PWM converter rating by compensating that large additional (leakage like) reactance ($\omega_p L_{m\sigma}$).
- When used as a generator at the power grid, facing grid voltage sags or swells, the short-circuiter opens and thus $\omega_p L_{m\sigma}$ limits the overcurrents in the machine.
- Developments on DFRM design, control, and applications are expected in the near future mainly due to its brushless ruggedness and reduced PWM converter KVA rating for up to 2/1 speed control range, not to mention that the machine may stand light starts as an induction cage motor by short-circuiting the control windings until the minimum speed for self-synchronization $\omega_{r\,\min} = (\omega_s - |\omega_{s\,\max}|) \cdot p_r$ is reached.

REFERENCES

1. A. R. W. Broadway, Cageless induction motor, *Proc. IEE*, vol. 1, no. 18, 1971, pp. 1593–1600.
2. L. J. Hunt, The cascade induction motor, *J-IEE*, vol. 52, 1914, pp. 406–426.
3. I. Boldea, *Reluctance Synchronous Electric Machines and Drives*, book, Ch. 4, Oxford University Press, 1996.
4. C. I. Heyne, A. M. El-Antably, Reluctance and doubly excited reluctance motors, Washington report, subcontract 86x-95013c, 1984.
5. L. Ling, L. Xu, T. A. Lipo, dq analysis of variable speed doubly a.c. excited reluctance motor, *EMPS (now EPCS) Journal*, vol. 19, no. 2, 1991, pp. 125–138.
6. L. Xu, Analysis of a doubly excited brushless reluctance machine by FEM, *Record of IEEE-IAS-1992*, vol. 1, pp. 171–177.
7. A. M. Knight, R. E. Betz, D. G. Dorrell, Design and analysis of brushless doubly fed reluctance machines, *IEEE Trans.*, vol. IA–49, no. 1, 2013, pp. 50–58.
8. M. F. Hsieh, I.-H. Lin, D. G. Dorrell, An analytical method combining equivalent circuit and magnetic circuit for BDFRG, *IEEE Trans.*, vol. Mag–50, no. 11, 2014, pp. 1–5.
9. M. Jovanovic, Sensored and sensorless control methods for brushless doubly fed reluctance machines, *IET*, vol. EPA–3, no. 6, 2009, pp. 503–513.
10. S. Ademi, M. G. Jovanovic, M. Hasan, Control of brushless doubly-fed reluctance generators for wind energy conversion systems, *IEEE Trans.*, vol. EC–30, no. 2, 2015, pp. 596–604.
11. M. Jovanovic, J. Yu, E. Levi, Encoderless direct torque controller for limited speed range applications of brushless doubly fed reluctance motors, *IEEE Trans.*, vol. IA–42, no. 3, 2006, pp. 712–722.
12. D. G. Dorrell, M. Jovanovic, On the possibilities of using a brushless doubly fed reluctance generator in a 2MW wind turbine, *Record of IEEE-IAS-2008 (IEEEXplore)*, pp. 1–8.
13. L. Xu, M. Liu, Rotor pole number studies for doubly excited brushless machine, *Record of IEEE-ECCE —2009*, pp. 207–213.
14. L. Xu, B. Guan, H. Liu, L. Gao, K. Tsai, Design and control of a high efficiency doubly fed brushless machine for wind generator application, *Record of IEEE-ECCE-2010*, pp. 2409–2016.
15. R. A. McMahon, X. Wan, E. Abdi-Jalebi, P. J. Tavner, P. C. Roberts, The BDFM as a generator in wind turbines *Record of EPE-PEMC*, 2006, pp. 1859–1865.

9 Switched Flux–Permanent Magnet Synchronous Motor Analysis, Design, and Control

9.1 INTRODUCTION

When, in 1943 [1] and 1955 [2] (Figure 9.1a and b), the outer and inner, respectively, rotor single-phase configurations for generator mode were invented, the "switched-flux" machine had PMs in the laminated stator together with 1(2) AC coils and a salient-pole passive laminated iron rotor.

The three-phase configuration of SF-PMSMs introduced in [3] provided radial magnets with tangential opposite polarity to make room for three phase coils and secure symmetric emfs (Figure 9.2).

The principle of FS-PMSM may be derived by the flux-modulation approach [6], which divides the SF-PMSM in three parts:

- PM (+ eventually) DC excitation (mmfs) on the stator (its speed is $\Omega_e = 0$) with p_e pole pairs
- The armature (AC) winding mmf on the stator (its speed is Ω_a and it has p_a pole pairs)
- The flux modulator (the rotor here) with speed Ω_m (mechanical) and p_m pole pairs; the rotor has p_r salient poles ($p_r = p_m$)

If only the fundamental of airgap permeance is considered, to provide synchronism between the AC winding mmf wave and the excitation (PM) produced airgap flux density, and thus secure non-zero average torque, two conditions have to be met [6] for maximum torque:

$$p_a = |p_e - p_m|; \quad \Omega_m = \frac{\Omega_a(p_m - p_e)}{p_m} \tag{9.1}$$

From the configurations in Figure 9.2, simply $P_e = N_s/2$, $P_m = N_r$ stator and rotor poles: $N_s = 12$, $N_r = 10$, $p_e = 12$, $p_m = 10$, $p_a = 2$, $f_a = p_m \cdot n$. The number of pole pairs of $p_a = 2$ means that even a distributed AC winding with four poles may be used. However, its larger end coils lead to larger copper losses unless the stator stack core length is large in comparison with the AC coil throws. The configurations in Figure 9.2b and c with 6 stator poles and 13 (respectively, 14) rotor poles operate on a similar principle, but use other harmonic of airgap permeance, again with the rotor as flux modulator, as explained later in more detail.

The flux modulator "matches" the different pole pairs (p_e and p_a) of excitation PMs and armature mmfs, such that, ultimately, FS-PMSMs may be assimilated to synchronous machines.

Many slightly different configurations claiming novelty have been introduced, such as:

- The outer rotor FS-PMSM (Figure 9.3) [7]
- The axial airgap FS-PMSM (Figure 9.4a and b) [8, 9]
- DC-assisted FS-PMSM (Figure 9.5a and b) [10, 11]

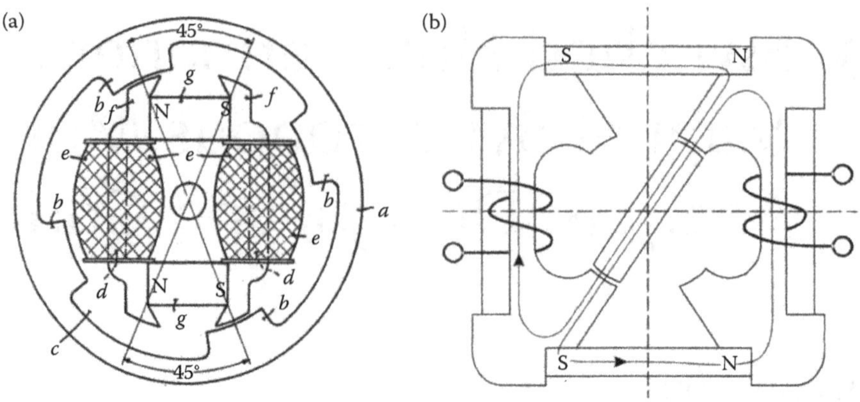

FIGURE 9.1 Switched-flux PM single-phase AC machine: (a) with outer rotor (Adapted from E. Binder and R. Bosch, Patent no. 741163, Class 21d1, Grorep 10, Sept 16, 1943 [1]); (b) with inner rotor. (Adapted from S. E. Raunch and L. J. Johnson, *AIEEE Trans.*, vol. 74, no 3, 1955, pp. 1261–1268. [2])

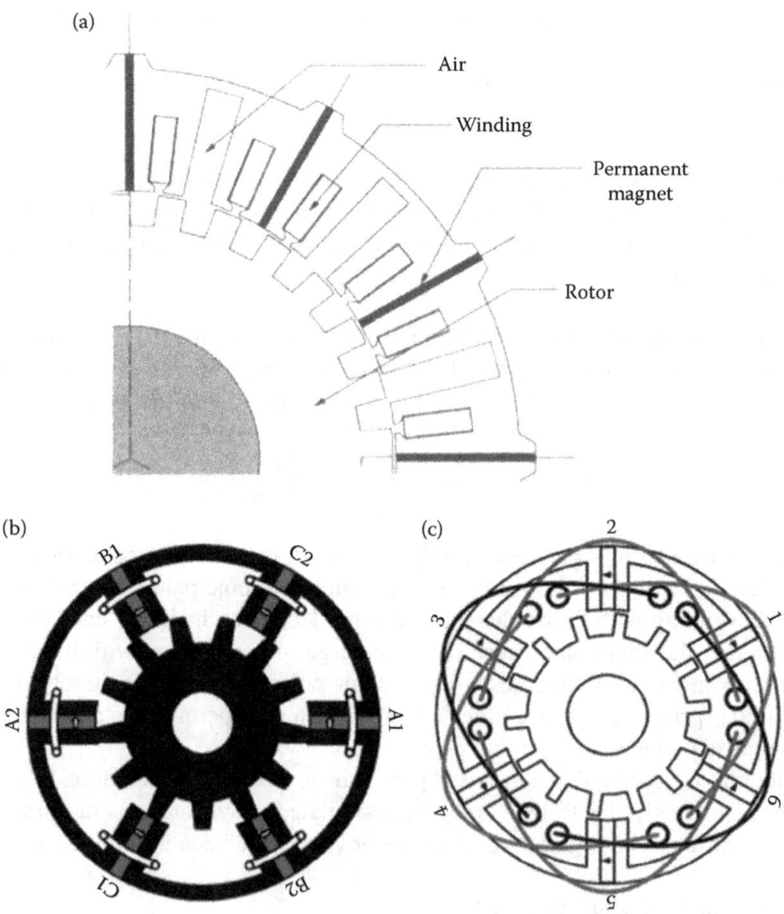

FIGURE 9.2 Typical three-phase SF-PMSM with inner rotor: (a) with regular stator core (12/10); (b) with C-type stator cores (6/13); (c) with distributed AC winding (6/14). (After E. Hoang, A. N. Ben Ahmed, J. Lucidarme, *Record of the European Conference on Power Electronics and Applications*, vol. 3, 1997, pp. 903–908; J. T. Chen et al., *IEEE Trans.*, vol. IA–47, no. 4, 2011, pp. 1681–1691.; D. Li et al., *IEEE Trans.*, vol. EC–31, no. 1, 2016, pp. 106–116. [3–5])

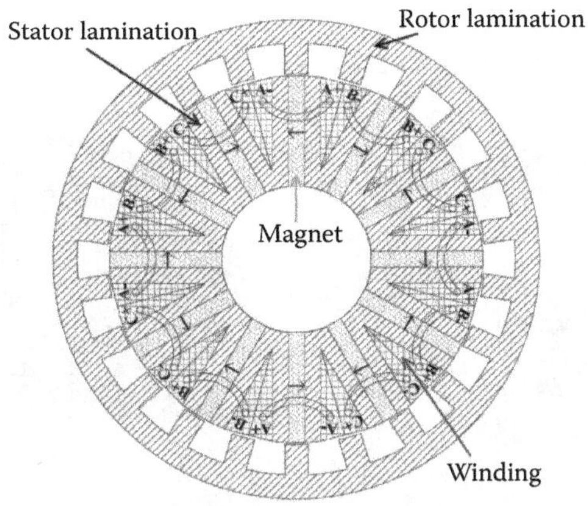

FIGURE 9.3 Outer rotor 12/22 FS-PMSM. (After W. Fei et al., *IEEE Trans.*, vol. IA–48, no. 5, 2012, pp. 1496–1506. [7])

A few general remarks are in order here:

- The main advantage of an SF-PMSM is that it retains torque density of regular PMSMs (in some conditions), while the placements of magnets on the stator leads to their easy cooling and to a passive salient pole rotor, so the machine may be considered thermally and mechanically rugged.
- However, as evident in Figure 9.2, only at best half of all coils throw is "visited" at a moment by the maximum PM flux (positive or negative); PM flux concentration is inherent to obtain the same torque in the same stator volume as for regular PMSMs. Still, "PM flux fringing" is so high with identically placed tangentially magnetized PMs that the mass of PMs per Nm

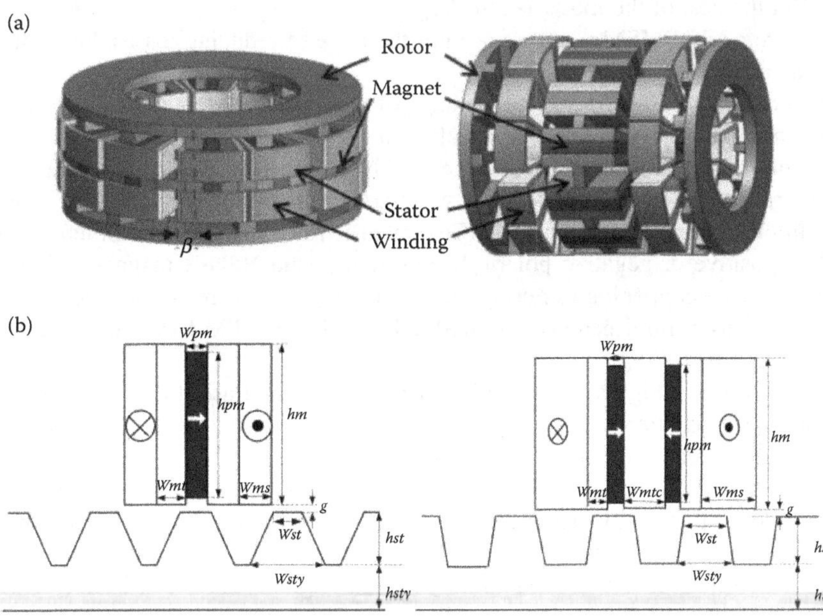

FIGURE 9.4 Axial airgap FS-PMSM: (a) with stator back iron. (After W. Zhao, T. A. Lipo, B.-I. Kwon *IEEE Trans.*, vol. MAG–51, no. 11, 2015, pp. 8112204 [8].); (b) without stator back iron. (After B. Zhang et al., *IEEE-Trans.*, vol. MAG–51, no. 11, 2015, pp. 8204804. [9])

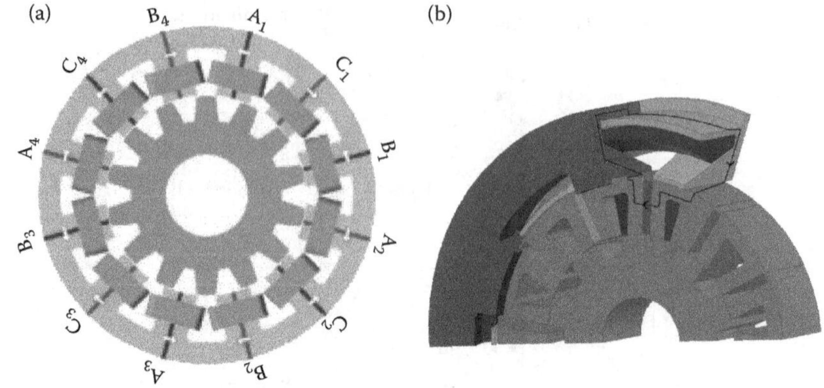

FIGURE 9.5 DC-assisted FS-PMSM: (a) with memory additional PMs. (After H. Yang et al., *IEEE Trans.*, vol. IA–52, no. 3, 2016, pp. 2203–2214. [10]); (b) with claw-pole stator-circular dc coil. (After A. Dupas et al., *IEEE Trans.*, vol. IA–52, no. 2, 2016 pp. 1413–1421. [11])

is almost twice as large with respect to regular PMSMs with same stator volume and same copper losses. The reduction of PM weight for the same emf at the same speed for the same stator volume is demonstrated in [4] (Figure 9.2b) with C stator cores and 6 (instead of 12) stator poles and 13 (instead of 10) rotor poles. However, at overloads, this performance improvement is not notable, a sign that the C core machine is more saturated.

- It has been demonstrated [12] that even at full load, with NdFeB magnets, the risk of demagnetization is high and the use of Ferrites in the standard configuration (Figure 9.2a) may lead to early PM demagnetization.
- However, the machine inductances $L_d \approx L_q$ are notably larger than for regular IPMSMs and thus the CPSR at full power is larger for SF-PMSMs for the same stator volume and copper losses. But the cost of the motor is also larger.
- The outer-rotor SF-PMSM was proposed in the hope of reducing losses, but it did not yet yield satisfactory results.
- The axial airgap TF-PMSM takes advantage of two rotor sides and is claimed to increase torque density (torque/volume) but not efficiency.
- The notably higher inductances make the SF-PMSM operate at a lower power factor as a motor (higher KVA in the inverter) and with large voltage regulation in generator mode.
- The additional memory magnets (AlNiCo), magnetized/demagnetized in tens of milliseconds (for positive or negative polarity), may reduce the NdFeB magnet weight per Nm and thus decrease copper losses during flux weakening, but the response in torque is marred by the time (tens of milliseconds) required to bring the total PM flux at the required torque level first.
- The configuration in Figure 9.5b—with DC claw-pole stator assisting excitation—requires reasonable DC excitation power, but about 35% of NdFeB flux is lost through the claw-pole structure that embraces the now-bulky stator.
- The rather large highly peaked airgap flux density in the airgap (1.4–1.7 [1.8] T) leads to large (local) radial forces, which increases noise and vibrations and thus at least the number of rotor poles p_r has to be even in order to avoid uncompensated radial forces.
- The "switched-flux" term was used initially [2] to designate, in fact, a PM flux "reversal" in the AC coil (coils) when maybe the second term ("reversal") would be more proper, because a "switch" is characterized by 0 and 1, not by $-1/+1$, situations.
- As "flux-reversal" PMSMs allow slightly different embodiments [13] from SF-PMSMs, they will be treated in Chapter 14.

- The doubly salient machines with DC stator coils (instead of PMs) in addition to AC coils may also be considered a variety of FS machines, but they will also be treated separately in a dedicated chapter, as they stem from the switched reluctance motor core.
- Given the myriad analyzed configurations, we will present here a synthesis of SF-PMSM principles and their modeling, then analyze a few key implementations with their control.

9.2 THE NATURE OF SWITCHED FLUX–PERMANENT MAGNET SYNCHRONOUS MOTORS

SF-PMSMs represent, in fact, synchronous machines with stator PMs and a flux-modulation rotor to couple the PM airgap flux density with the AC stator mmf to produce nonzero average torque.

The airgap permeance harmonics affect this magnetic coupling and, by using this method [5], we can not only explore different stator and rotor pole (N_s, N_r) combinations, but also calculate analytically, for preliminary design, the torque, inductances, and finally performance. More advanced modeling by MEC or by 2D (3D) FEM should follow to allow more precision for modeling the double saliency of the machine, considering magnetic saturation, and investigating PM demagnetization aspects. But, as shown in [6], the method of pole pair matching of rotor, stator PM mmf, and AC coil mmf may be instrumental in designing the AC coil windings with their connections [15].

The "airgap permeance" approach [15] will be followed primarily here. But first, let us notice the operating principle of PM flux linkage in the AC coil "switching" or (+ −) in Figure 9.6.

9.2.1 Airgap Permeance Harmonics

The SF-PMSM has doubly saliency, whose airgap permeance Λ_g (in the absence of magnetic saturation) may be approximated to [16]:

$$\Lambda_g(\theta_s, t) \approx \frac{g \cdot \Lambda_s(\theta) \cdot \Lambda_r(\theta_s, t)}{\mu_0} \tag{9.2}$$

where $\Lambda_s(\theta_s)$, $\Lambda_r(\theta_s, t)$ are permeances related to the stator and, respectively, rotor slotting.

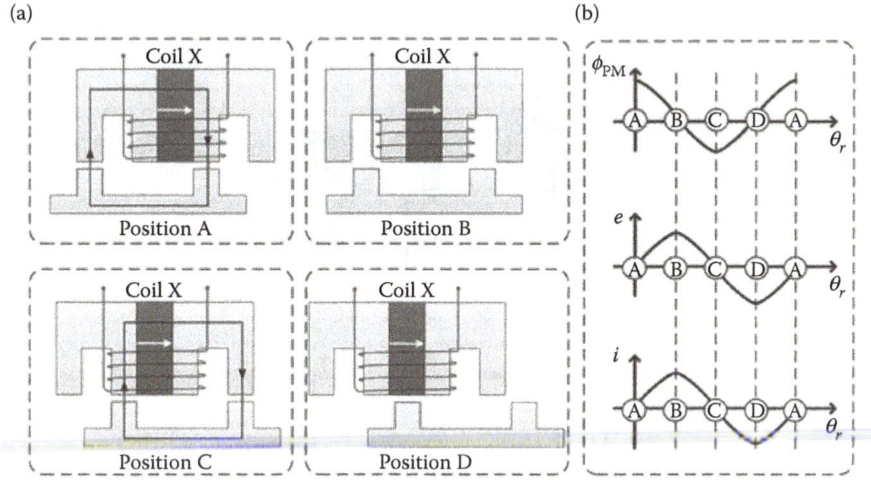

FIGURE 9.6 Principle of PM flux switching (a); and the ideal PM flux, emf and current in the AC coil versus rotor position (b). (After Y. Shi et al., *IEEE Trans.*, vol. 63, no. 3, 2016, pp. 1425–1437. [15])

For a single saliency $\Lambda(\theta)$—say, for the stator one—:

$$\Lambda_s(\theta_s) \approx \Lambda_0 + \Lambda_1 \cos(Z_s\theta_s) \tag{9.3}$$

if only the first harmonic is considered.

By conformal mapping [16]:

$$\Lambda_0 \approx \frac{\mu_0}{g_1}\left(1 - 1.6\beta\frac{s_o}{t_s}\right)$$

$$g' = g + L_m/\mu_r;\ \beta = 0.5 - \frac{1}{2\sqrt{1+\left(\frac{s_o}{2g'}\right)^2}} \tag{9.4}$$

$$\Lambda_1 = \frac{\mu_o}{g}\frac{4}{\pi}\beta\left[0.5 + \frac{(s_o/t_s)^2}{0.78125 - 2\left(\frac{s_o}{t_s}\right)^2}\right]\cdot\sin\left(1.6\cdot\pi\frac{s_o}{t_s}\right)$$

with s_o—slot opening, t_s—slot pitch, μ_r—recoil permeability of PMs, Z_s—number of stator slots (see Figure 9.7). After neglecting airgap permeance harmonics higher than $2Z_1$ in the stator and Z_2 in the rotor (Z_1, Z_2—stator teeth in the stator and in the rotor), from Equation 9.2:

$$\Lambda(\theta_s, t) = \frac{g}{\mu_0}(\Lambda_{s0} - \Lambda_{s1}\cdot\cos(2Z_1\theta_s))\cdot(\Lambda_{r0} - \Lambda_{r1}\cdot\cos(Z_2\theta_s - Z_2\Omega_r t)) \tag{9.5}$$

The constant component and the four harmonics in Equation 9.5, $2Z_1$, Z_2, $2Z_1 + Z_2$, $2Z_1 - Z_2$, with their speed Ω_r dependence are shown in Table 9.1 [15].

FEM verifications of Equation 9.5 for a 12/10 SF-PMSM are shown in Figure 9.8 [15].

Only the important harmonic 20 is missing in Equation 9.5, as it refers to the second harmonic ($2Z_2$) of the rotor component, which is missing.

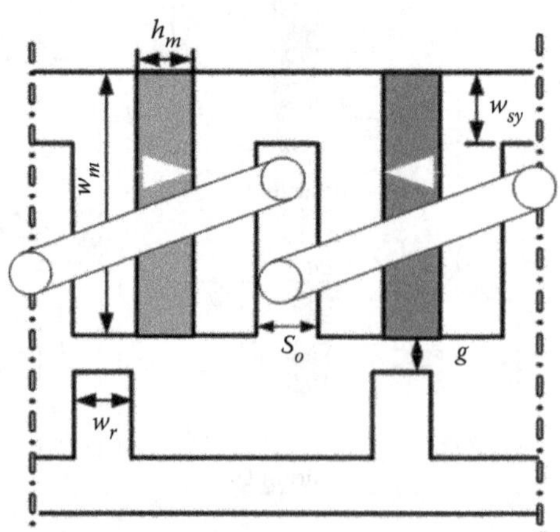

FIGURE 9.7 Two coils and their core section in SF-PMSM.

TABLE 9.1
Airgap Permeance Harmonics

Pole Pair	Amplitude	Speed	Source
Const.	$\frac{g}{\mu_0}\Lambda_{s0}\Lambda_{r0}$	0	Airgap
$2Z_1$	$\frac{g}{\mu_0}\Lambda_{r0}\Lambda_{s1}$	0	Stator slots
Z_2	$\frac{g}{\mu_0}\Lambda_{s0}\Lambda_{r1}$	Ω	Rotor slots
$2Z_1+Z_2$	$\frac{g}{2\mu_0}\Lambda_{s1}\Lambda_{r1}$	$\frac{Z_2}{2Z_1+Z_2}\Omega$	Stator and rotor slots
$2Z_1-Z_2$	$\frac{g}{2\mu_0}\Lambda_{s1}\Lambda_{r1}$	$-\frac{Z_2}{2Z_1-Z_2}\Omega^{a}$	Stator and rotor slots

[a] Negative value means opposite speed direction.
Source: Y. Shi et al., *IEEE Trans.*, vol. 63, no. 3, 2016, pp. 1425–1437. [15]

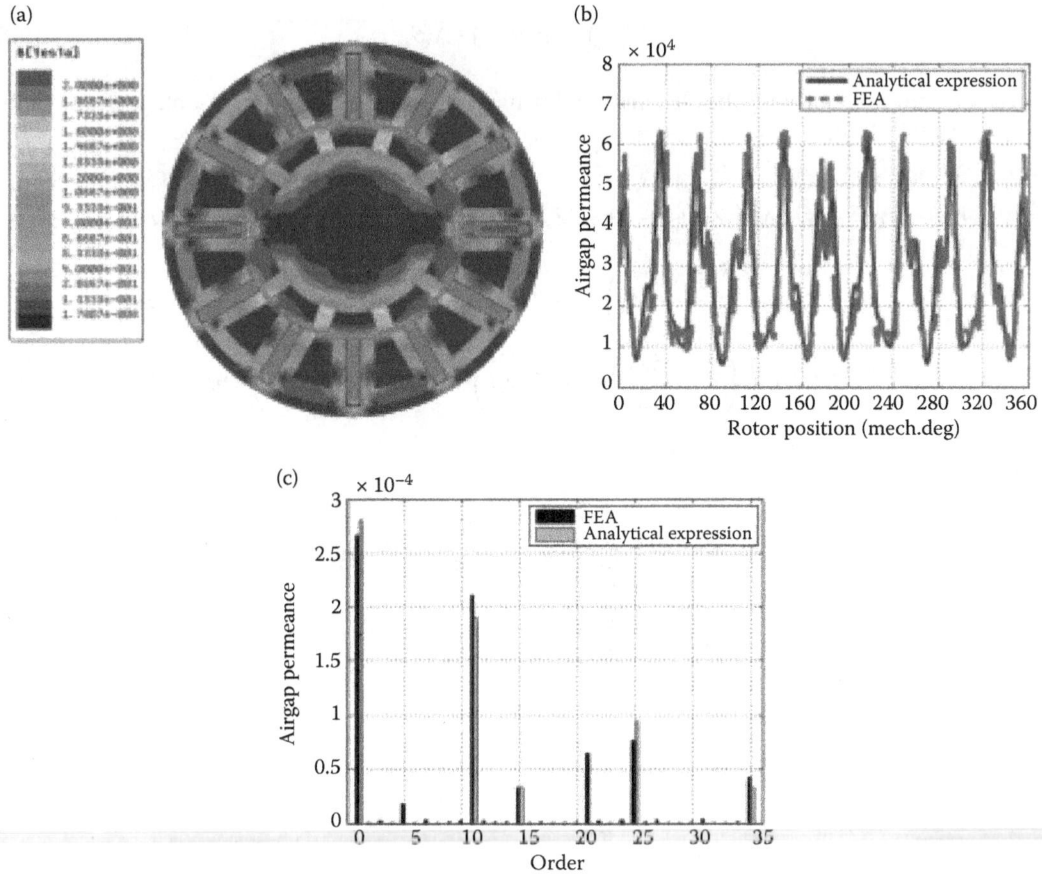

FIGURE 9.8 Airgap permeance harmonics amplitudes for a 12/10 SF PMSM (a) as calculated from Equation 9.5 (b), and with FEM (c). (After Y. Shi et al., *IEEE Trans.*, vol. 63, no. 3, 2016, pp. 1425–1437. [15])

9.2.2 No-Load Airgap Flux Density Harmonics

The magnet mmf may be described by its fundamental and third harmonic:

$$F_{1M}(\theta_s, t) = F_1 \sin\left(\frac{Z_1}{2}\theta_s\right) + F_3 \sin\left(\frac{3}{2}Z_1\theta_s\right) \tag{9.6}$$

Accounting for airgap by the product of Carter coefficients K_{cs}, K_{cr}:

$$\begin{aligned} F_1 &= \frac{4}{\pi}\frac{K_{\alpha 1}B_r h_{PM}}{\mu_o\mu_{rec}\left(2+\dfrac{h_{PM}}{gK_{cs}K_{cr}}\cdot\dfrac{\tau_{PM}}{l_{PM}}\right)} \\ F_3 &= \frac{1}{3}\frac{4}{\pi}\frac{K_{\alpha 3}B_r h_{PM}}{\mu_o\mu_{rec}\left(2+\dfrac{h_{PM}}{gK_{cs}K_{cr}}\cdot\dfrac{\tau_{PM}}{l_{PM}}\right)} \end{aligned} \tag{9.7}$$

where h_{PM}, l_{PM} are the magnet thickness and radial height, τ_{PM} is the peripheral distance between two adjacent magnets, and $K_{\alpha 1}$ and $K_{\alpha 2}$ are approximation coefficients.

The no-load airgap flux density $B_g(\theta_s,t)$ is thus:

$$B_g(\theta_s, t) = F(\theta_s, t)\cdot\Lambda(\theta_s, t) \tag{9.8}$$

Taking into account only the flux harmonics that will benefit the emf, we obtain (from Equations 9.5 to 9.7) [16]:

$$\begin{aligned} B_g(\theta_s, t) = &-B_1 \sin\left(\left(Z_2+\frac{Z_1}{2}\right)\theta_s - Z_2\Omega_r t\right) + B_2 \sin\left(\left(Z_2-\frac{Z_1}{2}\right)\theta_s - Z_2\Omega_r t\right) \\ &+ B_3 \sin\left(\left(5\frac{Z_1}{2}-Z_2\right)\theta_s + Z_2\Omega_r t\right) + B_4 \sin\left(\left(\frac{3}{2}Z_1-\frac{Z_1}{2}\right)\theta_s + Z_2\Omega_r t\right) \\ &+ B_5 \sin\left(\left(Z_2+\frac{5Z_1}{2}\right)\theta_s - Z_2\Omega_r t\right) + B_6 \sin\left(\left(Z_2+\frac{3Z_1}{2}\right)\theta_s - Z_2\Omega_r t\right) \end{aligned} \tag{9.9}$$

with

$$\begin{aligned} B_1 &= F_1\frac{g}{2\mu_0}\Lambda_{s0}\Lambda_{r1} + \frac{F_3\cdot g}{4\mu_0}\Lambda_{s1}\Lambda_{r1} \\ B_2 &= F_1\frac{g}{2\mu_0}\Lambda_{s0}\Lambda_{r1} + \frac{F_3\cdot g}{4\mu_0}\Lambda_{s1}\Lambda_{r1} \\ B_3 &= F_1\frac{g}{4\mu_0}\Lambda_{s1}\Lambda_{r1} \\ B_4 &= F_1\frac{g}{4\mu_0}\Lambda_{s1}\Lambda_{r1} + \frac{F_3\cdot g}{2\mu_0}\Lambda_{s0}\Lambda_{r1} \\ B_5 &= F_1\frac{g}{4\mu_0}\Lambda_{s1}\Lambda_{r1} \\ B_6 &= F_1\frac{g}{4\mu_0}\Lambda_{s1}\Lambda_{r1} + \frac{F_3\cdot g}{2\mu_0}\Lambda_{s0}\Lambda_{r1}; \end{aligned} \tag{9.10}$$

9.2.3 Electromagnetic Force and Torque

The emf expression, based on the winding function method, is:

$$e = -\frac{d}{dt}\sum \gamma_g l_{\text{stack}} \int_0^{2\pi} B_g(\theta_s, t) N_i(\theta_s) d\theta_s \tag{9.11}$$

where γ_g is the rotor outer radius, l_{stack} is the stack length, and N_i is distribution of stator coil turns of a phase along rotor periphery; m is the phases.

Finally, the electromagnetic torque T_e is simply:

$$T_e = \frac{e_A i_A + e_B i_B + e_C i_C}{\Omega_r} = \sum_{i=1.6} \left(3\sqrt{2} \cdot \frac{Z_2}{P_i} r_g l_{\text{stack}} I_{\text{RMS}} B_i W_s \right) \tag{9.12}$$

where $B_i = B_1, B_2, B_3, B_4, B_5, B_6$ as above and P_i is the order of flux density harmonics [15]:

$$(P_i)_{i=1,2,..,6} = Z_2 + Z_1/2, Z_2 - Z_1/2, 5Z_1/2 - Z_2, 3Z_1/2 - Z_2, Z_2 - 5Z_1/2, Z_2 + 3Z_1/2 \tag{9.13}$$

In Equation 9.12, the phase currents were considered in phase with emfs ($i_d = 0$), and N_s is the number of turns/phase.

9.2.4 Feasible Combinations of Stator and Rotor Slots: Z_1 and Z_2

In order to obtain a symmetric m phase winding:

$$\frac{Z_1}{\text{GCD}(P_i, P_a)} = mk; \; k\text{-integer positive} \tag{9.14}$$

P_i is the airgap flux density harmonic pole pair and P_a is the pole pair of armature (AC) winding mmf. Though there are many harmonics P_i, fortunately all have the same greatest common divisor (GCD) (Z_1, P_i) and thus P_a is easier to match in Equation 9.14, as the harmonic order may be "thrown out" in Equation 9.14 for compliance.

The minimum order of these harmonics may be chosen equal to P_a. In this case:

$$Z_1/Z_2 = k; \quad \frac{Z_1}{\text{GCD}(Z_1, P_a)} = mk \tag{9.15}$$

For single-layer windings, $Z_1/2$ will replace Z_1 in the second condition (Equation 9.15). Table 9.2 summarizes Z_1, Z_2 conditions for three-phase SF-PMSM.

Table 9.2 offers also P_a/pole pairs of armature mmf, K_w winding factor for nonoverlapping winding, SPP is slots/pole/phase and pole ratio PR $= Z_2/P_a$, which is a kind of "magnetic gear ratio" [15]:

$$\text{SPP}(g) = \frac{Z_1}{2mP_a} \tag{9.16}$$

The airgap flux density on no load for the 12/10 prototype in [5] appears in Figure 9.9.

The working airgap flux density harmonics that produce fundamental emf are the 4th, 8th, 16th, and 20th. The 6th and 18th harmonics are produced by the PM mmf third harmonic (Equation 9.16). Also, the second harmonic and 16th flux harmonic are produced by the fundamental PM mmf and the 20th ($2Z_2$) airgap harmonics and create a second-order emf harmonic that is ignored in Equation 9.9, but accounted for by FEM.

TABLE 9.2
Z_1 & Z_2 Conditions in Three-Phase SF-PMSM

Z_1	Z_2	2	4	5	6	7	8	10	11	12	13	14	16	17	19	20
6	Pa	1**	1**	2**		2	1**	1**	2		2	1**	1**	2	2	1**
	k_w	0.5	0.5	0.866	–	0.866	0.5	0.5	0.866	–	0.866	0.5	0.5	0.866	0.866	0.5
	SPP	1	1	0.5		0.5	1	1	0.5		0.5	1	1	0.5	0.5	1
	PR	2	4	2.5		3.5	8	10	5.5		6.5	14	16	8.5	9.5	20
12	Pa	4	2**	1		1	2**	4	5		5	4	2**	1	1	2**
	k_w	0.866	0.5	0.259	–	0.259	0.5	0.866	0.933	–	0.933	0.866	0.5	0.259	0.259	0.5
	SPP	0.5	1	2		2	1	0.5	0.4		0.4	0.5	1	2	2	1
	PR	0.5	2	5		7	4	2.5	2.2		2.6	3.5	8	17	19	10
18	Pa	7**	5**	4	3**	2	1**	1**	2	3**	4	5**	7**	8	8	7**
	k_w	0.9	0.735	0.62	0.5	0.617	0.167	0.167	0.617	0.5	0.62	0.735	0.9	0.945	0.945	0.9
	SPP	3/7	0.6	0.75	1	1.5	3	3	1.5	1	0.75	0.6	3/7	0.375	0.375	3/7
	PR	2/7	4/5	5/4	2	7/2	8	10	11/2	4	13/4	14/5	16/7	17/8	19/8	20/7

(light shading) Overlapping winding is preferred. k_w is the winding factor of FSPM machines with nonoverlapping winding.
(dark shading) Two slot-pitch winding is preferred. **means that the phase back emf waveform is asymmetrical.
Source: Y. Shi et al., *IEEE Trans.*, vol. 63, no. 3, 2016, pp. 1425–1437. [15]

9.2.5 Coil Connection in Phases

The standard "star of slot emfs" method used for regular AC symmetric windings may still be used here, but the electrical angle between the emfs in two adjacent slots α_e is:

$$\alpha_e = P_a \frac{2\pi}{Z_1} = P_a \cdot \alpha_{sgeometric} \tag{9.17}$$

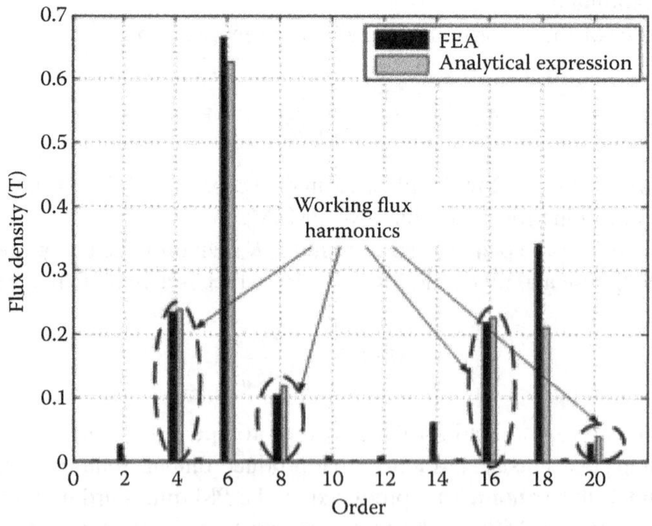

FIGURE 9.9 No-load airgap flux density harmonics spectrum. (After D. Li et al., *IEEE Trans.*, vol. EC–31, no. 1, 2016, pp. 106–116. [5])

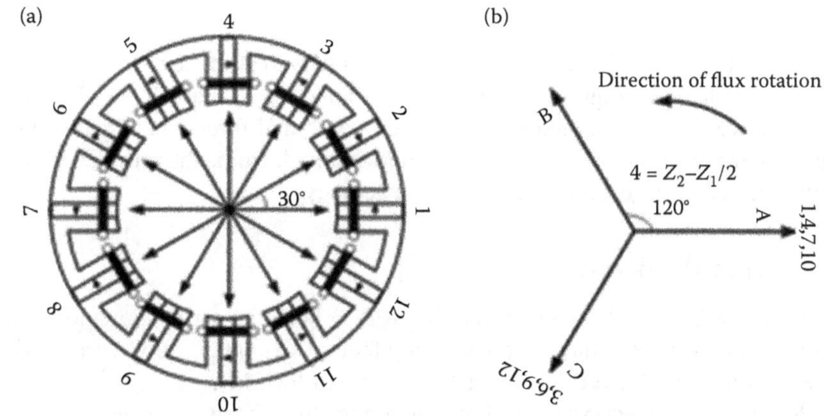

FIGURE 9.10 SF-PMSM, (a); and the star of slots, (b).

This is similar to conventional AC windings with P_a pole pairs. For $Z_1 = 12$ stator teeth, $Z_2 = 10$ rotor poles. $P_a = Z_2 - Z_1/2 = 10 - 12/2 = 4$ pole pairs. The star of slots (emfs)—Figure 9.10—is similar to that of a single-layer 12-slot/4 pole pair (8 poles) fractional slots (concentrated) winding regular PMSM (SPP = 0.5).

It has to be pointed out that all working flux harmonics (Equation 9.9) have the same star of slots!

9.2.6 Symmetrical Phase Electromagnetic Force Waveforms

By selecting a suitable combination of stator and rotor teeth (Z_1 and Z_2), the virtually asymmetric single-coil emfs (due to even time harmonics) may become symmetric in a phase emf when connected in series [17]. The even time harmonics in emf are produced by even flux harmonics (Equation 9.9).

Considering the machine periodicity, t_p (number of emfs per phase in phase with each other) is:

$$t_p = \text{GCD}(Z_1, P_a) \tag{9.18}$$

as in standard AC machines.

For odd, duplex-2, or duplex-odd phase machines, if t_p and $Z_1/m \cdot t_p$ are odd, the phase emf waveforms are asymmetrical. But if $Z_1/m \cdot t_p$ is even and

$$\frac{Z_2}{t_p} = 2k \pm 1, \tag{9.19}$$

the phase emf waveforms become symmetric.

Also, if t_p is even, but Z_1/t_p is odd, the number of star spokes is not a multiple of 2 m, but, still, if:

$$\frac{2Z_2}{t_p} = 2k \pm 1, \tag{9.20}$$

the phase emfs are symmetric.

Finally, for even or duplex-even (except for two-phase machines), if t_p and $2Z_1/mt_p$ are odd numbers, again, the phase emfs are asymmetric. But if $2Z_1/mt_p$ is even and Z_2 satisfies Equation 9.19, the phase emfs become symmetric again. However, with t_p even and $2Z_1/mt_p$ odd, Z_2 should satisfy Equation 9.20 to secure symmetric phase emfs.

9.2.7 High Pole Ratio $PR = Z_2/P_a$

For the 12/10 SF-PMSM PR = 10/4 = 2.5. A high PR is likely to produce higher torque density [18]. The C stator core SF machine (Figure 9.2b) with $Z_2 = 13$ and $P_a = 2$ has PR = 6.5 (!) and a winding factor of 0.866. However, it is plagued by uncompensated radial forces that create additional noise and vibration and core loss (due to $Z_2 = 13$, odd). The winding factor is the product of the distribution factor and chording factor, typical in standard AC machines.

9.2.8 Overlapping Windings?

In a high pole ratio, $PR = Z_2/P_a$ SF-PMSMs, the SPP(g) becomes large (Equation 9.16) and thus nonoverlapping windings tend to have a low winding factor ($K_{w1} = 0.5$, for C-core 6/14 SF-PMSM, PR = 14, $P_a = 2$, SPP = 1, $K_{w1} = 0.866$ for 12/10 standard PMSM).

For high PR (PR > 3), SF-PMSM distributed windings have been proposed:

- Integer q (for SPP > 1, integer) distributed AC windings (Figure 9.2c)
- Fractionary q (for SPP > 1) distributed windings
- 0.5 < SPP < 1–concentrated windings with two slot-pitch throw

The FEM-calculated single-coil emf harmonics spectrum in a C-core 6/14 SF-PMSM for overlapping and nonoverlapping windings is shown in Figure 9.11 [5].

It is evident that the overlapping type of winding for same number of turns/slot produces twice as much emf (and torque) for a given stator current. Unfortunately, the phase emfs of the C-core 6/14 SF-PMSM have asymmetric phase emfs that reflect themselves in third-order torque ripple. The 2/1 increase in emf with overlapping winding, for the same number of turns/slot and current (same current density), means more losses for the overlapping winding due to larger end turns, but also two times more torque.

With the C-core 6/14 SF-PMSM, $P_a = 1$ and thus the AC coils throw in the three-phase overlapping windings is $\pi D_r/2p_a = \tau_{coil}$. If $l_{stack} > \tau_{coil}$, the relevance of the end connections in copper losses for overlapping windings decreases. The two-slot pitch coil throw winding is a good compromise for the case in point. The 6/13 SF-PMSM with nonoverlapping windings falls in between the two overlapping windings for 6/14 SF-PMSM; it has symmetrical phase emfs but also

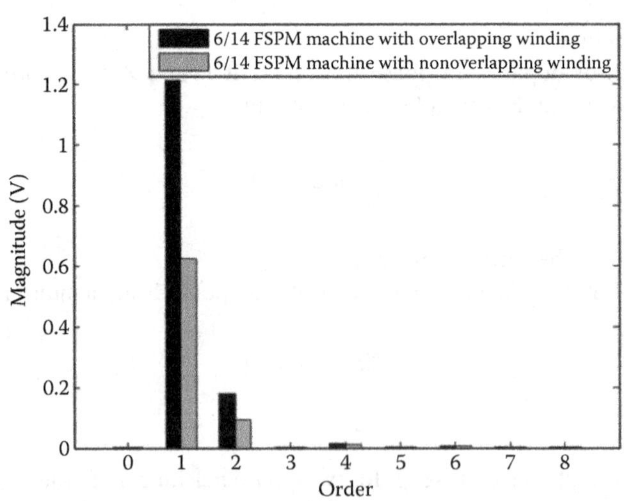

FIGURE 9.11 Single-coil emf for a 6/14 SF-PMSM with overlapping and nonoverlapping winding. (After D. Li et al., *IEEE Trans.*, vol. EC–31, no. 1, 2016, pp. 106–116. [5])

uncompensated radial forces. Also, the armature reaction field is larger for overlapping windings, and so is the mutual inductance between phases. Weaker fault tolerance is thus obtained. This only shows that choosing Z_1 and Z_2 is paramount; it has many consequences and should be selected application by application.

However, if the armature winding mmf pole pairs p_a are redefined to be the working harmonic pair, the star of slots (emfs) method may again be applied to allocate phase coils to slots, as in standard PMSMs with either nonoverlapping or overlapping AC windings. The number of slots/pole/phase is defined as in classical theory, $q = \text{SPP} = Z_1/2m \cdot P_a$, and then the winding factor may be calculated with same standard formulae for both nonoverlapping and overlapping AC windings [19]. When torque density is paramount, overlapping windings with C-core 6/13(14) and other SF-PMSMs are likely to be more suitable in terms of torque/volume, torque/losses, and torque/PM weight (cost).

Typical stars of emfs (slots), their allocation with phases, and the winding connections for the 12/11 and 12/26 standard SF-PMSMs are shown in Figure 9.12a through c [15].

The angle β of adjacent slot emfs is:

for 12/11 SF PMSM:

$$\beta = 150°, \text{ as } N_s = 12,\ P_a = N_r - N_s/2 = 5;\ \beta = \frac{360°}{N_s} \cdot P_a$$

for 12/13 SF PMSM:

$$\beta = 30°, \text{ as } N_s = 12,\ P_a = 13 - 12/2 = 7;\ \beta = \frac{360°}{12} \cdot 7$$

for 12/26 SF PMSM:

$$\beta = 240°, \text{ as } N_s = 12,\ P_a = 26 - 12/2 = 20;\ \beta = \frac{360°}{12} \cdot 20$$

While the 12/11 and 12/13 SF-PMSMs produce reasonable emfs and torque, the 12/26 configuration produces 20 times less [15] torque for the same geometry and is offered here just to illustrate the nature of SF-PMSMs. To further shed light on the nature of SF-PMSMs, [15] offers a parallel between the above-adopted classical (motor-oriented) theory and the generator-oriented theory to prove that, in fact, they are equivalent.

9.3 A COMPARISON BETWEEN SWITCHED FLUX–PERMANENT MAGNET SYNCHRONOUS MOTORS AND INTERIOR PERMANENT MAGNET SYNCHRONOUS MOTORS

SF-PMSM modeling starts from a method to calculate the no-load airgap-flux density, computes the emfs for then phases, then calculates L_{dm}, L_{qm}, L_{sl}, ψ_{PM} (the PM flux linkage per phase) [12]. Stator resistance R_s computation is straightforward, but core loss p_{iron} calculation, which is due to numerous flux harmonics (on load especially) in the stator and rotor (including PM eddy current), may be approached with adequate precision only by FEM. The flux density variation in time in all FEM elements is used in two (three) term core loss formulae initiated early by Steinmetz (Chapter 7).

This way, steady-state and transient performance (for control) may be approached by the dq model as for regular IPMSMs. Moreover, the magnetic saliency in SF-PMSMs is rather small and thus a wide CPSR is obtained just because the machine inductance in p.u. values is large, due to the small airgap and stator current mmf flux harmonics.

More and more, 2D-FEM is used to extract L_{dm}, L_{qm}, and ψ_{PM}, electromagnetic torque with all their variations, with rotor position (L_{dm} and L_{qm} vary little with rotor position, but enough to impede rotor position estimation in encoderless drives). A correction is needed, as known for IPMSMs.

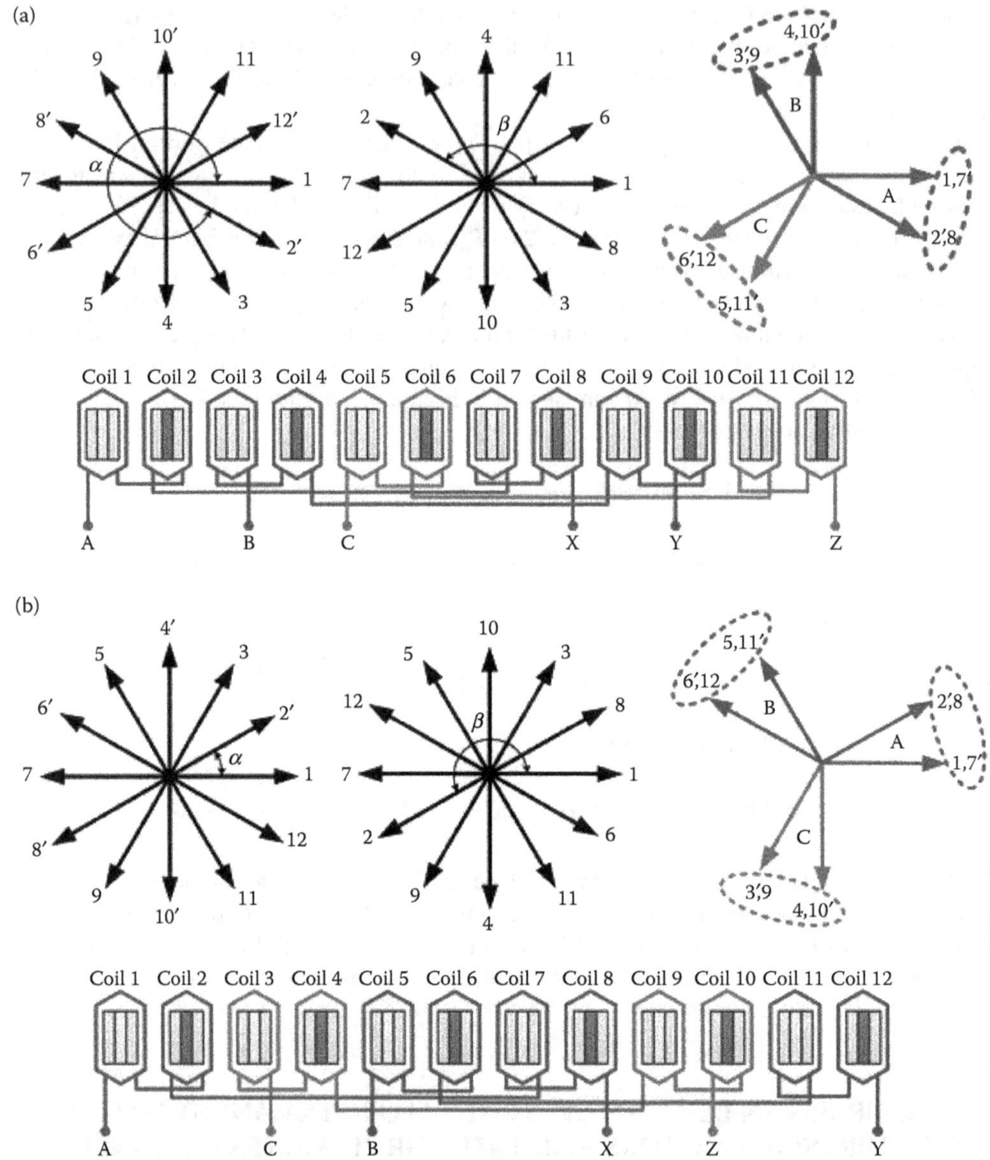

FIGURE 9.12 Coil (slot)—emfs, three-phase emfs and three-phase nonoverlapping standard 12 stator teeth (coils)/11, (a); 13, (b); rotor poles. (After Y. Shi et al., *IEEE Trans.*, vol. 63, no 3, 2016, pp. 1425–1437. [15]) *(Continued)*

As an example, starting with [12], a comparison between standard 12/10 SF-PMSMs and four different PMSMs with NdFeB or Ferrite magnets is unfolded here.

The four IPMSMs treated in (Figure 9.13) are:

- SPMSM (Figure 9.13a)
- IPMSM with flux-barrier-rotor (Figure 9.13b)
- IPMSM with spoke magnets (Figure 9.13c)
- IPMSM with V-shaped magnets (Figure 9.13d)

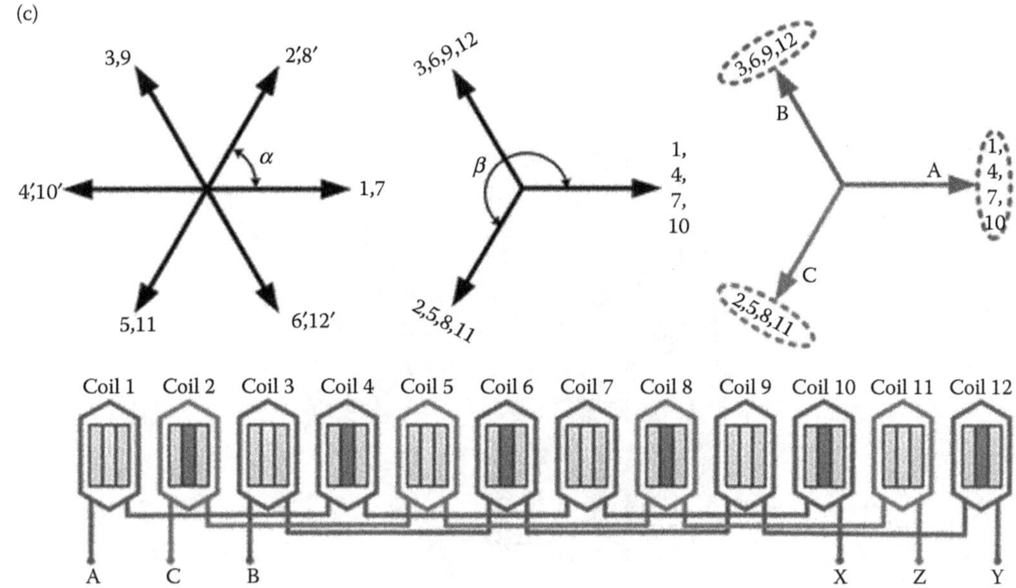

FIGURE 9.12 (Continued) Coil (slot)—emfs, three-phase emfs and three-phase nonoverlapping standard 12 stator teeth (coils)/26, (c) rotor poles. (After Y. Shi et al., IEEE Trans., vol. 63, no 3, 2016, pp. 1425–1437. [15])

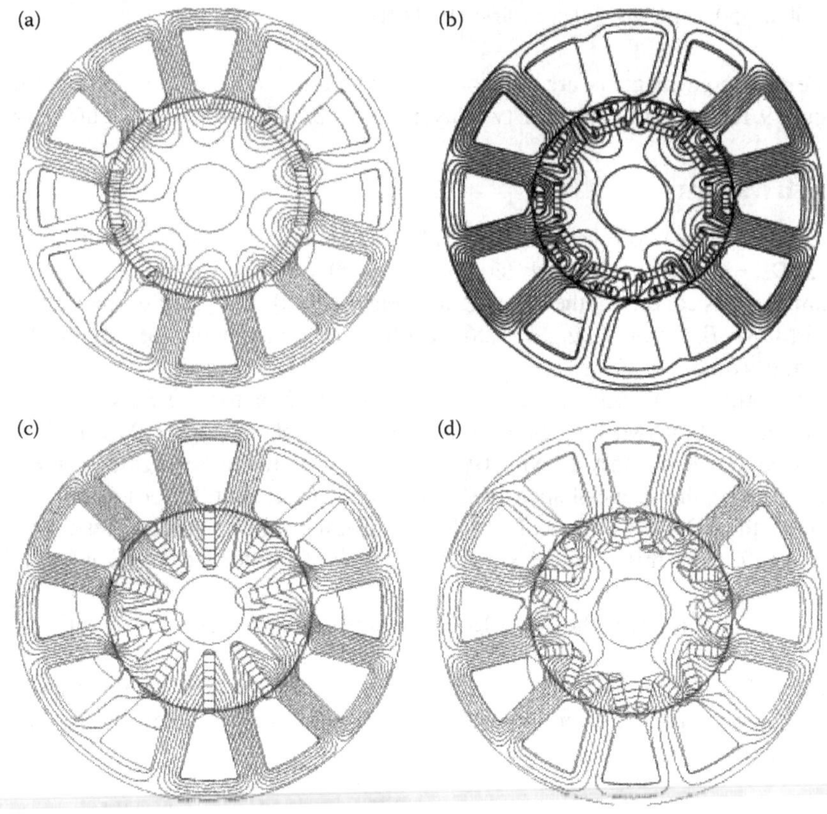

FIGURE 9.13 Four PMSMs: (a) SPMSMS; (b) flux-barrier-rotor IPMSM; (c) spoke-magnet IPMSM; (d) V-shaped magnet IPMSM. (After A. Fasolo, L. Alberti, N. Bianchi, *IEEE Trans.*, vol. IA–50, no. 6, 2014, pp. 3708–3716. [12])

FIGURE 9.14 12/10 SF-PMSM with magnetic flux lines, under load. (After A. Fasolo, L. Alberti, N. Bianchi, *IEEE Trans.*, vol. IA–50, no. 6, 2014, pp. 3708–3716. [12])

First, NdFeB-sintered magnets are considered, then Ferrite magnets are introduced in all four PMSMs (Figure 9.13) and in the regular two-layer main 12/10 SF-PMSM (Figure 9.14) [12].

9.3.1 NdFeB Magnets

All these machines have two-layer fractional identical winding. First, they all have same outer rotor diameter $D_{os} = 134$ mm and stack length $l_{\text{stack}} = 90$ mm, and the same inner stator diameter $D_{is} = 71.5$ mm. For this condition, the PM magnet weight was: 0.27 kg for flux-barrier-rotor IPMSM, 0.55 kg for SPMSM ($B_r = 0.97$ T, $\mu_{\text{rec}} = 1.06$ p.u.) for a nominal stator slot current of 1665 Aturns (peak)—9 A/mm^2 (rms).

The average torque (with pure i_q current) was 30 (31) Nm for the SPMSM and SF-PMSM and for the V-shaped IPMSM, but about 40 Nm for the spoke-shaped IPMSM. The flux-barrier IPMSM produced, however, only 15 Nm. By the torque/PM magnet weight, for the same copper losses and same stator, the spoke-shaped IPMSM is best, by far, but, unfortunately, as Figure 9.15 [12] shows, only the SPMSM is free of PM load demagnetization at full torque. The flux-barrier IPMSM is also safe from this point of view, but at half of the torque value, when all the others will also be okay.

As visible in Figure 9.16, even the SF-PMSM is not free of demagnetization at full load. Moreover, in the spoke IPMSM (Figure 9.15c) and SF-PMSM, the level of flux density along most of the PM length is smaller. This demagnetization inquiry was performed for 120° temperature (with NdFeB, 0.97 T, $\mu_{\text{rec}} = 1.06$ p.u. at 20°), when the intrinsic coercive force $H_{ic} = 450$ kA/m. $B_n > 0.4$ T was considered acceptable (free of demagnetization risk).

9.3.2 Ferrite Permanent Magnets

By putting Ferrite magnets in same geometry, torque has decreased and so has efficiency (for same current density). The average torque was 20.9 Nm for SF-PMSM, 10.3 Nm for flux-barrier IPMSM,

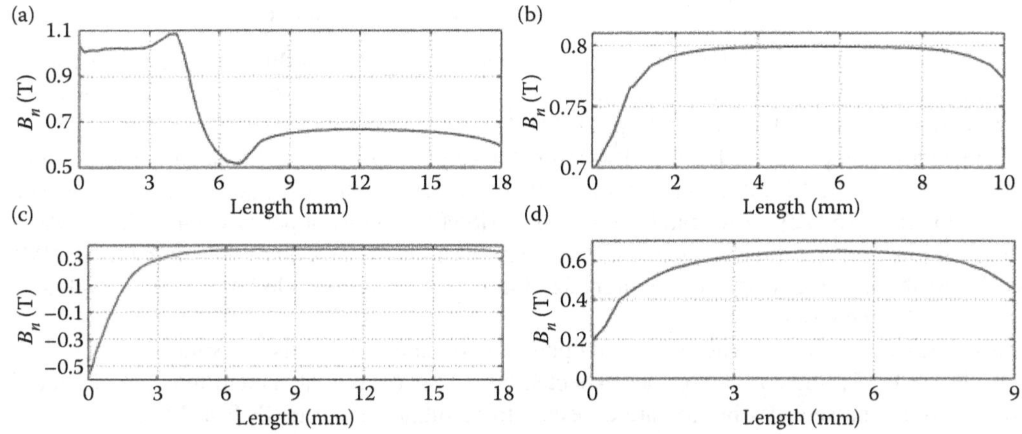

FIGURE 9.15 Flux density on the PM surfaces at full load along their length (a) SPMSM; (b) flux barrier IPMSM; (c) spoke-IPMSM; (d) V-IPMSM. (After A. Fasolo, L. Alberti, N. Bianchi, *IEEE Trans.*, vol. IA–50, no. 6, 2014, pp. 3708–3716. [12])

20.7 Nm for spoke IPMSM, 13.0 Nm for SPMSM, and 17.6 Nm for V-IPMSM [12], with the PM volumes in the ratio: 100%/210%/43%/43%/43%.

As Ferrite PMs get better with temperature rise, their demagnetization risk is checked at 20° (H_{ic} = 250–300 kA/m, μ_{rec} = 1.06 p.u.). $B_n > 0.1$ T was considered safe on the PM surface along its length.

Unfortunately, even this time the SF-PMSM is not demagnetization safe and neither are the spoke IPMSM and V-IPMSM. The flux-barrier IPMSM and SPMSM are safe, but at notably lower (50–65%) torque for the same stator geometry and lower losses. Better (or more) PMs and/or different PM placement are required to avoid demagnetization risk with both NdFeB and Ferrite magnets. However, the regular SF-PMSM produces least torque Nm/PM weight even with Ferrites.

The initial cost/torque of all 10 motors at 30 Nm (five with NdFeB and five with Ferrite PMs) was also calculated and found to be the largest for NdFeB-SF-PMSM (3.3 USD/Nm), but rather low for the Ferrite SF-PMSM (1.6 USD/Nm). Moreover, there is no guarantee that the whole magnet will survive demagnetization at full torque in both cases [12].

At about the same initial cost, both the NdFeB and Ferrite flux-barrier IPMSM are safe from this point of view, but only with NdFeB (2.0 USD/Nm) are they free from demagnetization at the lowest volume of all five machines. Not so for the Ferrite PM version of SPMSMS, which is not safe at full load (the price of magnetic materials was 60 USD/kg for NdFeB (0.97 T, μ_{rec} = 1.06 p.u.), 10 USD/Kg for Ferrite PMs, and 2.2 USD/Kg for iron laminations) [12].

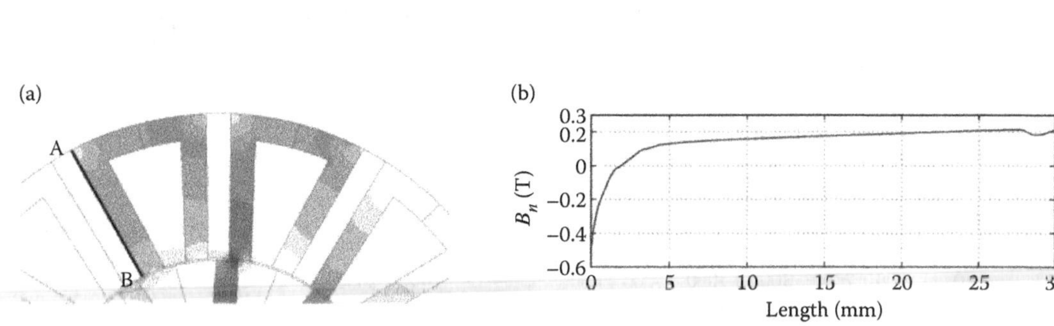

FIGURE 9.16 Flux density on magnet surface (line A-B) (a) in SF-PMSMS and flux density distribution (b). (After A. Fasolo, L. Alberti, N. Bianchi, *IEEE Trans.*, vol. IA–50, no. 6, 2014, pp. 3708–3716. [12])

9.3.3 Flux Weakening (Constant Power Speed Ratio) Capability

The same five motors with NdFeB magnets have been investigated for flux-weakening capability by imposing 30 Nm up to 1000 rpm and inverter current limit: 14 A; inverter voltage may be different to allow each machine 30 Nm at 1000 rpm [12].

To maintain voltage at speeds above 1000 rpm, the value of i_d increases (negative) or the current dq angle moves from the MTPA to maximum torque per voltage (MTPV). The results in power versus speed in Figure 9.16 [12] show that, here, the SF-PMSM is notably superior even to the V-IPMSM, not to mention the other three contenders. However, we should bear in mind that both the SF-PMSM and V-IPMSM face the danger of demagnetization at 30 Nm (up to 1000 rpm) and at high speeds (because $i_d < 0$ increases).

As the flux density in the stator yoke and upper and lower tooth zone is less sinusoidal in time with the SF-PMSM [12]; the core losses are expected be higher (in the latter) and thus the efficiency is lower. Also, the noise and vibration are expected to be higher with the SF-PMSM.

Finally, the torque/volume for the SF-PMSM (with NdFeB magnets) was found to be only 28.5 Nm/liter (at 9.00 A/mm^2), while it was 42.1 Nm/liter for the spoke PMSM; unfortunately, the latter is also under demagnetization risk. Torque pulsations for all motors studied here proved to be less than 6.4% [12] (Figure 9.17).

A few final remarks may be placed here:

- The regular SF-PMSM allows for a thermally and mechanically more rugged motor, but, when prime-quality NdFeB or SmCo magnets are used, it faces demagnetization risk, even at 9 A/mm^2 current density for only 28.5 N/liter at 30 Nm.
- The regular SF-PMSM is notably better in terms of CPSR once the demagnetization risk is eliminated, hopefully at moderate additional costs.
- It seems that there is hardly a way to make the regular 12/10 (or similar) SF-PMSM less expensive in terms of USD/Nm for the same copper losses as the SPMSM.
- Ferrite PMs should be checked carefully for demagnetization if ever tried for SF-PMSMs.
- The C-core (6/13, for example) SF-PMSM [4] has been proven to produce more emf (+40%) and torque (slot opening between the six poles is four to five times larger than PM thickness) in contrast to the 12/10 SF-PMSM of the same geometry, with half the PM (NdFeB) weight. The machine inductance also is increased 2.5 times, which may be good for wider SPSR, but at full (or over) load, the demagnetization risk is higher than in the regular SF-PMSM. The cost per torque Nm (in USD/Nm) will be decreased notably with respect to the regular SF-PMSM due to the use of about 50% PM weight.

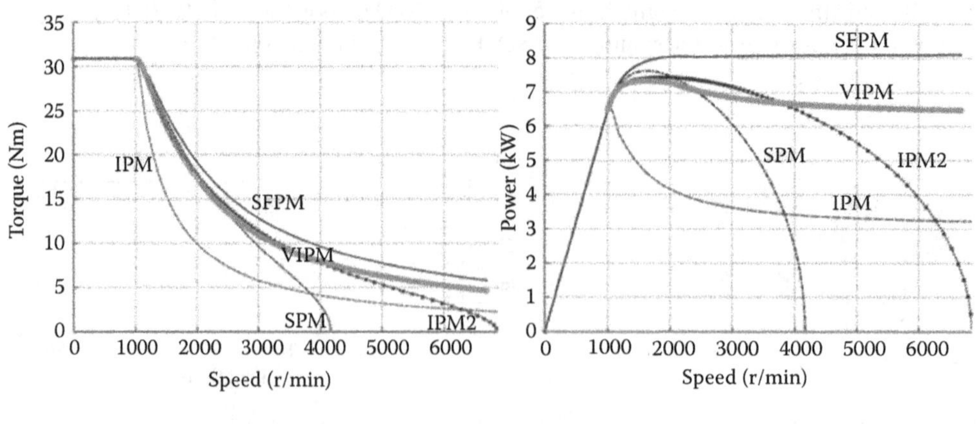

FIGURE 9.17 Flux weakening performance. (After A. Fasolo, L. Alberti, N. Bianchi, *IEEE Trans.*, vol. IA–50, no. 6, 2014, pp. 3708–3716. [12])

- Similar, better, results in terms of torque/PM weight have been claimed in [9] for a linear motor version of SF-PMSM with axial airgap and twin rotor where the stator in the middle has AC coils embracing two PMs (Figure 9.4b).
- Yet another attempt to increase torque/volume but also increase torque/PM weight [8] introduces an axial airgap twin-rotor 12/13 SF-PMSM, with dual stator winding (Figure 9.4a). Here, 3D-FEM is necessary (axial airgap). For an average torque of 70 Nm (torque ripple is about 5%) at 10 A/mm^2, a torque density of 13.5 Nm/liter is claimed for an efficiency of 0.8 at 450 rpm. Compared with the more than 45–60 Nm/liter in commercial EV and HEV propulsion at 50–100 KW for efficiency above 90%, with IPMSM or with copper cage IM, we feel that much more is to be done to make the FS-PMSM fully competitive and demagnetization safe in premium-quality (Nm/liter and efficiency) applications.

EXAMPLE 9.1: PRELIMINARY DESIGN

After elucidating to some degree the nature of SF-PMSMs and offering a rather complete comparison with four PMSMs, essentially using FEM for machine characterization, here we introduce a preliminary design example that may serve in feasibility studies or as an initial design for optimal design methodologies.

Let us design a $T_{en} = 100$ kNm direct drive SF-PMSM that works at $n_N = 30$ rpm and operates at $f_N = 50$ Hz.

The machine geometry, represented by two poles only, is shown in Figure 9.18.

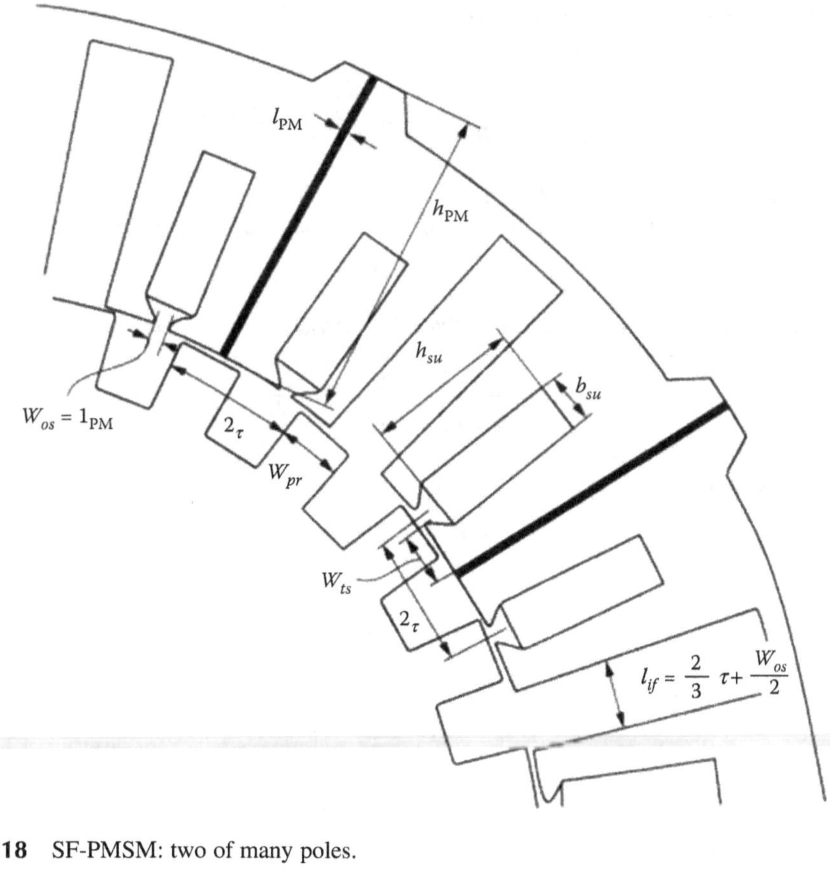

FIGURE 9.18 SF-PMSM: two of many poles.

A shear rotor stress $f_t = 5\ \text{N/cm}^2$ is adopted, and the ratio of stack length to interior stator diameter ratio is: $\lambda = l_{stack}/D_{is} = 0.15$. D_{is} may now be simply calculated as:

$$T_e = \frac{D_{is}}{2} \cdot f_t \cdot \pi \cdot D_{is} \cdot l_{stack}$$
$$D_{is} = \sqrt[3]{\frac{T_{en}}{\pi \cdot f_t \cdot \lambda}} = \sqrt[3]{\frac{100 \cdot 10^3}{\pi \cdot 5 \cdot 10^4 \cdot 0.15}} = 1.618\ \text{m} \tag{9.21}$$

So l_{stack} is:

$$l_{stack} = \lambda \cdot D_{is} = 0.15 \cdot 1.618 = 0.242\ \text{m} \tag{9.22}$$

The number of rotor salient poles for around 50 Hz is:

$$N_r = p_r = \frac{f_n}{n_n} = \frac{50}{30/60} = 100 \tag{9.23}$$

In reality, to simplify the design (Figure 9.18), here we magnetically separated the coils, and the extra available volume and area are used for cooling. Thus, if the rotor anisotropy has a periodicity of 2τ, one coil sector, which includes $2/3\tau$ shifting of coils (phases), is $(4\tau + 2\tau/3)$ long. The number of such sectors is 6 k and thus:

$$2N_r \cdot \tau = 6k\left(4\tau + \frac{2}{3}\tau\right) = 28k\tau \tag{9.24}$$

With $K = 7$; N_r becomes $N_r = 98$ rotor poles; $f_n = \frac{98}{100} \cdot 50 = 49\ \text{Hz}$. The airgap is $g = 2\ \text{mm}$ (assigned value), while the pole pitch:

$$\tau = \frac{\pi \cdot D_{is}}{2 \cdot N_r} = \frac{\pi \cdot 1.618}{2 \cdot 98} = 0.02592\ \text{m} = 25.92\ \text{mm} \tag{9.25}$$

The slot opening, equal to the PM thickness l_{pm}, is taken as $l_{pm} = 10\ \text{mm}$. So the stator tooth $w_{ts} = \tau - w_{0s} = 25.92 - 10 = 15.92\ \text{mm}$. The rotor pole span w_{pr} is in general:

$$w_{pr} = (0.7 - 1.0)\tau \tag{9.26}$$

We may now calculate the ideal no-load (PM) flux density in the airgap B_{gPMi}:

$$B_{gPMi} = \frac{B_r}{w_{ts}/h_{PM} + (\mu_{rec}/\mu_0) \cdot \frac{2g}{l_{PM}}} \tag{9.27}$$

which was derived based on:

$$\begin{aligned} B_m &= B_r + \mu_{rec} \cdot H_m \\ B_m \cdot h_{PM} &= B_g \cdot w_{ts} \\ H_m l_{PM} &+ \frac{B_g}{\mu_0} \cdot 2g = 0 \end{aligned} \tag{9.28}$$

With strong NdFeB magnets, $B_r = 1.2\ \text{T}$ and $\mu_{rec} = 1.05\,\mu_0$. Now, the PM flux fringing (K_{fringe}) and magnetic saturation (K_s) reduce B_{gPMi} to $B_{gPM} = 1.1\ \text{T}$.

$$B_{gPM} = B_{gPMi} \frac{1}{(1 + K_{fringe})(1 + K_s)} \tag{9.29}$$

With $K_{fringe} = 0.66$, $K_s = 0.2$, $B_{gPMi} = 2.19\ \text{T}$.

This high value is only theoretical, and from Equation 9.27:

$$2.19 = \frac{1.2}{15.92/B_{PM} + 1.05 \cdot 2.2/10} \tag{9.30}$$

the magnet radial height $h_{PM} = 124.6$ mm

Consequently, the maximum PM flux per coil turn φ_{PMmaxc} is:

$$\begin{aligned} \varphi_{PMmax} &= B_{gPM} \cdot w_{ts} \cdot l_{stack} \\ &= 1.1 \cdot 0.0159 \cdot 0.242 = 4.238 \cdot 10^{-3}\,\text{Wb} \end{aligned} \tag{9.31}$$

With $2K$ coils per phase (K = 7), the PM phase linkage ψ is:

$$\begin{aligned} \psi_{PMphase} &= 2K\varphi_{PMmax} \cdot n_c \\ &= 2 \cdot 7 \cdot 4.238.10^{-3} \cdot w_b \cdot n_c = n_c \cdot 59.33 \cdot 10^{-3}\,\text{Wb} \end{aligned} \tag{9.32}$$

n_c—number of turns per coil.

The emf per phase (peak value) is:

$$E_m = 2 \cdot \pi \cdot f_n \cdot \psi_{PMmax} = 2 \cdot \pi \cdot 49 \cdot 59.43 \cdot 10^{-3} \cdot n_c = 18.25 \cdot n_c \tag{9.33}$$

Now, the Ampere turns per coil (slot) can be calculated from torque T_e:

$$\begin{aligned} T_{en} &= 3\frac{E_m}{\sqrt{2}} I_n \frac{1}{2 \cdot \pi \cdot n_n} \\ n_c I_n &= \frac{2n_c}{3\sqrt{2}} T_{en} \cdot \frac{2\pi n_n}{E_m} = \frac{2 \cdot 100 \cdot 10^3 \cdot 2 \cdot \pi \cdot 60/60}{3\sqrt{2} \cdot 18.25} \\ &= 16.26 \cdot 10^3 \text{Aturns/slot(rms)} \end{aligned} \tag{9.34}$$

The slot width $w_{su} = (1 - 1.2)\tau_{PM} = 1.1 \cdot 25.92 = 28.512$ mm. The total available slot height $h_{su} = h_{PM} - (2/3)w_{ts} - 0.005 = 124.6 - (2/3) \cdot 15.92 - 5 = 109.09$ mm. Now, if we consider the coil cross-section rectangular and a slot filling factor $K_{fill} = 0.65$ (preformed coil), the current density j_{con} is:

$$\begin{aligned} j_{con} &= \frac{n_c I_n}{h_{su} w_{su} K_{fill}} \\ &= \frac{16.26 \cdot 10^3}{109.1 \cdot 28.52 \cdot 0.65} = 8.040\,\text{A/mm}^2\text{(rms)} \end{aligned} \tag{9.35}$$

With a turn length l_{coil} as:

$$\begin{aligned} l_{coil} &\approx 2l_{stack} + 4\tau_{PM} + \pi \cdot w_{su} \\ &= 2 \cdot 0.242 + 4 \cdot 0.02592 + \pi \cdot 0.0285 = 0.677\,\text{m} \end{aligned} \tag{9.36}$$

the stator phase resistance R_s is:

$$\begin{aligned} R_s &= 2K \cdot \rho_{Co} \cdot \frac{l_{coil}}{n_c I_n / j_{con}} n_c^2 \\ &= 2 \cdot 7 \cdot 2.1 \cdot 10^{-8} \cdot \frac{0.677}{16.26 \cdot 10^3 / 8.04 \cdot 10^6} \cdot n_c^2 \\ &= 9.571 \cdot n_c^2 \cdot 10^{-5}\Omega \end{aligned} \tag{9.37}$$

And thus the copper losses p_{copper} are:

$$p_{copper} = 3 \cdot R_s \cdot I_n^2 = 3 \cdot 9.571 \cdot (n_c \cdot I_n)^2 \cdot 10^{-5} = 75.913\ \text{W} \tag{9.38}$$

But the electromagnetic power P_{elm} is:

$$P_{elm} = T_e \cdot 2\pi \cdot n_n = 100 \cdot 10^3 \cdot 2\pi \cdot 60/60 = 628\ \text{kW} \tag{9.39}$$

So it seems that the copper losses are rather large (almost 13% of the electromagnetic power). This is mainly due to the rather large $f_t = 5\ \text{N/cm}^2$ allowed in the design. On the other hand, the design should continue, as apparently the machine inductance will be very large, so thicker PMs might be needed at a larger airgap, also to secure an acceptable ideal power factor of 0.707 when:

$$\frac{E_m}{\sqrt{2}} = \omega_n L_s I_n\ (\text{rms});\ \text{then: } V_{sn} \approx \frac{E_m}{\sqrt{2}} \cdot \sqrt{2}\ (\text{rms}) \tag{9.40}$$

V_{sn} is the rms phase stator rated voltage.

From these equations, the number of turns per coil n_c may be calculated (from V_{sn}). After that, the phase inductance $L_s \approx L_{sl} + 1.05L_{dm}$ is calculated using analytical expressions. If too large, L_s would make the solution of n_c infeasible (by not complying with the second equation in Equation 9.40); measures to reduce L_s have to be taken.

9.4 E-CORE HYBRID EXCITED SWITCHED FLUX–PERMANENT MAGNET SYNCHRONOUS MOTORS

DC assistance in SF-PMSMs is possible first in the regular 12/10 full reluctance stator core configuration (Figure 9.19a) [20, 21].

The use of assisting DC excitation allows for:

- Lower magnet volume, weight, cost
- Stronger flux weakening (FW) with zero stator i_d control and eventually lower total AC copper losses
- DC excitation-positive DC current at full torque below base speed and negative DC current for FW above base speed

Pertinent methods to optimize such a hybrid excitation SF-PMSM referring to N_r (number of rotor poles), PM thickness l_{PM}, slot width w_s, yoke depth h_{ys}, PM radial length h_{PM}, slot height h_{su}, stator (B_s) and rotor (B_r) tooth (pole) widths, and the stator split ratio D_{is}/D_{os} have been delivered [21].

The thin teeth embraced by the DC coil span may be also optimized.

The E-core (Figure 9.19b) 12/10 SF-PMSM keeps 12 stator poles, but 6 are narrow and have DC coils placed on them. This way, the configuration falls close to the C-core 6/10 SF-PMSMs. The C-core configuration has been proven to need about half the PM weight of the initial 12/10 configuration, but in a 6/13 slot/pole combination. The C core $N_s = 6$(coils)/$N_r = 13$ rotor poles SF-PMSM, also known to produce symmetric phase mmfs, is also chosen to increase torque (Nm/PM weight). As evident in Figure 9.19b for the E-core machine, the flux of DC coil II overcomes that of DC coil I because of the difference in magnetic reluctance. Consequently, the DC excitation mmf does not need to be so large, as the airgap is rather small (parallel hybrid excitation).

It has been proven that, even for $j_{dc} = \pm 15\ \text{A/mm}^2$ in the DC coils, the no-load flux linkage may be varied only by +8%, −13% while, with Ferrite PMs, the variation is +45%, −63% for the regular 12/10 SF-PMSM. The flux variation range for the E-core 6/11, 6/13 SF-PMSM is smaller than for the regular 12/10 SF-PMSM, but the average torque of the E-core 6/13 SF-PMSM is 25% larger.

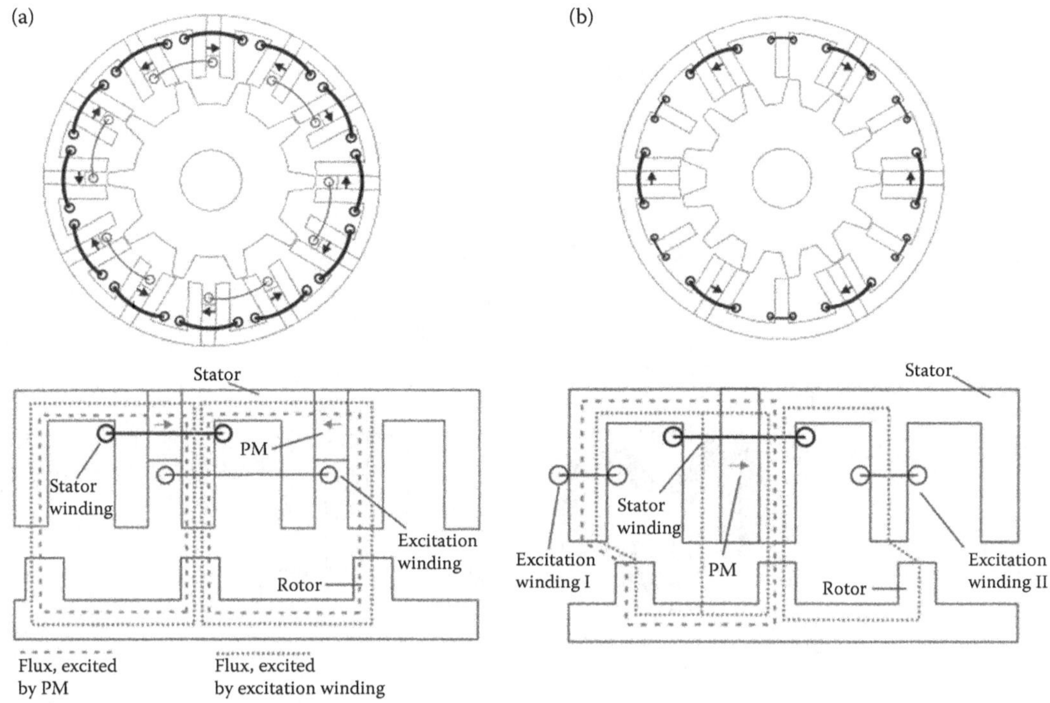

FIGURE 9.19 12/10 SF-PMSMs with hybrid excitation (a) with regular stator core; (b) with E-core stator. (After Y. Wang, Z. Deug, *IEEE Trans.*, vol. MAG–48, no. 9, 2012, pp. 2518–2527; J. Krenn, R. Krall, F. Aschenbrenner, *Optimization of Electrical and Electronic Equipment (OPTIM), 2014 International Conference on*, Brasov, Romania, 22–24 May 2014. [20,21])

The flux variation capability is larger for Ferrite PMs, but the average torque for similar geometry is notably smaller. It may be fairly stated that the flux variation capability with the SF-PMSM is still small, and new ways to improve it are needed.

9.5 SWITCHED FLUX–PERMANENT MAGNET SYNCHRONOUS MOTORS WITH MEMORY ALNICO ASSISTANCE FOR VARIABLE FLUX

Performing flux weakening in IPMSMs, as in SF-PMSMs, is accomplished by negative i_d current component control, which at high speeds means higher copper losses and thus lower efficiency than even IMs in same situation.

It should be ideal to control the SF-PMSM at $i_d = 0$ (maximum torque per current as $L_d \approx L_q$) but weaken the flux. This operation may be done through memory AlNiCo magnets with DC current coils, which may magnetize them positively or negatively "online" to increase/decrease the flux of the main (NdFeB) magnets over a wide range of flux adjustment.

This situation [10], already alluded to earlier in this chapter, is presented here in more detail, as the flux adjustment range may reach a 5/1 ratio eventually. The proposed configuration [10] is shown in Figure 9.20.

The 12/14 SF-PMSM configuration is chosen to mitigate between average torque, torque ripples, and compensated radial forces. To reduce torque ripples from 16.5% to 8.6%, tooth steps and rotor pole chamfering are added (Figure 9.20b), though the average torque is reduced by 7% (Figure 9.21).

To avoid too-heavy magnetic saturation in the stator core, care should be exercised in choosing the proportion (hybridization ratio HR) of AlNiCo and NdFeB magnets (HR = 0.85 for the case in

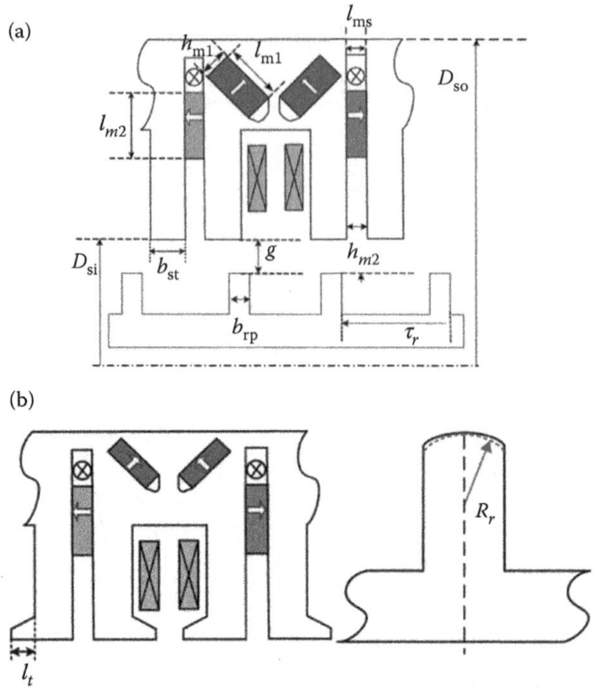

FIGURE 9.20 Hybrid 12/14 SF-PMSM with memory AlNiCo additional magnets—only one stator slot surrounding is shown. (a) barrier configuration; (b) tooth tip and rotor pole chamfer. (After H. Yang et al., *IEEE Trans.*, vol. IA–52, no. 3, 2016, pp. 2203–2214. [10])

point). The flux adjustment ratio increases with more AlNiCo area per pole, while the average torque is larger with more NdFeB area (which has a 6/1 higher coercive field H_c):

$$H_c = \frac{A_{\text{AlNiCo}}}{A_{\text{NdFeB}}} \tag{9.41}$$

A typical result of this compromise [10] is shown in Figure 9.22a.

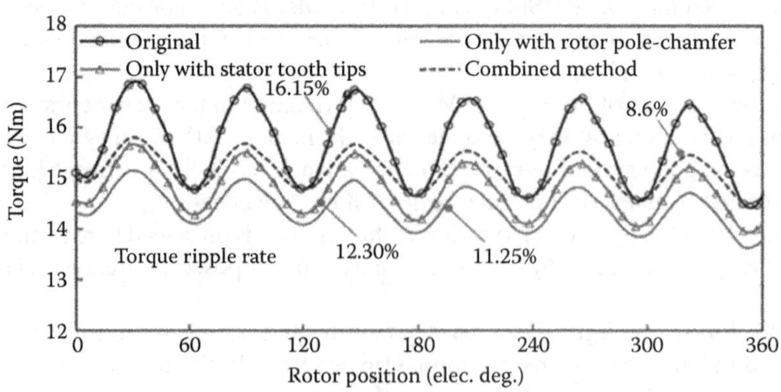

FIGURE 9.21 Full torque dependence on rotor position. (After H. Yang et al., *IEEE Trans.*, vol. IA–52, no. 3, 2016, pp. 2203–2214. [10])

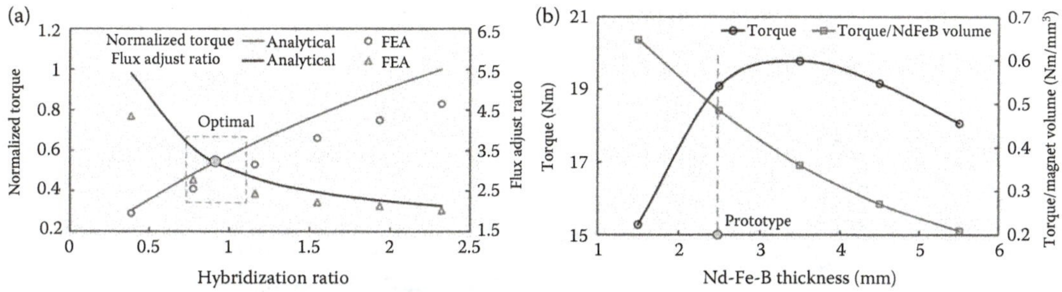

FIGURE 9.22 P.u. torque and flux adjustment ratio versus HR (hybridization ratio), (a); and torque/PM volume versus NdFeB magnet thickness, (b). (After H. Yang et al., *IEEE Trans.*, vol. IA–52, no. 3, 2016, pp. 2203–2214. [10])

Also, to reduce NdFeB thickness is important, but preserving low enough cogging torque and high torque/PM volume (weight)—Figure 9.22b [10]—is paramount. If the NdFeB magnets are too thick, they will simply demagnetize the AlNiCo magnets under load. This new problem needs careful scrutiny. Results in Figure 9.23 illustrate the need for a new compromise for 0.8 T no-load airgap flux density to secure a zero demagnetization ratio (in %) for the AlNiCo magnets when the NdFeB magnets are 2.5 mm thick [10].

It has to be remembered that the online magnetization/demagnetization of AlNiCo within tens of milliseconds by DC coils $\pm$ currents makes the torque response slower. But this may be acceptable in some applications with wide CPSR, such as traction, ventilation, and pumps.

As the efficiency improvement at high speed (FW) was the main goal of the AlNiCo memory magnets' presence, Figure 9.24 [10] shows the optimum wide operation torque-speed range where high efficiency is secured as it should be.

To conclude, we notice the following:

- The introduction of memory AlNiCo magnets to supplement NdFeB main magnets in a 12/14 SF-PMSM has raised the flux adjustment ratio to 5/1, for a hybridization (area) of 0.85 (AlNiCo area/NdFeB area), which means a notable reduction of NdFeB weight for a given peak torque.
- The design has to avoid too-thick NdFeB magnets to avoid AlNiCo PMs demagnetization under full load. However, too-thin NdFeB magnets will lead to demagnetization risk at high

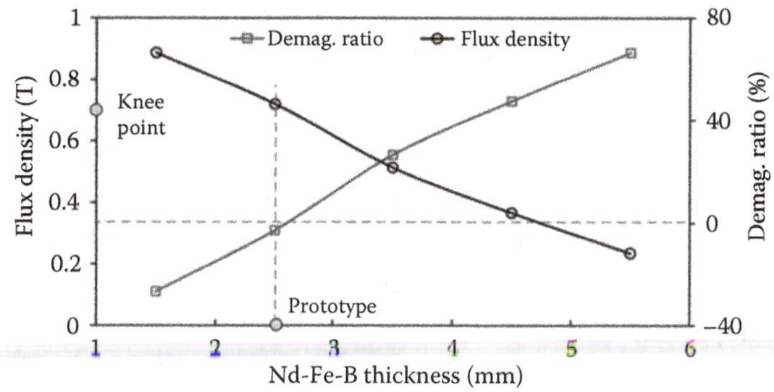

FIGURE 9.23 Airgap no-load flux density and AlNiCo demag. ratio versus NdFeB magnet thickness. (After H. Yang et al., *IEEE Trans.*, vol. IA–52, no. 3, 2016, pp. 2203–2214. [10])

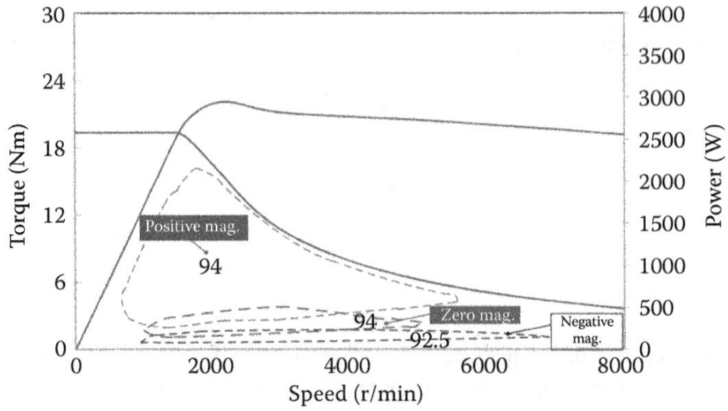

FIGURE 9.24 Optimum wide torque speed envelope with high efficiency for the AlNiCo SF-PMSM. (After H. Yang et al., *IEEE Trans.*, vol. IA–52, no. 3, 2016, pp. 2203–2214. [10])

loads; also, it will lead to a large inductance, which will reduce the power factor and thus increase the inverter KVA ratings.
- A few tens of milliseconds DC ±current pulses from dedicated DC coils are required to change the flux level, so the torque response is slower.
- The energy loss in DC coils to magnetize/demagnetize the AlNiCo magnets online is quite small.
- The higher efficiency at high speeds (for wide CPSR) by the method is demonstrated.
- All the above merits are paid for by additional hardware and space (larger outer stator diameter) and a four-quadrant DC-DC converter to handle the mag/demag current pulses. The peak KVA rating of the DC–DC converter should not be trivial (≈3%–5%).

9.6 PARTITIONED STATOR SWITCHED FLUX–PERMANENT MAGNET SYNCHRONOUS MOTORS

SF-PMSMs are characterized, as now known, by stator PMs that produce in the AC stator coils (nonoverlapping or overlapping) a bipolar (ideally sinusoidal) flux linkage due to the stator and the rotor magnetic saliency (open slots on both sides).

In general, nonoverlapping coil windings embrace about one rotor pole pitch in SF-PMSMs, while overlapping coils embrace around two or more rotor pole pitches to match the flux-modulator principle:

$$p_a = N_r - N_s/2 \tag{9.42}$$

p_a, N_r, N_s are the AC winding mmf pole pair, rotor, and stator tooth number.

Note: Flux-reversal PMSMs [13,14], in contrast, have multi-PM pole pitch nonoverlapping windings with PMs either on the stator or rotor and will be treated in a separate chapter.

Rather recently, inspired by the "flux reversal" concept [13], partition-stator SF-PMSMs have been proposed [22, 23]. In such machines, the external stator carries the AC nonoverlapping coil winding placed in N_s teeth with open slots, while the internal stator also carries N_s or $N_s/2$ PM poles of alternate polarity (with surface, consequent, or spoke-shaped PMs (Figure 9.25). In between, the rotor is made of N_r laminated iron segments that provide a magnetic saliency with N_r periods per periphery. So, here, only the PMs of the SF-PMSM have been placed in a separate stator.

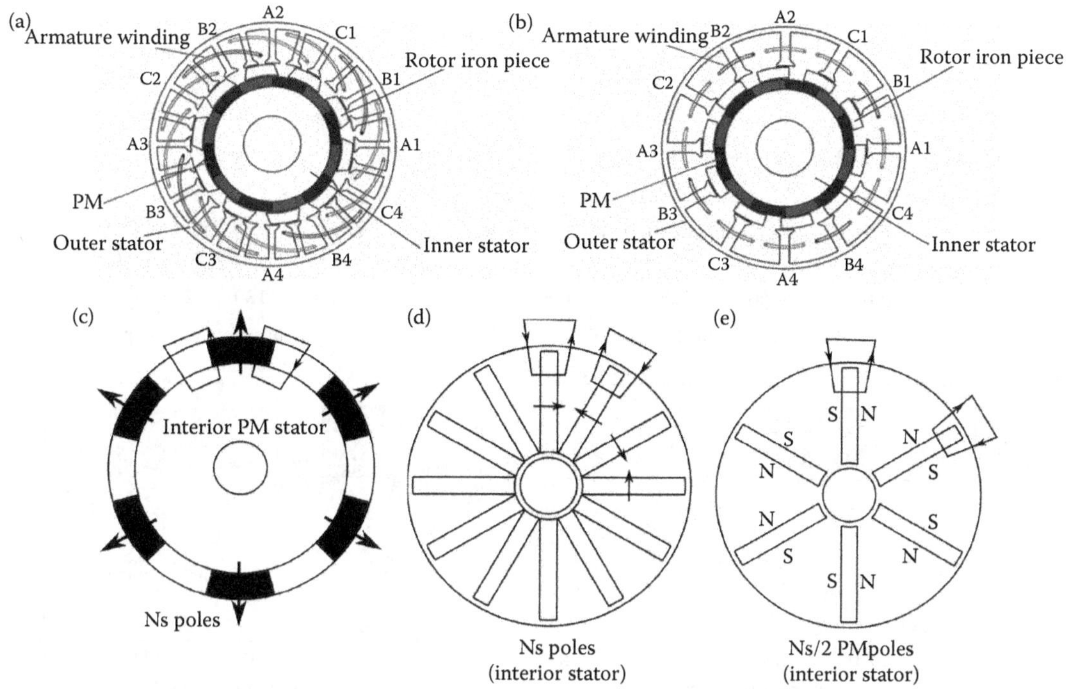

FIGURE 9.25 Partitioned-stator 12/10/12 SF-PMSM configurations with radial airgap and segmented rotor: (a) with overlapping (distributed: q > 1) winding and SPMs; (b). with tooth-wound winding and SPMs; (c) with consequent interior stator $N_s = 12$ PM poles (12/11/12). (Adapted from C. C. Awah et al., *IEEE Trans.*, vol. MAG–52, no. 1, 2016, pp. 9500310. [23]); (d) with spoke-shape $N_s = 12$ PM poles on the interior stator. (Adapted from C. C. Awah et al., *IEEE Trans.*, vol. MAG–52, no. 1, 2016, pp. 9500310. [23]); (e) with tooth-wound AC outer stator coils and $N_s/2$ interior stator PMs ($N_{so}/N_{si} = 12/6$).

According to the flux modulation principle (Equation 9.42) the pole pairs of stator mmf $p_a = 4$ (it is the fundamental one for the $q = 1$, 24-slot stator winding [Figure 9.25a] and the largest forward harmonic for the tooth-wound 12-slot winding [Figure 9.24b]).

Consequently, $N_s = 12$, $N_r = 10$, $p_a = 4$ in Equation 9.42. Now, for the PS-SF PMSMs, the inner stator pole pair $p_{PM} = N_s/2 = 6$ again, as it is for the regular SF-PMSM (12/10).

To reduce PM weight, consequent poles may be used in the interior stator, while, to increase mf and torque, spoke magnets are adequate.

Finally, it is possible to use only six PM poles on the interior stator (for the 12/10/6 combination), but only for a single layer tooth-wound winding, which has been proven to produce wide CPSR and good PM utilization at lower torque per outer stator volume [23].

Thus, only the PS-SF-PMSM with surface PMs on the interior stator (Figure 9.24a and b) will be treated in more detail here.

The interior stator may be shifted with respect to the outer stator poles to obtain a kind of flux weakening. However, the flux adjustment is only from 100% to about 60%, which hardly justifies the introduction of an even small power servo drive with a high-ratio latched gear to do the flux weakening. And, again, an analytical technical theory of PS-SF-PMSMs may be elaborated on for preliminary design and as a safe starting set of variables for optimal design codes. The segmented character of the rotor and the slotting in the stator introduce additional phenomena besides magnetic saturation that warrant the use of 2(3)D-FEM. 3D-FEM is imperative in short stack length l_{stack}/coil-through ratio (less than 1–1.5) machines. For comparison, two such machines (24/10/12 and 12/10/12),

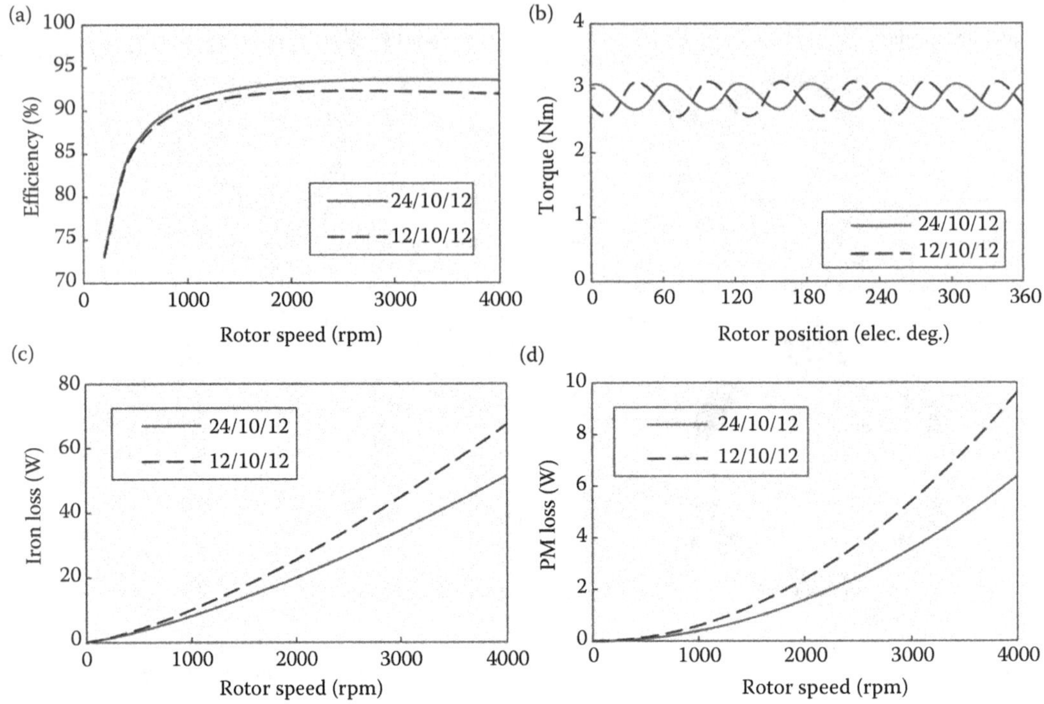

FIGURE 9.26 Design optimization result for 24/10/12 and 12/10/12 PS-SF-PMSMs with $D_{os} = 90$ mm, $l_{stack} = 25$ ms: (a) efficiency; (b) torque versus rotor position; (c) on load iron losses; (d) PM eddy current losses. (After I. Boldea and S. A. Nasar, *Electric drives*, 3rd edition, CRC Press, Taylor and Francis Group, New York, 2016. [25])

PS-SF-PMSMs with outer stator diameter $D_{os} = 90$ mm and stack length $l_{stack} = 25$ mm, are investigated, globally optimized, and compared for 20 W of ideal copper losses at rated speed $n_n = 400$ rpm.

The calculated electrical efficiency (mechanical losses are taken as zero, when they are notable with two small airgaps around the anisotropic laminated rotor) obtained (Figure 9.26a) [25] seems remarkable. The rated average torque for 20 W ideal copper losses for both machines investigated is around 2.87 Nm (Figure 9.26b). The core loss is larger for the 12/10/12 configuration because of the richer content of harmonics of the outer stator mmf (Figure 9.26c), and so are the PM eddy current losses (Figure 9.26d).

The 24/10/12 configuration has been shown to have higher phase self-inductance and lower mutual inductance (due to shorter flux lines, in essence) and thus it shows a smaller short-circuit current and slightly larger CPSR. Despite larger end coil connections, the slightly higher emf of the 24/10/12 configuration leads to slightly better (1%) efficiency because of lower core losses. But the difference is small and the machine frame is about 10% larger, heavier, and costlier (in copper and framing).

Figure 9.27 exposes the first prototype of these two machines [22]:

Note: Though the torque/volume of around 8 Nm/liter is, for an outer stator diameter $D_{os} = 90$ mm, rather acceptable, the mechanical complexity of the machine leads to increased fabrication costs and it is still to be determined if the PS-SF-PMSM will replace the regular SPMSMs and IPMSMs with tooth-wound AC stator windings. But if, as shown in [24], increasing the number of rotor poles to 13 (14) and reducing the outer rotor pole tips leads to a further 80% increase in torque at the price of thicker magnets, which also reduces demagnetization risk, high torque/volume may be expected.

FIGURE 9.27 The PS-SF-PMSM prototypes for $D_{os} = 90$ mm and $l_{stack} = 25$ mm (a) outer stator with 24 slots; (b) outer stator with 12 slots; (c) 10-pole cup-shaped rotor; (d) rotor laminations with bridges (for one piece); (e) 12-pole inner stator; (f) inner stator with bearing. (After Z. Z. Wu, Z. Q. Zhu, H. L. Zhan, *IEEE Trans*, vol. EC–31, no. 2, 2016, pp. 776–788. [22])

9.7 CIRCUIT *dq* MODEL AND CONTROL OF SWITCHED FLUX–PERMANENT MAGNET SYNCHRONOUS MOTORS

SF-PMSMs may be assimilated to synchronous machines where the number of pole pairs is the number of rotor salient poles N_r. There is some magnetic saliency ($L_d < L_q$), but not much. If a DC field stator winding is added, its equation should be added. Also, the emf (by PMs and DC coils) is rather sinusoidal. So the *dq* model in rotor coordinates $\omega_r = 2\pi N_r n$ for synchronous machines may be used to portray SF-PMSM behavior for transients and control design:

$$\begin{bmatrix} V_d \\ V_q \\ V_f \end{bmatrix} = \begin{bmatrix} R_s + pL_d & -\omega_r L_q & pM_f \\ \omega_r L_q & R_s + pL_q & \omega_r M_f \\ pM_f & 0 & R_f + pL_f \end{bmatrix} \cdot \begin{bmatrix} i_d \\ i_q \\ i_f \end{bmatrix} + \begin{bmatrix} 0 \\ \omega_r \psi_{PM} \\ 0 \end{bmatrix} \tag{9.43}$$

with

$$\begin{bmatrix} \psi_d \\ \psi_q \\ \psi_f \end{bmatrix} = \begin{bmatrix} L_d & 0 & M_f \\ 0 & L_q & 0 \\ \frac{3}{2}M_f & 0 & L_f \end{bmatrix} \cdot \begin{bmatrix} i_d \\ i_q \\ i_f \end{bmatrix} + \begin{bmatrix} \psi_{PM} \\ 0 \\ \psi_{mf} \end{bmatrix} \tag{9.44}$$

and torque T_e:

$$T_e = \frac{3}{2} p_r i_q (\psi_{PM} + M_f i_f + (L_d - L_q) i_d) \tag{9.45}$$

In the absence of DC stator excitation (with positive and negative excitation current i_f), the dq model in Equations 9.43 through 9.45 "degenerates," as expected, into that of regular IPMSMs. So all the control heritage of IPMSMs:

- Field-oriented control
- Direct torque and flux control
- Scalar V/f (or I-f) control with stabilizing loops

may be applicable here [25]. The inductances may be imported from finite-element analysis (FEA) or may have analytical formulae based on MEC modeling. Yes, cross-coupling saturation occurs, and this may cause errors in rotor position (and speed) observers for encoderless control.

As the SF-PMSM is similar to a regular IPMSM with small magnetic saliency $L_d/L_q = 0.8 - 0.85$, we refer the reader to the rich literature (papers and books) on the IPMSM control with and without encoder feedback [25], except for encoderless control with signal injection, which bears some peculiarities worth considering here. So attention will be placed on DC-excited SF-PMSMs for wide speed range control, where this machine shows definite merits.

9.7.1 Signal Injection Encoderless Field-Oriented Control of Switched Flux–Permanent Magnet Synchronous Motors

For sustainable operation at low speeds (a few rpms) and for nonhesitant starting, either encoder control or signal injection encoderless control may be applied. However, the magnetic saliency for signal injection (especially) is low in SF-PMSMs and thus position observers have difficulties in assessing rotor position correctly, especially as SF-PMSMs tend to have a large number of poles.

The FOC of SF-PMSMs may start with a "signal injection" rotor position (and speed) estimation and then "fuse in" an emf (or active flux [25])-based speed and position observer for wide speed control (with flux weakening). But this control for SM drives is already treated extensively in the literature [25], where, from MTPA to MTPV, controls are corroborated seamlessly.

However, as the saliency is small, the CPSR is acceptable (2/1 to 3/1) only due to large machine inductance in p.u.

Ref. 26 investigated four 12/10 SF-PMSMs as follows:

- Three-phase with double-layer tooth-wound AC coils (Figure 9.28a)
- Three-phase with one-layer tooth-wound AC coils (Figure 9.28b)

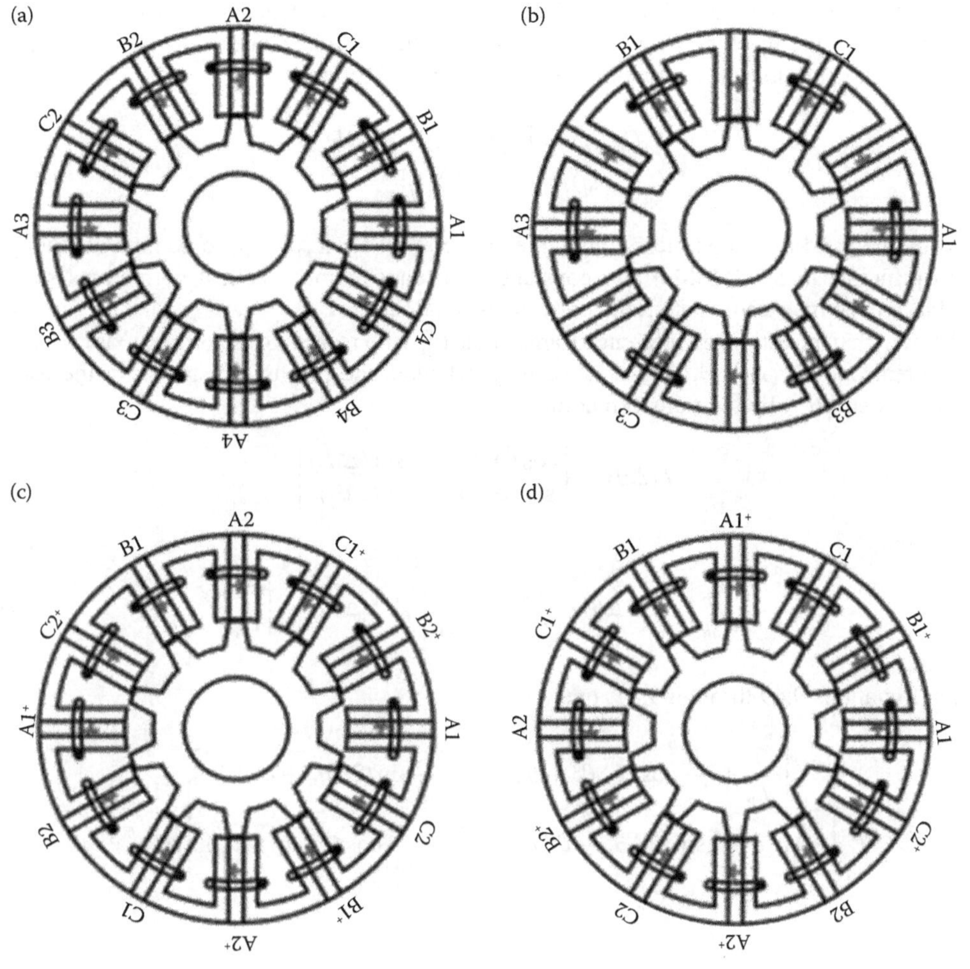

FIGURE 9.28 12/10 SF-PMSM (a) double layer; (b) single layer; (c) alternating 2 × 3 phases; (d) adjacent 2 × 3 phases. (After T. C. Lin et al., *IEEE Trans.*, vol. IE–63, no. 1, 2016, pp. 123–132. [26])

- 2 × 3 phases (phases A_1 and A_2 are in series) for fault tolerance with two 50% rating inverters (Figure 9.28c) or with one of 100% ratings with alternating (Figure 9.28c) and adjacent (Figure 9.28d) three-phase coil groups

For high-frequency ω_k injection at standstill or at low speed ($\omega_r/\omega_k < 0.05$), the dq model in synchronous dq coordinates is:

$$\begin{vmatrix} V_{dh} \\ V_{qh} \end{vmatrix} = \begin{vmatrix} L_{dh} & L_{dqh} \\ L_{qdh} & L_{qh} \end{vmatrix} \frac{d}{dt} \begin{vmatrix} i_{dh} \\ i_{qh} \end{vmatrix} \tag{9.46}$$

where the HF inductances—which include cross-coupling terms ($L_{qdh} \approx L_{dqh} \neq 0$)—are:

$$\begin{aligned} L_{dh} &= \left[\psi_d(i_d + \Delta i_d, i_q) - \psi_d(i_d, i_q)\right]/\Delta i_d \\ L_{qh} &= \left[\psi_q(i_d, i_q + \Delta i_q) - \psi_q(i_d, i_q)\right]/\Delta i_q \end{aligned} \tag{9.47}$$

$$\begin{aligned} L_{dqh} &= \left[\psi_d(i_d, i_q + \Delta i_q) - \psi_d(i_d, i_q)\right]/\Delta i_q \\ L_{qdh} &= \left[\psi_q(i_d, i_q + \Delta i_q) - \psi_q(i_d, i_q)\right]/\Delta i_d \end{aligned} \tag{9.48}$$

These inductances may be FEM calculated and shown to be dependent on both current i_d, i_q and smaller, with the machine heavily saturated, than the dq inductances calculated from FEM in zero-frequency conditions at standstill, with frozen permeabilities as:

$$\begin{aligned} L_d(i_d, i_q) &= \left[\psi_d(i_d, i_q) - \psi_{\text{PM}}\right]/i_d \\ L_q(i_d, i_q) &= \psi_q(i_d, i_q)/i_q \end{aligned} \tag{9.49}$$

It would be good for encoderless signal injection control for $L_{qh} - L_{dh}$ to be notable (though not large in principle in SF-PMSMs) and constant with load; if not, practical corrections should be applied to reduce the steady-state rotor position estimation error.

Assuming that the estimated reference frame angle θ_{er}^e with respect to the rotor d axis differs a little from the real one θ_{er}: $\Delta\theta_e = \theta_{er}^e - \theta_{er}$, we may calculate the machine response in the estimated reference system (θ_{er}^r) by the transformation:

$$T(\Delta\theta) = \begin{vmatrix} \cos(\Delta\theta_e) & -\sin(\Delta\theta_e) \\ \sin(\Delta\theta_e) & \cos(\Delta\theta_e) \end{vmatrix} \tag{9.50}$$

We obtain:

$$\begin{vmatrix} V_{dh}^e \\ V_{qh}^e \end{vmatrix} = T(\Delta\theta) \cdot \begin{vmatrix} L_{dh} & L_{dqh} \\ L_{qdh} & L_{qh} \end{vmatrix} .T^{-1}(\Delta\theta) \cdot \frac{d}{dt} \begin{vmatrix} i_{dh}^e \\ i_{qh}^e \end{vmatrix} \tag{9.51}$$

From Equations 9.49 through 9.51, one obtains:

$$\frac{d}{dt}\begin{bmatrix} i_{dh}^e \\ i_{qh}^e \end{bmatrix} = \begin{bmatrix} \frac{1}{L_p} + \frac{1}{L_m}\cos(2\Delta\theta_e + \theta_m) & \frac{1}{L_m}\sin(2\Delta\theta_e + \theta_m) \\ \frac{1}{L_m}\sin(2\Delta\theta_e + \theta_m) & \frac{1}{L_p} - \frac{1}{L_m}\cos(2\Delta\theta_e + \theta_m) \end{bmatrix} \begin{bmatrix} V_{dh}^e \\ V_{qh}^e \end{bmatrix} \tag{9.52}$$

with

$$\begin{aligned} &L_{s0} = (L_{dh} + L_{qh})/2,\ L_{s2} = (L_{qh} - L_{dh})/2 \\ &\theta_m = \tan^{-1}\left(\frac{-L_{dqh}}{L_{s0}}\right);\ L_p = \frac{L_{dh}L_{qh} - L_{dqh}^2}{L_{s0}};\ L_n = \frac{L_{dh}L_{qh} - L_{dqh}^2}{\sqrt{L_{dh}^2 + L_{qh}^2}} \end{aligned} \tag{9.53}$$

If the voltage injection into the virtual (estimated) reference frame has a pulsating character (only along axis d^e):

$$\begin{bmatrix} V_{dh}^e \\ V_{qh}^e \end{bmatrix} = V_h \begin{bmatrix} \cos\alpha \\ 0 \end{bmatrix};\ \alpha = \omega_h t + \varphi \tag{9.54}$$

where V_h, α are related to voltage injection parameters. Using Equation 9.54 in Equation 9.52, one obtains approximately:

$$\begin{bmatrix} i_{dh}^e \\ i_{qh}^e \end{bmatrix} = \begin{bmatrix} I_p + I_n\cos(2\Delta\theta_e + \theta_m) \\ I_n\sin(2\Delta\theta_e + \theta_m) \end{bmatrix} \sin(\omega_h t + \varphi) \tag{9.55}$$

with

$$I_p = V_h/(\omega_h L_p),\ I_n = V_h/\omega_h L_n \tag{9.56}$$

As expected, $I_n > I_p$ and its processing, after filtering the ω_h frequency component, will yield a raw rotor position estimation $\hat{\theta}_{er}$.

To obtain the amplitude of carrier (ω_h) current response, synchronous detection and low pass filter (LPF) are applied [26]:

$$\begin{bmatrix} i^e_{dh} \\ i^e_{qh} \end{bmatrix} = LPF\left(\begin{bmatrix} i^e_{dh} \\ i^e_{dh} \end{bmatrix} \cdot 2\sin(\omega_h t + \varphi)\right) = \begin{bmatrix} I_p + I_n \cos(2\Delta\theta_e + \theta_m) \\ I_n \sin(2\Delta\theta_e + \theta_m) \end{bmatrix} \tag{9.57}$$

The "road" from measured carrier $d^e q^e$ currents to their pulsating amplitudes and to i^e_{qh} response (according to Equation 9.57) is shown in Figure 9.29 [26].

It has been shown [26] that the amplitude of the i^e_q circle (Figure 9.29c) is higher for the two-phase 12/10 single-layer and 2 × 3-phase 12/10 double-layer windings. However, additional saliencies occur, which finally render the double layer 12/10 winding best for encoderless control! The rotor position extracted from Equation 9.57, even with a single PLL tracking observer (Figure 9.29a), is not improved enough in quality, unless an orthogonal PLL observer is applied [26]; see Figure 9.30b.

Even tracking observers based on motion equation, such as the PLL observer, may be used to improve the quality of rotor position and speed estimation ([25], Chapter 9), especially during speed transients. The entire FOC diagram is represented in Figure 9.31.

The 12/10 SF-PMSM investigated [26] has the data: $T_{en} = 2.7$ Nm, $n_n = 4000$ rpm, $L_d = 0.277$ mH, $L_2 = 0.377$ mH; digital simulations at 25 rpm and test results in transients from 0 to 25 rpm illustrated in Figure 9.32 [26] show remarkable quality performance.

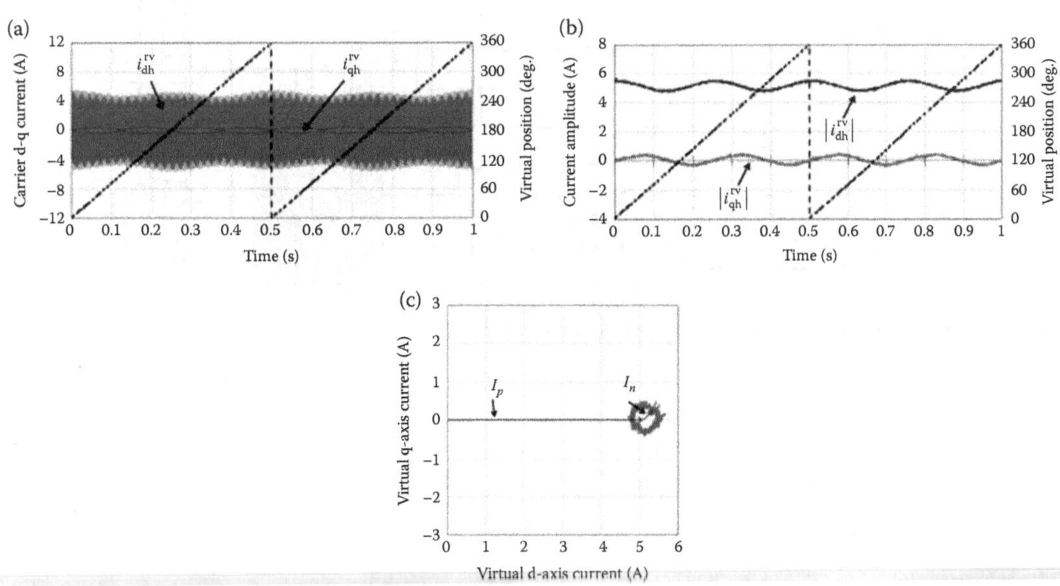

FIGURE 9.29 Measured HF (signal injection) currents in the estimated reference frame $d^e q^e$, (a); their amplitudes, (b); and i^e_q locus due to machine saliency, (c). (After T. C. Lin et al., *IEEE Trans.*, vol. IE–63, no. 1, 2016, pp. 123–132. [26])

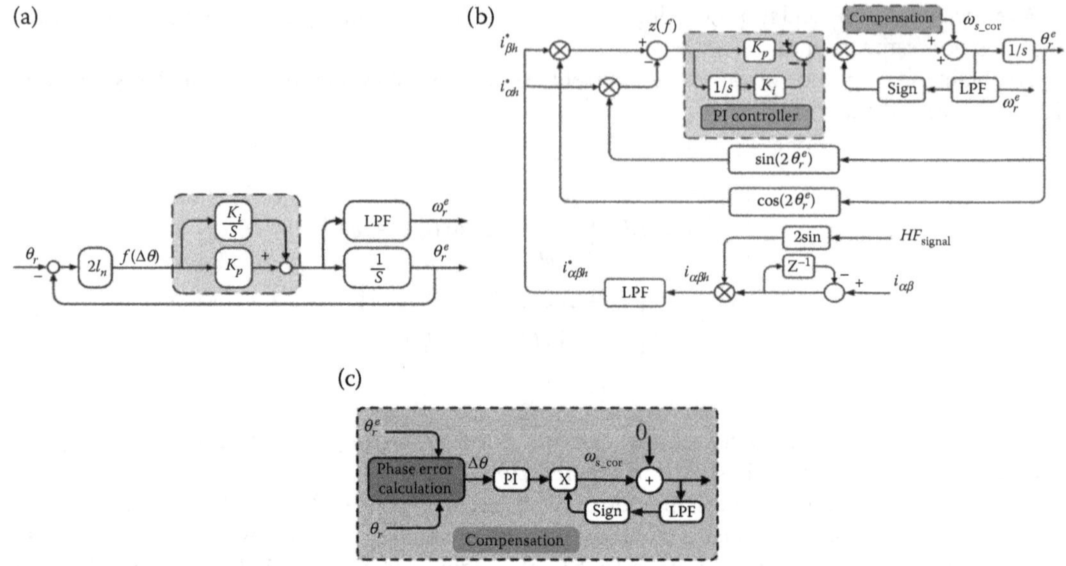

FIGURE 9.30 Single, (a); and orthogonal, (b); position tracking observers and compensator, (c). (After T. C. Lin et al., *IEEE Trans.*, vol. IE–63, no. 1, 2016, pp. 123–132. [26])

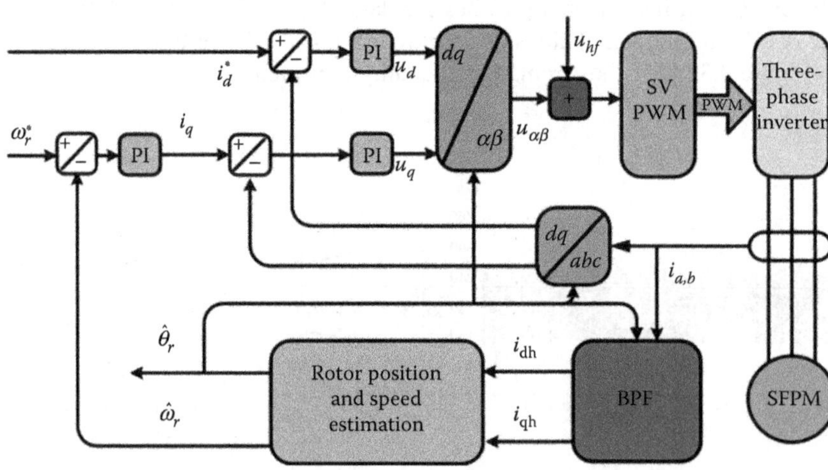

FIGURE 9.31 Encoderless signal injection system in SF-PMSM.

9.7.2 Direct Current–Excited Switched Flux–Permanent Magnet Synchronous Motors Field-Oriented Control for a Wide Speed Range

This time, the complete *dq* model (including the field circuit equation) is used. An axial-airgap topology with six stator E-cores (and DC-poles on the thinner teeth in the middle of E cores) and 11 rotor poles [27] is used for exemplification (Figure 9.33). The same topology, but with a radial airgap, was described earlier in the chapter as a good candidate for wide CPSR (Figure 9.33) and PM weight reduction by using positive (magnetizing) and negative (demagnetizing) DC excitation current.

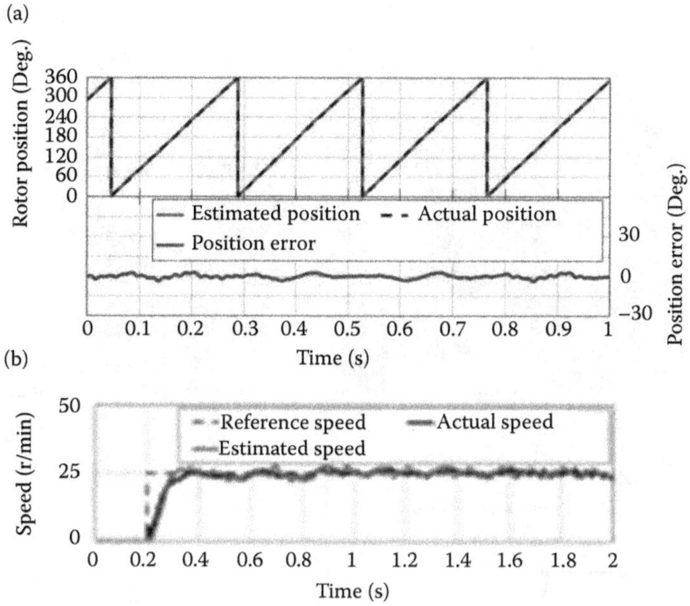

FIGURE 9.32 12/10 SF-PMSM: encoderless signal-injection FOC performance: (a) digital simulation at 25 rpm; (b) transients from zero to 25 rpm. (Adapted from T. C. Lin et al., *IEEE Trans.*, vol. IE–63, no. 1, 2016, pp. 123–132. [26])

For a wide torque speed control range envelope, $i_d = 0$ control with positive i_f below base speed (for more torque) and negative i_f above base speed (for less flux) may be described in three stages [27]:

$$\begin{aligned} &i_d^* = 0,\ i_F^* = \frac{2T_e - 3p_r\psi_{PM}i_{qn}}{3p_1 M_f i_{qn}} \quad \text{for } T_{load} > \frac{3}{2}p_r\psi_{PM}i_{qn};\ \omega_r < \omega_b \\ &i_d^* = 0,\ i_F^* = 0,\ i_q^* = \frac{2T_e}{3p_r\psi_{PM}} \quad \text{for } T_{load} > \frac{3}{2}p_r\psi_{PM}i_{qn} \end{aligned} \tag{9.58}$$

Above base speed ω_b, for flux weakening, there are two additional options:

$$\text{Stage 1}\begin{cases} i_F^* = \dfrac{\sqrt{\left(V_{dc}/\sqrt{3}\omega_r\right)^2 - (L_q i_q^*)^2} - \psi_{PM}}{M_F^*} > 0 \\ i_d^* = 0 \\ i_q^* = \dfrac{2T_e^*}{\dfrac{3}{2}p_r(\psi_{PM} + M_F i_F^*)}, \quad \text{if } V_s^* < V_{DC}/\sqrt{3} \end{cases} \tag{9.59}$$

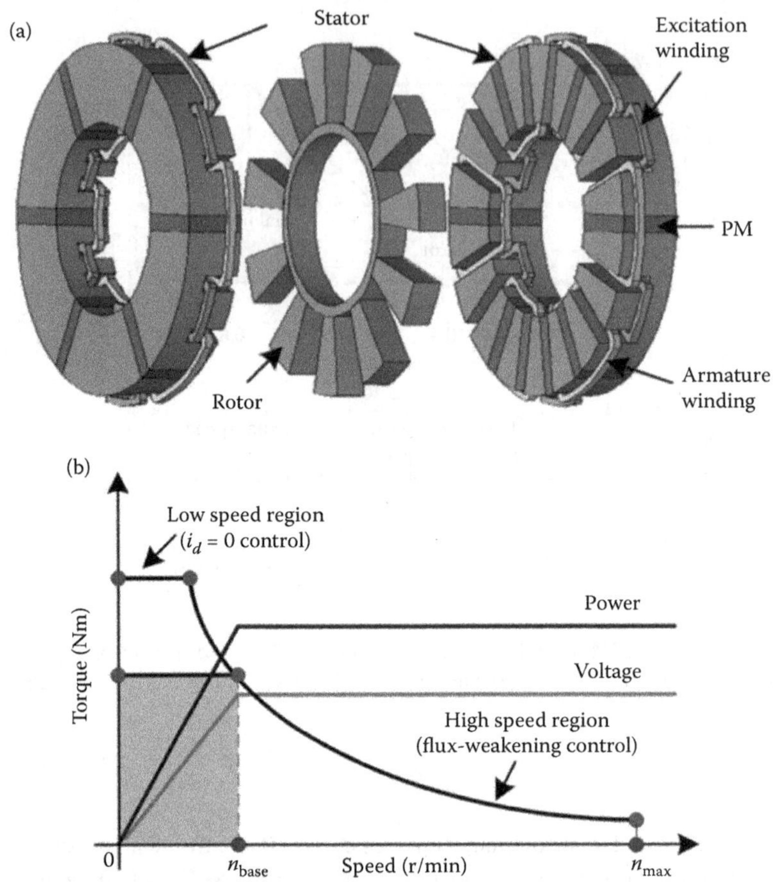

FIGURE 9.33 Axial airgap 12/10 dc excited SF-PMSM, (a); wide CPSR requirements, (b).

For higher speeds:

$$\text{Stage 2}\begin{cases} i_F^* = -|i_{Fn}| < 0 \\ i_d^* = \dfrac{\sqrt{(V_{dc}/\sqrt{3}\omega_r)^2 - (L_q i_q^*)^2} - (\psi_{PM} - M_f i_{Fn}^*)}{L_d} < 0 \\ i_q^* = \dfrac{2T_e}{3p_r[\psi_{PM} + (L_d - L_q)i_d^* - M_f i_{Fn}]} \end{cases} \tag{9.60}$$

Alternatively, above base speed, the machine may be run at unity power factor [25].

Particle swarm optimization is used online in [27] to optimally choose one of the four control strategies depicted in Equations 9.58 through 9.60. By optimal behavior, small torque, excitation current, and speed ripples are targeted.

For a 12/10 axial-airgap DC-excited SF-PMSM with the data: $V_{DC} = 200$ V, $n_n(n_b) = 750$ rpm, $T_n = 6$ Nm, $I_n = 4$ A, $i_{fn} = 3$ A, $\psi_{PM} = 0.1$ Wb, $R_s = 3.4\ \Omega$, $L_d = 10$ mH, $L_q = 15$ mH, $R_f = 7.9\ \Omega$, $M_f = 15$ mH, sample results such as in Figure 9.34a and b have been obtained through digital simulations and experiments [27].

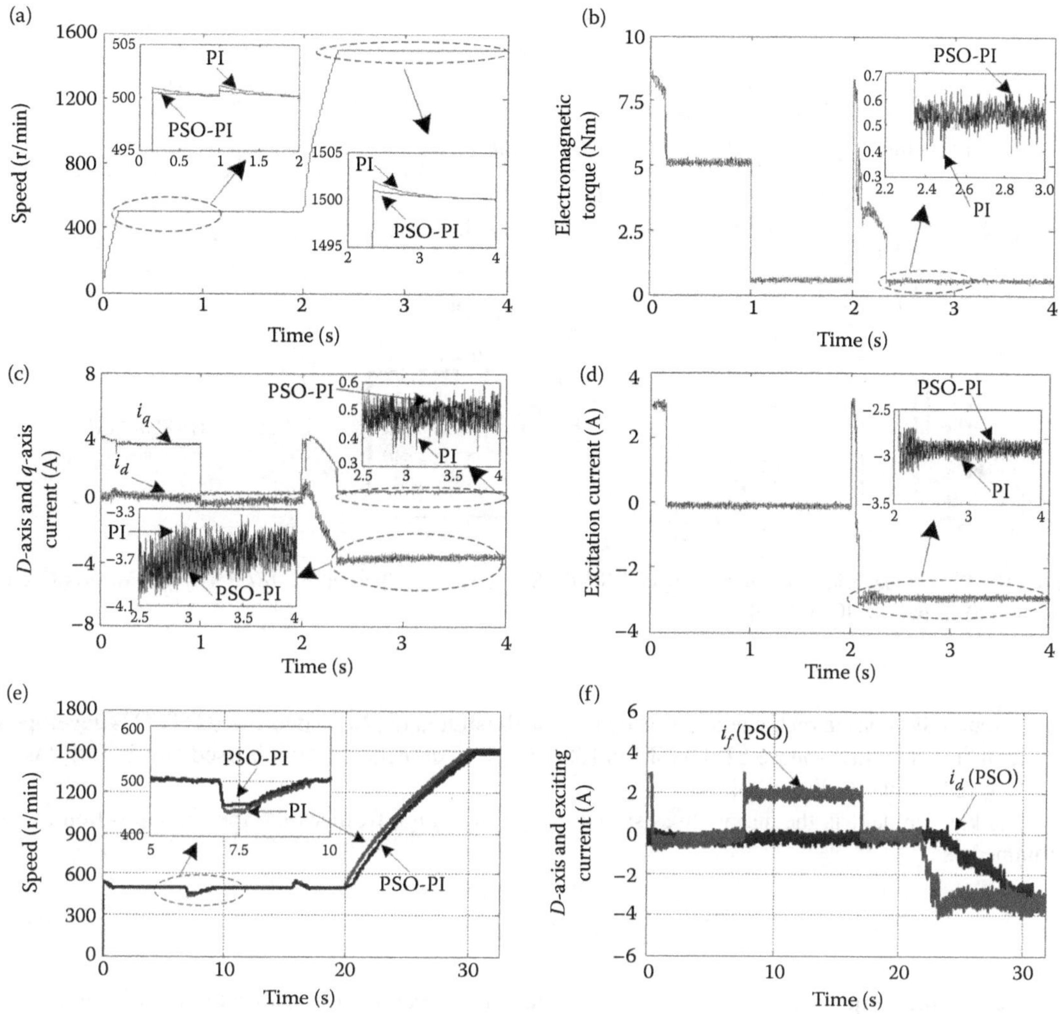

FIGURE 9.34 12/10 axial airgap DC-excited SF-PMSM digital simulation results for speed, (a); torque, (b); i_d, (c); field current i_F, (d); measured transients in speed, (e); in i_d and i_f, f). (After J. Zhao et al., *IEEE Trans.*, vol. MAG–51, no. 11, 2015, pp. 8204004. [27])

The results in Figure 9.34 trigger remarks such as:

- The DC-excited SF-PMSM is capable of a wide speed range with positive and negative DC field current contribution.
- i_d* is either 0, or, at the largest speeds, it is negative, together with a negative DC field current.
- PSO leads to lower ripples in torque, current, and speed.

The generic FOC system is shown in Figure 9.35.

9.7.3 Encoderless Direct Torque and Flux Control of Direct Current–Excited Switched Flux–Permanent Magnet Synchronous Motors

As already mentioned, being basically a cageless rotor synchronous motor, FOC and DTFC may be applied accordingly ([25], Chapter 14) to SF-PMSMs. While the encoderless FOC implementation of

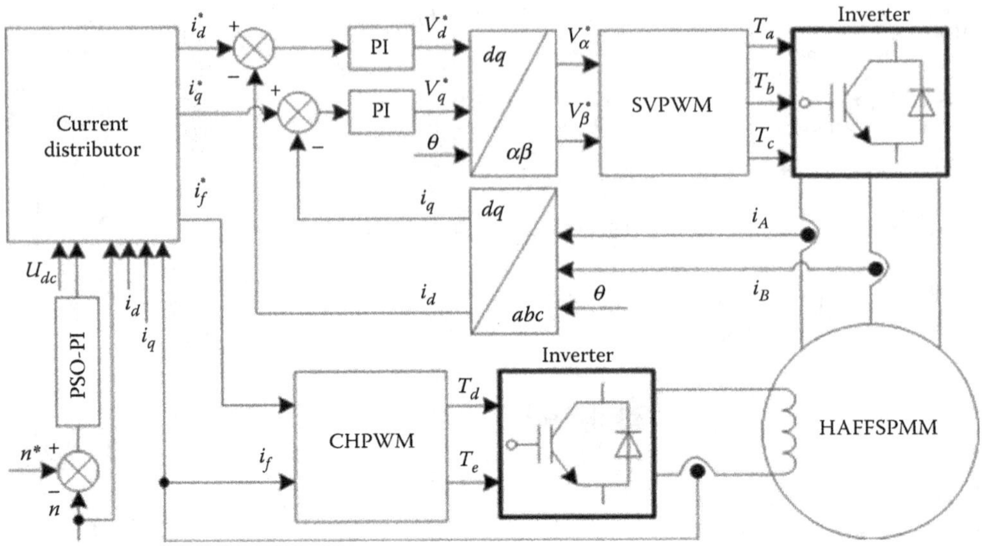

FIGURE 9.35 Generic FOC of DC excited SF-PMSM. (After J. Zhao et al., *IEEE Trans.*, vol. MAG–51, no. 11, 2015, pp. 8204004. [27])

DC-excited SMs based on the active flux concept is illustrated in [25], a dedicated DTFC is developed for a DC-excited three-phase SF-PMSM in [28]. Space vector modulation is used also in [28] (as in [25]) to reduce the torque ripple.

The key element is the stator flux estimator ($|\hat{\psi}_s|$, $\hat{\theta}_{\psi_s}$, $\hat{\omega}_1$), from which the rotor position $\hat{\theta}_{er}$ is obtained as:

$$\hat{\theta}_{er} = \hat{\theta}_{\psi_s} - \hat{\delta} \tag{9.61}$$

where $\hat{\delta}$ is the torque angle, the angle of the stator flux vector with rotor axis d (the axis of DC excitation and of PMs), calculable from torque:

$$\begin{aligned} \hat{T}_e &= \frac{3}{2} p_r (\psi_{s\alpha} i_{s\beta} - \psi_{s\beta} i_{s\alpha}) \\ &= \frac{3}{2} p_r |\hat{\psi}_s| (\psi_{\text{PM}} + M_f i_f) \sin \hat{\delta} \end{aligned} \tag{9.62}$$

A first-order and band-path (BP) filter are used to estimate stator flux from voltages e_α, e_β, while a multiple low pass filter of e_α, e_β is used to estimate the stator flux vector speed $\hat{\omega}_1$ and its position $\hat{\theta}_{\psi_s}$. So, a voltage emf model is used for stator flux estimation and is prone to notable errors at low speeds. The use of the active flux concept $\overline{\Psi}_d^a$ in a combined voltage/current model observer ([25] Chapter 14) leads to:

$$\bar{\psi}_d^a = \bar{\psi}_s - L_q \bar{i}_s \tag{9.63}$$

Aligning to the rotor d axis $\hat{\theta}_{\bar{\psi}_d^a} = \theta_{er}$ would simplify the speed estimator though a PLL observer acting on $\hat{\theta}_{\bar{\psi}_d^a}$ rather than on θ_{ψ_s}, which would allow operation at much lower speeds. The estimated $\hat{\theta}_{er} = \hat{\theta}_{\bar{\psi}_d^a}$ is used only in the stator flux observer at low speeds in DTFC. One more aspect to treat is how to control the DC field current i_F.

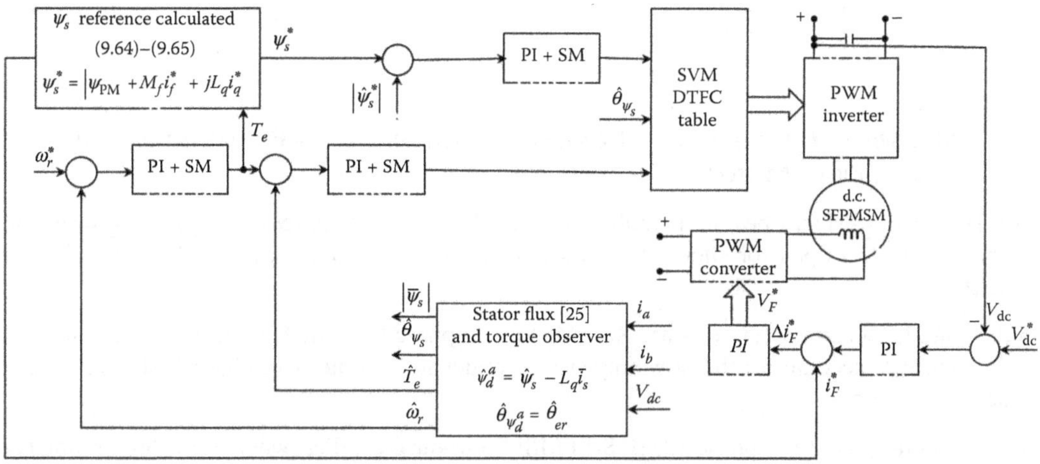

FIGURE 9.36 Generic encoderless DTFC of DC-excited SF PMSM using the active flux concept to simplify the rotor position and speed observer.

Equations 9.58 through 9.60 may be used for wide speed control in three stages, to produce stator flux $|\bar{\psi}_s^*|$, torque T_e^* reference values [27], and DC $\pm$field current i_F^*.

For autonomous generators, the DC output voltage in Equation 9.59 may be replaced by V_{dc}^*. But, in addition, to correct for machine parameter detuning, V_{dc} may be measured and a PI controller will add an additional DC field current reference Δi_F^*:

$$\Delta i_F^* = PI(V_{dc}^* - V_{dc}) \tag{9.64}$$

So, a generic DTFC based on the active flux model [25] could be used as in Figure 9.36.

For motoring (and grid operation), V_{dc} in Equation 9.57 could be replaced by the reference control voltage $V_s^* = V_{dc}/\sqrt{3}$, while the control system in Figure 9.36 holds.

Transients with SVM-DTFC encoderless DC-excited SF-PMSM specific control at high speeds is shown in Figure 9.37 [28]. The errors in position and speed estimation are large, but the response is

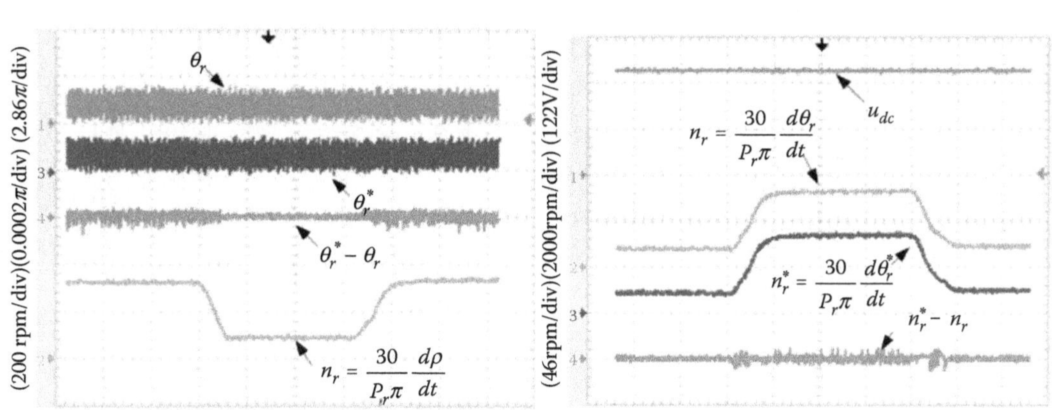

FIGURE 9.37 Generator mode specific DTFC of DC-excited SF-PMSM: high-speed transients with rotor position and speed estimators and their notable errors (in rad and in rpms, respectively). (After Yu Wang, Z. Q. Deng, *IEEE Trans.*, vol. EC–27, no. 4, 2012, pp. 912–921. [28])

still stable. Potentially, the scheme in Figure 9.36 should operate even below 10 rpm safely without signal injection in encoderless control.

9.7.4 Multiphase Fault-Tolerant Control of Switched Flux–Permanent Magnet Synchronous Motors

Note: The SF-PMSM may be built in multiphase (say, five or 2 × 3 phases) configurations when their control with 1, 2, 3 open (or shorted) phases is of interest for safety-critical AC drives [29]; see Figure 9.38.

Figure 9.38 [29] illustrates a five-phase SF-PMSM with phase C shorted and reconfigured analytical hybrid control to account for the noncompensated reluctance torque with phase C shorted; a small torque ripple is left.

Note: In contrast, the 2 × 3 phase 12/10 SF-PMSM introduced earlier, when 2% × 50% rating twin three-phase inverters are used, offers nonreconfigured control with one faulty phase at 50% torque by shutting down the three-phase group with the faulty phase(s).

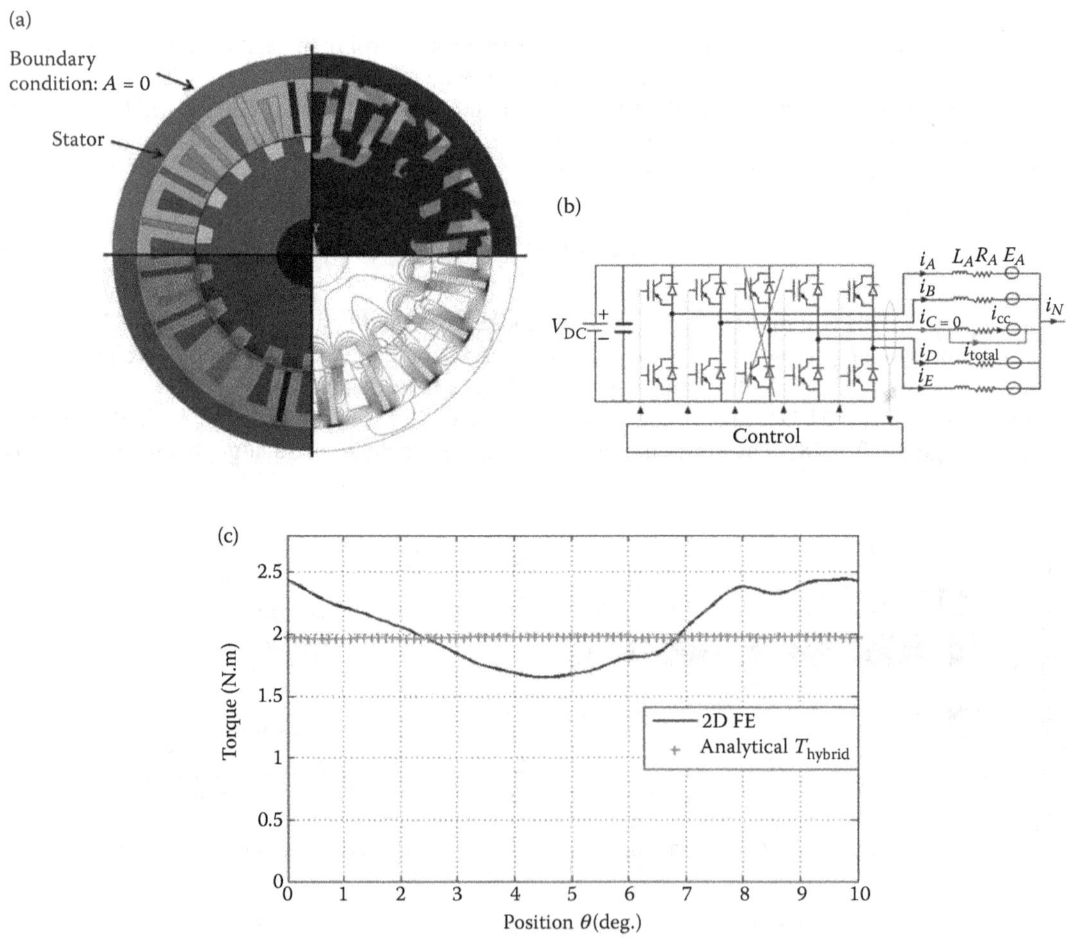

FIGURE 9.38 Five-phase SF-PMSM for fault-tolerant operation (a) adjacent phases; (b) five leg inverter with phase c shorted; (c) 2D-FEM + analytical torque with phase c shorted and reconfigured control. (After E. B, Sedrine et al., *IEEE-Trans.*, vol. EC–30, no. 3, 2015, pp. 927–938. [29])

9.8 SUMMARY

- Placing the PMs on the stator allows for easier fabrication and PM cooling, while a salient N_r pole passive rotor offers a rugged motor configuration. This is how switched-flux PMSMSs were born in one-phase topologies in 1943 [1] and "reborn" in 1955 [2].
- In essence, the AC coils on the stator have their throw (span) around the rotor saliency pole pitch. However, they comprise two stator teeth *that sandwich* a radially placed but tangentially magnetized PM, which implicitly produces PM flux concentration and thus the airgap no-load flux density is, in general, from 1.3 to 1.7 T for reasonable torque density (Nm/liter).
- The PM flux of the AC coils, modulated by stator and rotor saliency, changes (switches) polarity; this is why they are called "switched-flux" PMSMs.
- SF-PMSM meet the synchronization condition:

$$p_a = N_r - p_{PM}, p_{PM} = N_s/2; \quad N_r = p_r = p_m \tag{9.65}$$

where p_a, p_{PM}, N_s are the stator (armature) mmf max. forward space harmonic and number of PM flux pole pairs ($p_{\mathrm{PM}} = N_s/2$, in general), and, respectively, N_r is the number of flux modulator (rotor) salient poles equivalent to the pole pairs p_r (p_m).

- The closer N_s, N_r are, the better in terms of average torque pulsations, and so on. $N_s - N_r = 2$ is a preferred combination, but many more are feasible as long as they lead to symmetric $3(n)$ phase emfs.
- Unfortunately, the bipolar PM flux in the AC coils exploits at best half of coil span, and this is why good torque density implies heavy magnetic saturation.
- The frequency of stator currents $f_c = p_r \cdot n$; n—the rotor speed at synchronism.
- It has been proven that regular SF-PMSMs can develop in the same stator (outer diameter and stack length) for the same copper losses and similar torque as regular IPMSMs, but for less Nm/PM weight and with a stronger risk of PM demagnetization than in IPMSM. Still, in some configurations (say, instead of 12/10 combinations), C core 6/13 combinations produce similar Nm/PM weight.
- Despite double magnetic saliency, the *dq* model magnetization inductances L_{dm}, L_{qm} do not differ much, and $i_d{}^* = 0$ control is used below base speed for MTPA in most cases.
- However, the SF-PMSM has a larger synchronous inductance $L_s = L_{dm} \approx L_{qm}$ and thus may offer good CPSR operation.
- DC hybrid excitation may be added to reduce PM weight and extend the CPSR.
- Also, AlNiCo memory magnets (with mag/demag. DC coils) may be added to the stator to reduce NdFeB weight by at least 40%, though keeping the peak torque below base speed. AlNiCo magnets add to NdFeB magnet flux, but may be demagnetizing for flux weakening while keeping $i_d = 0$ for large CPSRs.
- The regular SF-PMSM may be treated almost as a nonsalient pole (surface) PMSMS, but with its large inductances, influenced by current, due to heavy magnetic saturation.
- The small differential (at AC) magnetic saliency may still be used for signal injection encoderless FOC where the popular 12/10 (and so on) configuration apparently fares best.
- Partitioned stator (exterior and interior) configurations have also been tried, where the segmented lamination-made rotor (flux modulator) is placed in between stators. This more complicated topology does not bring more torque density, but placing the PMs on a separate (interior) stator may allow easier replacement by replacing the entire interior stator. Also, the demagnetization risk of PMs may be reduced. Finally, a high power factor with such a PS-SF-PMSM has been reported recently [30]—around 0.96 for 2.7 Nm.
- The circuit model and encoder and encoderless control of SF-PMSMs is similar to that of small-saliency, strong-emf IPMSMs, or of DC-excited SMs (when SF-PMSM has a DC excitation assistance). So, FOC, DTFC, and scalar (V/f and I-f) with stabilizing loops

control, as known for regular AC synchronous motors/generators [25], is feasible for SF-PMSMSs.

- Multiphase fault-tolerant configurations with their reconfigurable control are also applicable to SF-PMSMs.
- Multiphysics optimal design, including electromagnetic, thermal, and mechanical aspects, is paramount in establishing SF-PMSM practicality.
- A hybrid excitation flux-switching PM motor with stator PM placed between the DC coils has recently been proven capable of producing competitive performance in experiments [32–33].
- Significant R&D effort is anticipated in this field [31], which seems close to industrialization both at low speeds and high torque and at medium speeds for medium and small torque applications, where fundamental frequency stays, in general, below 1 kHz.

REFERENCES

1. E. Binder and R. Bosch, Magnetoelectric rotary machine (magnetelektrische schwungradmaschine in German), Patent no. 741163, Class 21d1, Grorep 10, Sept 16, 1943.
2. S. E. Raunch and L. J. Johnson, Design principles of switching-flux alternators, *AIEEE Trans.*, vol. 74, no. 3, 1955, pp. 1261–1268.
3. E. Hoang, A. N. Ben Ahmed, J. Lucidarme, Switching flux permanent magnet polyphased synchronous machines, *Record of The European Conference on Power Electronics and Applications*, vol. 3, 1997, pp. 903–908.
4. J. T. Chen, Z. Q. Zhu, S. Iwasaki, R. Deodhar, Influence of slot opening in optimal stator rotor pole combinations and electromagnetic performance of S-F PM brushless machines, *IEEE Trans.*, vol. IA–47, no. 4, 2011, pp. 1681–1691.
5. D. Li, R. Qu, J. Li, W. Xu, L. Wu, Synthesis of flux-switching permanent magnet machines, *IEEE Trans.*, vol. EC–31, no. 1, 2016, pp. 106–116.
6. D. Li, R. Qu, J. Li, Topologies and analysis of flux modulation machines, *Record of IEEE-ECCE-2015*, pp. 2153–2160.
7. W. Fei, P. C. Kwong Luk, J. X. Shen, Y. Wang, A novel PM F-S machine with an outer rotor configuration for in-wheel light traction applications, *IEEE Trans.*, vol. IA–48, no. 5, 2012, pp. 1496–1506.
8. W. Zhao, T. A. Lipo, B.-I. Kwon, A novel dual-rotor axial-field fault tolerant flux-switching PM machine with high torque performance, *IEEE Trans*, vol. MAG–51, no. 11, 2015, pp. 8112204.
9. B. Zhang, M. Cheng, J. Wang, S. Zhu, Optimization and analysis of yawless linear flux switching PM machine with high torque density, *IEEE Trans.*, vol. MAG–51, no. 11, 2015, pp. 8204804.
10. H. Yang, H. Lin, Z. Q. Zhu, D. Wang, S. Fang, Y. Huang, A variable flux hybrid PM switched flux memory machine for EV/HEV applications, *IEEE Trans.*, vol. IA–52, no. 3, 2016, pp. 2203–2214.
11. A. Dupas, S. Hlioui, E. Hoang, M. Gabsi, and M. Lecrivain, Investigation of a new topology of hybrid-excited FS machine with static global winding: experiments and modeling, *IEEE Trans.*, vol. IA–52, no. 2, 2016 pp. 1413–1421.
12. A. Fasolo, L. Alberti, N. Bianchi, Performance comparison between SF and IPM machines with rare earth and Ferrite-PMs, *IEEE Trans.*, vol. IA–50, no. 6, 2014, pp 3708–3716.
13. R. Deodhar, S. Andersson, I. Boldea, T. J. E. Miller, The flux-reversal machines: a new brushless doubly-salient permanent magnet machine, *IEEE Trans.*, vol. IA–33, no. 4, 1997, pp. 925–934.
14. C. X. Wang, I. Boldea, S. A. Nasar, Characterization of 3 phase flux-reversal as autonomous generator, *IEEE Trans.*, vol. EC–16, no. 1, 2001, pp. 74–80.
15. Y. Shi, L. Jian, J. Wei, Z. Shao, W. Li, C. C. Chan, A new perspective on the operation principle of flux switching PM machines, *IEEE Trans.*, vol. 63, no. 3, 2016, pp. 1425–1437.
16. B. Heller, V. Hamata, *Harmonic Field Effects in Induction Machines*, book, Elsevier, Amsterdam, 1977, pp. 50–66.
17. W. Hua, M. Cheng, Z. Q. Zhu, D. Howe, Analysis and optimization of back emf waveform of a SF-PM motor, *IEEE Trans.*, vol. EC–23, no. 3, 2008, pp. 723–733.
18. J. T. Chen, Z. Q. Zhu, S. Iwasaki, R. Deodhar, Influence of slot opening in optimal stator rotor pole combinations and electromagnetic performance of S-F PM brushless machines, *IEEE Trans.*, vol. IA–47, no. 4, 2011, pp. 1681–1691.

19. J. Cros, P. Viarouge, Synthesis of high performance PM machines with concentrated windings, *IEEE Trans.*, vol. EC–17, no. 2, 2002, pp. 248–252.
20. Y. Wang, Z. Deug, Comparison of hybrid excitation flux-switching machines, *IEEE Trans.*, vol. MAG–48, no. 9, 2012, pp. 2518–2527.
21. J. Krenn, R. Krall, F. Aschenbrenner, Comparison of flux switching permanent magnet machines with hybrid excitation Optimization of Electrical and Electronic Equipment (OPTIM), 2014 International Conference on, Brasov, Romania, 22–24 May 2014.
22. Z. Z. Wu, Z. Q. Zhu, H. L. Zhan, Comparative analysis of partitioned stator flux reversal PM machines having fractional-slot non-overlapping and integer slot overlapping windings, *IEEE Trans.*, vol. EC–31, no. 2, 2016, pp. 776–788.
23. C. C. Awah, A. Q. Zhu, Z. Z. Wu, H. L. Zhan, J. T. Shi, D. Wu, X. Ge, Comparison of partitioned stator switched flux permanent magnet machines having simple and double layer windings, *IEEE Trans.*, vol. MAG–52, no. 1, 2016, pp. 9500310.
24. Z. Q. Zhu, Z. Z. Wu, D. J. Evans, W. Q. Chu, Novel electrical machines having separate PM excitation on stator, *IEEE Trans.*, vol. MAG–51, no. 4, 2015, pp. 8104109.
25. I. Boldea and S. A. Nasar, *Electric Drives*, book, 3rd edition, CRC Press, Taylor and Francis Group, New York, 2016.
26. T. C. Lin, Z. Q. Zhu, K. Liu, J. M. Liu, Improved sensorless control of switched flux PM SM based on different winding configurations, *IEEE Trans.*, vol. IE–63, no. 1, 2016, pp. 123–132.
27. J. Zhao, M. Lin, D. Xu, L. Hao, W. Zhang, Vector control of a hybrid axial field SF-PMSM based on particle swarm optimization, *IEEE Trans.*, vol. MAG–51, no. 11, 2015, pp. 8204004.
28. Y. Wang, Z. Q. Deng, A position sensorless method for direct torque control with SVM of hybrid excitation flux-switching generator, *IEEE Trans.*, vol. EC–27, no. 4, 2012, pp. 912–921.
29. E. B. Sedrine, J. Ojeda, M. Gabsi, I. Slama-Belkhodja, Fault tolerant control using GA optimization considering the reluctance torque of a five phase flux-switching machine, *IEEE Trans.*, vol. EC–30, no. 3, 2015, pp. 927–938.
30. D. J. Evans, Z. Q. Zhu, Novel partitioned stator switched flux PM machines, *IEEE Trans.*, vol. MAG–51, no. 1, 2015, pp. 8100114.
31. J. D. McFarland, T. M. Jahns, A. M. EL-Refaie, Analysis of torque production mechanism for flux-switching PM machines, *IEEE Trans.*, vol. IA–51, no. 4, 2015, pp. 3041–3049.
32. H. Nakane, Y. Okada, T. Kosaka, N. Matsui, Experimental study on windage losses reduction using two types of rotor for hybrid excitation flux-switching motor, *Record of ICEM-2016*, pp. 1709–1715.
33. Y. Maeda, T. Kosaka, N. Matsui, Design study on hybrid excitation flux switching motor with PMs placed at middle of field coil slot for HEV drives, *Record of ICEM-2016*, pp. 2524–2530.

10 Flux-Reversal Permanent Magnet Synchronous Machines

10.1 INTRODUCTION

As suggested in Chapter 1 [1], the flux-reversal (FR) PMSM may be considered a kind of flux modulation single-airgap (single-rotor) machine similar to the switched-flux PMSM.

However, at the time of FR-PMSM introduction [2–6], the main idea was to use it in doubly salient machine multiple alternate polarity PMs on wider stator poles that host nonoverlapping multiple phase windings made of 2mk coils (*m*—number of phases, *k*—an integer). Only the passive rotor has the function of magnetic saliency with N_r salient poles (Figure 10.1).

The large stator (coil) poles contain $2n_p$ PMs of alternate polarity located as surface PMs (Figure 10.2a) or as radially placed but tangentially magnetized PMs inside the former (Figure 10.2b).

Considering the PM pole pitch on the stator as τ_{PM}, allowing the rotor magnetic anisotropy to have the same period of $2\tau_{PM}$, and providing for $2\tau/3$ displacement between adjacent stator large poles to secure symmetric emfs, the following relationship is valid:

$$2\tau_{PM}N_r = \left(2n_p \cdot 6K + 6K \cdot \frac{2}{3}\right)\tau_{PM} \quad (10.1)$$

or

$$N_r = 6K \cdot n_p + 2K \quad (10.2)$$

It may be rightfully agreed that, this way, with same periodicity of $2\tau_{PM}$ in the PMs and of rotor saliency, a synchronous machine is targeted.

The increase of AC nonoverlapping coil span in FR-PMSM to 2,4,6,8 PM poles was intended for copper loss reduction (in a kind of magnetic gear effect), while torque increases (for given geometry and AC coil mmf) with the number of PM poles ($2n_p$) per coil span. But, at the same time, the PM fringing flux between poles increases with the reduction of the τ_{PM}. Eventually, at a certain (large, though) N_r, the two effects neutralize each other.

This fringing effect occurs in all flux-modulation machines when the pole pitch decreases, reducing torque magnification.

On top of that, with stator PMs, in general, only half the PMs (or half the AC coil span in SF-PMSMs) are active at any time. Still, the reduction of pole pitch $\tau_{PM} = 10$–12 mm (large N_r) leads to significant torque density. However, the PM weight per Nm has to be used to assess performance.

The price to pay is larger machine inductance in p.u. and thus a lower power factor for competitive torque density, which will lead to larger KVA ratings in the PWM converter that feeds the machine in a variable-speed drive. At the same time, the large inductance L_s is "good" for flux weakening, providing for wide CPSR.

But, as at low speeds, the electric machine is more expensive than the PWM converter (torque defines machine cost), the latter's KVA overrating might add reasonably to the total cost of the drive.

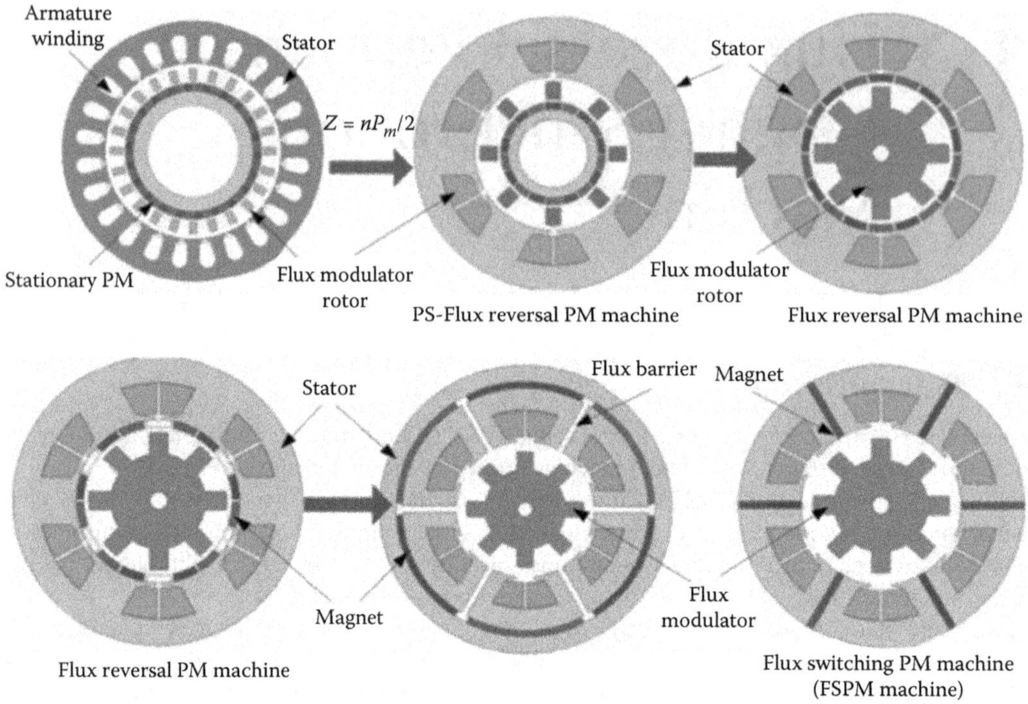

FIGURE 10.1 "Metamorphosis" of flux-modulation machines to "flux-reversal" and to "switched-flux" stator PMSMs. (After D. Li et al., *Record of IEEE-ECCE*, 2015, pp. 2153–2160. [1])

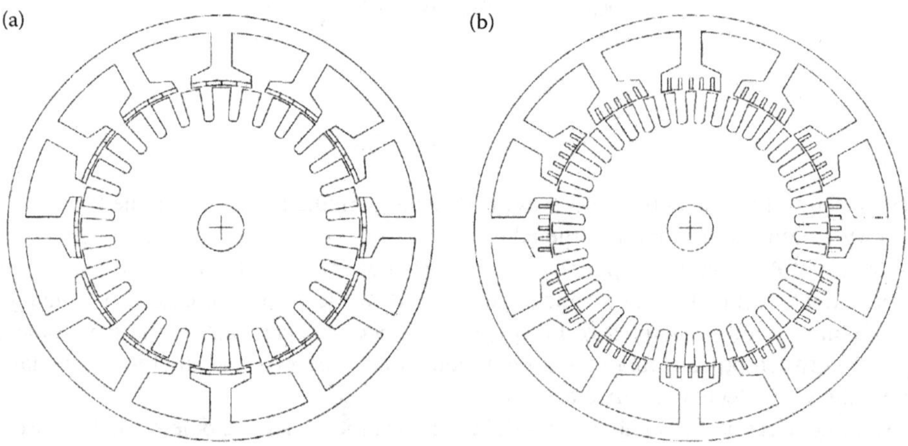

FIGURE 10.2 Typical three-phase FR-PMSMs: (a) with stator surface PMs ($N_s = 12$, $n_p = 2$, $N_r = 28$); and (b) inset radial stator PMs ($N_s = 12$, $n_p = 3$, $N_r = 40$). (After I. Boldea, *Variable Speed Generators*, book, 2nd edition, CRC Press, Taylor and Francis Group, New York, 2015 (Chapter 11: "Transverse flux and flux reversal permanent magnet generator systems", pp. 479–527). [7])

Another embodiment of FR-PMSM with stator-PMs is shown in Figure 10.3a and b, for one- and three-phase implementation for small power applications. It has Ferrite PMs that are "planted" in the stator after the AC coils are in place and thus PMs may be easily replaced and easily detachable. Here, all PMs are active all time, but still only half the stator coil span is active at any time.

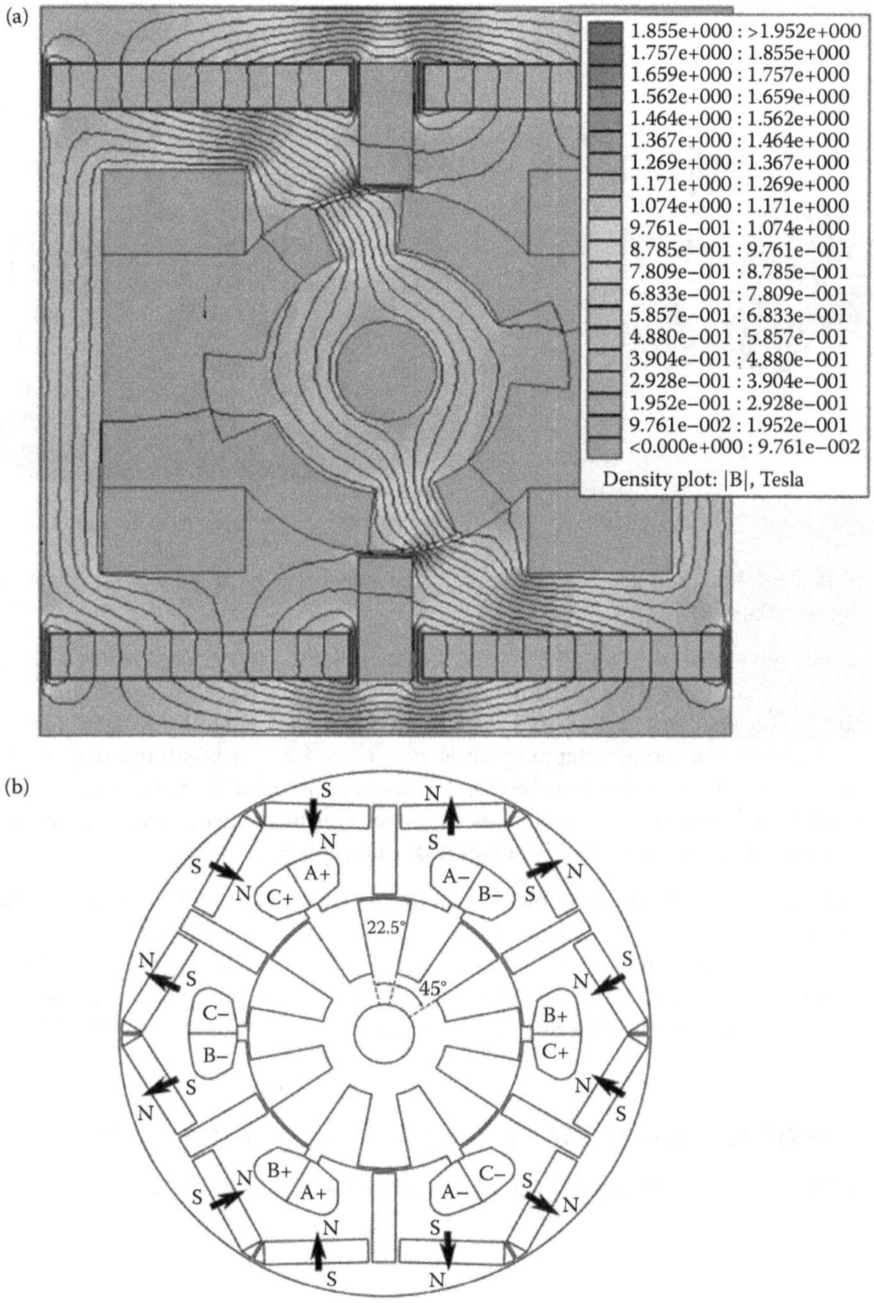

FIGURE 10.3 Detachable-stator PM FR machine: (a) one-phase; (b) three-phase.

In an effort to better use the PMs (to make them all active all the time) the spoke-shaped PMs may be placed in the rotor, sandwiched between two magnetically anisotropic stators that are phase shifted with each other by a PM pole pitch. Now the number of rotor pole pitches is $2N_r$, but the number of pole pairs is only N_r (Figure 10.4). Both stators may be provided with 2mk AC coils for more Nm/volume. To simplify the fabrication, only the outer stator bears AC coils on large poles that comprise $2np \times \tau_{PM}$ active spans. Consequently, the interior stator is simpler (Figure 10.4) and shorter axially, to allow a shorter glass-shaped rotor.

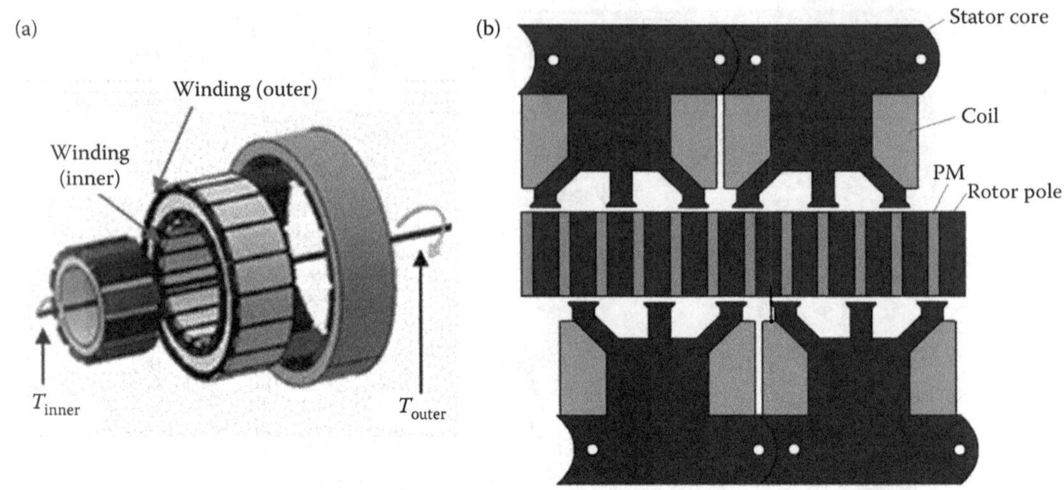

FIGURE 10.4 Rotor PM dual-stator three-phase FR PMSM (a) radial airgap; (b) axial airgap.

To simplify the fabrication process, an axial airgap version of the rotor-PM three-phase FRM is shown in Figure 10.4b [9].

Note: It may be argued that the rotor-PM FR-PMSM is similar to the Vernier PMSM with two stators and spoke-shaped rotor PMs. This is true, but the nonoverlapping three-phase winding in the stator makes the difference when 2, 4, 6 pole, $q = 2$ stator/pole/phase overlapping windings are used in the Vernier PMSM [10]. The nonoverlapping 2mK coil ($K = 1,2, \ldots$) windings may lead to lower copper losses per torque for same rotor pole pitch τ_{PM} and number of rotor pole pairs N_r ($2N_r$—PM poles). Also, the separation of coils leads to longer coil life between eventual repairs that are much easier to do (only the faulty coil is "extracted" and replaced).

The low power factor is the main problem of all flux modulation machines. Its alleviation through optimal design and rather high PM airgap flux density (1.4–1.5 T) to allow even 1 A/mm^2 current density in high-torque machines [10] and thus make $L_sI_n/\psi_{PM} = 0.7-0.8(\cos\varphi = 0.85)$. Finally, all "F-M" machines have small functional saliency ($L_d \approx L_q$) despite the physical double saliency, because they mainly use the first airgap magnetic permeance harmonic to couple PM and armature-current mmf waves.

10.2 TECHNICAL THEORY VIA PRELIMINARY DESIGN CASE STUDY

Let us consider a 60-Hz, 128.5-rpm, 200-Nm, three-phase FR-PMSM with surface stator PMs (Figure 10.5).

10.2.1 Solution

The operation at 60 Hz for a base speed of 128.5 rpm is good, since a regular off-the-shelf PWM inverter may be used. But the number of rotor poles N_r (pole pairs, in fact) should be large:

$$f_n = n_n \cdot N_r;\ N_r = \frac{60}{128.5/60} \approx 28 \tag{10.3}$$

However, for symmetric-phase emfs, Equation 10.2 has to be fulfilled:

$$N_s\left(n_p + \frac{1}{3}\right) = N_r \tag{10.4}$$

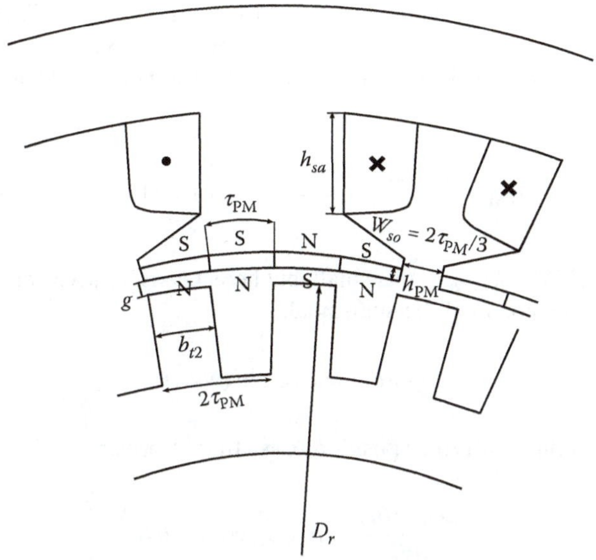

FIGURE 10.5 Three-phase FR-PMSM with stator surface PMs geometry.

with $n_p = 2$ ($2n_p = 4$) and $N_s = 12$ stator large poles (coils); also $N_r = 28$ rotor salient poles + inter poles ($2N_r$ pole spans of τ_{PM}). The magnets (Figure 10.5) are placed on the large stator pole shoes that make up $2n_p = 4$ of them along the periphery. The big stator slot opening is $2\tau_{PM}/3$.

The relationships between PM radial thickness h_{PM}, airgap g, and τ_{PM} are intricate and require an optimization calculus, say, for maximum PM flux in the coil. This would lead to an optimal $(h_{PM} + g)/\tau_{PM}$ ratio of less than ½, as in any variable reluctance structure.

To start the preliminary design, we adopt a shear rotor stress $f_t = 1.5\text{–}3\ \text{Nm}^2$ for the base torque T_{eb}:

$$T_{eb} = f_t \cdot \pi \cdot D_r \cdot l_{\text{stack}} \cdot \frac{D_r}{2} \tag{10.5}$$

D_r—rotor diameter, l_{stack}—axial length of stator core. To limit end-coil length, the stack length l_{stack} is:

$$\lambda_s = \frac{l_{\text{stack}}}{D_r} \geq 0.2 - 1.4 \tag{10.6}$$

$$D_r = \sqrt[3]{\frac{2T_e N_s}{f_{t_n} \pi^2 \cdot \lambda_s}} = \sqrt[3]{\frac{2 \cdot 270 \cdot 12}{2 \cdot 10^4 \cdot \pi^2 \cdot 1.055}} = 0.18\ \text{m} \tag{10.7}$$

The stack length $l_{\text{stack}} = \lambda_s \pi D_r = 1.05 \cdot 0.182 = 0.19\ \text{m}$.

The maximum value of PM flux in a coil with $2n_p$ PMs in its throw is:

$$\varphi_{\text{PM}} = l_{\text{stack}} \cdot (B_{g\,\text{PM}\,i} K_{\text{fringe}}) \cdot \tau_{\text{PM}} \cdot n_p \tag{10.8}$$

$B_{g\,\text{PM}\,i}$ is the ideal maximum flux density in the airgap:

$$B_{g\,\text{PM}\,i} = B_r \cdot \frac{h_{\text{PM}}}{h_{\text{PM}} + g} \tag{10.9}$$

with $h_{\text{PM}} = 2.5\,\text{mm}$, $g = 0.5\,\text{mm}$, and, for the PM with $B_r = 1.21\,\text{T}$, $k_{\text{fringe}} = 0.4$ (fringing coefficient: only 40% of the ideal PM airgap flux embraces the AC coils), $B_{g\,\text{PM}\,i} = 1.008\,\text{T}$, but $B_{g\,\text{PM}} = B_{g\,\text{PM}\,i} \cdot k_{\text{fringe}} = 0.4\,\text{T}$. This is a strong penalty for being able to reduce the PM pole pitch τ_{PM} to:

$$\tau_{\text{PM}} = \frac{\pi \cdot D_r}{2N_r} = \frac{\pi \cdot 0.180}{2 \cdot 28} = 10.1 \cdot 10^{-3}\,\text{m} \tag{10.10}$$

It is to be proven by FEM if k_{fringe} is this small, but here a conservative value is assigned. Approximately, the PM flux in the coils will be sinusoidal:

$$\varphi_{\text{PM}}(\theta_r) = \varphi_{\text{PM}} \sin{(N_r \theta_r)} \tag{10.11}$$

The derivative of this flux with rotor position θ_r is straightforward:

$$\frac{d\varphi_{\text{PM}}(\theta_r)}{d\theta_r} = N_r \varphi_{\text{PM}} \cos{(N_r \theta_r)} \tag{10.12}$$

But we have $N_s/3$ coils per phase (four coils in our case) which, if connected in series, will lead to the emf per phase peak value E_1:

$$E_1 = \frac{N_s}{3} \cdot n_c \cdot 2\pi n \cdot l_{\text{stack}} \cdot \pi \cdot D_r \cdot k_{\text{fringe}} \cdot \frac{n_p}{2} \cdot B_{\text{PM}\,i} \tag{10.13}$$

where n_c—the number of turns per coil. Note that πD_r in Equation 10.13 represents the entire rotor periphery and thus “hides” the torque magnification N_r.

Now the torque T_e with current I_1 in phase with emf E_1 ($i_d = 0$ in the FOC) is simply:

$$T_e = \frac{3}{2} E_1 I_1 \sqrt{2}/(2\pi n) \tag{10.14}$$

This way, from Equations 10.13 and 10.14, the number of amperturns per coil (peak value) is:

$$n_c I_1 \sqrt{2} = n_c \frac{2}{3} T_e \frac{2\pi n}{E_1} = 1032\,\text{Aturns/coil (peak value)} \tag{10.15}$$

Choosing a rated current density $j_{\text{con}} = 3.57\,\text{A/mm}^2$ and a slot filling factor $k_{\text{fill}} = 0.5$, with two coils per slot (double side-by-side layer winding), the active slot area is:

$$A_{\text{slot}} = \frac{2n_c I_1}{j_{\text{con}} \cdot k_{\text{fill}}} = \frac{\sqrt{2} \cdot 1032}{35 \cdot 10^6 \cdot 0.5} = 831.5 \cdot 10^{-6}\,\text{m}^2 \tag{10.16}$$

Now the slot geometry should be easy to settle, as the big stator slot pitch $\tau_{\text{stator}} = \pi D_r/N_s = \pi \cdot 0.18/12 = 0.0471\,\text{m}$. Even with a 30-mm-deep stator slot (h_{sa}), a slot geometry fulfilling Equation 10.16 may be found.

As the slot sizing completion, airgap, and leakage inductance calculations are straightforward, we will not insist on these aspects here, but proceed with FEM characterization to check the technical oversimplifying theory put forward here to provide a starting geometry.

10.3 FINITE-ELEMENT MODEL GEOMETRY

At no load, for the now known geometry, 2D-FEM has been used ($l_{stack}/D_r = 1.055!$), and the PM flux per pole and cogging torque have been calculated first (Figure 10.6).

Figure 10.6 shows clearly that the PM flux per stator poles in phases A, B, C are rather symmetric and sinusoidal and that the cogging torque may be drastically reduced to (1.5%!) for a 1.8–mechanical degree skewing.

This advantage is paid for by a 7% reduction in EMF (and in torque for given stator current); see Figure 10.7.

The PM flux derivative (with rotor position) shows a large fundamental and tiny third harmonic (Figure 10.8).

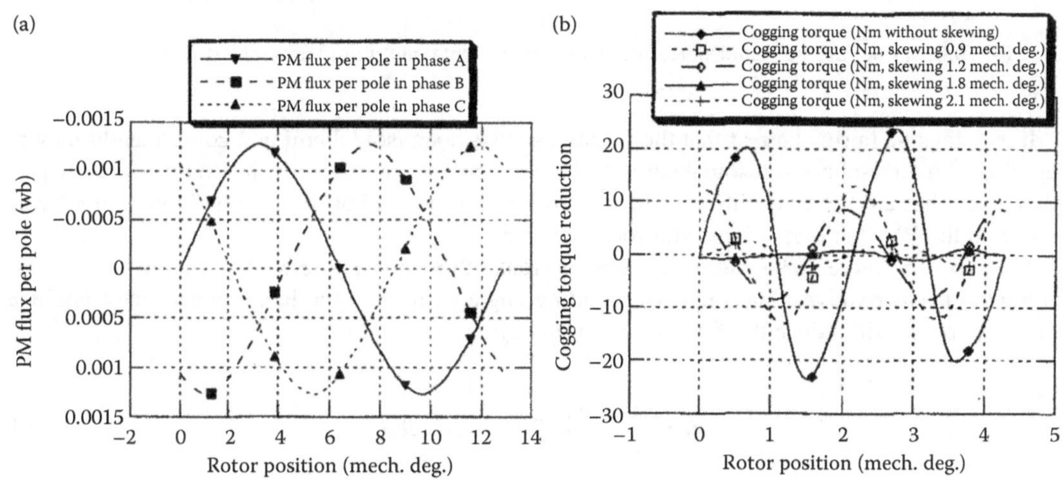

FIGURE 10.6 PM flux per stator pole versus rotor position, (a); and cogging torque (without and with rotor skewing), (b). (After I. Boldea, *Variable Speed Generators*, book, 2nd edition, CRC Press, Taylor and Francis Group, New York, 2015 (Chapter 11: "Transverse flux and flux reversal permanent magnet generator systems", pp. 479–527). [7])

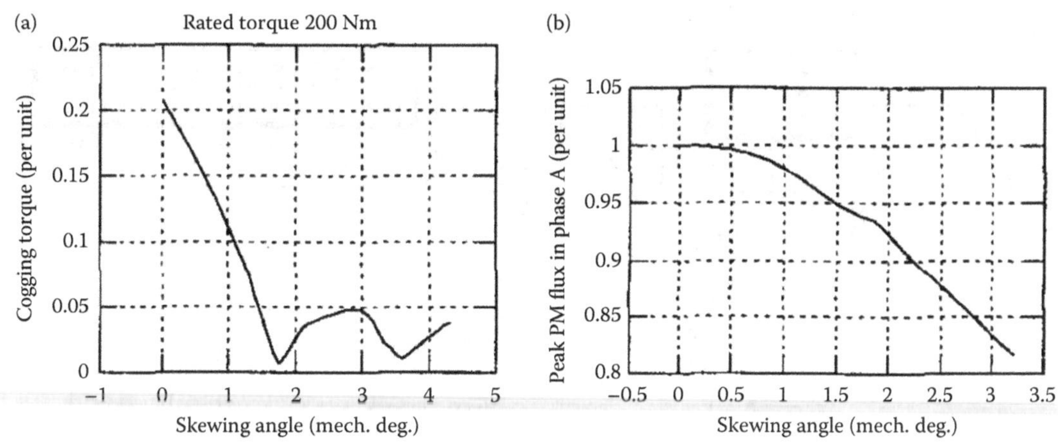

FIGURE 10.7 Reduction of cogging torque with skewing angle, (a); and the corresponding peak flux per stator coil reduction, (b).

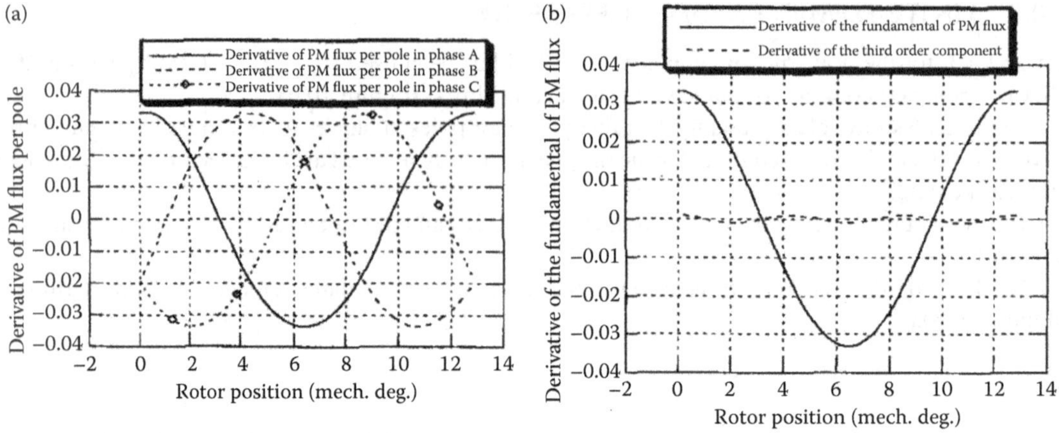

FIGURE 10.8 $d\varphi_{PM}/d\theta_r$ per stator pole, (a); and its fundamental and third harmonic, (b).

It may thus be fairly claimed that the machine shows a sinusoidal emf and consequently may be supplied with three-phase sinusoidal currents, here in phase with their emfs, not to total flux per-pole derivatives that are not only different (in amplitude from emf), but, as expected, are shifted with respect to the PM flux derivatives with rotor position.

As visible in Figure 10.9a and b, the phase currents (though in phase with emfs) are not in phase with total flux derivatives, and the corresponding angle is, in fact, the ideal power factor angle φ_i (machine losses will reduce it a few percent in motoring).

The total emf:

$$E_t = \frac{d\varphi_t}{d\theta_r} \cdot \frac{N_s}{3} \cdot 2\pi n \,(\text{peak value}) \tag{10.17}$$

Now the peak voltage equation during steady state V_A is:

$$V_A \cong \sqrt{\left(E_{t\max}/\sqrt{2} + R_s I_n \cos\varphi_i\right)^2 + R_s^2 I_n^2 (\sin\varphi_i)^2}, \text{ (rms)} \tag{10.18}$$

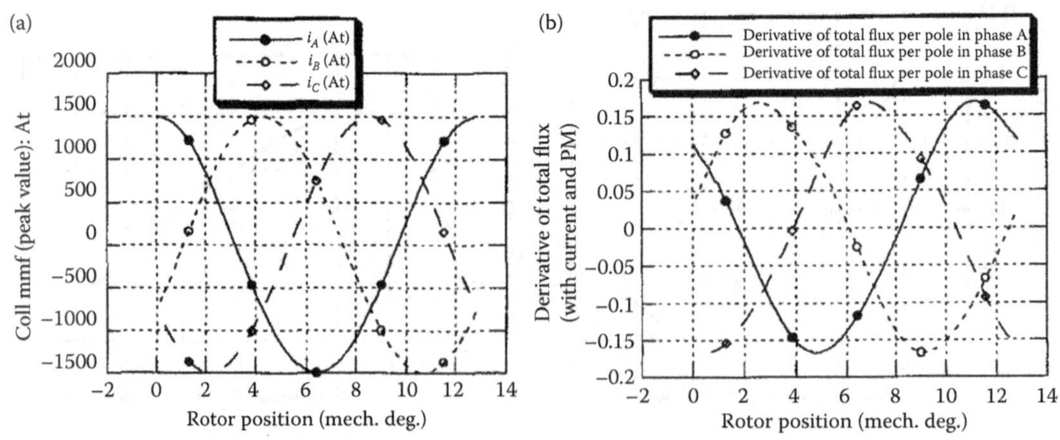

FIGURE 10.9 Phase currents mmfs (a) and total flux/stator pole derivatives with rotor position (b).

From here we may calculate the number of turns per coil with given V_A (phase voltage [rms]) and stator phase resistance voltage R_sI_n as:

$$R_sI_n = \rho_{co}\frac{l_{\text{coil}}\cdot j_{\text{con}}}{n_cI_n}\cdot\frac{N_s}{3}\cdot n_c^2I_n$$
$$= 2.1\cdot 10^{-8}\cdot 0.458\cdot 3.5\cdot 10^{-6}\cdot 4\cdot n_c = 0.1346\cdot n_c \tag{10.19}$$

$$l_{\text{coil}} \approx 2\cdot\left(l_{\text{stack}} + 2\cdot n_{pp}\tau_{\text{PM}}\right) = 0.458\,\text{m} \tag{10.20}$$

$E_{t\,\max}$ can also be expressed as $E_{t\,\max} = K_e\cdot n_c$ and thus the number of turns/coil n_c comes simply from Equation 10.18. In practical designs, a notable voltage reserve has to be left to allow for "chopping" the sinusoidal current properly at rated (base) speed (frequency).

Further, back to FEM, we may calculate directly, from the total flux per stator pole derivatives (Figure 10.9b), the instantaneous torque T_e:

$$T_e = 3\cdot\frac{N_s}{3}\left[\begin{array}{l}\dfrac{d\varphi_A^t(\theta_r)}{d\theta}\cdot n_c\cdot i_A(\theta_r) + \dfrac{d\varphi_B^t(\theta_r)}{d\theta}\cdot n_c\cdot i_B(\theta_r)\\ +\dfrac{d\varphi_C^t(\theta_r)}{d\theta}\cdot n_c\cdot i_C(\theta_r)\end{array}\right] \tag{10.21}$$

The results for full torque versus rotor position, for sinusoidal currents in phase with their emfs, depicted in Figure 10.10a–d, lead to remarks such as:

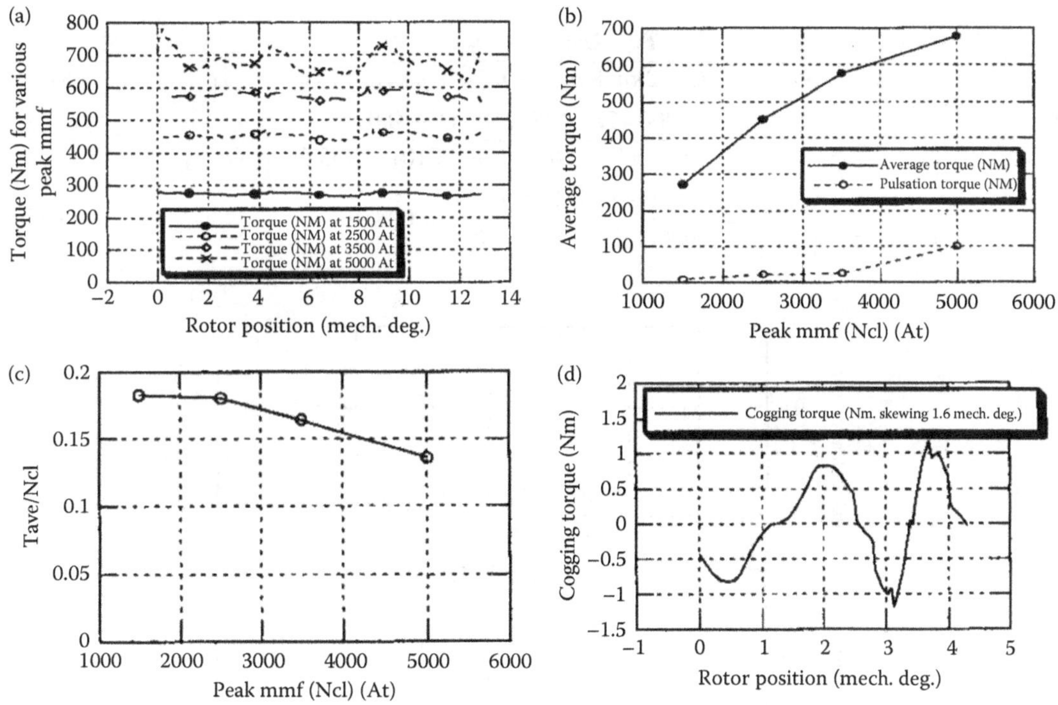

FIGURE 10.10 Torque versus position, (a); average torque and torque ripple, (b); average torque coefficient reduction versus coil mmf (peak value), (c); cogging torque, (d). (After I. Boldea et al., *Proc. IEE*, EPA-146, no. 2, 1999, pp. 139–146 [4]; I. Boldea et al., *Elec. Mach. Power Syst. Journal*, vol. 27, 1999, pp. 848–863 [5]; I. Boldea et al., *IEEE Trans.*, vol. IA-38, no. 6, 2002, pp. 1544–1557. [6])

- A torque of 200 Nm may be obtained for 1032 Aturn/coil/peak value; for 1500 Aturns, 270 Nm is obtained. This means that the large fringing coefficient $K_{\text{fringe}} = 0.4$ was, after all, realistic.
- The machine may be overloaded up to 5000 Aturns/coil (peak value) when an average torque of 680 Nm is still obtained; yes, for a large overcurrent.
- The torque pulsations are small at rated torque (a few percent) but increase steadily with overload.
- The question arises if the PMs risk demagnetization at peak torque (680 Nm). To verify this, the radial flux density on the PM surface was calculated in these conditions by FEM (Figure 10.11).

As expected, only the N/S polarity magnets are demagnetized, but not to endanger their self-remagnetization. Part of this happy ending is the large flux fringing and part the large total (magnetic) airgap.

The copper losses. It is important to calculate the copper losses p_{con}:

$$p_{\text{con}} = 3R_s I_1^2 = 3(R_s I_1) \cdot \frac{I_1 n_c}{n_c}$$
$$= 3 \cdot 0.1346 \cdot n_c \cdot \frac{1032}{n_c\sqrt{2}} = 295.5\,\text{W} \tag{10.22}$$

For 200 Nm, the electromagnetic power and efficiency, with core and mechanical loss lumped up to 1.5% Pelm, and $P_{\text{mec}} = 0.5\%\text{P}_{\text{elm}}$ (128.5 rpm only!):

$$\eta_n = \frac{T_{en} \cdot 2\pi n \cdot (1 - p_{\text{mec}})}{T_{en} \cdot 2\pi n + p_{\text{con}} + p_{\text{core}} + p_{\text{mec}}}$$
$$= \frac{200 \cdot 2\pi \cdot 1285/60 \cdot (1 - 0.015)}{200 \cdot \pi(128.5/60)(1 + 0.010) + 295.5}$$
$$= \frac{2690(1 - 0.005)}{2690(1 + 0.01) + 295.5} = 0.9046 \tag{10.23}$$

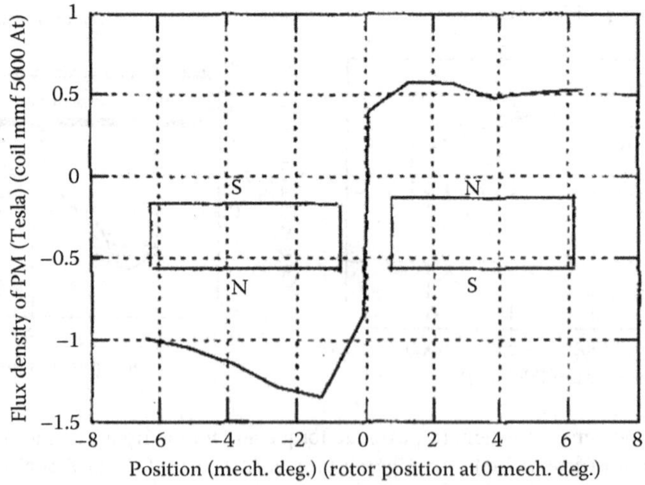

FIGURE 10.11 Flux density on the PM surface for peak torque conditions.

The efficiency looks satisfactory, but the power factor is in the vicinity of 0.7 only.

The rated torque/volume is:

$$\frac{T_{\text{en}}}{(\pi \cdot D_{os}^2/4) \cdot l_{\text{stack}}} = \frac{200}{\pi \cdot (0.27^2/4) \cdot 0.19} = 18.394\,\text{Nm/l} \tag{10.24}$$

with D_{os}—stator outer diameter.

It reaches 60 Nm/l for the peak of 680 Nm. While the performance may be considered acceptable, the large fringing PM flux, the torque density, efficiency, and Nm/kg of PM (cost) are notable.

10.4 COMPARISON BETWEEN FLUX-REVERSAL PERMANENT MAGNET SYNCHRONOUS MACHINES AND SURFACE PERMANENT MAGNET SYNCHRONOUS MOTOR

Reference 11 unfolds a complete comparison between a 6/14 FRM with $2n_p = 4$ PM/stator pole and a 24-slot/28–rotor pole SPMSM with $D_r = 160$ mm, $D_{os} = 270$ mm, stack length $l_{\text{stack}} = 150$ mm (close to our design in the previous section), airgap $g = 0.5$ mm, PM height $h_{\text{PM}} = 2.5$ mm. Both motors that are compared operate at the same frequency for the same speed, so core losses should be also comparable. The average torque (for the same coil mmf) was 114 Nm for FR-PMSM and only 73 Nm for the SPMSM.

But phase self and mutual inductances have been found to be five times larger in FR-PMSMs. They hardly vary with rotor position (almost no saliency: $L_d \approx L_q$)—Figure 10.12 [11].

So the power factor was only 0.73 for FR-PMSM but 0.94 for the SPMSM. However, the efficiency, from the tests run (for FR-PMSM with series capacitor compensators of the machine large reactance $[\omega_1 L_s]$ in generating on a resistive load) have proved about the same efficiency, but the generator power was extended by 50%.

Finally, the torque-speed envelope was checked for the same inverter max. voltage, with 6/14 FR-PMSM exceeding spectacularly the 24/28 SPMSM (Figure 10.13).

Wide CPSR applications seem to favor FR-PMSM.

10.5 THREE PHASE FLUX-REVERSAL PERMANENT MAGNET SYNCHRONOUS MACHINES WITH ROTOR PERMANENT MAGNETS

As alluded to earlier in this chapter (Figure 10.5), in order to increase the axial flux in the airgap, PM flux concentration (through spoke-shaped rotor magnets) with all the magnets active all the time is required. The PMs are placed on the rotor, which is sandwiched between two stators with their stator saliency phase shifted by a rotor PM pole pitch. The inner stator is not wound to simplify fabrication. The two twin stators are, however, both active in an axial airgap configuration.

To grasp the nature of this machine and acquire a feeling of the magnitudes, let us perform again a preliminary design followed by key FEM validation checks [12].

Let us now consider the configuration in Figure 10.14, which now has two airgaps and double saliency in both. Again, the stator's saliency period is $2\tau_{\text{PM}}$, while the large stator slot opening is $2\tau_{\text{PM}}/3$, scaled for diameter.

There are six large stator poles (six AC coils, two per phase), which carry four smaller salient poles (and three interpoles) $2np_{pr} = 4$. So the number of rotor PM poles is $2p_{pr} = 46 = (7 \cdot 6 + (2/3) \cdot 6)$ and, at 50 Hz, the speed n_b is:

$$n_b = \frac{f_b}{p_{pr}/2} = 50/23\,\text{rps} = 130.43\,\text{rpm} \tag{10.25}$$

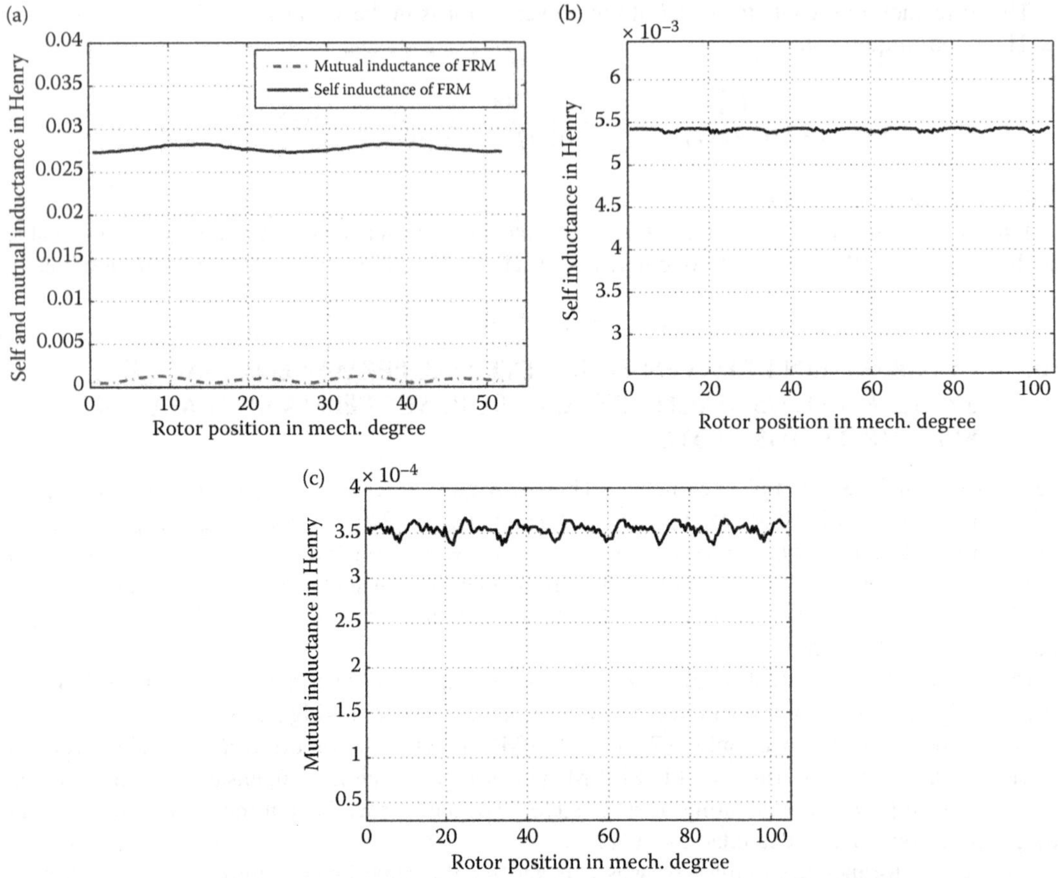

FIGURE 10.12 Self and mutual inductances: (a) for the 6/14 FR-PMSM; (b) and c) for the 24/28 SPMSM. (After D. S. More et al., *Record of IEEE-IECON*, 2008, pp. 1131–1136. [11])

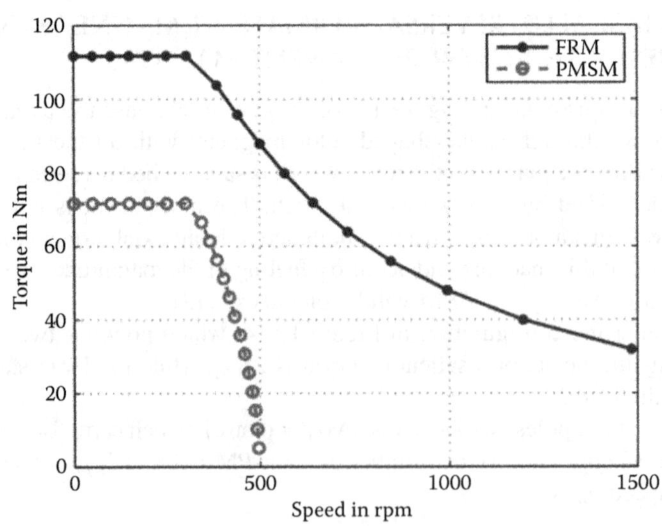

FIGURE 10.13 Torque speed envelopes: (a) for 6/14 FR-PMSM; (b) for 24/28 SPMSM. (After D. S. More et al., *Record of IEEE-IECON*, 2008, pp. 1131–1136. [11])

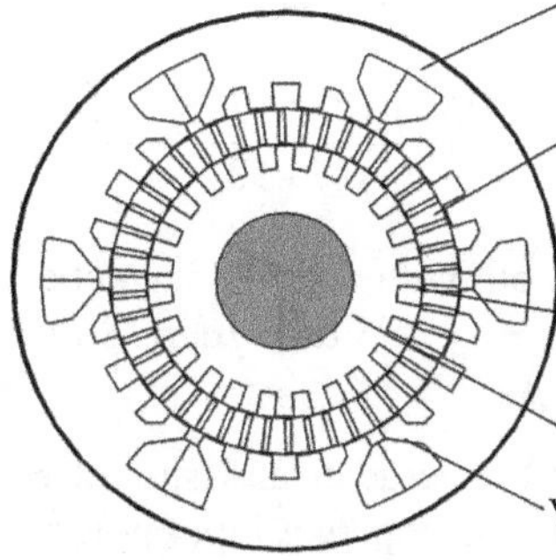

FIGURE 10.14 Three-phase FR-PMSM with dual stator and PM rotor (radial airgap).

A much higher rotor sheer stress $f_t = 6\,\text{N/cm}^2$ for a $T_{eb} = 150\,\text{Nm}$ motor with $l_{stack} = 70\,\text{mm}$ and stator outer diameter $D_{os} < 300\,\text{mm}$, with $j_{con} < 6\,\text{A/mm}^2$ and a line voltage of 380 V, efficiency above 90%, and a power factor above 70% are expected.

Using the same technical theory as in section 10.2, we obtain in sequence:

$$D_r \approx \sqrt{\frac{2T_{eb}}{\pi f_t \cdot l_{stack}}} = 176\,\text{mm} \tag{10.26}$$

$$\tau_{PM} = \frac{\pi D_r}{p_{pr}} = \frac{\pi \cdot 180}{46} = 12.02\,\text{mm} \tag{10.27}$$

The airgap is considered here as $g = 0.7$ mm, and the rotor stack length is 73 mm (in contrast to $l_{stack} = 70$ mm in the stator) to avoid large axial forces on the bearings.

The leakage inductance will reduce this value further.

We chose here NdFeB PMs with $B_r = 1.2$ T and $\mu_{rec} = 1.05\,\mu_0$ at 20°C. So the ideal radial height of the PM spokes h_r will be:

$$\frac{h_r}{\tau_{PM}} = \frac{B_{PM}/B_r}{2K_{fringe}\left(1 - (B_{PM\,g}/B_r)\cdot(2g\mu_{rec}/\mu_0)/(h_{PM})\right)} \tag{10.28}$$

The rotor core is made of one-piece regular laminations and thus two 0.5-mm-thick iron bridges are left above each encapsulated magnet whose thickness $h_{PM} = 4$ mm (the pole pitch τ_{PM} is only 12 mm!). The PM flux fringing coefficient $K_{fringe} = 0.42$. It is slightly larger than usual because all magnets are active, but here a conservative design is targeted.

Now, from Equation 10.29, the PM radial height/length $h_r = 1.8\tau_{PM} = 21.6$ mm.

The coil amperturns (rms) $n_c I_c$:

$$n_c I_c = \frac{T_{eb}\sqrt{2}}{3\cdot\varphi_{PM\,max}\cdot p_{pr}\cdot(N_s/3)} = 325.8\,\text{Aturns/coil (rms)} \tag{10.29}$$

$$\varphi_{PM\,max} = B_{g\,PM}\cdot\tau_{PM}\cdot l_{stack}\cdot 2\cdot n\cdot p_{pr};\quad 2\cdot p_{pr} = 4 \tag{10.30}$$

Let us now consider the two coils per phase in parallel and correct the voltage (due to end-coil inductance) by a factor $K_{cor} = 0.27$:

$$n_c\left(\frac{d\varphi_{PM}}{d\theta_r}\right)\cdot 2\pi n\sqrt{1+\left(\frac{B_{ag}}{B_{PM(-)}}\right)^2}(1+K_{cor}) = V_1\sqrt{2} \tag{10.31}$$

The airgap flux density produced by all $|i_q|$ stator rated current is:

$$B_{ag} \approx \frac{\mu_0 n_c I_c \sqrt{2}}{2g} \approx \frac{1.256 \cdot 10^{-6} \cdot 325 \cdot \sqrt{2}}{2 \cdot 0.7 \cdot 10^{-3}} = 0.5\,\text{T} \tag{10.32}$$

This leaves room for a 2.0-p.u. current, but with notable magnetic saturation.

With $V_1 = 220\,\text{V}$ from Equation 10.31, $n_c = 155$ turns/coil. So the coil current $I_{cn} = n_c I_c / n_c = 325/155 = 2.10$ A, phase current $I_n = 4.2$ A.

Note: Now the assumption of $B_{ag} = 1.0\,\text{T}$ is crucial and should be carefully checked, best by FEM; supposing it holds, we may calculate the apparent input power S_n:

$$S_n = 3V_1 I_n = 3 \cdot 220 \cdot 4.2 = 2772\ \text{VA} \tag{10.33}$$

The electromagnetic power P_{en} writes:

$$P_{en} = T_{eb} \cdot 2\pi n_s = 2041\ \text{W} \tag{10.34}$$

So:

$$\frac{P_{en}}{S_n} = \eta_n \cos\varphi_n \approx 0.736 \tag{10.35}$$

To complete the machine general sizing, we first calculate the outer stator back iron h_{yos}:

$$h_{\text{yos}} = \frac{\varphi_{\text{PM pole}}}{2 \cdot B_{ys} \cdot l_{\text{stack}}} \approx 22.4\ \text{mm} \tag{10.36}$$

With a slot filling factor $K_{\text{fill}} = 0.5$ and rated current density $j_{\text{con}} = 3\ \text{A/mm}^2$, the slot active area A_{slot} is:

$$A_{\text{slot}} = \frac{2n_c I_{cn}}{K_{\text{fill}} \cdot j_{\text{con}}} = 433\ \text{mm}^2 \tag{10.37}$$

The copper wire diameter $d_{co} = 0.9238$ mm.

As the sizing of the slot is straightforward with D_r given and six large stator poles, we leave it out here. But, for copper loss computation, the average turn length coil is required:

$$\begin{aligned} l_{\text{coil}} &\approx 2 \cdot l_{\text{stack}} + 2(4+3)\tau_{\text{PM}} + \tau_{\text{PM}} \\ &\approx 2 \cdot 0.07 + 14 \cdot 0.012 + 0.012 = 0.32\,\text{m} \end{aligned} \tag{10.38}$$

The coil resistance R_{coil}:

$$\begin{aligned} R_{\text{coil}} &= \rho_{\text{coil}} \frac{l_{\text{coil}} \cdot n_c \cdot j_{\text{con}}}{I_{cn}} \\ &= 1.68 \cdot 10^{-8} \frac{0.32 \cdot 155 \cdot 3 \cdot 10^6}{2.1} = 1.19\,\Omega \end{aligned} \tag{10.39}$$

So the copper losses:

$$p_{\text{copn}} = 3 \cdot 2 \cdot R_{\text{coil}} \cdot I_{cn}^2 = 31.49\,\text{W} \tag{10.40}$$

(Again, the two coils per phase are connected in parallel.)

At 50 Hz, the core losses should not be significant, but let us suppose they are 75 W (with about 30 kg of stator core weight); the mechanical losses are taken as 1% P_{elm}. Thus, the rated efficiency is:

$$\eta_n = \frac{P_{en} - p_{mec}}{P_{en} + p_{core} + p_{copn}} = \frac{2041 - 20}{2041 + 75 + 87.65} = 0.9411 \quad (10.41)$$

So, from Equation 10.34, the power factor should be:

$$\cos\varphi_n = \frac{\eta_n \cos\varphi_n}{\eta_n} = \frac{0.736}{0.917} = 0.782 \quad (10.42)$$

The total motor active weight was calculated at 40 kg, so the torque/weight is 3.75 Nm/kg or 30 Nm/l. This is about twice as much as for the surface-rotor-PM FR machine of about the same specifications. The added difficulties in its fabrication have, though, to be weighed carefully.

The FEM verifications done for the scope show the results visible in Figure 10.15a–c.

It is confirmed by FEM that:

- The cogging torque is reduced, for 0.5 mechanical degrees skewing, to 2 Nm (1.5% of rated torque of 150 Nm).

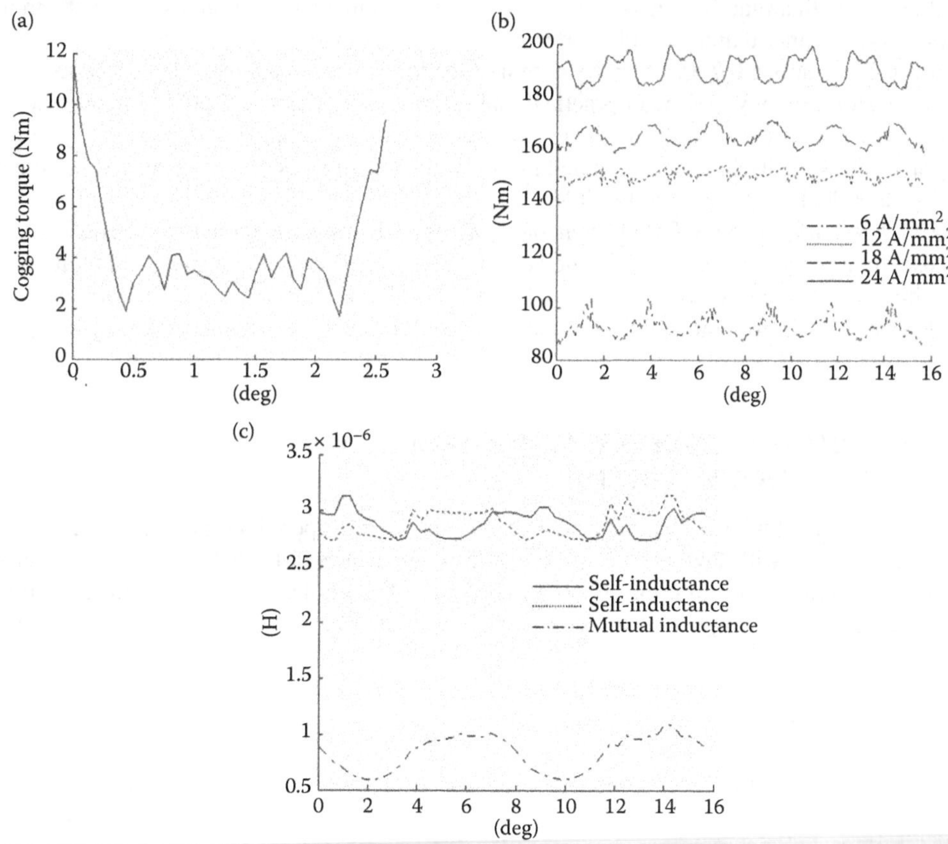

FIGURE 10.15 FEM verification for the three-phase FR-PMSM with spoke-rotor-PMs and dual stator: (a) cogging torque versus skewing angle; (b) torque versus rotor position angle; (c) self and mutual inductance for one-turn coil.

- The machine may develop an average torque $T_{eb} \approx 150\,\mathrm{Nm}$ at 4.2 A/phase (rms) with 155 turns/coil and two coils in parallel per phase.
- It is seen that, in contrast to surface-stator PM FR-PMSM, the machine saturates rather heavily. Apparently, overloading is not an asset of this machine.
- With 155 turns/coil, the self-inductance (Figure 10.15c) may be written as:

$$L_{AA} \approx L_0 + L_2 \cos\theta_{er} \tag{10.43}$$

with

$$L_0 = 0.06826H, L_2 = 0.03688H \tag{10.44}$$

End-coil inductance is not considered here.

- The mutual inductance is much smaller than the self-inductances, but it varies more with rotor position and it has lower-order harmonics. This property may be used for rotor position estimation in encoderless FOC (DTFC) in variable speed drives.
- The phase voltage drop on the machine inductance is approximately: $V_{Ls} = \omega_1 L_s i_n = 2\pi \cdot 50 \cdot (3/2)0.0684 \cdot 4.2 = 135.0$ V (rms); this seems to validate the power factor calculated analytically.
- The copper losses for 150 Nm (FEM validated) torque are confirmed.

The conclusion is that FEM-based optimal design is needed to extract more power per volume at good efficiency with airgap flux density $B_{g\,\mathrm{PM}} = 1.4$–1.5 T and a large airgap (here $g \approx 0.7$ mm), for larger outer stator inner diameter and shorter l_{stack} (pancake) shape topologies.

A power factor above 0.7–0.8 may be obtained (as for Vernier machines [10]). The axial airgap twin active stator topology may be of practical interest also. Still, the machine PWM converter initial cost, plus the capitalized cost of the losses for a most-frequent torque and speed pair, per system functional years (hours in all), has to be included in the optimal design code to see if and where the rotor-PM dual-stator FR-PMSM drive is feasible.

Note on three (m)-phase FR-PMSM control. Again, as the emf is sinusoidal and the dq inductances are almost equal to each other, the machine behaves like an SPMSM but is easy to saturate magnetically (with a large synchronous inductance). So FOC, DTFC with and without encoder feedback, or scalar (V/f or I-F) control with stabilizing loops, typical for synchronous motors, may be applied.

10.6 ONE-PHASE FLUX-REVERSAL PERMANENT MAGNET SYNCHRONOUS MACHINE

The one-phase PMSM with self-starting facility (by, in general, tapered airgap) may be implemented in FR configurations with PMs on the stator, both with surface PMs (NdFeB) or with detachable Ferrite-PMs with flux concentration. Small power (<100 W) applications may profit from this thermally and mechanically rugged motor topology. It may also be used at high speeds with a lower flux density.

A typical configuration is shown in Figure 10.16.

If the surface PM configuration (Figure 10.16) is characterized by a small maximum PM flux density ($B_{g\,\mathrm{PM}} \leq 0.4$ T, even with high-quality NdFeB sintered magnets), $B_{g\,\mathrm{PM\,max}}$ in the airgap can reach values above 0.7 T with Ferrite-PMs bellow the active two rotor poles (Figure 10.17 [17]).

By tapering the airgap, it is possible to obtain a stator/rotor geometry that allows for nonzero (positive) total torque (above a certain current level) for any rotor position. This is to say that safe self-starting is obtained. For a 35 W, 0.2 A, 290 V_{DC} (a single-phase inverter is needed), at 1600 rpm motor, such a self-starting torque/position level with trapezoidal bipolar current is shown in Figure 10.18 (by 2D-FEM) [17].

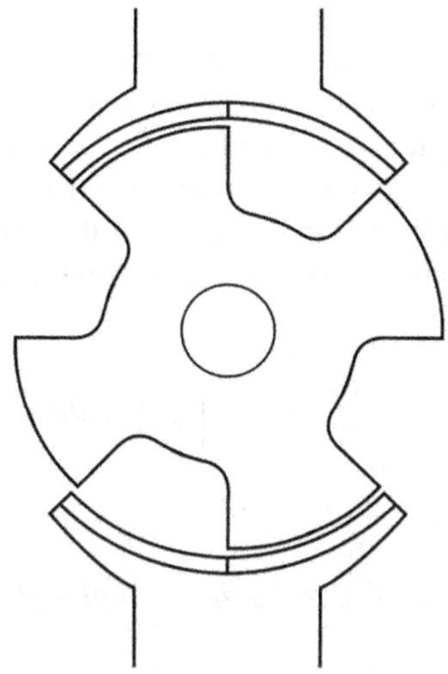

FIGURE 10.16 One-phase stator-PM FR-MSMS: with surface PMs.

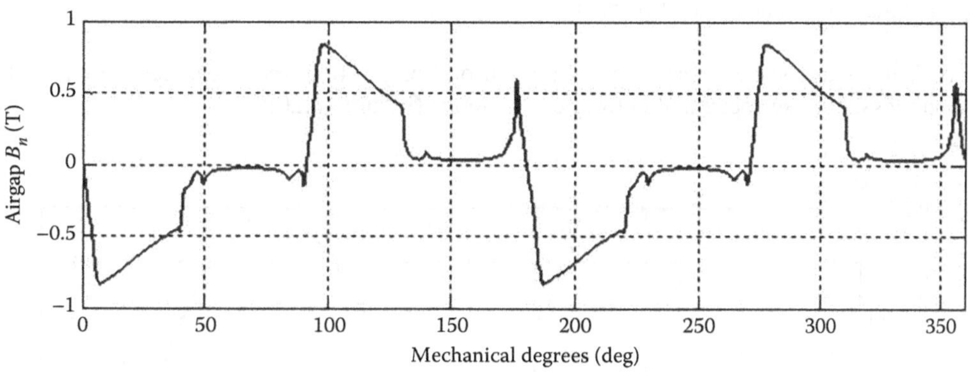

FIGURE 10.17 Airgap flux density for maximum PM flux rotor position in one-phase FR-PMSM. (After Y. Gao et al., *IEEE Trans.*, vol. IA–53, no. 5, 2017, pp. 4232–4241. [17])

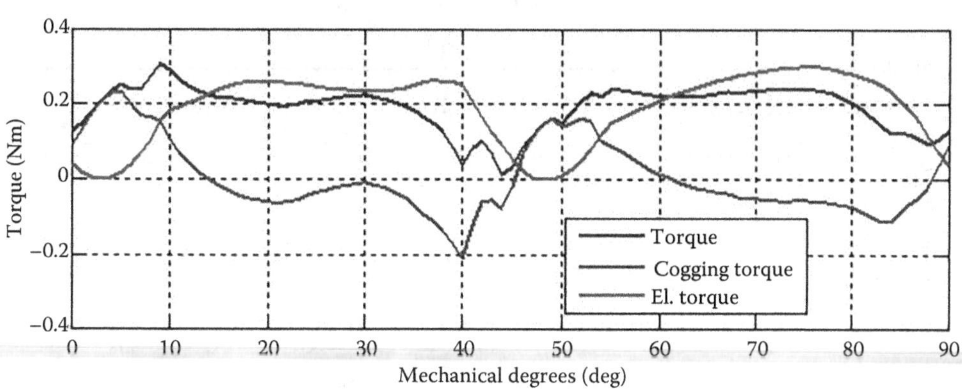

FIGURE 10.18 Electromagnetic cogging and total torque of a 35-W, 1600-rpm, one-phase Ferrite FR-PMSM with safe starting. (After Y. Gao et al., *IEEE Trans.*, vol. IA–53, no. 5, 2017, pp. 4232–4241. [17])

FEM-only–based optimization might be run with a penalty function as min(*torque*) > 0.1 Nm (in our case) to secure nonhesitant large self-starting torque (yes, in a single direction of motion).

The machine inductance does vary with rotor position (Figure 10.19).

The PM-emf waveform, cogging torque, and inductance dependence on rotor position and current (due to magnetic saturation) are a clear indication that FEM results (eventually after curve fitting) should be exported to a phase coordinate circuit model of this motor.

The circuit plus the motion equations may thus be summarized as:

$$\begin{bmatrix} \dfrac{di}{dt} \\ \dfrac{d\omega_r}{dt} \end{bmatrix} = \begin{bmatrix} -\dfrac{R}{L(\theta_r)} & -K_e(\theta_r) \\ \dfrac{K_e(\theta_r)}{J} & 0 \end{bmatrix} \begin{bmatrix} i \\ \omega_r \end{bmatrix} + \begin{bmatrix} \dfrac{1}{L(\theta_r)} & 0 \\ 0 & \dfrac{1}{J} \end{bmatrix} \begin{bmatrix} V \\ T_{cogg}(\theta_r) - T_{load} \end{bmatrix} \quad (10.45)$$

$$\frac{d\theta_r}{dt} = \omega_r; \; T_{em} = K_e(\theta_r) \cdot i; \; \text{emf} = K_e(\theta_r)\omega_r \quad (10.46)$$

where

$L(\theta_r)$ is the phase inductance versus rotor position (θ_r) and current

T_{load}—load torque (Nm)

I—phase current (A)

ω_r—angular mechanical speed (rad/s)

A typical control system may use a Hall sensor (for position and speed feedback) or it may be encoderless, but use speed and current closed loops (Figure 10.20).

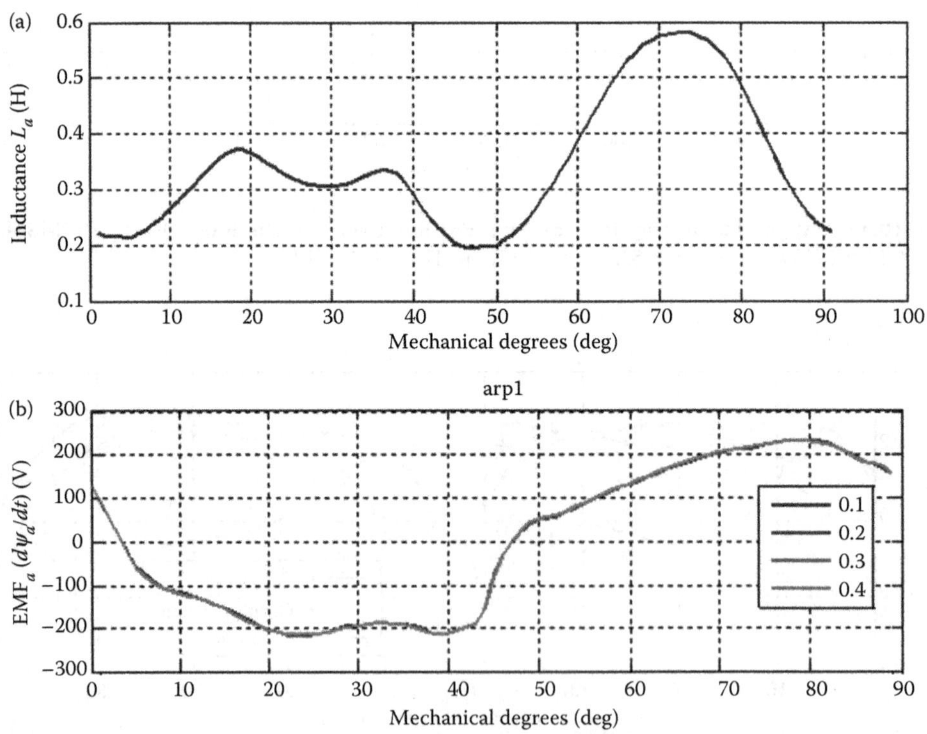

FIGURE 10.19 One-phase Ferrite FR-PMSM: phase inductance versus rotor position at rated current (a); and PM emf versus rotor position, (b). (After Y. Gao et al., *IEEE Trans.*, vol. IA–53, no. 5, 2017, pp. 4232–4241. [17])

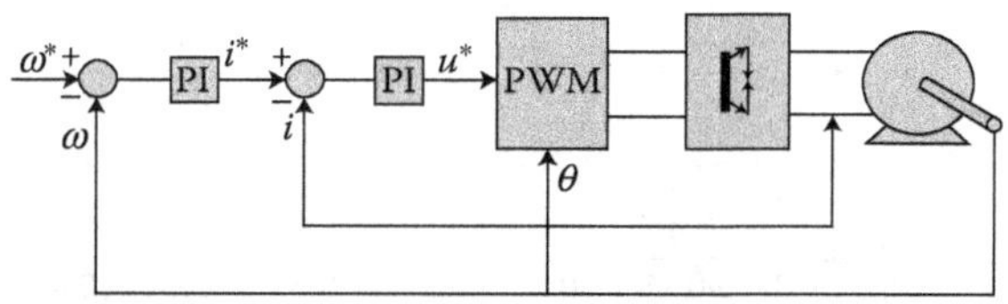

FIGURE 10.20 Generic encoder (Hall sensor) one-phase FR-PMSM control system.

Let us consider here a small compressor load whose torque model and motor data are described in Tables 10.1 and 10.2. The compressor torque expressions are [8]:

$$V_c = [\cos(\theta_r + \pi/6) + 1] \cdot R_{\text{exc}} \cdot S_{\text{pist}} + V_m \tag{10.47}$$

$$P_c = \begin{cases} P_d\left(\dfrac{V_m}{V_0}\right)^{n_{\text{gas}}}, & \text{if } V_c > 0 \text{ and } V_c < V_m\left(\dfrac{P_d}{P_s}\right)^{1/n_{\text{gas}}} \\ P_s, & \text{if } V_c > 0 \text{ and } V_c < V_m\left(\dfrac{P_d}{P_s}\right)^{1/n_{\text{gas}}} \\ P_s\left(\dfrac{V_{\max}}{V_c}\right)^{n_{\text{gas}}}, & \text{if } V_c \le 0 \text{ and } |V|_c > V_{\max}\left(\dfrac{P_s}{P_d}\right)^{1/n_{\text{gas}}} \\ P_d, & \text{if } V_c \le 0 \text{ and } |V_c| \le V_{\max}\left(\dfrac{P_s}{P_d}\right)^{1/n_{\text{gas}}} \end{cases} \tag{10.48}$$

TABLE 10.1
Compressor Torque Model

R_{exc} (Axis eccentricity)	10×10^{-3} m
S_{pist} (Piston surface)	0.45×10^{-3} m^2
V_m (Cylinder dead volume)	120×10^{-9} m^3
n_{gas} (Gas constant)	1.1
P_d (Discharge pressure)	5.3 bar
P_s (Suction pressure)	0.3 bar

TABLE 10.2
Motor Control Data

Parameters	
R—is the resistance (Ω)	37.85
J—is the inertia constant—(kg · m^2)	1×10^{-4}
Number of poles	4
Switching frequency (kHz)	10
DC link voltage (V)	300
Speed regulator proportional gain	2
Speed regulator integral gain	1×10^{-2}
Current regulator proportional gain	1
Current regulator integral gain	1×10^{-4}
Sapling frequency (kHz)	100

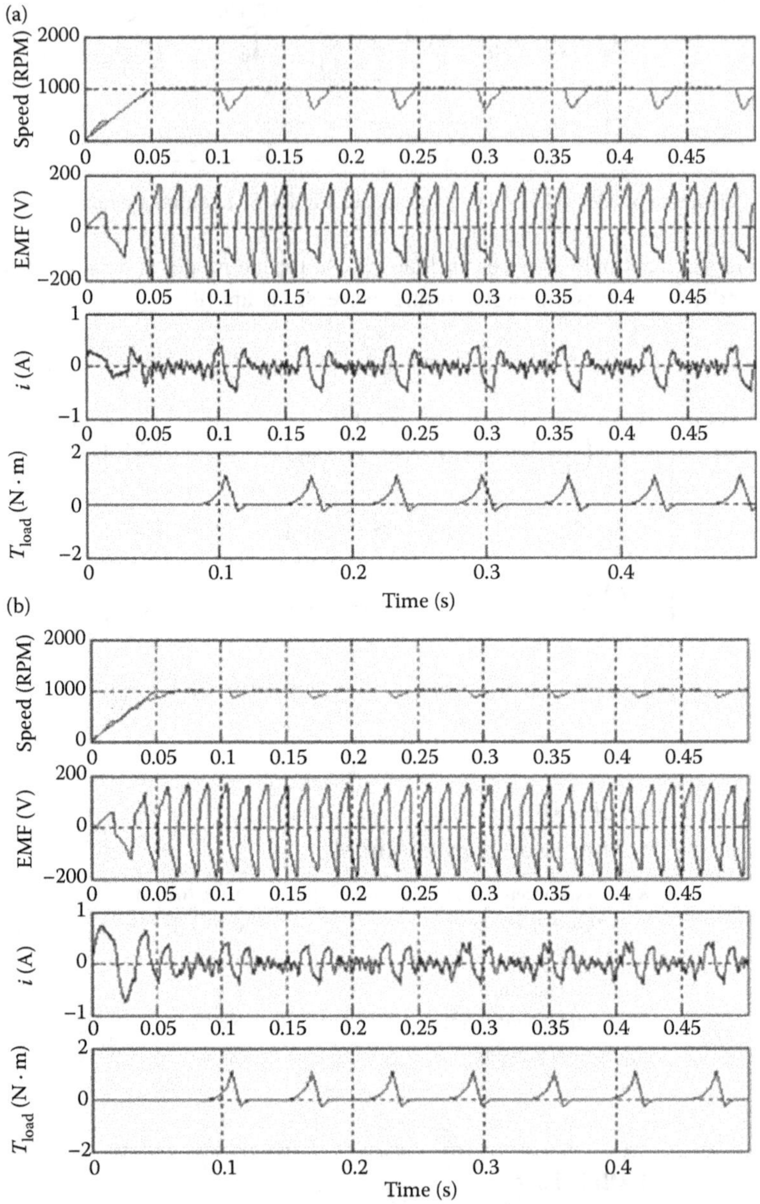

FIGURE 10.21 Starting and operation with an average of 0.2-Nm compressor load: (a) small inertia ($J = 1\ 10^{-4}\ \text{Kg m}^2$); (b) large inertia ($J = 3\ 10^{-4}\ \text{Kg m}^2$). (After F. J. H. Kalluf et al., *Record of ACEMP-OPTIM*, 2015, pp. 406–411. [8])

$$T_{\text{load}} = \sin\left(\theta_r + \frac{\pi}{6}\right) \cdot R_{\text{exc}} \cdot (P_c - P_s) \cdot S_{\text{pist}} \tag{10.49}$$

Digital simulation results (for $J = 1\ 10^{-4}\ \text{Kg m}^2$ and for $J' = 3\ 10^{-4}\ \text{Kg m}^2$) are shown in Figure 10.21a and b [8].

It is evident that only with an enlarged inertia is the compressor variable-with-rotor-position-load-torque handled with a small speed ripple.

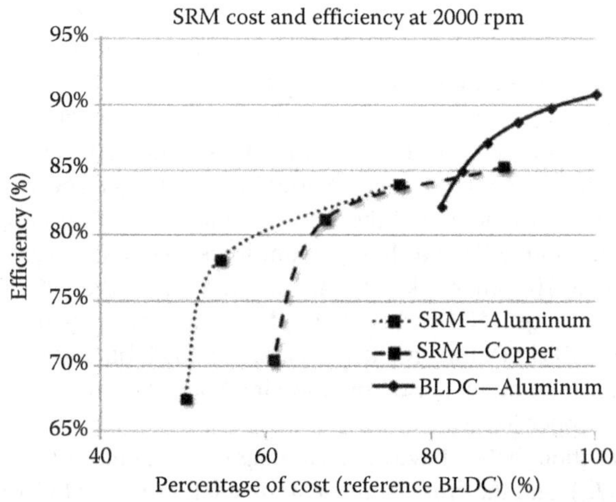

FIGURE 10.22 Electrical efficiency–cost Pareto fronts. (After F. J. H. Kalluf et al., *Record of ACEMP-OPTIM*, 2015, pp. 406–411. [8])

Design optimization (efficiency versus percentage costs of a three-phase BLDC PM motor) show that good performance is expected for up to 100 W in such small motors [9], where cost may be traded for electrical efficiency with both copper and aluminum windings (Figure 10.22).

Considerable talent is expected to be further devoted to small motor (actuator) drives within 100 (1000) W, with one-phase inverter control, for household, vehicular-auxiliary, info-gadget, and industrial applications.

10.7 SUMMARY

- Flux-reversal PMSMs use multi-PM pole nonoverlapping AC coils on the stator and salient pole rotors (N_r poles) such that for a three-phase machine: $N_s = (k_p + (1/3))N_r$; $k_p = 1, 2, \ldots$. In general, each of the N_s stator poles has $2n_p$ PM of alternate polarity.
- This may be considered a particular case of N_s, N_r combinations that secures symmetric three-phase emfs in the AC coils; the multiple AC coils are supposed to reduce copper loss for given torque, machine geometry, and PM pole pitch τ_{PM}. The stator slot openings between the N_s poles are $(2/3\tau_p)$ wide to secure symmetrical emfs.
- Alternatively, the PMs may be placed inside the stator pole, as, for PM flux concentration, behind the stator coils, when the stator poles are magnetically nonsalient.
- Finally, the PMs may be planted on a rotor with $2p_{PM}$ poles (instead of N_r salient poles and interpoles with $2\tau_{PM}$ periodicity) in a spoke shape to allow:
- PM flux concentration
  - Dual-stator operation when all PMs are active all the time.
  - The two stators are shifted by one pole pitch τ_{PM}: the interior stator may be moved by a servomotor to produce mechanical flux weakening. Axial airgap topologies are also feasible.
  - So, in regard to the above, FR-PMSM is somewhat different from SF-PMSM.
  - Allowing PM pole pitch of down to 10–12 mm for an up to 1–2 mm airgap, the FR-PMSM leads to notable torque magnification as all F-M machines do, and thus has suitable low and medium speeds when the fundamental frequency remains below 1 kHz in the KW power range and less for larger power/unit.
  - However, as the PM pole pitch decreases, the PM flux fringing increases, reducing the real torque magnification; finally, the machine inductance ends up larger in p.u. values

and the power factor remains low (below 0.75) in surface-stator PM FR-machines when there is an optimum ratio $(g + h_{PM})/\tau_{PM} \leq 0.5$. Still, the machine allows overloading and in contrast to tooth-wound coil SPMSMs (of the same τ_{PM} and geometry) may produce up to 50% more torque and power, but at smaller power factor (around 0.7–0.75).

- The spoke-PM-rotor dual stator FR-PMSM, with PM flux concentration and at 1.4 T PM flux density in the airgap, is capable of doubling the torque (power) with respect to the surface-stator-PM FR-machine of the same geometry, at lower copper losses.
- For this configuration, a 21-mm-high, 4-mm-thick magnets, airgap $g = 0.7$ mm, $\tau_{PM} =$ 10–12 mm, for a 70-mm stack length, 180-mm inner diameter of outer stator, the machine may produce 150 Nm at 22 Nm/l for an efficiency of 91.7%, power factor 0.80 at 130 rpm, 50 Hz PM magnetization at 2 p.u. load, but has to be checked carefully; the power factor may be increased by allowing some $I_d < 0$ in the control, to the detriment of larger copper losses.
- As FR-PMSMs, though they have physical magnetic saliency, operate without functional saliency ($L_d = L_q$), and the emf is almost sinusoidal, FOC, DTFC, or scalar control strategies typical to SPMSMs may be used [14], with the observation that heavy saturation may occur and has to be accounted for.
- FR-PMSMs operate at very low cogging and resultant full torque ripple less than 3%–5%
- Stator-Ferrite PM one-phase PMSMs, with dual-saliency and tapered airgap (for safe starting), are offering PM flux concentration and all PMs working at all times (yes, on only half of the stator coil throw, as in SF-PMSMs, etc); 0.75 T max. Ferrite PM flux density in the airgap was obtained with two stator poles ($2n_p = 2$) and four rotor salient poles. PMs have cubicle shape and may be inserted (or extracted) with AC coils in place. By airgap tapering with geometrical variables changed and FEM calculation, positive-only total torque is obtained, above a certain trapezoidal-shaped current level, to provide for self-starting; yes, only in one direction of motion. Preliminary results have shown efficiency above 0.84 for 100 W, 3000 rpm. Small power moderate cost actuators for home appliances, info gadgets, vehicular auxiliaries, and industrial process motion controllers below 100 W (1 kW) may profit from this rugged motor of moderate to low initial cost and good efficiency.
- As an example, an inner stator with DC excitation multiphase FR-machine (without PMs) has been proposed recently for wind power generation [13–14].
- New stator/rotor slots/pole combinations are investigated for lower torque pulsations [15].
- Also, transverse flux-reversal motors/generators with SMC cores have recently been proposed at 1.2 kHz fundamental frequency for 3000-rpm drives in the sub and kW range, for better efficiency but still at a small power factor [16].

- New design methods for FRMs are still being proposed [17].

REFERENCES

1. D. Li, R. Ou, J. Li, Topologies and analysis of flux modulation machines, *Record of IEEE-ECCE*, 2015, pp. 2153–2160.
2. I. Boldea, E. Serban, R. Babau, Flux reversal stator-PM single phase generator with controlled d.c. output, *Record of OPTIM-1996*, vol. 4, pp. 1124–1134.
3. R. Deodhar, S. Anderssen, I. Boldea, T. J. E. Miller, The flux reversal machine: a new brushless doubly-salient PM machine, *IEEE Trans.*, vol. IA-33, no. 4, 1997, pp. 925–934.
4. I. Boldea, C. X. Wang, S. A. Nasar, The flux reversal machine. *Proc. IEE*, EPA-146, no. 2, 1999, pp. 139–146.
5. I. Boldea, C. X. Wang, S. A. Nasar, Design of a 3 phase flux reversal machine, *Elec. Mach. Power Syst. Journal*, vol. 27, 1999, pp. 848–863.
6. I. Boldea, J. Zhang, S. A. Nasar, Theoretical characterization of flux reversal machine in low speed drives. The pole-PM configuration, *IEEE Trans.*, vol. IA-38, no. 6, 2002, pp. 1544–1557.

7. I. Boldea, *Variable Speed Generators*, book, 2nd edition, CRC Press, Taylor and Francis Group, New York, 2015 (Chapter 11: "Transverse flux and flux reversal permanent magnet generator systems", pp. 479–527).
8. F. J. H. Kalluf, A. D. P. Juliani, I. Boldea, L. Tutelea, A. R. Laureano, Single phase stator Ferrite-PM double saliency motor performance and optimization, *Record of ACEMP-OPTIM*, 2015, pp. 406–411.
9. I. Boldea, L. Tutelea, L. Parsa, D. Dorrel, Automotive electric propulsion systems with reduced or no PMs: an overview, *IEEE Trans.*, vol. IE-61, no. 10, 2014, pp. 5696–5711.
10. D. Li, R. Ou, T. A. Lipo, High power factor Vernier machines, *IEEE Trans.*, vol. IA-50, no. 6, 2014, pp. 3669–3674.
11. D. S. More, H. Kalluru, B. G. Fernandes, Comparative analysis of flux reversal machines and fractional slot concentrated winding PMSM, *Record of IEEE-IECON*, 2008, pp. 1131–1136.
12. I. Boldea, L. Tutelea, M. Topor, Theoretical characterization of three phase flux reversal machine with rotor-PM flux concentration, *Record of OPTIM 2012*, pp. 472–476 (IEEEXplore).
13. C. H. T. Lee, K. T. Chau, C. Liu, Design and analysis of cost effective magnetless multiphase flux-reversal DC-field machine for wind power generation, *IEEE Trans.*, vol. EC-30, no. 4 2015, pp. 1565–1573.
14. T. H. Kim, J. Lee, Influences of PWM mode on the performance of flux reversal machine, *IEEE Trans.*, vol. MAG-41, no. 5, 2005, pp. 1956–1959
15. Y. Gao, R. Qu, D. Li, J. Li, Design procedure of flux reversal PM machines, *Record of IEEE-ECE-2015*, pp. 1586–1592.
16. V. Dumitrievski, V. Prakht, S, Sarapulov, D. Askerov, A multiphase single phase SMC flux reversal motor for fans, *Record of IEEE ECCE-2015*, pp. 55–61.
17. Y. Gao, D. Li, R. Qu, J. Li, Design procedure of flux reversal permanent magnet machines, *IEEE Trans.*, vol. IA–53, no. 5, 2017, pp 4232–4241.

11 Vernier PM Machines

11.1 INTRODUCTION

Vernier reluctance machines were apparently introduced in 1960 [1], while their PM versions have been developed since 1995 [2–4]. Vernier PM machines are characterized by larger torque density and higher efficiency in low-speed drives. Being a flux-modulation machine (Chapter 1) with an open-slotted, stator with its slots containing a distributed ($q > 1$) winding with p_a pole pair mmf and $2p_r$ PM-rotor poles, the two conditions for nonzero average torque (and synchronous operation) are (Chapter 1):

$$\pm\, p_a = N_s - p_r;\ \omega_a = p_r \cdot 2\pi n;\ PR = p_r/p_a \tag{11.1}$$

Fundamentally, the first harmonic of the flux modulator (the slotted stator airgap magnetic permeance of N_s periods per the entire periphery) "couples" the rotor-PM mmf p_r pole pair mmf with the armature AC winding p_a pole pair mmf. As already demonstrated for switched-flux and flux-reversal PM machines, numerous stator slot (N_s)/rotor poles ($2p_r$) combinations are eligible, but the one in Equation 11.1 produces large torque density, with $p_a = 1,2,3$ and $q_1 = 1,2$ (slots per pole per phase).

The topology may vary from the surface PM rotor (Figure 11.1a) [5] to the spoke-shaped PM-rotor sandwiched between two stators (Figure 11.1b) [6–9], or a combination of stator and rotor PM structure (Vernier + flux-reversal concept)—Figure 11.1c [10].

In an effort to increase the torque PM weight and reduce the end-coil length (and their copper losses), the consequent PM pole outer rotor with toroidal (Gramme-ring) AC windings on the interior stator ($2p_a = 2$) was also investigated (Figure 11.2) with partial success.

We rank the success chance of Vernier PM machines as moderate because of the fabrication complexity of the interior Gramme-ring winding stator while, for similar complexity, the spoke-shaped PM rotor dual stator (Figure 11.1b) produces better torque/volume, torque/PM weight, and efficiency. Since all PMs are active all the time, ($2p_a = 4$, $q = 1,2$), and the emf is large with two airgaps (airgap flux density is as high as 1.4–1.45 T), the spoke-shaped PM-rotor dual stator Vernier machine ends up having a high power factor (above 0.8) with a rather large torque/volume index. It may be practical to use thicker and deeper Ferrite PMs on such a rotor dual-stator machine to preserve high torque density [13].

Axial–airgap versions of three-/multiphase Vernier PM machines look more manufacturable, as the two stators are identical. Also, one-phase Vernier PM machines with tapered airgap (on the stator part) may be imagined to allow safe self-starting conditions, as for FR-PMSMs (Chapter 10).

As the number of the stator slots $N_s \geq 12$, the number of rotor PM pole pairs $p_r > 10$ (with $p_a = 2$), the frequency for a given speed tends to be rather large. So, if used at $n_n = 3000$ rpm, the minimum fundamental frequency of the AC stator current would be $f_a = n_n p_r > 50 \times 10 = 500$ Hz.

This already implies costly high-quality thin lamination cores. So, in general, Vernier PM machines are mostly suitable for low and moderate speeds and high-torque applications (tens of *Nms* and more).

As already noted in Chapter 10, on flux-reversal PM machines with spoke-shaped PM–rotors, the Vernier PM machine counterparts are similar, but the windings use overlapping three-/multiphase coils (with $2p_a = 2,4,6$), while FR-PM machines use nonoverlapping AC windings with four equivalent mmf poles.

As Vernier PM machines in variable speed motor/generator drives have not yet reached commercialization, here we will proceed with technical design methodologies with FEM validation [12–14]

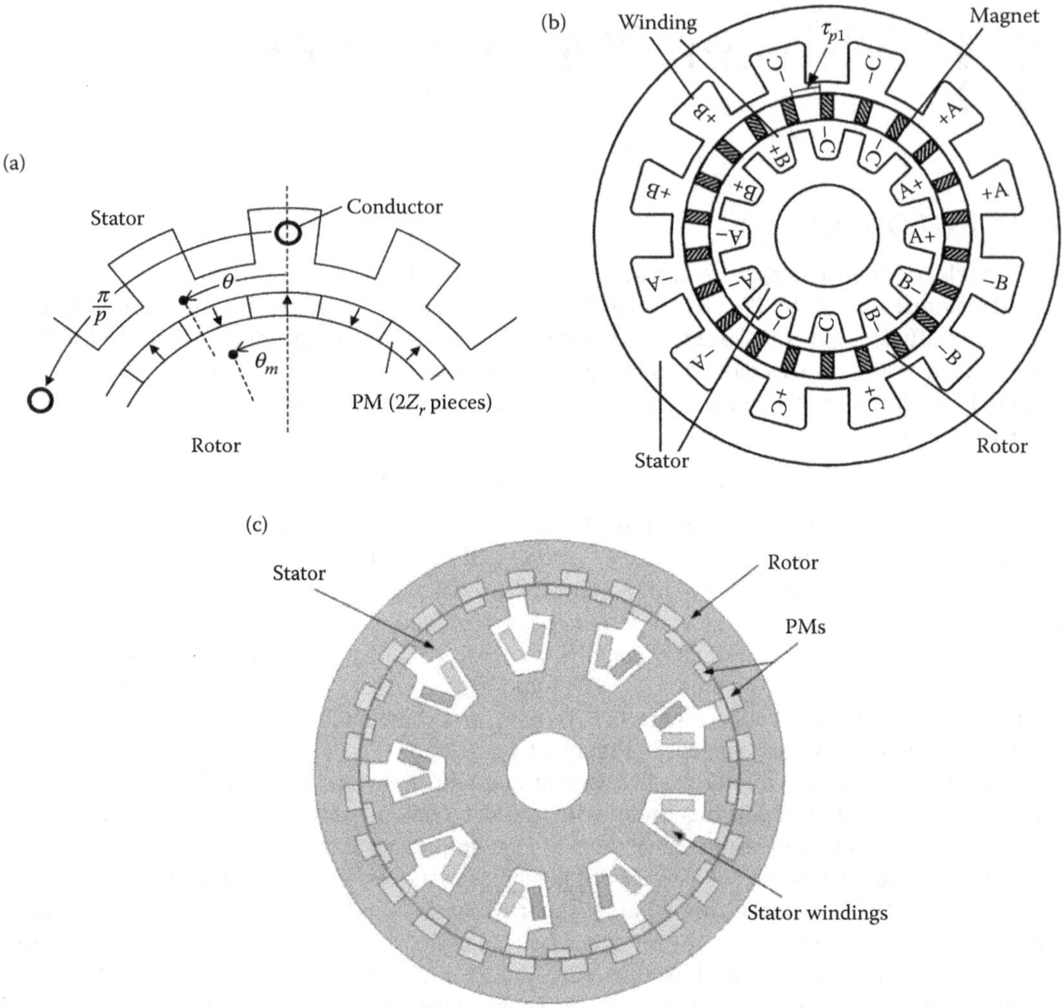

FIGURE 11.1 Basic three-phase Vernier PM machine topologies: (a) with surface-PM rotor; (b) with spoke-shaped PM rotor and stator; and (c) with combined rotor and stator PMs. (After Z. S. Du, T. A. Lipo, *Record of IEEE-ECC*, 2015, pp. 6082–6089; S. Niu, S. L. Ho, W. N. Fu, *IEEE Trans. on MAG*, vol. 50, no. 2, 2014, pp. 7019904. [7,10])

and case studies and control principles with some test results to offer the reader a feeling of the magnitude of his or her choice of a drive suitable for a specific application.

11.2 PRELIMINARY DESIGN METHODOLOGY THROUGH CASE STUDY

Let us consider a 5000-Nm, 30-rpm, line voltage <380 V(rms) spoke-shaped PM-rotor dual-stator Vernier PM machine [as in 8]; see Figure 11.1b.

The preliminary design methodology includes quite a few stages, as follows.

11.2.1 Twin Alternating Current Windings and Their Stator Phase Shift Angle

Let us, for a start, imagine that we have already chosen from Equation 11.1: $p_a = 2$ stator mmf pole pairs, $N_s = 24$ slots ($q_1 = 2 = \text{SPP}$), and thus $p_r = 22$ PM pole pairs (44 poles) on

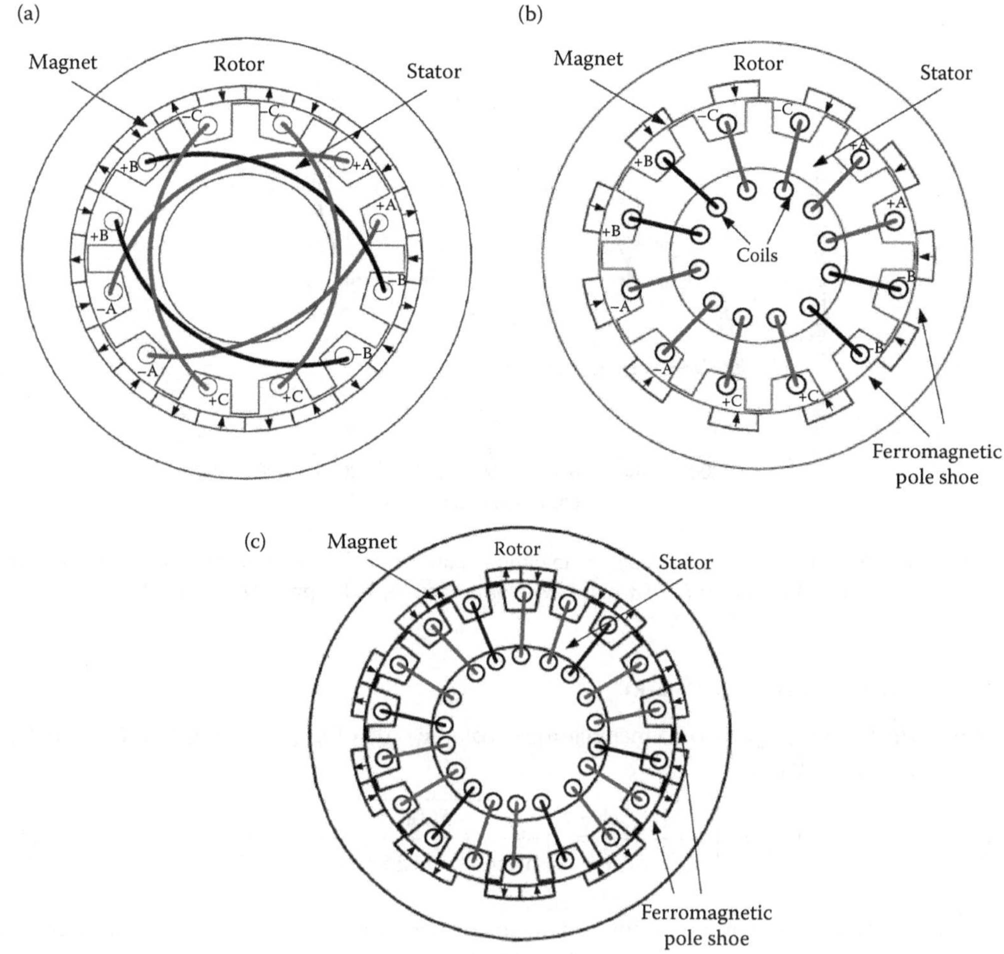

FIGURE 11.2 Vernier PM machine topologies with: (a) surface-PM outer rotor and regular distributed stator windings; (b) consequent N-N PM outer rotor and toroidal winding stator; and (c) consequent N-S PM outer rotor and toroidal winding stator. (After Z. S. Du, T. A. Lipo, *Record of IEEE-ECC*, 2015, pp. 6082–6089; S. Niu, S. L. Ho, W. N. Fu, *IEEE Trans. on MAG*, vol. 50, no. 2, 2014, pp. 7019904. [7,10])

the rotor ($PR = p_r/p_a = 22/2 = 11$). Later, this decision's motivation will be examined (Figure 11.3) [7].

A single-layer regular symmetric winding ($q_1 = 2$ slots/pole/phase) is chosen to yield a fundamental winding factor $K_{w1} = 1$. The two windings should be connected in series, as their stator cores are shifted by the geometrical angle α_0:

$$\alpha_0 = \frac{p_a \pi}{p_r + p_a} \tag{11.2}$$

The angle α_0 corresponds to the "slot versus tooth" relative position of the two stators and allows the PMs on the rotor to streamline their flux lines. Theoretically, the emfs in the two windings should be in phase, but, with rectangular cross section PMs, the outer and inner airgap permeances are not identical. However, this angle is smaller than $\alpha_{es}/2 = \pi/2\,\mathrm{mq} = \pi/(2 \times 3 \times 2) = \pi/12 = 15^\circ$ [8].

A 2D-FEM analysis is mandatory for enough precision, but an initial geometry is first required.

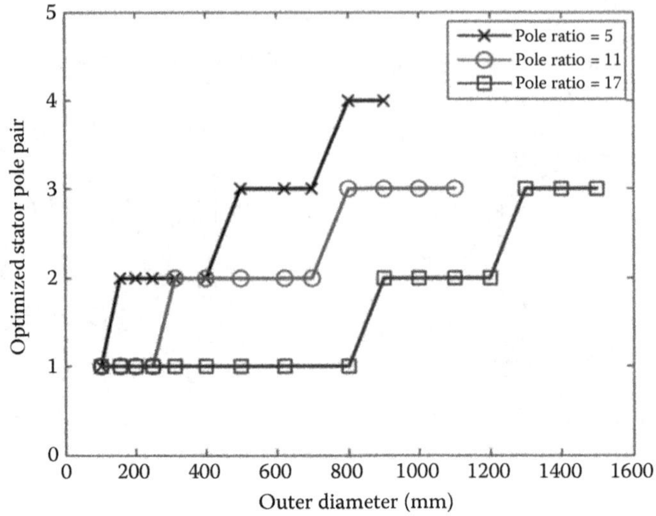

FIGURE 11.3 Optimal stator pole pair (pa) for maximum emf versus Vernier PM machine outer diameter for PR = 5, 7, 11. (After D. Li et al., *IEEE Trans. on IA*, vol. 51, no. 4, 2015, pp. 2972–2983. [8])

11.2.2 General Stator Geometry

Adopting a stack length $l_{stack} = 0.25$ m and torque/volume ratio of $t_{evn} = 66$ Nm/liter (at outer stator diameter D_{0so}), the latter is:

$$D_{0so} = \sqrt{\frac{T_{en}}{t_{evn}\frac{\pi}{4}l_{stack}}} = \sqrt{\frac{5000}{66 \times \frac{\pi}{4} \times 0.25 \times 10^3}} \approx 0.62 \text{ m} \tag{11.3}$$

The ratio between the inner and outer diameter of the outer stator is assigned the value $D_{ios}/D_{0so} = 0.77$. It is, in general, between 0.65 and 0.7 for a four-pole stator winding, but is larger here, as 40% of the mmf is produced in the interior stator.

Thus

$$D_{i0s} = 0.77 \cdot D_{0so} = 0.77 \cdot 0.62 \approx 0.470 \text{ m} \tag{11.4}$$

11.2.3 Permanent Magnet Stator Sizing

With magnet radial length l_{PM}, thickness (tangentially) h_{PM}, and PM pole pitch τ_{PM}, the following equations are valid on no load:

$$\frac{B_{gPMi}}{\mu_0}2g + H_m h_{1M} = 0; \quad l_{PM}B_m \approx B_{gPMi}\tau_{PM} \tag{11.5}$$

$$B_m = B_r + \mu_{rec}H_m; \; H_m < 0 \tag{11.6}$$

Then, considering flux fringing through the airgap permeance and the flux-bridges (thickness: h_{bridge}) in the rotor lamination that holds the magnets, we may calculate the actual airgap flux density that really produces emf B_{gPM} is:

$$B_{gPM} = B_{gPMi}K_{fringe}K_{bridge} \tag{11.7}$$

Before FEM verification, K_{fringe} may only be guessed by appearance; $K_{\text{fringe}} = 0.55$ is a safe value for the case in point, while K_{bridge} should modify B_r; in fact:

$$K_{\text{bridge}} = \left(1 - \frac{2B_{\text{sat}}h_{\text{bridge}}}{l_{\text{PM}}}\right) \tag{11.8}$$

with $B_{\text{sat}} = 2$ T, $h_{\text{bridge}} = 0.7$–0.8 mm, and $l_{\text{PM}} = 20$–40 mm, $K_{\text{bridge}} = 0.85$–0.92.

Now, adopting $B_{g\text{PM}} = 0.8$, $B_{g\text{PM}i}$ (Equation 11.7) is:

$$B_{g\text{PM}i} = \frac{0.8}{0.55 \times 0.95} \approx 1.414\ \text{T} \tag{11.9}$$

This value is needed to calculate the rotor PM geometry.

This is a rather high value, which should be available in the rotor, so the rotor core will be saturated, especially under load; thus, Equations 11.5 and 11.6 are only approximate ($\mu_{\text{rec}} = \infty$). With an airgap $g = 1.5$ mm and the PM thickness $h_{\text{PM}} = 0.40 \cdot \tau_{\text{PM}} = 0.4 \cdot 34.04 = 13.62\ \text{mm} \approx 14$ mm, $B_r = 0.3$ T, $\mu_{\text{rec}} = 1.25\mu_0$ (at high temperatures [100°C]), finally, from Equations 11.5 and 11.6, the magnet radial height l_{PM} is calculated in the sequence below:

$$\begin{aligned} H_m &= -\frac{B_{g\text{PM}i}}{\mu_0}\frac{2g}{h_{\text{PM}}} \\ &= \frac{1.414 \times 2 \times 1.5}{1.256 \times 10^{-6} \times 14} = -0.2412 \times 10^{-6}\ \text{A/m} \\ B_m &= B_r + \mu_{\text{rec}} H_m \\ &= 1.13 + 1.20 \times 1.256 \times 10^{-6} \times (-0.2412 \times 10^{-6}) = 0.665\ \text{T} \end{aligned} \tag{11.10}$$

$$l_{\text{PM}} = \frac{B_{g\text{PM}i}}{B_m}\frac{\tau_{\text{PM}}}{2} = \frac{1.414 \times (34/2)}{0.6652} \approx 36\ \text{mm} \tag{11.11}$$

11.2.4 Permanent Magnet Flux Linkage and Slot Magnetomotive Force

The PM flux per one-turn AC coil (per stator coil span) ψ_{PM} is:

$$\varphi_{p\max} = B_{g\text{PM}} \frac{\pi D_{i0s}}{2p_a}(1 - K_{s01}) l_{\text{stack}} \tag{11.12}$$

The slot opening ratio $K_{s01} = 0.6 - 0.65 = b_{0s1}/(b_{0s1} + b_{0t1})$; b_{0s1}, b_{0t1} are the outer stator slot and tooth spans.

The total (peak) PM flux linkage per phase ψ_{PM1} is:

$$\psi_{\text{PM1}} = \varphi_{\text{PM}av} k_{w1} p_a q_1 n_s \tag{11.13}$$

where $q_1 = \text{slots/pole/phase} = 2$ (in our case) and $n_s = n_c$ is the number of turns/slot (or per coil in the adopted single-layer winding).

The torque expression is simple if the reluctance torque is neglected (as it is small):

$$T_{\text{en}} = \frac{3}{2} p_a \psi_{\text{PM1}} I_{1\text{peak}} \cos\gamma_i \tag{11.14}$$

For simplicity, we consider here $\gamma_i = 0$ to compensate for not including the reluctance torque, which may account for 7–10%.

From Equations 11.3 and 11.14 with ψ_{pmax} from Equation 11.12, the peak total slot mmf $n_s I_{peak}$ is:

$$n_s I_{peak} = \frac{2T_e}{3k_{w1}p_a q_1 B_{gPM}\dfrac{\pi D_{i0s}}{2p_a}(1-k_{as1})l_{stack}p_r}$$
$$= \frac{2\times 5000}{3\times 1\times 2\times 2\times 1.8\times \pi \times \dfrac{0.477}{4}\times 0.4\times 0.25\times 22}$$
$$= 1264.325 \text{ Aturns peak} \tag{11.15}$$

Thus, the total electric average loading is:

$$A_{tpeak} = \frac{N_s(n_s I_{peak})}{\pi D_{i0s}} = \frac{24\times 1264.375}{\pi \times 0.477} = 20260 \text{ Aturns/m} \tag{11.16}$$

Distributing this electric loading between the outer and inner stator is now rather straightforward, as the outer diameter of inner stator D_{i0s} is:

$$D_{0is} = D_{i0s} - 2(l_{PM} + 2h_{bridge}) - 2g$$
$$= 477 - 2(36 + 2\times 0.8) - 2\times 1.5 = 398.8 \text{ mm}$$

$$\frac{A_{outer}}{A_{inner}} \approx \frac{D_{i0s}}{D_{0is}} = \frac{477}{398.8} = 1.196 \tag{11.17}$$

11.2.5 Armature Reaction Flux Density B_{ag} and Ideal Power Factor

A quick check of armature reaction airgap flux density B_{ag} gives:

$$B_{ag} = \frac{\mu_0 6 n_s I_{peak}}{p_a \pi 2g} = \frac{1.256\times 10^{-6}\times 6\times 1264}{\pi \times 2\times 221.5\times 10^{-3}} = 0.5068 \text{ T} \tag{11.18}$$

This is a notable value, but in comparison with $B_{gPM} = 0.8$ T, it leads to a peak flux density in the stator teeth $B_{tmax} = \sqrt{B_{gPM}^2 + B_{ag}^2} = 0.946$ T, which is an acceptable value.

However, at 2 p.u. current, the machine core may notably saturate and thus the airgap may be increased to cope with a strong overload application with high PM demagnetization risks. It is now feasible to calculate approximately the ideal power factor $\cos\varphi_i$:

$$\cos\varphi_i \approx \frac{B_{PMg}}{\sqrt{B_{gPM}^2 + B_{ag}^2(1+K_{leakage})^2}} \tag{11.19}$$

where $K_{leakage} = L_{leakage}/L_{qm}$. A value of $K_{leakage} = 0.2$ is assigned here and thus:

$$\cos\varphi_i \approx \frac{0.8}{\sqrt{0.8^2 + 0.5068^2\times 1.2^2}} \approx 0.7966 \tag{11.20}$$

11.2.6 Rated Current and Number of Turns/Coil

Even the number of turns per slot (and per phase $W_s = p_a q_1 n_s = 4n_s$ in both stators) may be computed by using the expression:

$$\psi_{PM1}\omega_1/\cos\varphi_i \approx V_s;\quad V_s = \frac{V_L\sqrt{2}}{\sqrt{3}} = 220\sqrt{2} \tag{11.21}$$

with $\omega_{1n} \approx 2\pi n p_r = 2\pi(30/60) \times 22 = 69.08(\text{rad/s})$:

$$\psi_{\max} = \frac{V_s \cos\varphi_i}{\omega_{1n}} = \frac{220\sqrt{20.89}}{69.08} = 4.026 \text{ Wb} \tag{11.22}$$

So, from the torque expression (11.14), the peak current per phase is:

$$I_{1\text{peak}} = \frac{2 \cdot T_{\text{en}}}{3 \cdot p_r \cdot \psi_{\max}} = \frac{2 \times 5000}{3 \times 22 \times 3.76} = 37.625 \tag{11.23}$$

Consequently, the number of turns per "slot pair" n_s is:

$$n_s = \frac{n_s I_{1\text{peak}}}{I_{1\text{peak}}} = \frac{1264.375}{37.625} = 33.58 \approx 33 \text{ turns} \tag{11.24}$$

So $I_{1\text{peak}} \approx 38\text{A}$ (peak value).
For a phase:

$$W_s = p_a \cdot q_1 \cdot n_s = 2 \cdot 2 \cdot 33 = 132 \text{ turns/phase} \tag{11.25}$$

Note: By "slot pair," we mean inner and outer stator corresponding slots.

11.2.7 Copper Loss and Electromagnetic Power

The copper loss may be calculated by adopting an average AC coil length l_{coil} and a current density $j_{\text{con}} = 1.5 \text{ A/mm}^2$:

$$l_{\text{coil}} \approx 2 l_{\text{stack}} + \pi\left(\frac{\pi D_{i0s}}{2p_a}\right) = 2 \times 0.25 + \frac{\pi^2 0.477}{4} = 1.675 \text{ m} \tag{11.26}$$

The end coil still amounts to two times more than the active coil length. So, the rated copper losses p_{con} are:

$$\begin{aligned} p_{\text{con}} &= 3\rho_{\text{co}} \frac{l_{\text{coil}}(4n_s)}{\left(\frac{I_{1\text{peak}}}{\sqrt{2} j_{\text{con}}}\right)} \frac{I_{\text{peak}}^2}{2} \\ &= \frac{3 \times 2.1 \times 10^{-8} \times 1.675 \times 4 \times 33 \times 38^2}{2 \frac{38}{\sqrt{2} \times 1.5 \times 10^6}} = 563.1 \text{ W} \end{aligned} \tag{11.27}$$

The electromagnetic power P_{en} is simply:

$$P_{\text{en}} = T_{\text{en}} 2\pi n_n = 5000 \times 2 \times \pi \times \frac{30}{60} = 15700 \text{ W} = 15.7 \text{ kW} \tag{11.28}$$

The ratio of copper losses per electromagnetic power is only $p_{\text{con}}/P_{\text{en}} = 563/15700 = 3.586$ %. An efficiency well above 0.92 is expected, as core and mechanical losses should be moderate, too.

11.2.8 Stator Slot Area and Geometry

The active slot area of the outer stator slot (with 18 turns in the outer stator slots and 15 turns in the inner stator slot; 33 turns per "slot pair" as in Equation 11.24) is:

$$A_{\text{slot0}} \approx \frac{n_s(I_{\text{peak}}/\sqrt{2}) \times 18/33}{j_{\text{con}}K_{\text{fill}}} = \frac{(1264.375/\sqrt{2})18/33}{1.5 \times 0.45} = 724 \text{ mm}^2 \tag{11.29}$$

For a slot pitch of $\tau_{\text{slot0}} = \pi D_{i0s}/N_s = \pi \times (0.477/24) = 0.062$ m and a slot opening of 35 mm, the outer stator slot will be only at most 24 mm deep (the active part).

Note: We should remember, though, that such a long stack (0.25 m) is not easy to handle in fabrication, as the rotor, sandwiched between the stators, will be that long, and so will be that of the inner stator plus the axial length of coils ends. It might be more practical to "plant" the entire winding in the outer stator slots, which will end up 45 mm + 5 mm = 50 mm deep (the aspect ratio will still be low to ensure a low slot leakage inductance). In this case, fabrication complexity and the effort of cooling of windings may be reduced.

11.3 HARD-LEARNED LESSONS FROM FINITE-ELEMENT METHOD ANALYSIS OF VERNIER PERMANENT MAGNET MACHINES

A few recent pertinent FEM case studies on Vernier PM machines and some with experimental verifications have led to conclusions like the following:

- The optimal stator pole pair (p_a) increases with the outer stator external diameter, but is smaller if the pole ratio $PR = p_r/p_a$ increases (Figure 11.3) [8].
- A larger pole ratio *PR* leads to larger torque, but also may lead to a lower power factor (Figure 11.4) [8].
- The $p_a = 2$ (with $q_1 = 1$) ($N_s = 12$ slots with $N_r = 10$ PM rotor pole pairs) seems a practical compromise between high torque density and good power factor.
- The spoke-PM rotor Vernier machine (12/10/2) with two stators may produce twice as much torque as the conventional four-pole surface PMSM in the same volume [12],

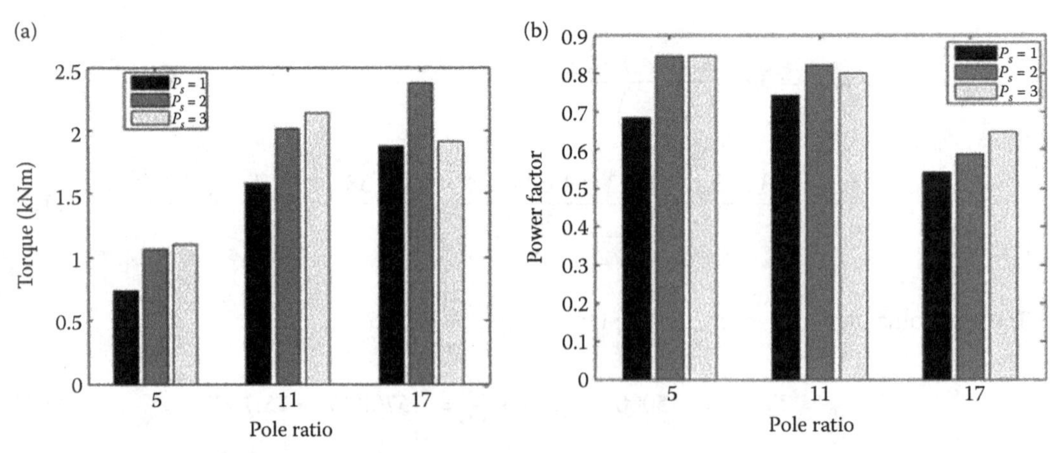

FIGURE 11.4 Influence of PR $= p_r/p_a$ on torque, (a); and on power factor, (b) for $p_a = 1,2,3$. (After D. Li et al., *IEEE Trans. on IA*, vol. 51, no. 4, 2015, pp. 2972–2983. [8])

but the Nm/kg of PM is about the same, and the power factor is smaller (though still satisfactory); however, the machine topology is notably more complex, increasing the fabrication cost.

- There is an optimal stator slot opening ratio (0.6–0.65) to maximize torque (Figure 11.5) [8].
- $r_g^+/g^+ = r_g^-/g^-$(ratios of radius to respective airgap) is required to fully exploit the dual stator spoke PM rotor.
- The surface PM-rotor single-stator Vernier machine may produce about the same torque as the spoke-PM-rotor with dual stators but at a notably smaller power factor [12].
- There is an optimal PM pole ratio $(1 - h_{PM}/\tau_{PM}) \approx 0.58 - 0.64$ [8] for maximum torque, but even for the PM-like ratio of 0.71, the machine may develop high enough torque at $B_{gPMi} = 1.5$ T (ideal flux density in the airgap).
- A large PM airgap ideal flux density $B_{gPMi} = 1.41$ T in the airgap before considering fringing is mandatory for good torque density (66 Nm/liter, here [8]).
- Larger stator slot opening ratios S_{01} and S_{02} (0.65–0.68) lead to higher power factors.
- High-torque machines (2000 Nm, 30 rpm in [8] and 5000 Nm, 30 rpm in our example) are designed at 1.2 (1.5) A/mm^2 current density, which means that more copper is used, as the useful active PM airgap flux density is not $B_{g1} = 1.43$ T but $B_{g1}K_{fringe}$ (fringe factor $K_{fringe} \approx 0.55$), which is only $B_{gPM} = 0.8$ T.
- Longer stator stacks $l_{stack}/(\pi D_{0so}/2p_a) \approx 1$ may be tempting to further reduce copper weight (and losses), but then the longer inner stator length, including end-coil axial length, would pose serious fabrication problems. This only shows that the torque was obtained largely by increasing the airgap diameter.
- The peak radial forces, which are determined by $B_{gPMi} \approx 1.41$ T, are large and thus noise and vibration will be important.
- The cogging torque is not so small, especially with thick spoke-shaped Ferrite-magnets (thick magnets are needed to avoid their demagnetization at 2 + p.u. current). Rectangular-area magnets influence the flux distortion in the inner airgap differently and thus torque ripple may be as high as 20%. However, very good performance has been reported (with notably longer [radially] magnets [l_{PM} longer]) [13].
- The surface single-stator Vernier PM machines [14], even in consequent PM pole stator and rotor configuration [10], may not match the dual stator spoke-shaped PM rotor Vernier machine in the torque density, though they make better use of the PMs and may reduce the machine initial cost via a simplified topology. Power factor is also to be investigated here.

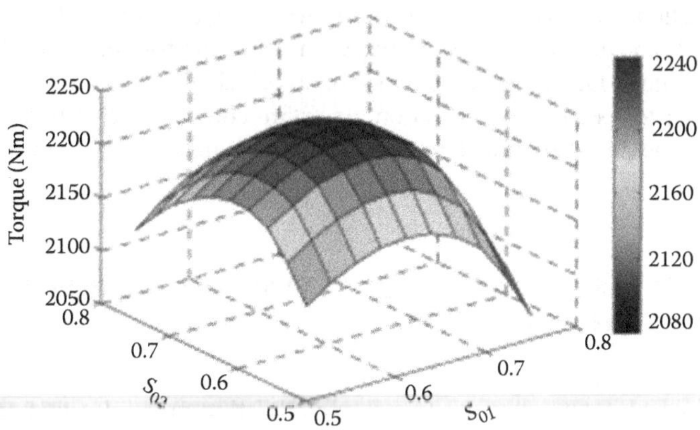

FIGURE 11.5 Torque versus slot opening rotors S_{01} and S_{02} (for $B_{gPMi} = 1.43$ T). (After D. Li et al., *IEEE Trans. on IA*, vol. 51, no. 4, 2015, pp. 2972–2983. [8])

- The axial-airgap alternative configuration of spoke-shaped PM (even with Ferrites)-rotor twin-stator Vernier machine might be easier to manufacture.
- The power factor was already raised to a reasonable value as well as efficiency, but the machine initial cost (fabrication cost included) has to be carefully addressed, as Vernier PM machines are not otherwise likely to be less expensive (in Nm/USD for a given efficiency and power factor) than multiple-pole SPMSMs with tooth-wound coils for PM pole pitches above $\tau_{PM} > 30$ mm, even in large torque applications, as already demonstrated for SF-PMSMs [15]. The same is true for consequent Vernier PM motors [11] in terms of power factor and efficiency, though at higher torque for the same geometry.

11.4 VERNIER PERMANENT MAGNET MACHINE CONTROL ISSUES

The rather sinusoidal emf (as for flux reversal PM machines; see Chapter 10) and the small dq (orthogonal) magnetic anisotropy ($1 > L_{dm}/L_{qm} > 0.85$) allows for dq-modeling for transients and control. The machine may be treated almost like an SPMSM, with a synchronous inductance $L_s \approx L_d \approx L_q$ easier to saturate [10].

The machine reactance $\omega_1 L_s$ in p.u. is larger than in SPMSM, so these machines show better flux weakening capability (wider but not excessive CPSR) and may sustain 2 + p.u. overcurrents with no demagnetization risk, if properly designed, even with Ferrite-PMs. This is partly due to the rather large PM flux fringing and the fact that most armature reaction mmf field lines go parallel to the spoke-shaped PMs in the rotor.

The FOC, DTFC, and scalar (V/f, I-f) control science heritage of AC (SPMSM-like) motor/generator drives, with stabilizing loops, may be used here. As saliency is small and load dependent, encoderless control should be carefully pursued to avoid notable errors in the rotor position observers.

11.5 VERNIER PERMANENT MAGNET MACHINE OPTIMAL DESIGN ISSUES

Analytical preliminary design methodologies such as the one in this chapter have been introduced to produce a design start-up geometry not very far from the optimal one.

With enough multiframe computer hardware available, FEM-only-based optimal design cases that integrate dq circuit models with FEM imported parameters for transients and control design, for a wide speed range and using multiobjective optimization algorithms, are the way to go, though it still takes tens of parallel computation hours.

For less computation time (10 times less), nonlinear analytical model-based optimization, followed by a few FEM runs for the optimal geometry, should be performed to check key performance; if not met, adequate fudge factors (based on regression principles) are introduced to adjust the analytical model in order to speed up optimization algorithm convergence. It is too soon to tell if and when Vernier PM machine drives will reach international markets, but strong R&D efforts are still underway.

11.6 COMBINED VERNIER AND FLUX-REVERSAL PERMANENT MAGNET MACHINES

Reference 10 describes “a novel stator and rotor dual PM Vernier motor.” In fact, it combines, in a single interior stator and outer rotor, the two flux-modulation principles that animate the Vernier machine (stator modulator-consequent pole PM rotor) and the flux-reversal (flux-switching) effect by consequent PMs on the 3k large stator poles that host 3k nonoverlapping coils (Figure 11.6). The machine may also be built in interior rotor topology.

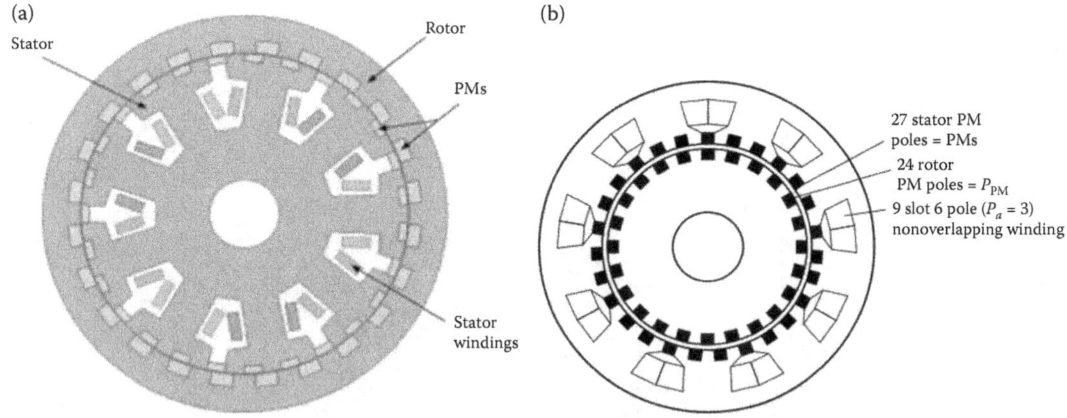

FIGURE 11.6 Combined Vernier + flux-reversal: (a) with outer rotor; (b) with inner rotor.

The machine fabrication is notably simplified, as it has a single stator and the windings use non-overlapping coils (as in flux-reversal or FS-PMSMs), allowing shorter frame length for a given airgap diameter while still showing moderately long end coils.

The two conditions for synchronous operation and nonzero average torque of flux-modulation machines (Chapter 1) still hold:

$$\begin{aligned} p_a &= p_m - p_{PM} \\ p_{PM}\omega_{PM} - p_m\omega_m + (p_m - p_{PM})\omega_a &= 0 \end{aligned} \quad (11.30)$$

where:

p_a, ω_a—pole pair and mechanical speed of stator winding mmf
p_m, ω_m—pole pairs and mechanical speed of flux modulator
p_{pm}, ω_{pm}—pole pairs and mechanical speed of PM excitation

With ω_m (the stator is the flux modulator), a first working frequency ω_{a1} is:

$$\omega_{a1} = -p_{PMr}\omega_{PMr}/(p_m - p_{PM}) = p_{PMr}\omega_{PMr}/p_a \quad (11.31)$$

In this machine, however, dual flux modulation is exercised by the consequent pole PM/teeth structures of the rotor and the stator.

Looking from the rotor side (Vernier principle), we have already obtained Equation 11.31. But, from the stator side (flux-reversal principle), the p_m pole pair stator consequent pole PMs are modulated by the p_r consequent pole pairs of anisotropic rotor, where ω_{pm} (stator PMs produce a fixed magnetic field).

A second working frequency ω_{a2} is obtained:

$$\omega_{a2} = p_{mr}\omega_{mr}/(p_{mr} - p_{PMs}) \quad (11.32)$$

with $p_{mr} = p_{PM}$, $\omega_{mr} = \omega_r$, $p_{PMs} = p_{ms}$. It is evident that ω_{a1} and ω_{a2} will have the same pole pair ($p_{PMr} - p_{PMs}$), and thus the two flux-modulation parts add the emfs and torques.

There are, for example, 27 PMs (and 27 teeth) in the stator and 24 PMs (and 24 teeth) in the rotor. The nonoverlapping winding forms a six pole ($p_a = 3$) mmf (9/6) typical combination in tooth-winding coil-PMSMs.

As visible in Equation 11.31, the machine speed ω_{PMr} is low, since $p_a < p_{PMr}$ and thus the stator fundamental frequency is $f_a = p_{PMr}n$ (n-speed in rps), with ω_a, ω_m, ω_{PM} as mechanical angular speeds.

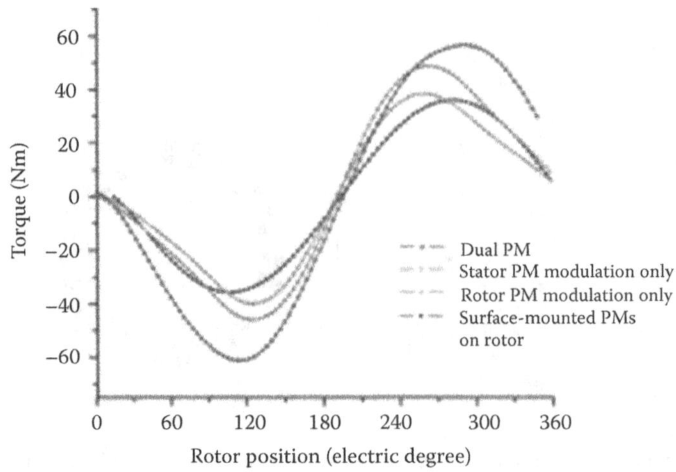

FIGURE 11.7 Static torque waveform for dual PM (Vernier + FR) machine. (After S. Niu, S. L. Ho, W. N. Fu, *IEEE Trans. on MAG*, vol. 50, no. 2, 2014, pp. 7019904. [10])

For an outer rotor outer diameter of 213.4 mm and 50 mm stack length, airgap $g = 0.6$ mm, $B_r = 1.1$ T, 1.5-kW, 250-rpm, 75-V (100-Hz) machine, calculations by 2D-FEM with sinusoidal pure I_q current ideal control show torque results as in Figure 11.7 [10].

The dual-PM (Vernier + FR) machine shows a notably enlarged torque (up to 60 Nm).

Also, the no-load airgap flux density in the airgap with PMs only in the rotor and stator flux-modulation (Figure 11.8a) (Vernier machine) and the one with PMs only in the stator and rotor flux-modulation (flux-reversal machine)—Figure 11.8b—are in phase, adding up contributions to torque [10].

As Figure 11.7 shows, the maximum torque with dual PM design is 50% larger than for only stator-PM-modulation, only 8.2% more than for only rotor-PM-modulation, but 60% larger than for a conventional surface PM-rotor motor machine. For 60 Nm and an outer stator diameter of 0.21 m and stack length 0.05 m, a torque density of 34.66 Nm/liter is reported [10].

11.7 SUMMARY

While the results are encouraging, further steps are required to prove the eventual practicality of the dual PM assessment of Vernier machine:

- PM demagnetization risk at 2.0 p.u. pure i_q current.
- Evaluation of efficiency and power factor against machine initial cost.
- Optimal design methodologies integrating time-stepping FEM and circuit model with field-oriented control [10] for a given torque/speed envelope, and an assigned maximum inverter voltage vector $V_{s\max} = V_1\sqrt{2}$.
- To reduce copper losses, fractionary-slot Vernier PM machines have been proposed; while the torque production is acceptable, the power factor is still low with surface PM rotors [16].
- Directly driven wind generators have been targeted for Vernier PM machines. For 3 kW/350 rpm/163.3 Hz a $N_s = 36$ slots, $p_s = 8$, $p_r = 28$, such a machine, with spoke-shaped Ferrite-rotor, good performance in terms of torque density and losses has been proven, but at a notably smaller power factor than a conventional tooth-wound PMSG [17].
- Sensorless vector or direct torque control of concentrated (or distributed) winding Vernier PM machines is very similar to that of standard permanent magnet synchronous generators (PMSMs) [18].
- Gramme-ring (toroidal) winding dual-rotor axial-airgap Vernier machines for $p_s = 2$ stator mmf pole pairs ($p_r = 22$, $p_s = 24$ stator slots) have been proposed very recently in order to

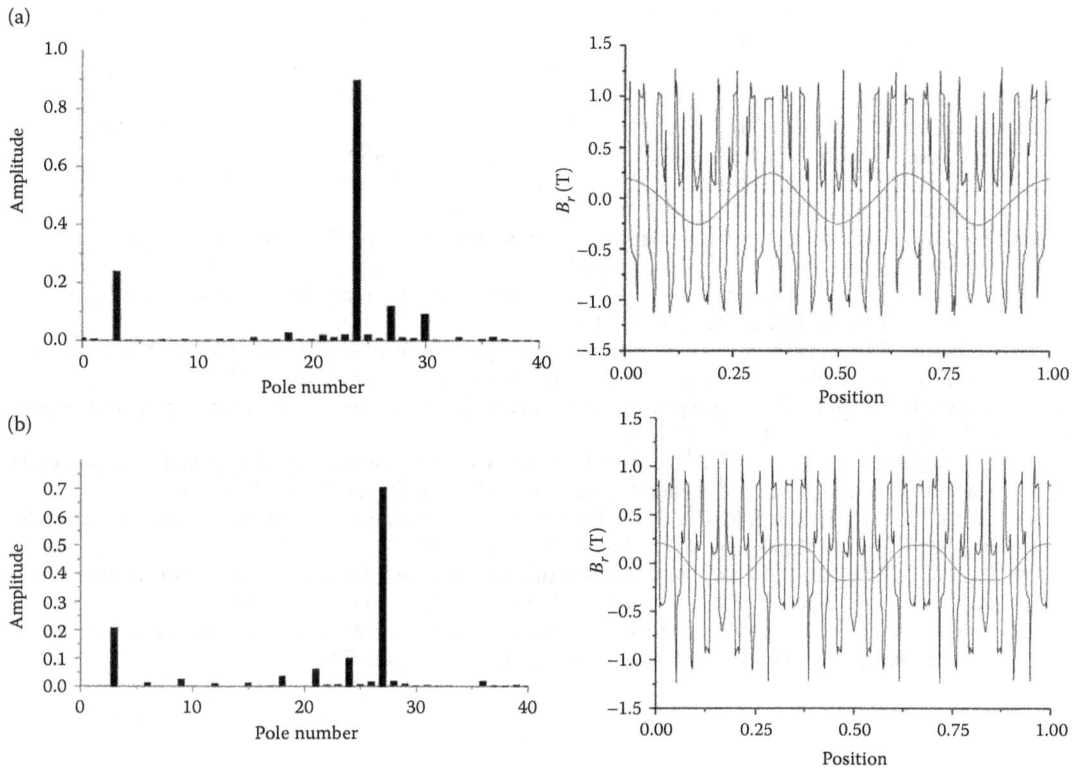

FIGURE 11.8 Airgap flux density at no load: (a) with rotor PMs only and stator FM; (b) with stator PM only and rotor FM only. (After S. Niu, S. L. Ho, W. N. Fu, *IEEE Trans. on MAG*, vol. 50, no. 2, 2014, pp. 7019904. [10])

reduce copper losses [19], but the power factor remains low (less than 0.6), and the efficiency is 0.86 for a 100-Nm, 320-rpm machine at 32 Nm/l torque density. Also, a spoke-Ferrite-rotor dual-stator Vernier machine has been claimed very recently to produce 700 Nm at 400 rpm for 96% efficiency (power factor 0.72) at 23 Nm/l for a 0.355-m stator outer diameter, without PM demagnetization at 3.0 p.u. load current [20].

REFERENCES

1. C. H. Lee, Vernier motor and its design, *IEEE Trans. on PAS*, vol. 82, no. 66, 1960, pp. 343–349.
2. A. Ishizaki, T. Tanaka, K. Takasaki, S. Nishikata, Theory and optimum design of PM Vernier motor, *Record of IEEE-ICEMD*, 1995, pp. 208–212.
3. A. Toba, T. A. Lipo, Generic torque-maximizing design methodology of surface permanent-magnet Vernier machine, *IEEE Trans. on IA*, vol. 36, no. 6, 2000, pp. 1539–1546.
4. D. Li, R. Qu, Sinusoidal back-EMF of Vernier permanent magnet machines, *Record of IEEE-ICEMS*, 2012, pp. 1–6.
5. K. Okada, N. Niguchi, K. Hirata, Analysis of Vernier motor with concentrated windings, *IEEE Trans. on MAG*, vol. 49, no. 5, 2013, pp. 2241–2244
6. B. Kim, T. A. Lipo, Operation and design principles of a PM Vernier motor, *IEEE Trans. on IA*, vol. 50, no. 6, 2014, pp. 3656–3663.
7. Z. S. Du, T. A. Lipo, High torque density Ferrite permanent magnet Vernier motor analysis and design with demagnetization consideration, *Record of IEEE-ECC*, 2015, pp. 6082–6089.
8. D. Li, R. Qu, W. Xu, J. Li, T. A. Lipo, Design procedure of dual-stator spoke-array Vernier permanent-magnet machines, *IEEE Trans. on IA*, vol. 51, no. 4, 2015, pp. 2972–2983.

9. D. Li, R. Qu, Z. Zhu, Comparison of Halbach and dual side Vernier permanent magnet machines, *IEEE Trans. on MAG*, vol. 50, no. 2, 2014, pp. 7019804.
10. S. Niu, S. L. Ho, W. N. Fu, A novel stator and rotor dual PM Vernier motor with space vector pulse width modulation, *IEEE Trans. on MAG*, vol. 50, no. 2, 2014, pp. 7019904.
11. D. Li, R. Qu, J. Li, W. Xu, Consequent-pole toroidal-winding outer-rotor Vernier permanent magnet machines, *IEEE Trans. on IA*, vol. 51, no. 6, 2015, pp. 4470–4480.
12. B. Kim, T. A. Lipo, Analysis of a PM Vernier motor with spoke structure, *IEEE Trans. on IA*, vol. 52, no. 1, 2016, pp. 217–225.
13. Z. S. Du, T. A. Lipo, High torque density ferrite permanent magnet Vernier motor analysis and design with demagnetization consideration, *Record of IEEE-ECCE*, 2015, pp. 6082–6089
14. B. Kim, T. A. Lipo, Design of a surface PM Vernier motor for a practical variable speed application, Record of IEEE-ECCE, 2015, pp. 776–783.
15. A. Fasolo, L. Alberti, N. Bianchi, Performance comparisons between SF and IPM machines with rare-earth and Ferrite PMs, *IEEE Trans. on IA*, vol. 50, no. 6, 2014, pp. 3708–3716.
16. T. Zoa, R. Qu, D. Li, D. Jiang, Synthesis of fractional-slot Vernier permanent magnet machines, *Record of IEEE-ICEM*, 2016, pp. 911–917.
17. J. F. Kolzer, T. P. M. Bazzo, R. Carlson, F. Wurtz, Comparative analysis of a spoke ferrite permanent magnets Vernier synchronous generator, *Record of IEEE-ICEM*, 2016, pp. 218–224.
18. T. Zoa, X. Han, D. Jiang, R. Qu, D. Li, Modeling and sensorless control of an advanced concentrated winding Vernier PM machine, *Record of IEEE-ICEM*, 2016, pp. 1112–1118.
19. T. Zou, D. Li, R. Qu, J. Li, D. Jiang, Analysis of dual-rotor, toroidal-winding, axial-flux Vernier permanent magnet machine, *IEEE Trans. on IA*, vol. 53, no. 3, 2017, pp. 1920–1930.
20. Z. S. Du, T. A. Lipo, Torque performance comparison between a ferrite Vernier vector and an industrial IPM machine, *IEEE Trans. on IA*, vol. 53, no. 3, 2017, pp. 2088–2097.

12 Transverse Flux Permanent Magnet Synchronous Motor Analysis, Optimal Design, and Control

12.1 INTRODUCTION

Transverse flux PMSMs were apparently introduced in 1986 [1] in an effort to increase torque density in low and medium variable-speed motors, by separating the magnetic and electric circuit design and thus reducing the pole pitch τ_{PM} down to 10–20 mm even in large torque applications.

The initial single-sided topologies [1–3] are shown in Figure 12.1a and b.

The I-shaped cores between U-lamination stator cores allow for all magnets to be active all the time in a flux line with four airgap crossings. The IPM rotor allows for PM flux concentration, that is, higher torque density. The I-shaped core advantage is paid for in a larger leakage inductance of the machine. The main problem is PM flux fringing, while the main advantage is torque magnification by increasing the number of poles $2p_{PM}$.

Double-sided stator configurations, especially with IPMs, are favorable in reducing the PM flux fringing by making all the magnets active all the time and increasing torque/volume (Figure 12.2).

Further, PM flux fringing may be notably reduced by planting additional lateral magnets on the IPM rotor poles of a single-sided TF-IPMSM (Figure 12.3).

TF-PMSMs operate as synchronous motors with the number of stator U-shaped cores (N_s) equal to the number of rotor PM pole pairs (p_{PM}). They have a modular structure with each phase being independent, and thus notable eddy current losses occur in the magnets and rotor core by the inverse component of the stator mmf field. The fundamental frequency $f_n = p_{PM} \cdot n$ limits the core losses, while dividing the pole PMs into a few pieces circumferentially (and/or axially) will reduce PM eddy current losses drastically. TF-PMSM was included (Chapter 1) in the category of flux-modulation PM machines, where the pole pairs of stator mmf $p_a = 0$:

$$p_a = N_s - p_{PM} = 0 \tag{12.1}$$

The torque magnification effect with the circular coil (phase) is evident if we only imagine increasing the number of stator U-shaped cores and the number of poles in the rotor for a given machine general geometry.

The maximum PM flux in the coil remains about the same with increasing p_{PM}, but its polarity reversal in the coil is performed in a shorter time (angle) and thus a larger emf and torque (for same coil mmf) are produced.

As the pole pitch τ_{PM} decreases (with increasing p_{PM}), PM fringing is enlarged such that there is an optimum in terms of maximum emf (torque) for a given machine geometry and copper loss for a certain (large) number of rotor poles.

When p_{PM} increases, for a given speed, frequency f_n increases and thus core losses increase. All these aspects have to be considered carefully in designing TF-PMSMs, as it seems that, again, $g + h_{PM}/\tau_{PM} < 1/2$, to secure high torque density. TF-PMSMs suffer from the same drawback of a

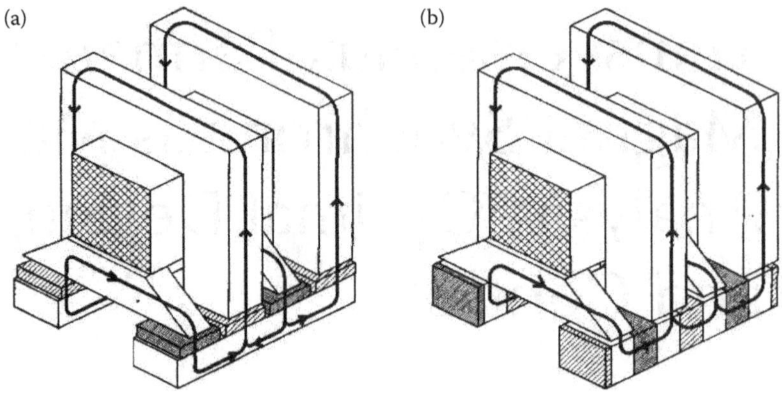

FIGURE 12.1 Single sided TF-PMSMs (one phase is shown) (a) with surface PMs; (b) with interior (spoke) PMs.

lower power factor for higher efficiency (due to coils' circular shape), mainly because of PM fringing. In spoke-shaped PM-rotor configurations, the power factor may be kept within reasonable values.

Besides rotor-PM topologies, TF-PMSM was investigated for stator-PM configurations also, with a variable reluctance passive rotor (Figure 12.4) [4].

Also, the outer rotor configuration [4] may be preferred; see Figure 12.5.

The stator-PM configurations (Figure 12.4) pose fabrication problems. SMC cores are used because the flux lines are 3D in the variable reluctance rotor.

However, in low-speed applications, when rated frequency is below 50 (60) Hz, a stator-PM configuration with radial-airgap and shifted regular laminations in the stator may be used to make the machine more manufacturable (Figure 12.6).

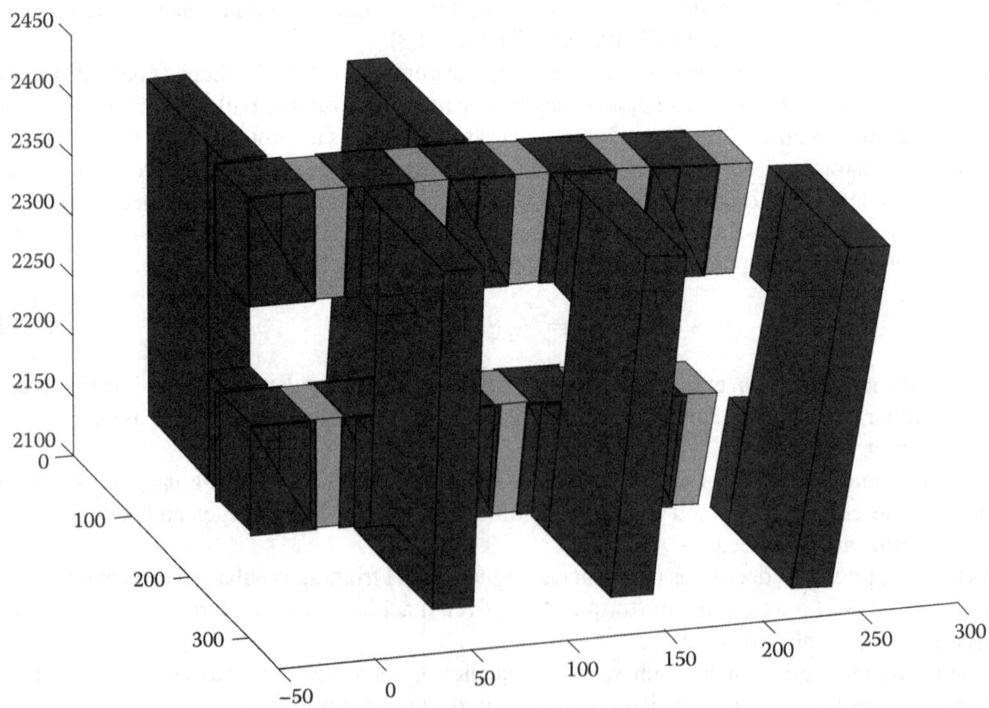

FIGURE 12.2 Double sided TF-IPMSM (one phase is shown).

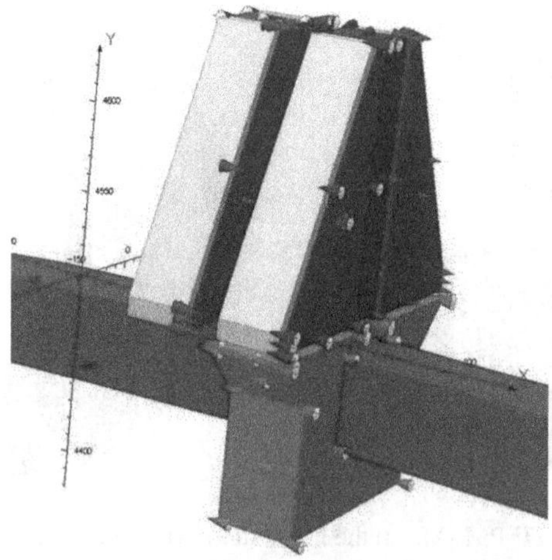

FIGURE 12.3 Single sided TF-IPMSM with antifringing additional PMs.

The stator flux barriers filled with PMs also play the role of slits that reduce the eddy currents produced by magnetic field lines at 90° with respect to laminations (mainly in the yoke zone). Also, the variable reluctance rotor that experiences axial and radial flux line portions, though made of axial laminations rolled over one another, has thin slits to reduce the core eddy currents (see view of C in Figure 12.6). Heavy PM flux concentration with all PMs active at any time is obtained, as in TF-IPMSMs with dual stators (Figure 12.2). The configuration in Figure 12.6 is suitable for easier (less costly) fabrication.

It is also possible to use a modular claw-pole SMC stator configuration with an ordinary surface PM rotor (2, 3 phases [modules] are placed axially), especially for small powers, in order to produce a compact motor of low and medium speed (less than 3000 rpm, in general [5]); see Figure 12.7.

Reference 5 shows that for 20-tone, force-produced, low-density (5.8 g/cm^3) soft magnet composites, stator core fabrication costs and core losses are reasonable. At 60 W, 3000 rpm, the average core losses are already 6 W (10%) of rated power, so efficiency is below 90%.

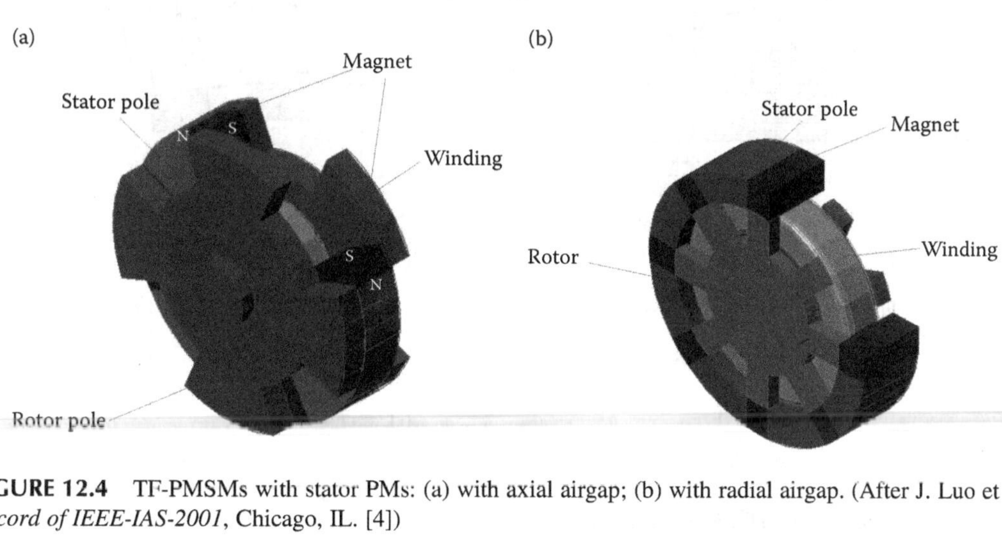

FIGURE 12.4 TF-PMSMs with stator PMs: (a) with axial airgap; (b) with radial airgap. (After J. Luo et al., *Record of IEEE-IAS-2001*, Chicago, IL. [4])

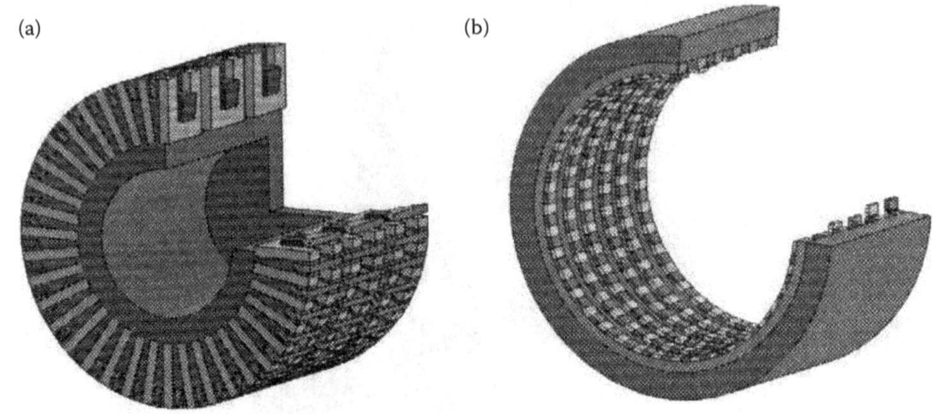

FIGURE 12.5 The outer rotor TF-PMSM: (a) the inner stator; (b) outer rotor. (After I. Boldea, *Variable Speed Generators*, 2nd edition, book, CRC Press, Taylor and Francis, New York, 2015 (Chapter 11). [2])

Efforts to reduce cogging torque in TF-PMSMs refer to increasing the number of phases, which are anyway shifted by $2\pi/m$ radians (in the stator or rotor) or by skewing PM placement on the rotor or stator.

Using herringbone teeth [6] on a double C-hoop stator [7] may also lead to torque ripple reduction for 7%–8% of average torque reduction, though [8–10].

In yet another design (Figure 12.8, after [8]), the IPM rotor with axially straight spoke-shaped magnets interacts with stator U-shaped cores that are skewed by one pole pitch.

The stator may be made of rectangular distinct axial laminations packs for stator slot walls (teeth) and yokes. Still, the stator and rotor need to be encapsulated in nonmagnetic frames, in contrast with the stator-PM TF-IPMSM in Figure 12.6. One important aspect of TF-PMSM is that it is made of m ($m \geq 2$) axially placed (and shifted) single-phase modules. So, even if a multiphase TF-PMSM has a total small torque ripple (5% or so), each phase-part (module) of the machine experiences higher

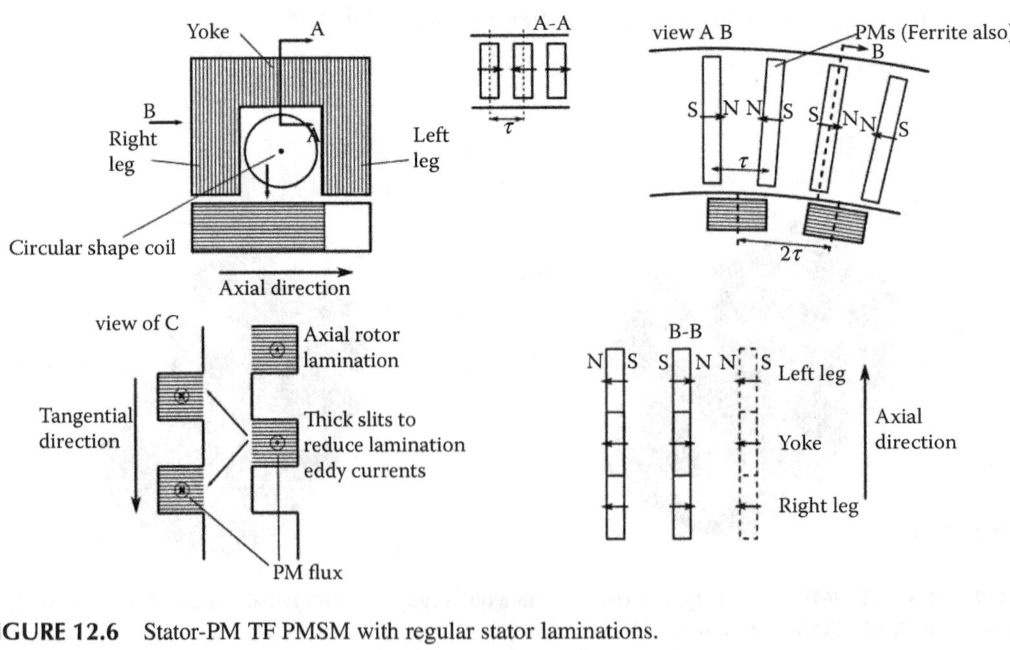

FIGURE 12.6 Stator-PM TF PMSM with regular stator laminations.

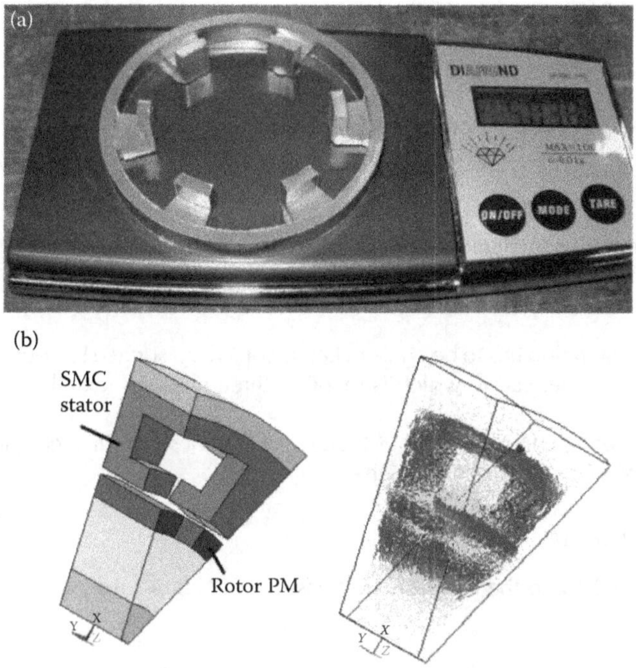

FIGURE 12.7 Claw-pole-SMC-stator TF-SMPSM (one phase is shown): (a) molded SMC claw pole stator (half of it); (b) 3D view.

torque ripple. So, the noise and vibration aspects are to be given special attention. In what follows, we will treat in more detail:

- A preliminary nonlinear analytical design methodology (to produce a start-up geometry for eventual more involved MEC modeling)
- Magnetic-equivalent circuit modeling
- MEC-based optimal design methodology via a case study
- 2D and 3D FEM characterization
- TF-PMSM control issues

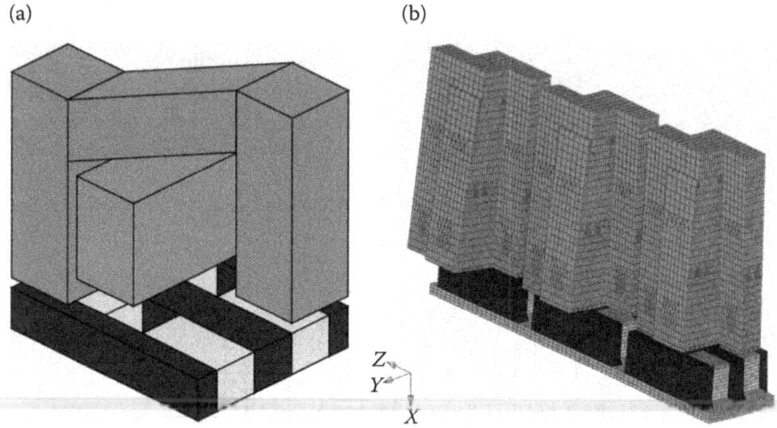

FIGURE 12.8 Twisted-stator cores straight-IPM-rotor TF-PMSM (a) 3D-view—one phase; (b) 3D mesh for FEM analysis. (After E. Schmidth, *IEEE Trans.*, Vol. MAG-47, no. 5, 2011, pp. 982–985. [8])

12.2 PRELIMINARY NONLINEAR ANALYTICAL DESIGN METHODOLOGY

EXAMPLE 12.1

Let us consider a set of specifications of a TF-PMSM:

$T_{en} = 5000$ Nm
$n_n = 75$ rpm, $f_n = 50$ Hz
$V_{nL} = 380$ V (rms)—line voltage (star connection of the 3 phases)

Solution

First, a suitable configuration has to be chosen. Let us consider a surface PM outer rotor with U- and I-shaped cores in the inner stator, which is made of three modules (axially), one for each phase (Figure 12.9).

The I-shaped cores help in making all PMs active at all times and thus reducing the PM fringing flux to keep the PM airgap flux density high.

12.2.1 Stator Geometry

With $\lambda_{st} = l_{\text{stack}}/D_{is} = 0.1$ and the rotor shear stress $f_{tn} = 4\ \text{N/cm}^2$, the outer stator diameter D_{is} is:

$$D_{is} = \sqrt[3]{\frac{T_{en}}{3\pi \cdot \lambda_{st} \cdot f_{tn}}};\ T_{en} = \pi \cdot D_{os} \cdot f_{te} \cdot \frac{D_{is}}{2} \cdot 3 \cdot 2 \cdot l_{\text{stack}} \tag{12.2}$$

Three modules (phases) have been considered, and l_{stack} is the axial length of one leg of the U-shaped stator.

So:

$$D_{is} = \sqrt[3]{\frac{5000}{3 \cdot \pi \cdot 0.1 \cdot 4 \cdot 10^4}} = 0.51\ \text{m} \tag{12.3}$$

Thus, the U-shaped leg axial length $l_{\text{stack}} = \lambda_{st} \cdot D_{is} = 0.1 \cdot 0.51 = 0.051$ m.

With a pole pitch $\tau_{\text{PM}} = 0.020$ m, the number of rotor poles $2p_m$ is:

$$2p_m = \frac{\pi D_{is}}{\tau_{\text{PM}}} = \frac{\pi \cdot 0.51}{0.02} = 80;\ p_{\text{PM}} = 40 \text{ pole pairs} \tag{12.4}$$

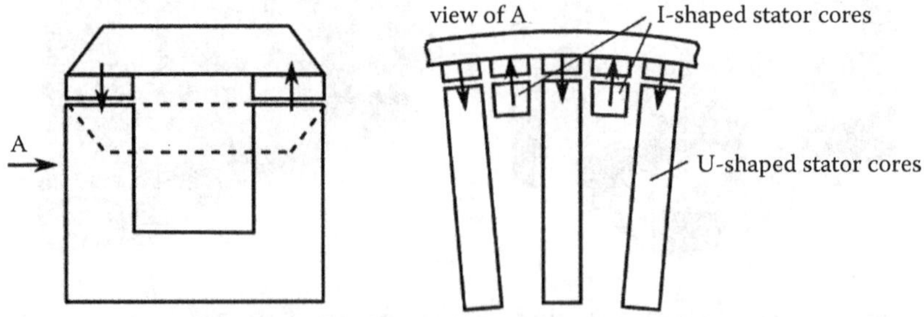

FIGURE 12.9 Surface-PM outer-rotor TF machine with U- and I-shaped stator cores.

So the frequency is indeed:

$$f_n = p_{\mathrm{PM}} \cdot n = 40 \cdot \frac{75}{60} = 50 \text{ Hz} \tag{12.5}$$

We already know from FR-PMSMs (Chapter 10) that $h_{\mathrm{PM}} + g < \tau_{\mathrm{PM}}/2$. So, if we choose $g = 1.5$ mm (airgap), $h_{\mathrm{PM}} = 6$ mm would be a good choice.

With medium-quality PMs, $B_r = 1.13$ T, $\mu_{\mathrm{rec}} = 1.07\,\mu_0$ (still sintered NdFeB). Consequently, the ideal PM airgap flux density $B_{g\,\mathrm{PM}\,i}$ is:

$$B_{g\,\mathrm{PM}\,i} = B_r \frac{h_{\mathrm{PM}}}{h_{\mathrm{PM}} + g} = 1.13 \cdot \frac{6}{6 + 1.5} = 0.904 \text{ T} \tag{12.6}$$

Now, assuming that PMs span $b_{\mathrm{PM}}/\tau_{\mathrm{PM}} = 0.9$ of the pole pitch and the U (and I) stator cores cover $b_{us}/\tau_{\mathrm{PM}} = 0.8$ of the pole pitch, with PM flux fringing coefficient $K_{\mathrm{fringe}} = 0.25$ (surface PMs with I-shaped additional cores) and neglecting magnetic saturation, the maximum PM flux/PM pole $\varphi_{\mathrm{PM\,max}}$ is:

$$\begin{aligned}\varphi_{\mathrm{PM\,max}} &= B_{g\,\mathrm{PM}\,i} \cdot \frac{b_{\mathrm{PM}} + b_{us}}{2} \cdot l_{\mathrm{stack}} \frac{1}{1 + k_{\mathrm{fringe}}} \\ &= 0.904 \cdot \frac{1.7 \cdot 0.02}{2} \cdot 0.051 \cdot \frac{1}{1.25} = 6.27 \cdot 10^{-4} \text{ Wb}\end{aligned} \tag{12.7}$$

As with other F-M PM machines, the PM flux in the coil (phase) switches polarity from $+\,n_c \cdot p_{\mathrm{PM}} \cdot \varphi_{\mathrm{PM\,max}} (= \psi_{\mathrm{PM\,max}})$ to $-\,n_c \cdot p_{\mathrm{PM}} \cdot \varphi_{\mathrm{PM\,max}}$ over one PM pole pitch angle:

$$\theta_{\tau_{\mathrm{PM}}} = 2\pi/2p_{\mathrm{PM}} \tag{12.8}$$

We may suppose at this stage that the PM flux linkage $\psi_{\mathrm{PM}}(\theta_r)$ in a coil (phase) varies as:

$$\psi_{\mathrm{PM}a}(\theta_r) = \psi_{\mathrm{PM\,max}} \cdot \sin(p_{\mathrm{PM}}\theta_r) \tag{12.9}$$

12.2.2 Coil Peak Magnetomotive Force

A third-space harmonic may be added for a flatter variation of $\Psi_{P\,\max}$ with rotor position θ_r, if needed. Now, with three phases and phase currents in phase with their emfs (pure i_q control), the torque T_{en} is:

$$\begin{aligned}T_{en} &= \frac{3}{2} I_{\mathrm{speak}} \left| \frac{\psi_{\mathrm{PM}}(\theta_r)}{d\theta_r} \right| \\ &= \frac{3}{2} I_{\mathrm{speak}} \cdot p_{\mathrm{PM}} \cdot \psi_{P\,\max} \\ &= \frac{3}{2} n_c I_{\mathrm{speak}} p_{\mathrm{PM}}^2 \varphi_{P\,\max}\end{aligned} \tag{12.10}$$

Consequently, the coil peak mmf $n_c I_{\mathrm{speak}}$ is:

$$\begin{aligned}n_s I_{\mathrm{speak}} &= \frac{2}{3} \frac{T_{en}}{p_{\mathrm{PM}}^2 \varphi_{P\,\max}} \\ &= \frac{2}{3} \frac{5000}{40^2 \cdot 6.27 \cdot 10^{-4}} = 3322 \text{ Aturns/coil}\end{aligned} \tag{12.11}$$

The active stator slot area is thus:

$$A_{\text{slot}} = \frac{n_c I_{\text{speak}}/\sqrt{2}}{j_{\text{con}} \cdot k_{\text{fill}}} \tag{12.12}$$

A slot fill factor $k_{\text{fill}} = 0.6$ may be assigned, as there is only one circular coil in slot. Also, we consider $j_{\text{con}} = 3\ \text{A/mm}^2$ and thus (from Equation 12.13) A_{slota} is:

$$A_{\text{slota}} = \frac{3322/\sqrt{2}}{3 \cdot 0.6} \approx 1309\ \text{mm}^2 \tag{12.13}$$

The torque per machine cylinder volume is thus 54 Nm/liter (Figure 12.10).

The I-shaped core radial height $L_{ci} = l_{\text{stack}}/2 = 0.0250$ m. Finally, leaving $h_{si} = 5$ mm distance between the I cores and the coil, the total slot radial weight h_{st} is:

$$h_{st} = h_{sa} + h_{si} + h_{ci} \tag{12.14}$$

The slot aspect ratio w_s/h_{su} has to remain small to yield a small slot leakage inductance to compensate for the large leakage flux inductance corresponding to tangential armature coil field lines between the U- and I-shaped cores (the distance between them is only $(1 - (b_{\text{suli}}/\tau_{\text{PM}}))\tau_{\text{PM}} = 0.2 \cdot \tau_{\text{PM}} = 4$ mm).

Fixing the 40 U- and I-shaped cores via nonmagnetic nonconducting material pieces to the machine frame is an art in itself, as there are strong attraction (normal) forces between the PMs and the latter.

With slot axial width, $W_s \approx 0.8 l_{\text{stack}} = 40$ mm, the total stator slot depth h_{st} Equation 12.14 is:

$$h_{st} = \frac{A_{\text{slot}\,a}}{w_s} + h_{si} + h_{ci} = \frac{1309}{40} + 5 + 25.5 = 63\ \text{mm} \tag{12.15}$$

Note: The twice-deeper-than-necessary slot height, to leave room for I-shaped cores and thus make all PMs active all times and the machine look like it has four airgaps for the PM field, but mainly with

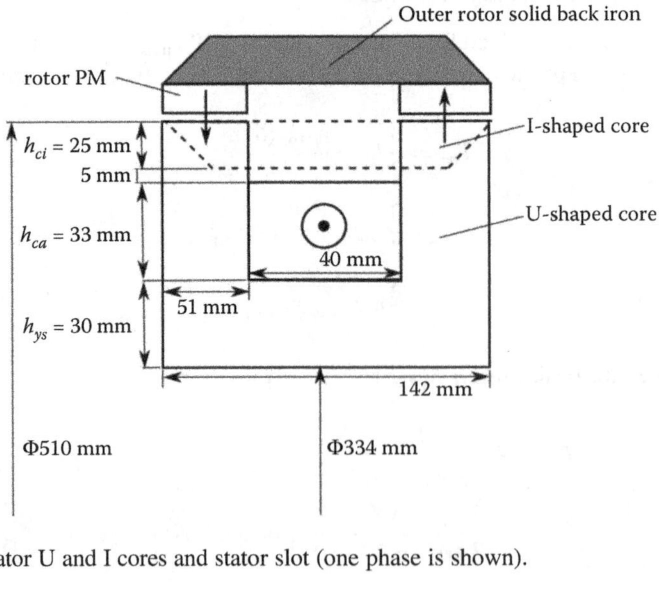

FIGURE 12.10 Stator U and I cores and stator slot (one phase is shown).

only two airgaps for the armature field, should pay off and finally provide a small enough synchronous inductance that allows for a moderate power factor. Also, keeping the airgap diameter the same, an inner stator–outer rotor configuration could have been adopted to further lower the machine volume.

12.2.3 Copper Losses

The copper losses (with sinusoidal current) are:

$$\begin{aligned} p_{\text{con}} &= \frac{3}{2} R_s I_{\text{speak}}^2 = \frac{3}{2} \rho_{co} \frac{l_{\text{coil}} \cdot n_c \cdot I_{\text{speak}}}{1/(\sqrt{2} \cdot j_{\text{con}})} \\ &= \frac{3}{2} \cdot 2.1 \cdot 10^{-8} \cdot 1.4 \cdot 3322 \cdot \sqrt{2} \cdot 3 \cdot 10^6 = 620 \text{ W} \end{aligned} \tag{12.16}$$

with

$$l_{\text{coil}} = \pi(D_{is} - h_{st}) = \pi(0.51 - 0.063) = 1.4 \text{ m} \tag{12.17}$$

But the electromagnetic power P_{en} is:

$$P_{en} = T_{en} \cdot 2 \cdot \pi \cdot n_n = 5000 \cdot 2\pi \frac{75}{60} = 39.250 \text{ kW} \tag{12.18}$$

Consequently, the copper losses represent only about 1.5% of electromagnetic power, which should secure high efficiency (97%) at 75 rpm, as mechanical and core losses (at 50 Hz) may not together surpass 1.5% P_{en}.

The power factor remains the main issue. A simple way to calculate it is to determine an approximate value of the number of turns n_c per coil (phase). For that, we have to notice that the inter-U- and I-shaped core leakage inductance, as the distance between the I and U cores is only 4 mm, while the magnetic airgap $g_m = g + h_{PM} = 7.5$ mm.

So, the synchronous inductance includes three types of flux lines representing it (straight through two airgaps), wondering through four airgaps and the space between the U- and I-shaped cores (tangentially).

So, the armature reaction flux density inductance with the most part as leakage, will be approximately:

$$L_s = L_{lss} + L_{slui} + L_{m2gm} \approx L_{slui} \cdot 1.5 \tag{12.19}$$

$$\begin{aligned} L_s &\approx \mu_0 n_c^2 p_{\text{PM}} \cdot \frac{(h_{ci} \cdot l_{\text{stack}}/2)}{2g_{ui}} \cdot 1.5 \\ &= 1.256 \cdot 10^{-6} \cdot 40 \cdot \frac{0.025 \cdot 0.051}{2 \cdot 4 \cdot 10^{-3}} \cdot n_c^2 \\ &\approx 6 \cdot 0.05 \cdot 10^{-6} n_c^2 (H) \end{aligned} \tag{12.20}$$

As known, the ideal power factor is:

$$\cos\varphi_i = \frac{\psi_{P\max}}{\sqrt{\psi_{P\max}^2 + (L_s i_{\text{peak}})^2}} \tag{12.21}$$

$$\psi_{P\max} = n_c p_{\text{PM}} \varphi_{P\max} = n_c \cdot 40 \cdot 6.27 \cdot 10^{-4} = n_c \cdot 25.08 \cdot 10^{-3} \tag{12.22}$$

But

$$L_s I_{\text{peak}} = 6.005 \cdot 10^{-6} \cdot n_c \cdot 3.322 = 19.99 \cdot 10^{-3} \cdot n_c \tag{12.23}$$

Consequently

$$\cos \varphi_i \approx \frac{25.08}{\sqrt{25.08^2 + 19.99^2}} = 0.782 \tag{12.24}$$

The losses in the machine may drive this value up to $\cos \varphi \approx 0.8$.

Note: Let us remember that about 2/3 of the armature flux will flow in the stator and thus will not contribute to the PM demagnetization. Thus, the presence of I-cores may become practical. A rather conservative design was approached here to serve as a solid start-up for eventual optimal design attempts.

The voltage equation in steady state (with pure I_q control) is:

$$V_{s\max} = \frac{\psi_{P\max}}{\cos \varphi_i} \cdot \omega_1 = \frac{25.08 \cdot 10^{-3} n_c \cdot 3.14}{0.782} = 220\sqrt{3} \tag{12.25}$$

So $n_c = 30.8$ turns/coil ≈ 30 turns/coil.
The rated phase RMS current is then $I_n = ((n_c I_{\text{speak}})/\sqrt{2} n_c) = (3322/\sqrt{2} \cdot 30) = 78.53A$.
As a verification:

$$\eta \cos \varphi_i \approx \frac{P_{en} - P_{\text{mec}}}{3V_{\max} \cdot I_n} \tag{12.26}$$

with $\eta_n = 0.965$ (1.5% for core and mechanical loss has to be fulfilled). On the other hand, the apparent input power (motoring) is:

$$S_n = \frac{3V_{s\max}}{\sqrt{2}} \cdot I_n = 3 \cdot 220 \cdot 78.53 = 51829 \text{ VA} = 51.289 \text{ kVA} \tag{12.27}$$

So, from Equation 12.26, $\cos \varphi_i$ is

$$\cos \varphi_i = \frac{39.520(1 - 0.005)}{0.965 \cdot 51.829} = 0.782 \tag{12.28}$$

QED.

12.3 MAGNETIC EQUIVALENT CIRCUIT METHOD MODELING

Let us consider now the case of an axial-airgap TF-IPMSG for wind energy conversion, with the specifications:

- $P_n = 3$ MW, generated electric power
- $f_n = 50$ Hz, rated output energy
- $n_n = 11$ rpm, rated speed
- $n_{n\,\min} = 9$ rpm, minimum active speed
- $n_{n\,\max} = 13$ rpm, maximum rated speed
- $V_{\text{line}} = 4.2$ kW, nominal line voltage

12.3.1 Preliminary Geometry

The investigated topology (Figure 12.11) has spoke-shaped PMs and dual stator, dual coil, and dual-rotor active sectors per phase.

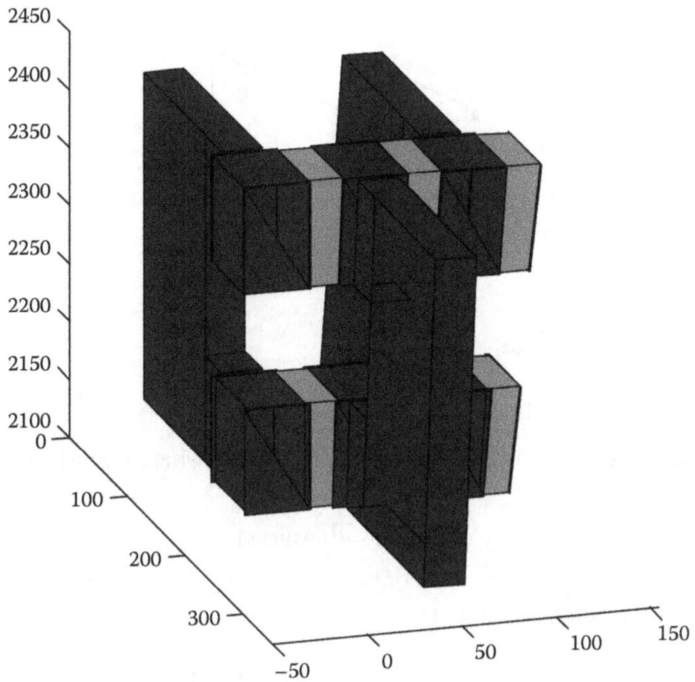

FIGURE 12.11 Axial airgap TF-IPMSM (G).

Making use of a preliminary analytical design methodology—as described in the previous section—a start-up geometry is provided:

- Core outer external diameter $D_{es} = 4900$ mm
- Radial height of stator leg along one airgap: $w_{st} = 95$ mm
- Coil slot width $w_c = w_{\text{slot}} = 40$ mm
- Coil height (axial) $H_c = 51$ mm
- Axial stator yoke $H_{sy} = 60$ mm
- PM axial length $l_{\text{PM}} = 65$ mm
- Airgap $g = 3$ mm

To reduce core weight, the stator and rotor poles represent only 60% of the pole pitch $\tau_{P\,\max} = 26.87$ mm.

- The average PM thickness $h_{\text{PM max}} = 13$ mm, with six stator modules, four of them back to back, in a three-phase machine, the generator total length $l_{ax} = 0.88$ m
- Interior stator diameter $D_{is} = 4440$ mm

Additional geometrical initial data is needed for MEC modeling [11].

12.3.2 Magnetic Equivalent Circuit Topology

MEC modeling is based on flux tubes whose permeance is characterized by uniform flux density distribution. Their magnetic permeance G_m (reluctance R_m) is:

$$G_m = \frac{\Delta\Phi}{\theta}\left(\frac{W_b}{\text{Aturns}}\right) = \frac{1}{R_m} \tag{12.29}$$

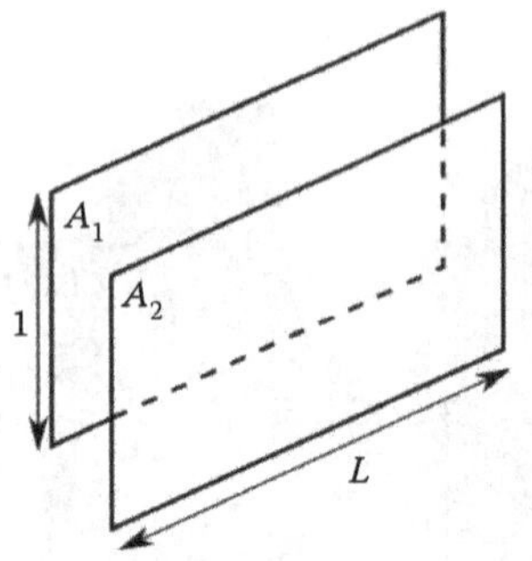

FIGURE 12.12 The generic flux tube.

where $\Delta\Phi$ is the magnetic flux between two chosen flux lines, where θ represents the mmf:

$$\theta = \int \overline{H} \cdot \overline{dl}\,(\text{Aturns}) \tag{12.30}$$

For a magnetostatic problem, the flux is:

$$\Delta\Phi = (A_1 - A_2) \cdot L;\ \theta \approx \Phi \frac{l}{\mu S} \tag{12.31}$$

with A_1, A_2 as vector potential values and l the radial length of U cores (Figure 12.12).

The MEC refers in our case to one pole pair along axis d, to calculate the maximum PM flux density linkage in the coil correctly (Figure 12.13).

12.3.3 Magnetic Equivalent Circuit Solving

MEC contains 16 nodes, which means 15 algebraic equations. The node magnetic potentials will be computed using Thevenin's theorem.

Part of magnetic reluctances is constant, but part of them is influenced by magnetic saturation and thus intensive computation is required. An algorithm to solve the magnetic circuit model is, in fact, needed. The unknowns are the flux densities (and fluxes) in the magnetic tubes.

The problem is with the iron flux tubes, which experience magnetic saturation. The computation process starts with assigned values of flux densities in the iron flux tubes. With these values, iron magnetic initial permeabilities are calculated (the iron magnetic curve is given as a table for interpolation through spline approximation). Then all magnetic permeances are calculated. With known mmfs (PMs are represented by mmfs), the MEC is solved and now flux densities are calculated. For those in iron, new permeabilities are found. For the speed-up computation convergence, an under-relaxation coefficient $K_{sr} = 0.2$ was adopted to define the permeabilities for the new computation cycle:

$$\mu_{\text{new}} = 0.8\mu_{\text{old}} + 0.2\mu_{\text{current}} \tag{12.32}$$

The iterative process continues until sufficient error-limit–based convergence in fluxes is obtained. With known fluxes, $\Phi_{P\,\text{max}}$ and $\Psi_{P\,\text{max}}$ are calculated (for zero currents), but calculations with nonzero currents may also be performed (as F_{mst}; refer to stator mmfs); see Figure 12.13.

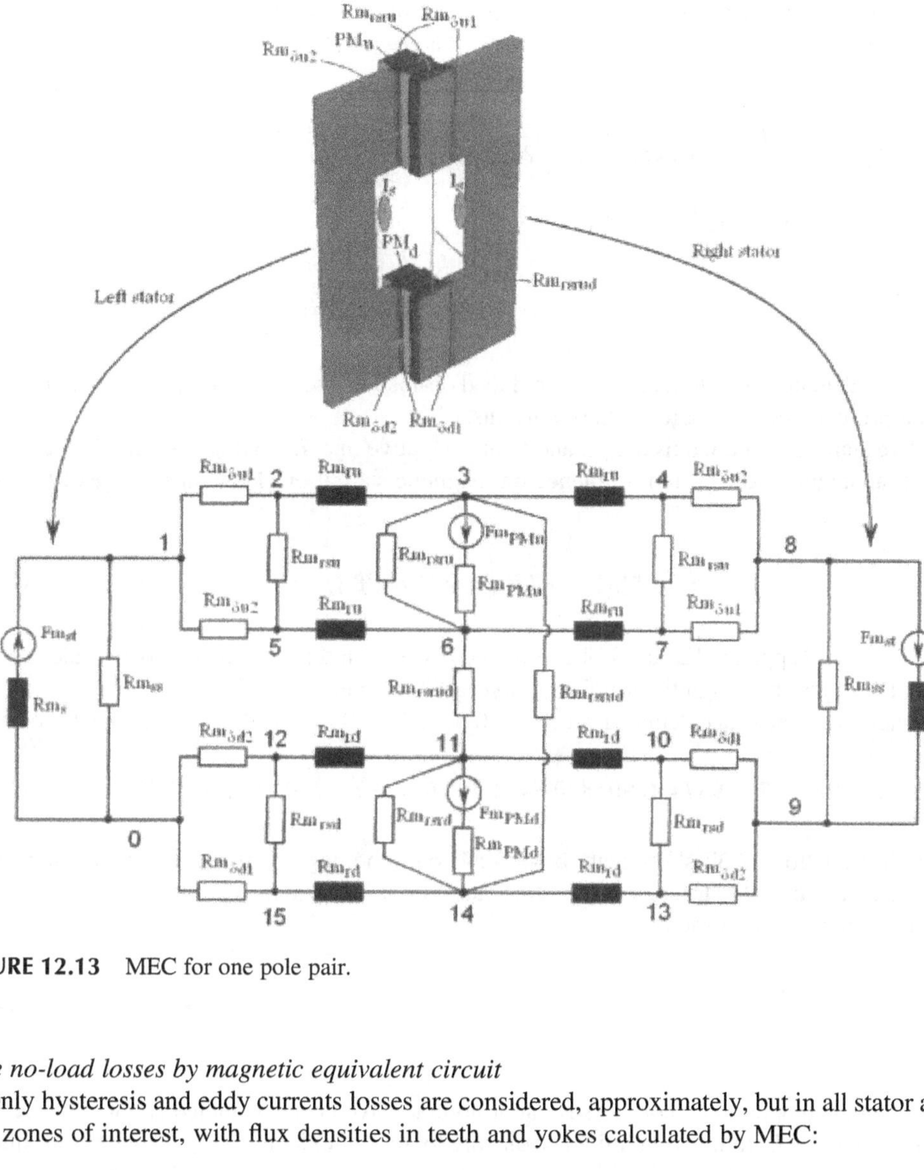

FIGURE 12.13 MEC for one pole pair.

Core no-load losses by magnetic equivalent circuit

Only hysteresis and eddy currents losses are considered, approximately, but in all stator and rotor core zones of interest, with flux densities in teeth and yokes calculated by MEC:

$$\begin{aligned} p_{Fe0,\text{tooth}} &= \left(\frac{B_{\text{stat tooth}}}{1.5}\right)^2 \cdot m_{\text{teeth}} \left[\left(\frac{f_n}{60}\right)^{1.5} \cdot p_{w1.5,60}\right] \cdot K_{\text{fabrication teeth}} \\ p_{Fe0,\text{yoke}} &= \left(\frac{B_{\text{stat yoke}}}{1.5}\right)^2 \cdot m_{\text{yoke}} \left[\left(\frac{f_n}{60}\right)^{1.5} \cdot p_{w1.5,60}\right] \cdot K_{\text{fabrication yoke}} \\ p_{Fe0s} &= 2m(p_{Fe0,\text{tooth}} + p_{Fe0,\text{yoke}}) \end{aligned} \tag{12.33}$$

similarly, in the rotors (on rotors, each with two active sections).

For the case in point (with $p_{w1.5,60} = 3.82$ W/Kg, $K_{\text{fabrication teeth}} = 1.7$, $K_{\text{fabrication yoke}} = 1.3$), the total stator and rotor iron losses in the entire three-phase machine were calculated as: $p_{Fe0s} = 21.82$ kW and $p_{Fe0r} = 27.98$ kW.

The no-load PM flux (for one turn coil) in all poles of a phase: $\Psi_{s10} = \Phi_s \cdot p_{\mathrm{PM}} = 0.4955$ Wb. With an assumed electrical efficiency of $\eta_n = 0.97$, the electromagnetic torque T_{en} is:

$$T_{en} = \frac{P_n}{\eta_n 2\pi n_n} = \frac{3 \cdot 10^6}{0.9 \cdot 2 \cdot \pi(11/60)} = 2.686 \cdot 10^6 \text{ Nm} \tag{12.34}$$

But the electromagnetic torque is also:

$$T_{en} = \frac{m_1}{2} \cdot p_{\mathrm{PM}} \cdot \Psi_{s10} \cdot n_c I_{\mathrm{peak}\ n} K_{\mathrm{red}} \tag{12.35}$$

As Ψ_{s10} is calculated at no load, it may be reduced on-load due to magnetic saturation and changed thus by the presence of, say, pure I_q stator currents.

A positive stator current in axis d, I_{d+}, and then a negative one, I_{d-}, are introduced in the MEC in axis d to simulate the q axis current influence on magnetic saturation. Then an average of the two is used:

$$((\Psi_{s1})_{I_{d+}} + (\Psi_{s1})_{I_{d-}})/2 = (\Psi_{s1})_{I_q} \tag{12.36}$$

At start, K_{red} in Equation 12.35 will be assigned a value, and then it is iteratively calculated by comparing $(\Psi_{s1})_{Iq}$ with Ψ_{s10} until sufficient convergence is met.

The reduction factor K_{red} calculated for (0.25, 0.5, 1, 1.25, 1.5, 1.7, 2)I_{peak} was thus found to be:

$$K_{\mathrm{red}} = [0.9857, 0.9874, 0.9058, 0.8444, 0.7786, 0.7097, 0.6412, 0.5783] \tag{12.37}$$

Results in Equation 12.37 show that above 50% load, the neglect of K_{red} turns into substantial errors, because in this IPM structure, magnetic saturation is important.

The output power P_{out} is then:

$$\begin{aligned} P_{\mathrm{out}} &= \Psi_{s10\ \mathrm{peak}} I_{\mathrm{speak}} \cdot K_{\mathrm{red}} \cdot \omega_1 - m_1 \cdot R_{s1}(I_{\mathrm{speak}})^2 - p_{\mathrm{Feload}} \\ \omega_n &= 2\pi f_n \end{aligned} \tag{12.38}$$

The problem is that only the core losses under no load have been calculated so far. But the MEC allows, again, with $\pm I_d$, mimicking influence on core losses under load. P_{Feload} may be consequently calculated simultaneously with K_{red}. I_{speak} corresponds here to one turn/coil. The number of turns/coil has not been calculated yet here. The rated peak mmf in a phase is $I_{s1n} = 9118$ Aturns/coil (Figure 12.14). There are two coils per phase.

The q-axis magnetization coil inductance is calculated as explained earlier.

$$L_{mq} = \frac{L_{md+} + L_{md-}}{2} = L_{m1n} \tag{12.39}$$

$$L_{md+} = (\Psi_{s1d+} - \Psi_{s10})/I_{d+};\ \ L_{md-} = (\Psi_{s1d-} - \Psi_{s10})/I_{d-}; \tag{12.40}$$

For one turn coil $(L_{mq})_{\mathrm{rated}} = 6.25 \cdot 10^{-5} H$, while the slot leakage inductance $L_{sl} = 0.247 \cdot 10^{-5} H$. The coil inductances for one turn versus its mmf are shown in Figure 12.15.

Again, magnetic saturation may not be neglected. The rated copper losses have been already calculated, as they do not depend on the number of turns/coil, but on coil mmf: $p_{\mathrm{con}} = 86.151$ kW.

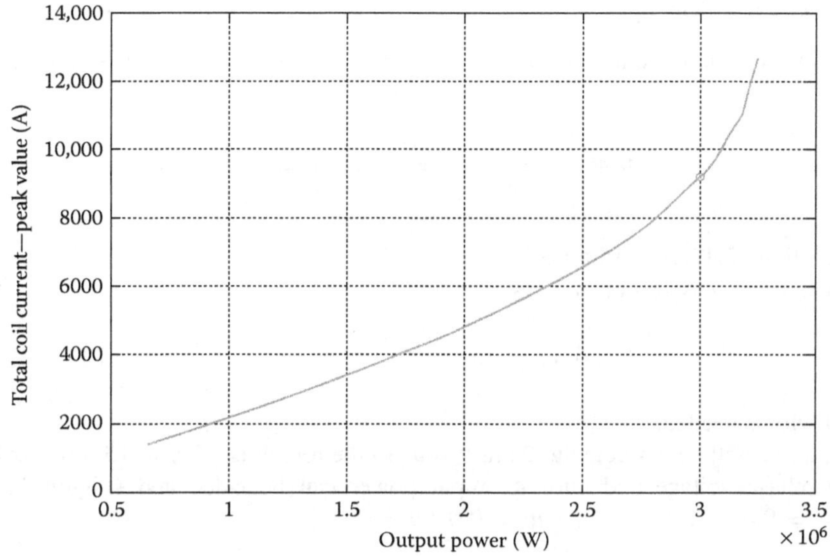

FIGURE 12.14 Coil peak mmf (two coils per phase) versus output power for the 3-MW, 11-rpm AA-TF-IPMSG.

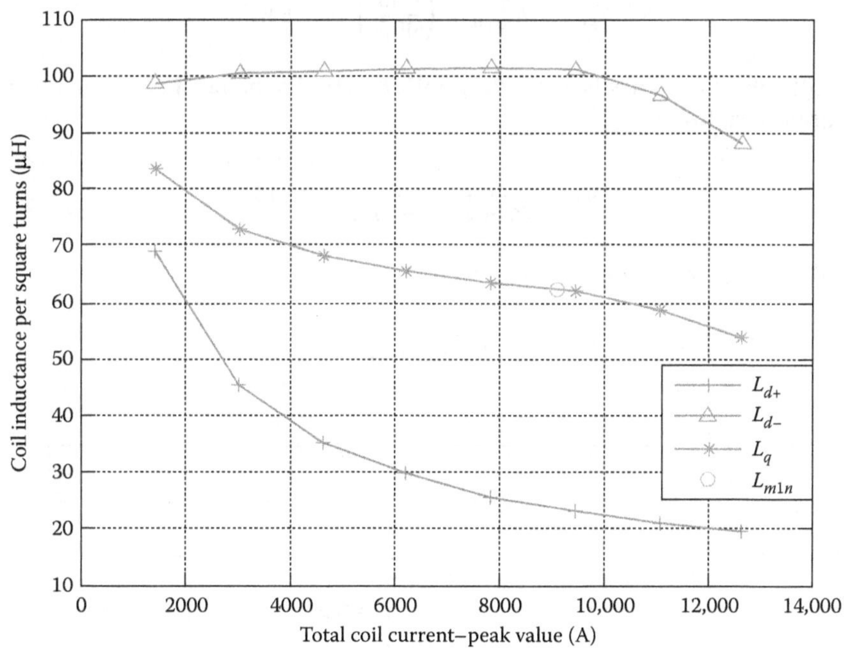

FIGURE 12.15 One-turn coil inductances versus its peak mmf (pure i_q control).

So the rated reduction coefficient K_{red} Equation 12.37 is:

$$K_{\text{red}\ n} \approx \frac{P_n + m_1 R_{s1} I_{s1n\ \text{peak}}^2}{m_1 \omega_1 \Psi_{s10} \cdot I_{s1n\ \text{peak}}} = 0.724 \tag{12.41}$$

12.3.4 Turns per Coil n_c

Approximately, considering only the copper losses, the number of coil turns is simply (the two coils are in parallel):

$$n_c = \frac{\sqrt{2}V_n}{\sqrt{(\omega_1 \Psi_{s10} K_{\text{red }n})^2 + (\omega_1 L_{s\,\ln} I_{s1\text{peak}})^2}} \tag{12.42}$$

From Equation 12.42, $n_c \approx 15$ turns/coil.

The copper wire cross-section A_{wire} is:

$$A_{\text{wire}} = K_{\text{fill}} \cdot \frac{w_s \cdot h_s}{n_c} = 68 \text{ mm}^2 \tag{12.43}$$

Stranded wire is needed.

The voltage V_n will be corrected to 2448 V (due to the rounding of n_c to 15 turns/coil).

Now the phase voltage and current output power can be calculated (Figure 12.16). Also, $R_s = R_{s1} \cdot n_c^2 = 0.077\ \Omega$, $L_{sn} = L_{s1n} \cdot n_c^2 = 0.0159$ H.

12.3.5 Rated Load Core Losses, Efficiency, and Power Factor

A factor K_{pFe} is introduced here to multiply no-load core losses for load conditions:

$$K_{pFe} = K_{\text{red }n}^2 + \left(\frac{L_{m1n}}{\Psi_{s10}}\right)^2 = 1.8467 \tag{12.44}$$

Consequently, the core losses become $P_{Fen} = 91.98$ KW. So the rated electrical efficiency η_{en} is (with zero mechanical losses):

$$\eta_{en} = \frac{P_n}{P_n + p_{\text{con}} + p_{Fen}} = 0.944 \tag{12.45}$$

But then the rated power factor cos φ_n yields:

$$\cos\varphi_n = \frac{P_n}{2m_1 V_{en} I_{fn\text{ coil}}} = 0.4752 \tag{12.46}$$

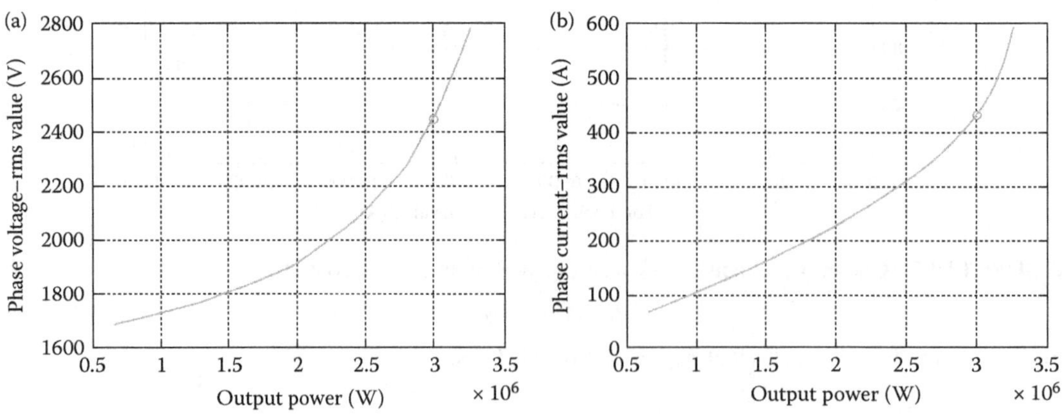

FIGURE 12.16 Phase voltage (a); and coil current (b) versus output power for pure I_q control (two coils in parallel per phase).

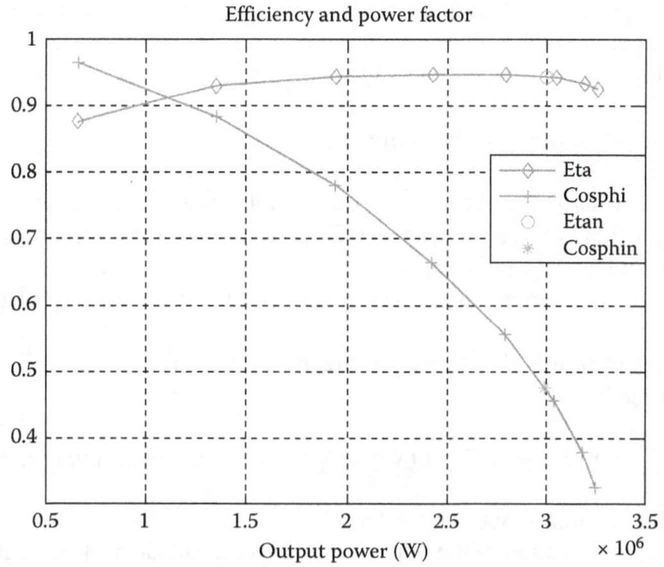

FIGURE 12.17 Power factor and efficiency versus output power.

The variation of power factor with output active power is shown in Figure 12.17.

Note: It is evident that the rated power factor is inadmissibly small. So a last attempt to save the situation would be optimal design in the hope of increasing performance. This was done as explained below.

By optimal design (to follow), electrical efficiency was thus increased to 0.9669, while the power factor was increased to 0.665. It has to be remembered that though this leads to notably larger KVA ratings in the inverter, the latter in this low-speed, super-high-torque application is less expensive than the electric machine (not so in regular or high-speed machines). Also, the greatest part of machine inductance is stator leakage, which does not demagnetize the magnets.

12.4 OPTIMAL DESIGN VIA A CASE STUDY

Let us develop an optimal electromagnetic design methodology and apply it to the case study in the previous section.

Such a methodology implies quite a few stages:

- Specifications
- The electric machine model
- Variables vector with its range
- The optimization algorithm (a modified Hooke–Jeeves algorithm is used here)
- The global objective cost function(s)
- The computer code

12.4.1 The Hooke–Jeeves Direct Search Optimization Algorithm

The main goal of an optimization algorithm is to calculate the chosen objective function $f(\bar{X})$ and find its global minimum, where $\bar{X}$ the variable vector is varied in a given domain.

A point in the pattern search is considered a new point $(\bar{X})$ if $f(\bar{X})$ is smaller than in the previous point. At each step, points around the current one are polled through an a priori number of iterations.

The poll stops when a point with a smaller $f(\bar{X})$ is found. This point is called successful and becomes the new current point at next iteration, while the current mesh size (for search) is increased. If the search fails, the poll is called unsuccessful and the current point stays the same for the next iteration, while the mesh size is decreased.

The Hooke–Jeeves algorithm may be summarized as:

- *Phase a*: Define a starting point $X^{(k-1)}$ in the feasible region and start with a large step size (the entire algorithm is to be run from 15–25 random initial starting points to increase the probability of finding a global optimum).
- *Phase b*: Exploratory moves in all coordinates (of x) are made to find the current base point $X^{(k)}$.
- *Phase c*: A pattern move $X^{(k+1)}$ along a direction from the previous $X^{(k-1)}$ to current $X^{(k)}$ point is performed:

$$X^{(k+1)} = X^{(k)} + a(X^{(k)} - X^{(k-1)}) \text{ with acceleration a} < 1 \tag{12.47}$$

- *Phase d*: The base point is set to $X^{(k-1)} = X^{(k)}$.
- *Phase e*: Tests [computation of the objective function are made to check if improvements in $f(x)$ took place].
  - If $X^{(k+1)}$ yields a smaller $f(x)$, then a new pattern move is performed by jumping at phase c.
  - If $X^{(k+1)}$ is not better ($f(X)$ is not smaller), the algorithm goes to phase *f*.
- *Phase f*: A check is made to see if the search step size should be reduced.
  - If the step size cannot be reduced further (its lower limit was fixed a priori), the algorithm is stopped and $X^{(k)}$ becomes the optimal vector of variables.
  - If the search step size may be further reduced, it will be reduced and the algorithm returns to Phase b.

Feasible initial and range values for the variables vector are required to run the Hooke–Jeeves algorithm.

12.4.2 The Optimal Design Algorithm

The electric generator (it could be a motor) designed in our case is modeled by MEC, and the optimization algorithm stages are presented here, for the sake of simplicity, in a few steps [11].

- *Step 1*: Choose the constants and variables vector $\bar{X}$ to optimize:

$$\bar{X} = [2p_{\text{PM}}, D_{\text{ext}}, w_{st}, w_c, H_c, H_{sy}, l_{\text{PM}}, l_{m\_s}, l_{m\_r1}, l_{m\_r2}] \tag{12.48}$$

  where:
  $2p_{\text{PM}}$ (-)—rotor pole number
  D_{ext} (mm) outer machine core diameter (it is a three-phase axial airgap TF-IPMSG with two circular coils per phase)
  w_c (mm)—stator slot width
  H_c (mm)—stator slot weight
  H_{sy} (mm)—stator U core yoke thickness
  l_{PM} (mm)—axial length (spoke-shaped PM rotor with dual active zone: two airgap zones)
  l_{m_s} ()—ratio between stator pole width and average pole pitch τ_{PM}
  l_{m_r1} ()—ratio between airgap rotor pole width and average pole pitch τ_{PM} at airgap 1
  l_{m_r2} ()—ratio between airgap rotor pole width and average pole pitch τ_{PM} at airgap 2

- *Step 2*: The minimum and maximum values of variables are selected:
  In our case:

$$\begin{aligned}\bar{X}_{\min} &= [44, 4900, 10, 5, 5, 20, 5, 0.2, 0.2, 0.2]^T \\ \bar{X}_{\min} &= [800, 6500, 250, 250, 250, 250, 0.9, 0.9, 0.9]^T\end{aligned} \tag{12.49}$$

  Geometrical and technological constraints are defined as:

$$\bar{g}_1(\bar{X}) = \bar{0}, \ \bar{g}_2(\bar{X}) \leq 0 \tag{12.50}$$

- *Step 3*: The objective scalar function represents the total cost (initial machine + converter + capitalized loss cost plus penalty cost). Its detailed content will be given later in his section, while here only a general description of $f(\bar{X})$ is given:

$$\begin{aligned}& f(\bar{X}) = f_1(X) + f_p(X) \\ & f_1(X) - \text{initial cost} + \text{losses cost} \\ & f_p(X) - \text{penalty cost}\end{aligned} \tag{12.51}$$

  with

$$f_p = \sum_{i=1}^{p+q} f_{p_i}(g_i(\bar{X})) \tag{12.52}$$

  and

$$f_{p_i} = \begin{cases} 0, & \text{if } 1 \leq i \leq p \quad \text{and} \quad g_i(\bar{X}) = 0 \\ 0, & \text{if } i > p+1 \quad \text{and} \quad g_i(\bar{X}) \leq 0 \\ & \text{monotonic positive} \end{cases} \tag{12.53}$$

- *Step 4*: The initial variable vector with its initial ($d\bar{X}_0$) and final ($d\bar{X}_{\min}$) steps of variation:

$$\begin{aligned}\bar{X}_0 &= [546, 4900\text{ mm}, 95\text{ mm}, 40\text{ mm}, 51\text{ mm}, 60\text{ mm}, 60\text{ mm}, 65\text{ mm}, 0.6, 0.5, 0.6] \\ d\bar{X}_0 &= [44, 100, 10, 5, 5, 10, 5, 0.1, 0.1, 0.1] \\ d\bar{X}_{\min} &= [2, 1, 0.1, 0.1, 0.1, 0.1, 0.1, 0.01, 0.01, 0.01]\end{aligned} \tag{12.54}$$

- *Step 5*: Based on the MEC, all geometrical dimensions and performance are calculated, based on which the objective function f_0 is evaluated. The objective function includes three big terms:

$$f(\bar{X}) = C_i(\bar{X}) + C_e(\bar{X}) + C_p(\bar{X}) \tag{12.55}$$

  with:
  C_i—initial machine and PWM converter cost in USD
  C_e—machine and converter energy loss for a given number of years, x hours of most frequent duty cycle of the machine
  C_p—penalty function related to, in general:

  a. Stator and rotor temperature limitation, based on a simplified thermal model and equivalent heat transmission coefficient $\alpha_{\text{heat}} = (14 - 100)\text{W/m}^2/^\circ C$

b. Demagnetization avoidance in most critical operation points

The penalty function components (Equation 12.52) should all be zero in the optimal design $\bar{X}_{\text{final}}$.

- *Step 6*: A first search along each variable in $|\bar{X}|$, with initial step $d\bar{X}$, up and down, in search of a set of points called a mesh around the current point, which is the point computed at the previous computation step. The objective function and its gradient $|\bar{h}|$ are calculated.

$$|\bar{h}| = \left[\frac{\partial f}{\partial x_1}\frac{\partial f}{\partial x_n}\cdots\frac{\partial f}{\partial x_n}\right] \tag{12.56}$$

The partial derivatives in Equation 12.56 are done numerically, by using a three-point evaluation:

$$\frac{\partial f}{\partial x_k} = \begin{cases} \dfrac{f_k - f_0}{dx_k}, & \text{if } f_{-k} \geq f_0 \geq f_k \ (1) \\ \dfrac{f_0 - f_{-k}}{dx_k}, & \text{if } f_{-k} < f_0 \leq f_k \ (2) \\ \dfrac{f_k - f_{-k}}{dx_k}, & \text{if } f_{-k} < f_0 > f_k \ (3) \\ 0, & \text{if } f_{-k} > f_0 > f_k \ (4) \end{cases} \tag{12.57}$$

f_k is the evaluation of f in X_k, obtained from X_0 by moving along the k direction with dX_k, respectively, $-dX_k$ for f_{-k}. Point $\bar{X}_0$ is on the slope for cases 1 and 2 in Equation 12.57, but is worse in case 3, with a better point in case 4.

We should mention that in cases 3 and 4 (in Equation 12.57), the derivative is not strictly valid mathematically, but they proved good enough to decide between leaving the worst point or staying near a good point.

The step along gradient $\bar{\Delta}$ is also calculated:

$$|\bar{\Delta}| = \left(\frac{h_1 dx_1}{|\bar{h}|}, \frac{h_2 dx_2}{|\bar{h}|}, \ldots, \frac{h_n dx_n}{|\bar{h}|}\right) \tag{12.58}$$

- *Step 7*: The variable vector is changed with step $|\bar{\Delta}|$ until the objective function decreases. That is, point P_i is reached in the two-variable vector example in Figure 12.18.
- *Step 8*: The search movement in step 5 is then repeated to find a new gradient direction and then the gradient movement is repeated (step 6) until the search movement is not able to find better points around the current point. This situation corresponds regularly to $|\bar{h}| = 0$, point P_j in Figure 12.18.
- *Step 9*: The variable variation step is reduced by ratio $r < 1$ and the previous steps are done again until the minimum step is reached and the gradient norm vanishes. Then the algorithm is run again from 20 to 30 random starting variable vectors to find rapidly the global optimum with better probability.

Sample results from the MATLAB optimal design code developed for this application are given as optimum design output file and machine characteristics. A total of 73 computation steps per run have been necessary, for 57 minutes of computation time on a contemporary desktop computer (Table 12.1).

The power factor was improved notably (from 0.47 to 0.665); the electrical efficiency increased by 2% (to 0.967), but the outer diameter increased from 4900 to 6500 mm, while the total active weight was reduced to 9345 Kg (from 9910 Kg). Also, the total cost function was reduced by optimal design from 639,100 USD to 466,600 USD (initial cost of machine, energy loss, and inverter cost are included). The prices of copper, laminations, PMs, energy loss, frame cost influence, inverter cost,

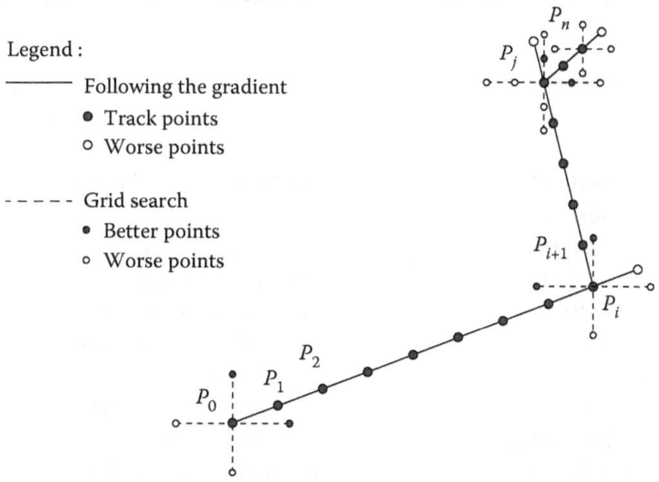

FIGURE 12.18 Search and gradient pattern movement for a two-variable vector in the Hooke–Jeeves algorithm.

hours per years, and number of years in operation were: 10 USD/kg, 5 USD/kg, 50 USD/kg, 0.1 USD/KWh, 7 USD/kg (at active materials weight), 25 USD/KVA of PWM converter, 1500 equivalent hours per year, and 10 years.

The penalty functions for stator and rotor over-temperature and PM demagnetization have used dimensionless coefficients applied to initial machine materials cost C_c: 1, 1.2, 2.0.

Complete comparison tables with MEC assessment design and optimal MEC-based design are offered in Tables 12.2 through 12.4 [11].

12.4.3 Final Remarks on Optimal Design

- It is based on MEC nonlinear model so key FEM validation is still applicable (as proposed in the next section).
- It uses a single composite objective function that includes initial costs (of the machine and PWM converter [per KVA, thus including the extra cost for a lower power factor in the machine]).
- Penalty functions have to be zero in the optimal solution, as proof of the temperature limitation and PM nondemagnetization in the critical operation point (2 p.u. I_q current, for example).
- The optimal design machine geometry, weights, initial costs, efficiency, and power factor depend heavily on the objective function (see [12], where the inverter cost (per KVA) is not included), but the objective function was first the global cost and then maximum efficiency.
- The less than 10 tons of machine active weight here for 3 MW, 11 rpm is dramatically less than the 16.8 tons for 3 MW, 16 rpm [13] with a regular SPMSG with three slots/pole distributed AC windings at a smaller outer diameter, similar efficiency but at above 0.9 power factor!
- As expected, the power factor may be increased further in the optimal design code by simply artificially increasing the price/KVA in the PWM converter. However, in low-speed drives, the electric machine is more expensive than the PWM inverter and thus the objective global function should be carefully chosen, depending on application.
- GA, PSO, ant-colony, and bee-colony optimization algorithms have been used against the Hooke–Jeeves algorithm only to see that the final (optimal) solutions are similar but at less

TABLE 12.1
Optimum Design Output File

Parameter	Value	Description
Electrical Rated Parameters		
V_{cn}	2491.857	[V] Phase rated voltage, rms value
I_{fn}	301.7328	[A] coil rated current, rms value
f_1	57.93333	[Hz] rated frequency
I_{s1n}	6827.434	[A] coil rated mmf, peak value
P_{sipm0}	7.19402	[n] PMs flux no load, peak value
P_{sipmn}	6.515624	[Wb] PMs flux at rated load
R_s	0.066177	[Ohm] Phase resistance
l_{isn}	0.016944	[H] Phase q axis rated inductance
P_{cun}	36149.74	[W] Rated copper loss
P_{fe0}	41917.96	[W] No load total iron losses
P_{fe0s}	22555.56	[W] No load total stator iron losses
P_{fe0r}	19362.41	[W] No load total rotor iron losses
P_{fen}	66502.24	[W] total iron losses
P_{fesn}	35784.22	[W] total stator iron losses
P_{fern}	30718.59	[W] total rotor iron losses
Etan	0.966915	[−] Rated efficiency
Cosphin	0.665004	[−] Rated power factor
Optimization Variables		
poles	632	[−] number of poles per primitive machine
D_{ext}	6500	[mm] core outer diameter
W_{st}	80.57489	[mm] radial height of stator along one airgap
W_C	96.97418	[mm] coil width
h_c	37.57053	[mm] coil height
h_{sy}	66.20056	[mm] axial stator yoke length
l_{pm}	39.07571	[mm] permanent magnet axial length
l_{m_s}	0.431178	ratio between stator pole width and average pole pitch
l_{m_r1}	0.391324	ratio between rotor pole body width and average pole pitch
l_{m_r2}	0.433279	ratio between airgap rotor pole width and average pole pitch
l_{st}	38.57053	[mm] axial stator tooth length
W_{sp}	13.3784	[mm] stator pole width
W_{pm}	18.88575	[mm] PM width
W_{rp}	13.44359	[mm] rotor pole width
Rotor Pole Dimensions		
$W_{rp\ \text{ext}}$	13	[mm] (upper) outer rotor teeth tangential length
W_{rp2}	13	[mm] lower outer rotor teeth tangential length
W_{rp3}	11.3	[mm] upper inner rotor teeth tangential length
$W_{rp\ \text{int}}$	11.3	[mm] (lower) inner rotor teeth tangential length
Stator Pole Dimensions		
W_{rs1}	51.24243	[mm] stator slot upper outer tangential length
W_{rs2}	50.90232	[mm] stator slot lower outer tangential length
W_{rs3}	46.46093	[mm] stator slot upper inner tangential length
W_{rs4}	46.11009	[mm] stator slot lower inner tangential length
r_{h2}	0.4	[mm] part of rotor length
l_r	47.87571	[mm] rotor axial length

(Continued)

TABLE 12.1 (*Continued*)
Optimum Design Output File

Parameter	Value	Description
s_b	16	[−] turns per coil
l_{turns}	19609.43	[mm] average length of one turn
q_w	113.8554	[mm^2] wire cross-section
Weights		
$m_{s\ teeth}$	204.961	[kg] stator teeth mass
$m_{s\ yoke}$	563.4766	[kg] stator yoke mass
$m_{st\ iron}$	768.4376	[kg] one stator iron mass
m_{cu}	318.9994	[kg] one coil copper mass
m_s	1087.437	[kg] one stator mass
m_{rp}	462.0967	[kg] rotor iron mass
m_{pm}	578.7335	[kg] PM mass per rotor
m_r	1040.83	[kg] one rotor total mass
m_t	9345.147	[kg] generator total mass
$r_{a\ pint}$	1736.201	[kg] total PM mass
m_{cut}	1913.997	[kg] total copper mass
m_{fet}	5694.949	[kg] total iron mass
Costs		
cu_c	19139.97	[USD] copper cost
lam_c	28474.75	[USD] lamination cost
PM_c	86810.03	[USD] PM cost
i.cost	134424.7	[USD] initial active material cost
pmw_c	65416.03	[USD] passive material cost
inverter_c	12781.28	[USD] inverter cost
energy_c	153977.5	[USD] energy cost
temp.cost	0	[USD] over-temperature penalty cost
demag.cost	0	[USD] PM demagnetization penalty cost
t_cost	466599.511234	[USD] total cost
i_trace	73	optimization steps
t1	57	[min] optimization time

Source: D. F. Andonie, "Axial airgap flux concentrated TF-PMSG for direct drive wind turbines". Licence Thesis, University Politehnica Timisoara, 2008 (in English), supervisor I. Boldea. [11]

than 50% of total computation time (even with 20 random initial starts with the H-J algorithm): $57 \times 20 = 1140$ minutes (in our case here) on a single desktop computer.

- Computationally efficient FEM optimal design methodologies are more precise and have been introduced recently, but in general, they still need tens of computation hours on multi-frame computation hardware [14].

12.5 FINITE-ELEMENT MODEL CHARACTERIZATION OF TRANSVERSE FLUX PERMANENT MAGNET SYNCHRONOUS MOTORS

Ideally, 3D-FEM should be used to analyze flux distribution, emfs, cogging torque, and total torque [8] (Figure 12.19). L_{dm}, L_{qm}, inductances, core loss, and, finally, performance may all be calculated by 3D-FEM.

TABLE 12.2
Geometry and Weight/3 MW, 11 rpm AA-TF-IPMSG

Mechanical Parameters	Analytical Model Value	Optimized Value
External diameter: D_{ext}	4900 [mm]	6500 [mm]
Internal diameter: D_{int}	4440 [mm]	5984 [mm]
Number of poles: $2P$	546	632
Generator axial length: l_{ax}	880.44 [mm]	754.7 [mm]
Height of stator (and PM) along one airgap: W_{st}	95 [mm]	80.5 [mm]
Stator coil width: W_c	40 [mm]	97 [mm]
Stator coil height: H_c	51 [mm]	37.5
Width of PM: w_{pm}	13 [mm]	18.9
Length of PM: L_{pm}	65 [mm]	39 [mm]
Width of stator pole: W_{sp}	16 [mm]	13.4
Width of rotor pole: W_{rp}	16 [mm]	13.4 [mm]
Number of turns per coil: N	15	16
Total copper mass: m_{cut}	801.8	1913
Total iron mass: m_{Fet}	7083.8	5694
Total PM mass: m_{pm}	648	578.7
Total generator mass: m_t	9910.5	9345

Source: D. F. Andonie, "Axial airgap flux concentrated TF-PMSG for direct drive wind turbines". Licence Thesis, University Politehnica Timisoara, 2008 (in English), supervisor I. Boldea. [11]

TABLE 12.3
Electrical Parameters/3 MW, 11 rpm AA-TF-IPMSG

Electrical Parameters	Analytical Model Value	Optimized Analytical Value
Phase rated voltage: V_{cn}	2448 [V]	2491 [V]
Coil rated rms current: I_{fn}	430 [A]	302 [A]
Rated frequency: f_n	50 [Hz]	58 [Hz]
Rated coil mmf, peak: I_{s1n}	9118 [AmpTurns]	6287 [AmpTurns]
Phase resistance: R_s	77.72 [m Ω]	66.17 [m Ω]
Phase q axis rated inductance: L_{sn}	15.93 [mH]	17 [mH]
Total copper losses: P_{cun}	86151 [W]	36150 [W]
Total iron losses: P_{Fen}	91981 [W]	66502 [W]
Rated efficiency: η_n	0.944	0.967
Rated factor: cos φ_n	0.475	0.665

It has been inferred that 2D-FEM may not be used because of the 3D shape of the PM (and stator mmf) flux lines. However, since TF-PMSMs have many poles, a sector of the machine with at least 10 poles may be made into a corresponding linear motor where the stator cores add additional fictitious parts. 2D-FEM, may be used then, within limited computation time to calculate flux distribution, say, in axis d and q on no load (Figure 12.20a–d) for the optimally designed machine in section 12.4.

The PM flux linkage in a one-turn coil Ψ_{s10} was also calculated to be $\Psi_{s10} = \Phi_s \cdot p_{\text{PM}} = 0.4955$ (Wb), which is very close to the MEC calculated value (Figure 12.21)—0.75 Wb.

The cogging torque was also calculated (Figure 12.22).

TABLE 12.4
Costs (3 MW, 11 rpm AA-TF-IPSMG)

Cost Parameters	Analytical Model Value	Optimized Analytical Value
Total copper cost: cu_c	8018 [USD]	19140 [USD]
Total iron laminations cost: lam_c	35419 [USD]	28474 [USD]
Total active material cost: i · cost	144684 [USD]	134424 [USD]
Total passive material cost: pmw_c	69373 [USD]	65416 [USD]
Inverter cost: inverter_c	157842.6 [USD]	112781 [USD]
Energy cost: energy_c	267197 [USD]	153977 [USD]
Total generator costs: t_cost	639100 [USD]	466600 [USD]

Source: D. F. Andonie, "Axial airgap flux concentrated TF-PMSG for direct drive wind turbines". Licence Thesis, University Politehnica Timisoara, 2008 (in English), supervisor I. Boldea. [11]

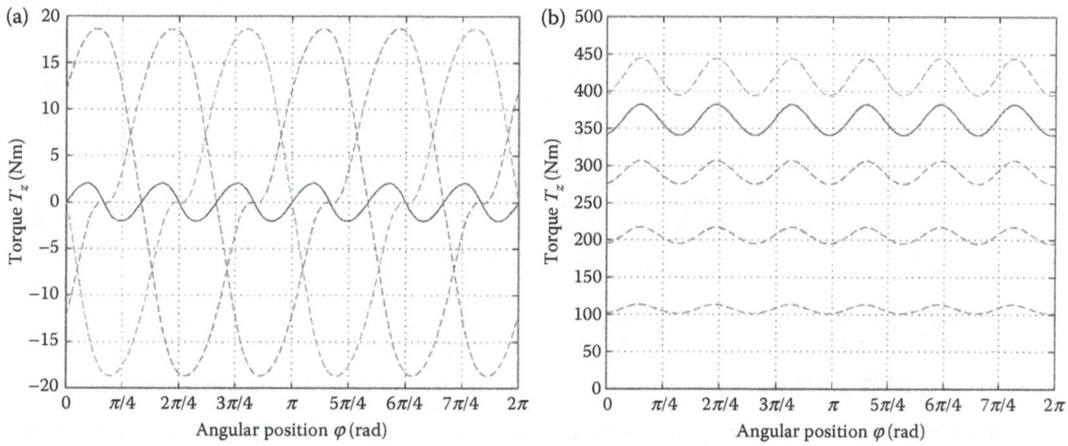

FIGURE 12.19 Cogging torque of each phase and of three-phase (a); and total torque (30, 60, 90, 120, 150 A), (b) waveforms for a 400 rpm, 400 Nm TF-IPMSM with single stator and rotor. (After E. Schmidth, *IEEE Trans.*, vol. MAG–47, no. 5, 2011, pp. 982–985. [8])

The total cogging torque was calculated at about 1.35% of this $2.68 \cdot 10^{-6}$ Nm three-phase machine, which is quite acceptable.

12.5.1 Six-MW, 12 rpm ($m = 6$ phase) Case Study by Three-Dimensional Finite-Element Model

The 3D-FEM was used to directly investigate a six-phase 6-MW, 12-rpm wind generator with a radial airgap in an interior-stator-outer-rotor configuration. The outer rotor uses U-shaped cores with radially placed, axially magnetized 1.3 T NdFeB magnets (Figure 12.23). From the analytical design, the main geometry was obtained, with the stator-outer (airgap) diameter, $D_{es} = 9$ m and $j_{con} = 6$ A/mm^2, efficiency 0.96, power factor 0.843, $l_{PM} = 150$ mm (radially), $h_{PM} = 30$ mm (axially), l_{stack} (per stator U-shaped leg) = 81 mm, stator slot area: 50·50 mm^2, airgap $g = 5$ mm, rotor pole pitch $\tau_{PM} = 56$ mm, pole pairs $p_p = 250$ pole pairs.

The element chosen to be studied by 3D-FEM corresponds to a period ($2\tau_{PM}$); see Figure 12.24a with the mesh in Figure 12.24b.

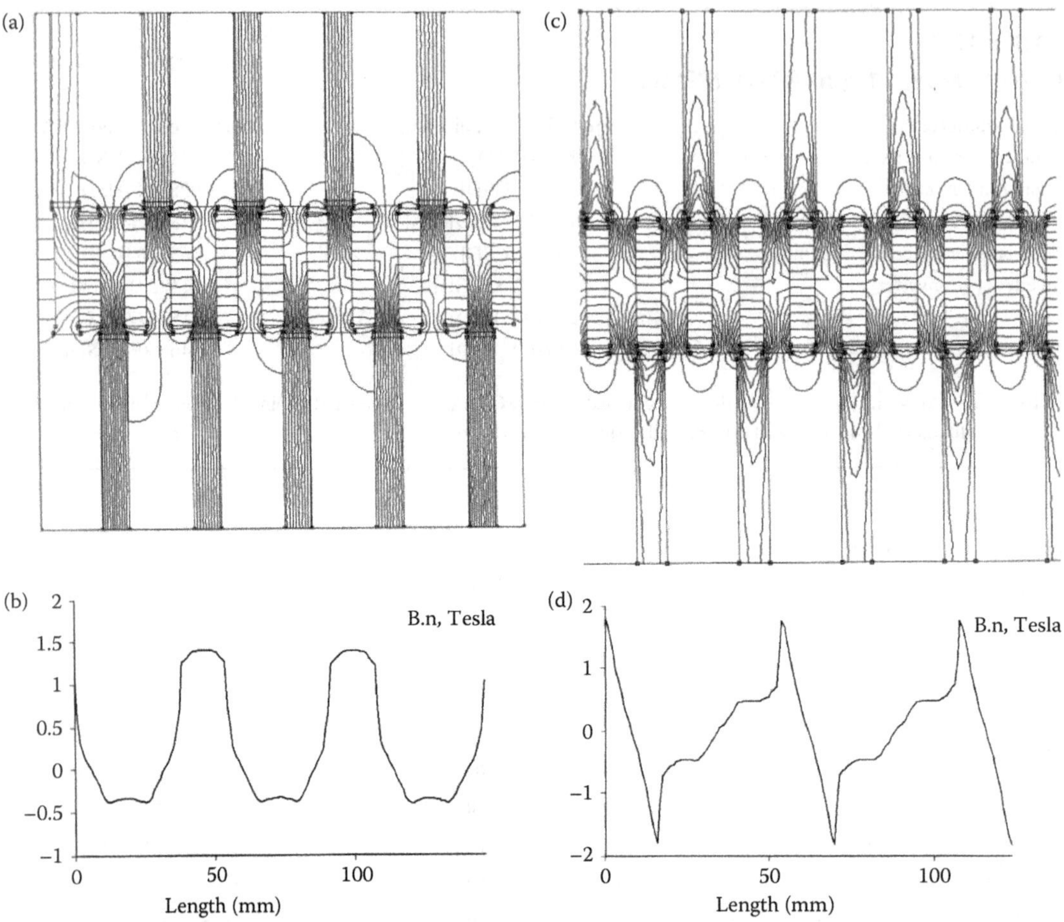

FIGURE 12.20 Flux lines and airgap flux distribution in axis *d*, (a), (b); and in axis *q*, (c), (d): 3 MW, 11 rpm, AA-TF-IPMSG. (After D. F. Andonie, "Axial airgap flux concentrated TF-PMSG for direct drive wind turbines". Licence Thesis, University Politehnica Timisoara, 2008 (in English), supervisor I. Boldea. [11])

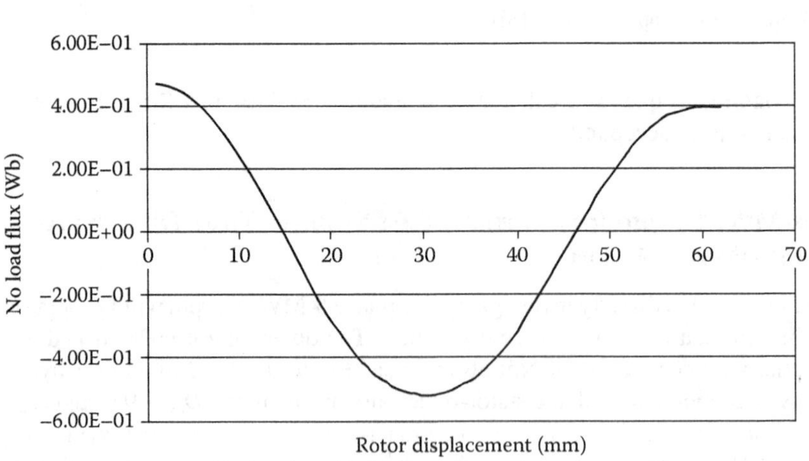

FIGURE 12.21 2D-FEM no-load flux linkage of a one-turn coil: 3 MW, 11 rpm, AA-TF-IPMSG. (After D. F. Andonie, "Axial airgap flux concentrated TF-PMSG for direct drive wind turbines". Licence Thesis, University Politehnica Timisoara, 2008 (in English), supervisor I. Boldea. [11])

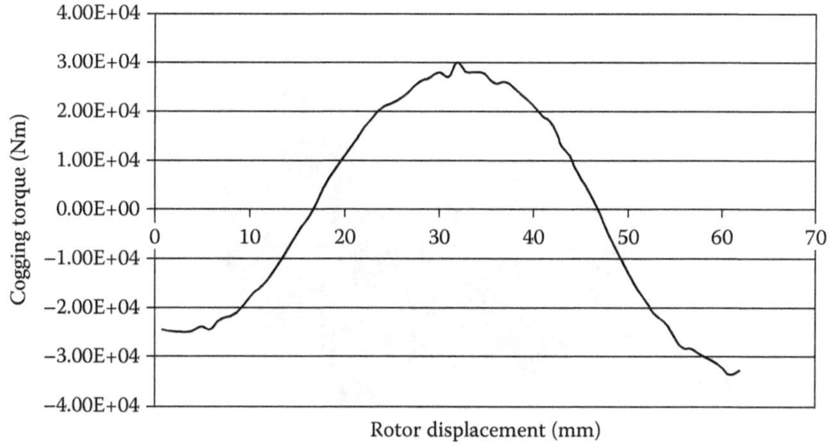

FIGURE 12.22 2D FEM cogging torque of the considered sector (primitive machine), calculated from the linear version cogging force, neglecting end-effect: 3 MW, 11 rpm, AA-TF-IPMSG. (After D. F. Andonie, "Axial airgap flux concentrated TF-PMSG for direct drive wind turbines". Licence Thesis, University Politehnica Timisoara, 2008 (in English), supervisor I. Boldea. [11])

Subsequently, to increase torque, additional PMs are placed in the lateral faces of rotor poles in order to reduce PM flux fringing of main IPMs (Figure 12.25) between rotor poles [12].

The torque/phase increases notably with the additional magnet thickness (Figure 12.26).

With six phases, the torque ripple is small (Figure 12.27), and with 5-mm-thick additional magnets, the machine produces the required average torque of around $5 \cdot 10^{-6}$ Nm. Similar results are reported in [12].

Errors of three electrical (3/250 mechanical) degrees in properly shifting one module (by 120° electrical) already produces an increase in torque pulsations to 4%–5%.

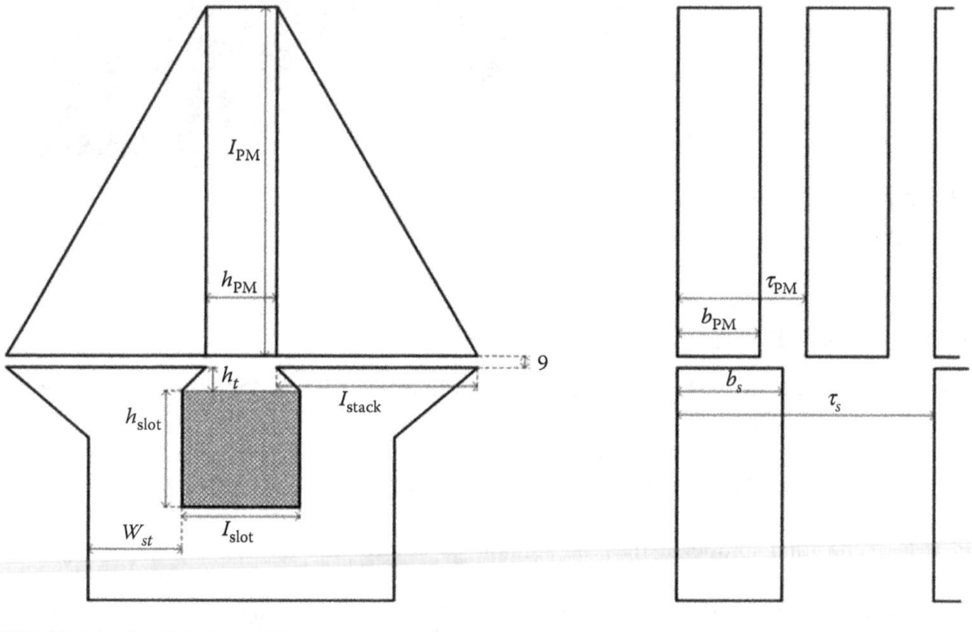

FIGURE 12.23 Radial airgap TF-PMSM(G) with outer IPM rotor.

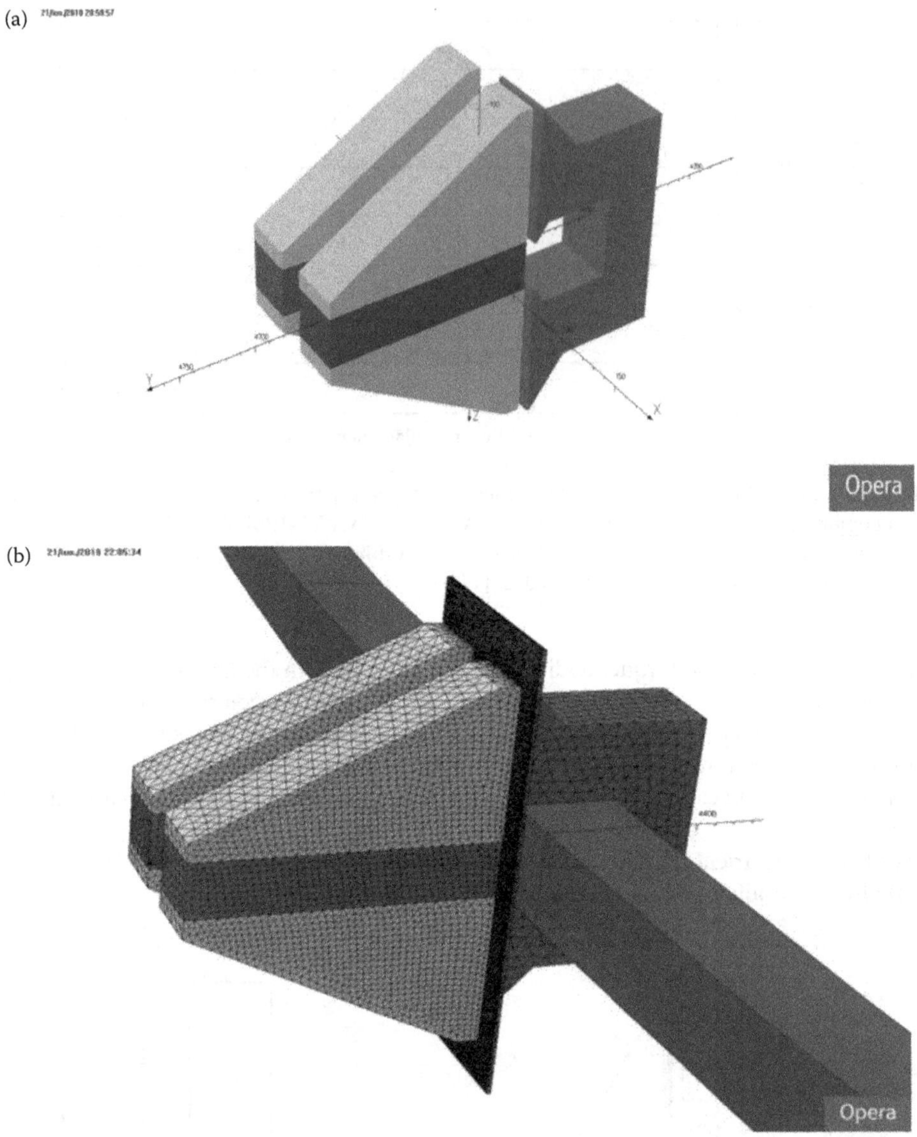

FIGURE 12.24 3D model (one period), (a); and its 3D mesh, (b).

12.6 CONTROL ISSUES

As Figure 12.21 shows, the PM flux linkage in a coil (phase) of a TF-PMSM is sinusoidal (or may be made so, by carefully carving the stator and rotor saliencies).

Also, with negative i_d, L_{md} is constant with increasing current, larger than L_{mq}, and much larger than L_{d+0}. At rated current (mmf) $L_{md} > L_{mq}$ and thus there is the same positive saliency, which for negative i_d leads to negative (moderate, though) reluctance torque. As this reluctance torque would reduce the total torque, it may be suitable to control the machine with pure i_q^* and zero $i_d^*(i_d^* = 0)$, unless flux weakening is required, when lagging power factor control may be in place. So, $i_d^* = 0$ control is used unless the control-designed reference voltage V_s^* is larger than $V_{s\ \max}$ (of the inverter) when, through a PI loop, a negative i_s^* is prescribed; an almost-unity power factor

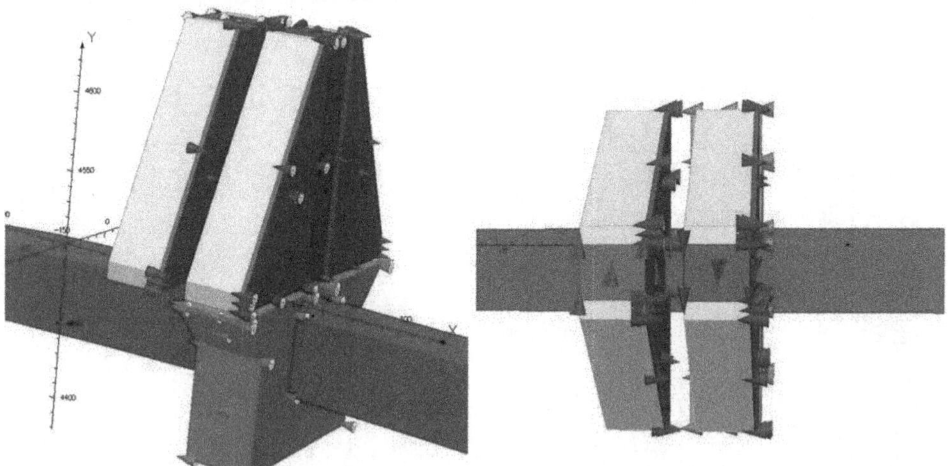

FIGURE 12.25 Adding additional magnets on rotor pole sides to reduce flux fringing of main IPMs.

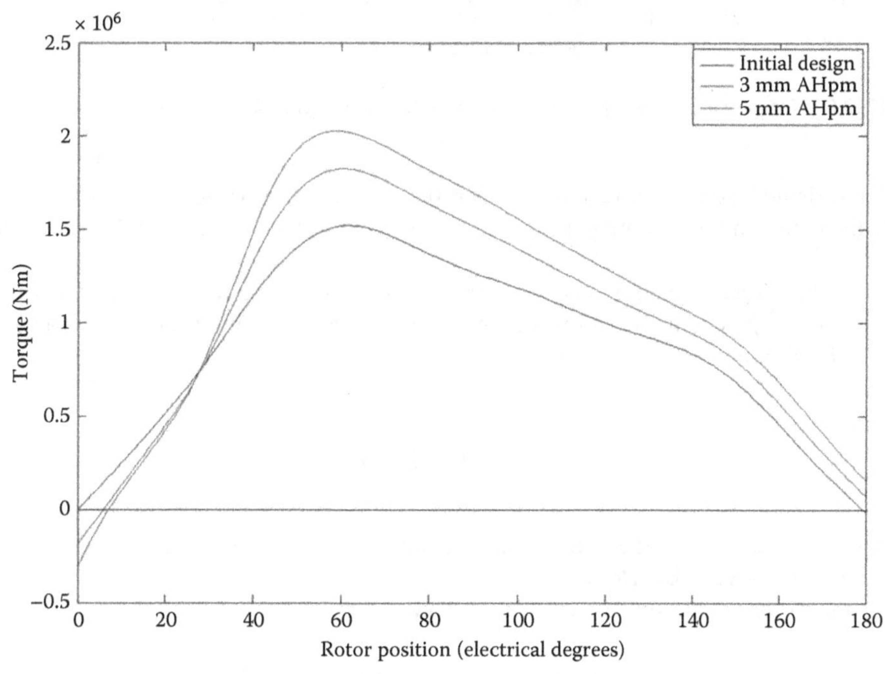

FIGURE 12.26 Torque/phase versus rotor position without and with 3- and 5-mm-thick additional magnets.

condition may add a Δi_d^* for precision:

$$
\begin{aligned}
&i_q^* \text{ control with } i_d^* = 0 \quad \text{for } V_s^* < V_{s\max} \\
&i_d^* = -\text{PI}(V_s^* - V_{s\max}) \quad \text{for } V_s^* \geq V_{s\max} \\
&\text{and} \\
&\Delta i_d^* = -\text{PI}\left(\varphi^* - \hat{\varphi}_1\right) \quad \text{for } V_s^* \geq V_{s\max}
\end{aligned}
\tag{12.59}
$$

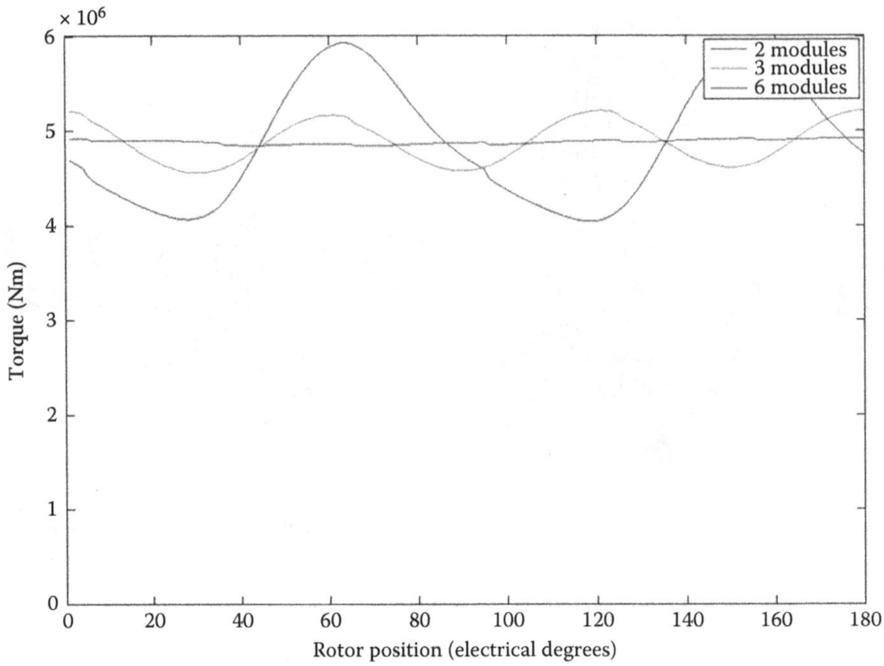

FIGURE 12.27 Torque versus rotor position, with 2, 3, 6 modules (phases) and 5 mm thick additional PMs.

The field-oriented control, direct torque (or power) and flux control or scalar (V/f or I-f) control with stabilizing loops may be applied for controlling TF-PMSM(G)s with sinusoidal currents.

$P(\theta_r)$ is the Park transformation, and PI + SM means PI + sliding mode (robust) controller. The power factor angle $\hat{\varphi}_1$ has to be estimated and used above a certain speed (frequency) and thus may be calculated online as:

$$\hat{\varphi}_1 = \tan^{-1}\left(\frac{V_q^* i_d - V_q^* i_d}{V_d^* i_d - V_q^* i_q}\right) \tag{12.60}$$

Equation 12.60 stems from the ratio of reactive to active input powers and should work properly with little filtering above a few Hertz.

For generator control, the speed loop may be replaced by a power loop:

$$-(\mathrm{PI} + \mathrm{SM}) \text{ of } (P^* - \hat{P}) = i_q^* < 0 \tag{12.61}$$

Where P^* is the reference active power and $\hat{P}$ is the actual machine active power; the rest of the control scheme in Figure 12.28 holds.

The speed of the prime mover that rotates the generator is controlled by the prime mover and influenced by generator-delivered power (braking the rotor). It is also feasible to have the optimal prime mover reference power (say, determined by wind speed), which leads to a certain reference speed ω_r^* when, this way, the speed reference is obtained. Again, the FOC system in Figure 12.28 holds.

The unity power factor operation may be limited (in the speed/torque envelope) and may be eliminated outside it to avoid instabilities in control.

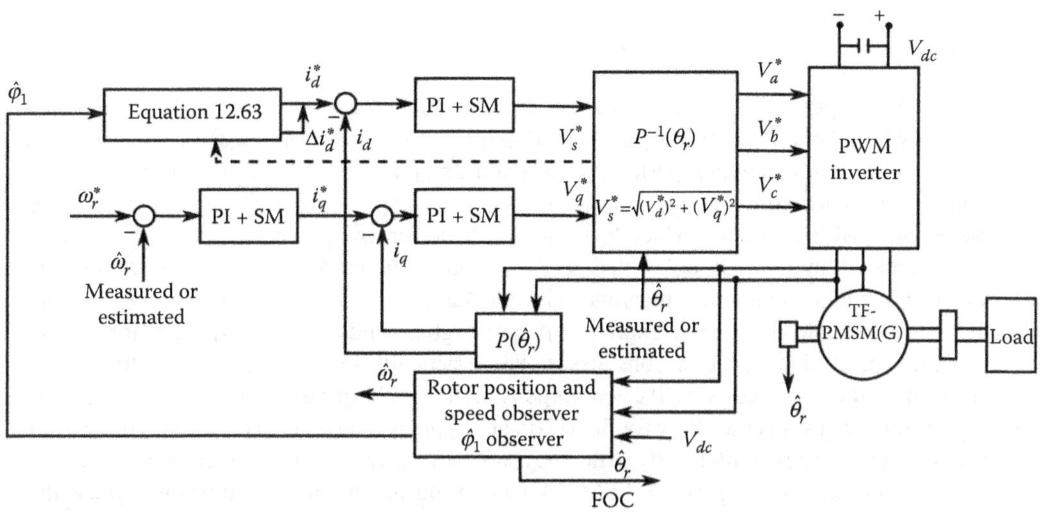

FIGURE 12.28 Generic FOC of TF-PMSM(G).

For autonomous generators, with DC-controlled voltage output, the DC voltage error $V_{dc}^* - \hat{V}_{dc}$ dictates the reference current i_q^*:

$$i_q^* = -(\text{PM} + \text{SI}) \text{ of } (V_{dc}^* - \hat{V}_{dc}) < 0 \tag{12.62}$$

The speed loop will be eliminated.

Note: $L_d > L_q$ in some TF-IPMSMs, mainly due to magnetic saturation, may not be the case for TF-SPMSMs when $L_d = L_q$, but the control in Figure 12.28 may still hold, though the rotor position and speed estimator should be adapted to the SPMSM if the drive is encoderless.

Note: With individual phases (separated coils), higher voltage is feasible in TF-PMSMs, also while fault tolerance is high, especially in multiphase ($m > 3$ phases) topologies.

12.7 SUMMARY

- Transverse-flux PMSMs were apparently introduced in 1986 [1] in order to increase torque density in low-speed high-torque electric motor drives.
- TF-PMSMs have a modular structure with $1(k)$ circular shape stator-coils per phase. The U-shaped stator cores embrace the coil, forming a flux modulator with the same periodicity as the PMs on the rotor ($2p_{\text{PM}}$). Phases (modules) are separated. Moreover, the magnetic circuit and electric circuit design are decoupled. Small pole pitch, down to 10–12 mm in high-torque (tens, hundreds of Nm) machines, is feasible, with large mmf in the coils, producing high torque density.
- Moreover, as the pole pitch goes down, the PM flux fringing goes up and thus counteracts the torque magnification effect by increasing the number of pole pairs for a given stator circular coil diameter.
- Moreover, the available space to house large coil mmfs despite a small pole pitch leads inevitably to a lower power factor in the machine.
- However, the circular stator multipole coils lead to the lowest torque/copper losses ever, especially in interior stator–outer rotor radial-airgap topologies.

- To further increase torque density, spoke-shaped PM dual stator configurations have been proposed; here all PMs are active all the time, so the emf per given machine configuration is increased notably.
- The spoke-shaped PMs may be placed on the stator but also embedded in the U-core walls and back iron. When the rotor is passive and magnetically anisotropic with p_{PM} salient poles and p_{PM} interpoles, saliency periodicity is again $2\tau_{PM}$, equal to that of the stator.
- Finally, the stator coils may be placed in claw-pole-shaped soft magnetic composite one-piece cores, with a regular surface PM rotor, suitable for low-power compact motor drives.
- The increased number of poles that leads to torque magnification (TF-PMSM may be viewed as a flux-modulation machine with the stator winding mmf pole pair $p_a = 0$, and thus the PR $= p_{PM}/p_a = \infty$!), implicitly means higher fundamental frequency for a given machine speed, that is, larger core losses. The drastically lower copper losses (due to the "blessing" of circular coils) will overcompensate for the larger iron losses in good designs.
- A preliminary design case study for 5000 Nm at 75 rpm produced a machine with outer-rotor radial airgap at 55 Nm/liter, 97% efficiency, and 0.78 power factor, which is encouraging.
- MEC nonlinear modeling of a TF-PMSM accounting for the 3D flux tubes was presented in detail with results on a 3-MW, 11-rpm wind generator case study: the important role of magnetic saturation was proven in an axial-airgap dual core and coil per phase stator and dual active area spoke-shaped-PM rotor per phase topology. But the power factor was too small (0.47) for an efficiency of ≈94.4%.
- Optimal design based on MEC and the Hooke–Jeeves algorithm was presented in detail for same case study as above, improving performance across the board (power factor rise from 0.47 to 0.67, efficiency from 0.944 to 0.967, weight and overall cost were also reduced); the still-low power factor resulted in a higher KVA in the converter cost, which was also considered in the objective function together with the main initial machine cost, energy loss costs over machine operation life (hours), and penalty functions.
- Penalty functions in the optimal design composite function referred to stator and rotor over-temperature limitation (by simplified thermal model, but real machine losses) and PM demagnetization avoidance at 2 p.u. pure i_q current.
- 3D-FEM is typically required to fully characterize TF-PMSMs, but 2D versions with a more than 10-poles sector translated into linear motor counter parts that add fictitious cores to close typical axial flux lines have been used with good results, saving time on orders of magnitudes.
- A pure 3D-FEM investigation on a given geometry for a 6-MW, 12-rpm, 50-Hz wind generator with a radial airgap single-sided stator and rotor and an outer IPM rotor is discussed in detail. Additional magnets on the core sides of the IPM rotor lead to an increase of 30% in torque, thus meeting the $5 \cdot 10^{-6}$ Nm torque range in a 9-m airgap diameter, total length 0.5 m, 97% efficiency, about 0.84 power factor, 5 mm airgap, $50 \cdot 50$ mm^2 slot, 0.5 m^3 PM volume design.
- With almost sinusoidal PM flux linkage variation in the circular coils, the TF-PMSM control may use the entire FOC, DTFC, scalar (V/f and I-f) control with stabilizing loop heritage of PMSM drives. Some TF-PMSMs resemble surface PMs and other IPM topologies (where L_{dm} > or < L_{qm}) depending on configuration and level of magnetic saturation. FEM computation of L_{dm}, L_{qm} under load is suggested in TF-PMSMs.
- A generic FOC of TF-PMSMs is presented where $i_d^* = 0$ control is switched to $i_d^* < 0$ when the reference voltage surpasses the maximum inverter voltage $V_{s\,max}$.
- The same FOC scheme (Figure 12.28) may be easily adapted to:
- DC grid constant voltage generating by replacing the speed regulator by an active power regulator to yield the reference current i_q^*
- DC output autonomous generator operation, when the speed regulator is replaced by a DC voltage regulator that yields, again, the reference current i_q^*

- TF-PMSM still has a way to go before reaching the international markets niche, but the R&D efforts so far [15] and in the near future may provide for such an outcome; the lowest copper losses (due to circular AC coils) seem too good of an asset not to exploit in industry.

REFERENCES

1. H. Weh, H. May, Achievable force densities for PM excited machines in new configurations, *Record of ICEM-1986*, no. 3, pp. 1107–1111.
2. I. Boldea, *Variable Speed Generators*, 2nd edition, book, CRC Press, Taylor and Francis, New York, 2015 (Chapter 11).
3. G. Henneberger, I. A. Viorel, *Variable Reluctance Electrical Machines*, Shaker Verlag, Aachen, 2001 (Chapter 6).
4. J. Luo, S. Huang, S. Chen, T. A. Lipo, Design and experiments of a novel circumferential current PM machine (AFCC) with radial airgap, *Record of IEEE-IAS-2001*, Chicago, IL.
5. Y. Guo, J. Zhu, D. G. Dorrell, Design and analysis of a claw pole PM motor with molded shift magnetic composite core, *IEEE Trans.*, vol. MAG–45, no. 10, 2009, pp. 4582–4585.
6. H. Ahn, G. Jang, J., Chang, Sh. Chung, D. Kang, Reduction of the torque ripple and magnetic force of a rotary two-phase TF machine using herringbone teeth, *IEEE Trans.*, vol. MAG–44, no. 11, 2008, pp. 4066–4069.
7. Z. Jia, H. Lin, Sh. Fang, Y. Huang, Cogging torque optimization of novel TF PM generator with double C-hoop stator, *IEEE Trans.*, vol. Mag–51, no. 11, 2015, pp. 8028104.
8. E. Schmidth, Finite element analysis of a novel design of a three phase TF machine with external rotor, *IEEE Trans.*, vol. MAG-47, no. 5, 2011, pp. 982–985.
9. E. Schmidt, 3D-FEM analysis of cogging torque of a TF machine, *IEEE Trans.*, vol. MAG–41, no. 5, 2005, pp. 1836–1839.
10. R. Blissenbach, Investigation of PM transverse flux machines for high torque electric drives, PhD thesis, RWTH Achen, Germany, 2002 (in German).
11. D. F. Andonie, Axial airgap flux concentrated TF-PMSG for direct drive wind turbines. Licence thesis, University Politehnica Timisoara, 2008 (in English), supervisor I. Boldea.
12. L. Strete, L. Tutelea, I. Boldea, C. Martis, I. A. Viorel, Optimal design of a rotary transverse flux motor (TFM) with PMs on the rotor, *Record of ICEM 2010*, Rome, Italy (IEEEXplore).
13. H. Polinder, F. F. A. van der Pijl, G. D. De Vilder, P. J. Tavner, Comparison of direct drive and geared generator concepts for wind turbines, *Record of IEEE-IEMDC-2005*, pp. 543–550 (IEEEXplore).
14. A. Fatemi, D. M. Ionel, N. A. O. Demerdash, T. W. Nehl, Fast multiobjective CMODE type optimization of electric machines for multicore desktop computers, *IEEE-ECCE-2015*, pp. 5593–5600 (IEEEXplore).
15. S. Jordan, N. J. Baker, Design and build of a mass-critical air cooled transverse flux machines for aerospace, *Record of ICEM-2016*, pp. 1455–1460.

13 Magnetic-Geared Dual-Rotor Reluctance Electric Machines *Topologies, Analysis, Performance*

13.1 INTRODUCTION

Magnetic-geared dual-rotor reluctance machines are also flux-modulation machines. Their two rotors provide for a magnetic-gear effect that may increase the torque/volume in low-speed high-torque applications, but at the inevitable cost of notable additional PM weight/Nm, though at acceptable efficiency.

But the elimination of mechanical transmission by using a magnetic gear capable of more than 100 Nm/liter is considered encouraging in pursuing the development of compact low-speed high-torque pseudodirect drives with MG-REMs. A typical high torque density (100 Nrm/liter) magnetic gear is shown in Figure 13.1 [1].

The MG has here an inner high-speed PM rotor with $p_{h\mathrm{PM}}$ pole pairs ($p_{h\mathrm{PM}} = 4$), an outer low-speed PM rotor with $p_{l\mathrm{PM}}$ pole pairs ($p_{l\mathrm{PM}} = 22$), and in between them there is a stationary flux modulator made of n_{F-M} ($n_{F-M} = 26$) laminated segmented poles and interpoles.

The principle of *F–M* machines holds here also:

$$n_{h\mathrm{PM}} = n_{F-M} - n_{l\mathrm{PM}} \tag{13.1}$$

Other pole pair combinations are feasible.

The rotor speed relationship of $F - M$ machines holds here also:

$$\Omega_{\mathrm{low}} = \frac{p_{h\mathrm{PM}}}{p_{h\mathrm{PM}} - n_{F-M}}\Omega_h = -\frac{1}{5.5}\Omega_h \tag{13.2}$$

Typical resultant radial airgap flux densities in the airgap adjacent to high-speed and, respectively, low-speed rotors (Figure 13.2a and b) [1] show the flux-modulation effect of the flux-modulator magnetic permeance first-space harmonic.

The torques versus rotor position for the two rotors (Figure 13.3) emphasize the gear ratio $G = T_{e\mathrm{low}}/T_{e\_\mathrm{high}} = \Omega_h/\Omega_l = 5.5$ [1].

With 1.2 T ($\mu_{\mathrm{rec}} = 1.05\,\mu_0$) NdFeB magnets, a 100 Nm/liter torque density is demonstrated [1]. There are eddy and hysteresis losses in the cores (especially in the flux modulator) and in the PMs, which, however, amount in general to less than 20–25% of total losses. This means that both the torque density and the efficiency of mechanical gears may be hoped for from magnetic gears. MG is fail-safe in the sense that it trips in case of large over-torque.

It is to be seen how many such accidental stops the MG may survive, especially if it is performance/price competitive. If a PMSM stator with windings is added to drive the high-speed rotor of MG, the MG-PMSM with dual rotors is obtained. The MG-PMSM may be built with the PMSM on the inside [2–3] or outside [4] (Figure 13.4a and b).

The tooth-coil high-speed SPMSM is placed on the inner side of the assembly in Figure 13.4a and on the outside in Figure 13.4b. Also, while in Figure 13.4a, the flux modulator is stationary and the two rotors are provided with PMs, in Figure 13.4b, only the high-speed rotor has PMs, and the flux modulator represents the low-speed rotor.

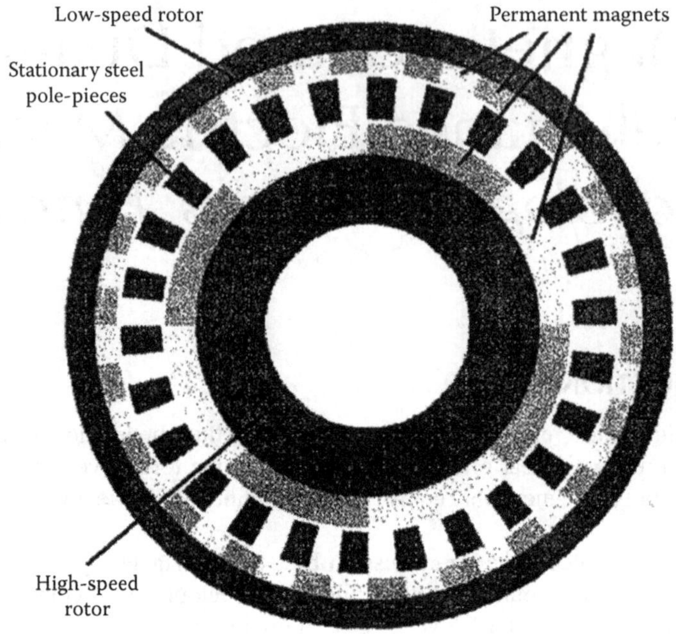

FIGURE 13.1 Magnetic gear cross-section.

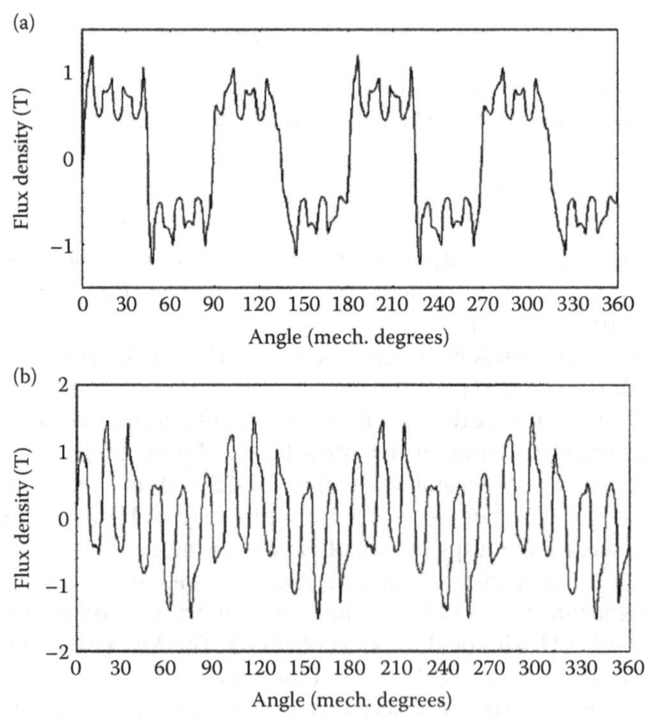

FIGURE 13.2 Resultant airgap flux density in the MG: (a) close to high-speed rotor (four pole pairs); (b) close to low-speed rotor (22 pole pairs). (After K. Atallah, D. Howe, *IEEE Trans.*, vol. MAG–37, no. 4, 2001, pp. 2844–2846. [1])

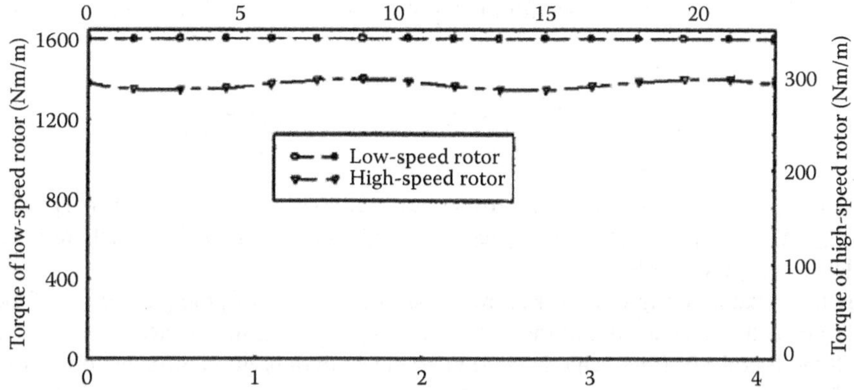

FIGURE 13.3 Torques of low- and high-speed rotors. (After K. Atallah, D. Howe, *IEEE Trans.*, vol. MAG–37, no. 4, 2001, pp. 2844–2846. [1])

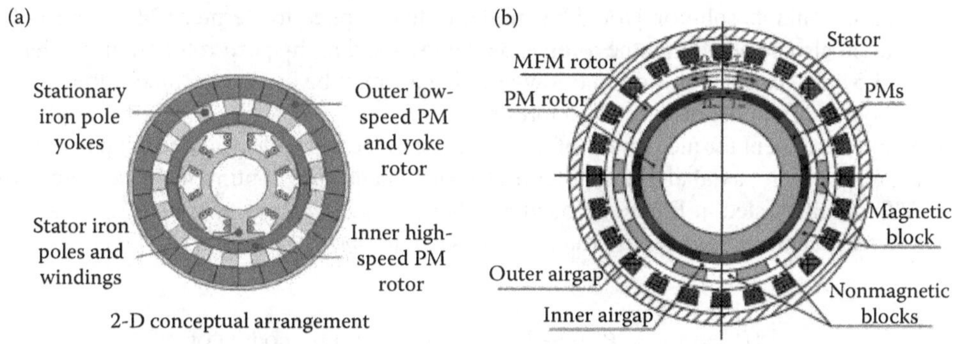

FIGURE 13.4 Magnetic geared REM: (a) with inner high-speed motor; (b) with outer motor.

The two conditions of flux modulation are met here, too:

$$\text{Fig.13.4a:}\quad \begin{cases} p_{h\mathrm{PM}} = |P_{\mathrm{PM}} - P_{F-M}|;\ \text{stationary flux modulator :} \\ \Omega_{F-M} = 0 \\ \Omega_a = \Omega_{h\mathrm{PM}} = \dfrac{P_{l\mathrm{PM}} \cdot \Omega_{\mathrm{PM}}}{P_{l\mathrm{PM}} - P_{F-M}} + \dfrac{P_{F-M}\Omega_{F-M}}{P_{F-M} - P_{l\mathrm{PM}}}; \\ p_{h\mathrm{PM}} = 2;\ P_{l\mathrm{PM}} = 17;\ p_{F-M} = 19 \\ \Omega_{h\mathrm{PM}} = \Omega_a \left(\begin{array}{l} \text{high-speed PM rotor pole pairs } p_{h\mathrm{PM}} \text{ is} \\ \text{the same as that of stator winding mmf} \end{array} \right) \end{cases} \tag{13.3}$$

$$\text{Fig.13.4b:}\quad \begin{cases} P_a = P_{F-M} - P_{\mathrm{PM}};\ P_a = 4,\ P_{\mathrm{PM}} = 17;\ P_{F-M} = 21 \\ \Omega_a = \dfrac{P_{\mathrm{PM}} \cdot \Omega_{\mathrm{PM}}}{P_{\mathrm{PM}} - P_{F-M}} - \dfrac{P_{F-M}\Omega_{F-M}}{P_{\mathrm{PM}} - P_{F-M}};\ \Omega_{\mathrm{PM}} \neq 0;\ \Omega_{F-M} \neq 0 \\ \Omega_a - \text{speed of stator mmf wave} \\ \omega_a \text{ angular frequency of stator currents.} \end{cases} \tag{13.4}$$

From the equilibrium of powers, say, for the topology in Figure 13.4b, the torque relationships are obtained:

$$T_{F-M} = -\frac{P_{F-M}T_{\text{stator}}}{P_a}; \qquad T_{\text{PM}} = -\frac{P_{\text{PM}}}{P_a}T_{\text{stator}} \tag{13.5}$$

For both cases, the MG part reduces speed but increases torque of the low-speed PM rotor (Figure 13.4a) and, respectively, the low-speed flux modulator (made of P_{F-M} salient magnetic segmented poles; see Figure 13.4b).

As in the inner stator configuration (Figure 13.4a), the number of pole pairs $p_{h\text{PM}}$ of the adjacent high-speed rotor is the same as that of the stator mmf ($p_a = p_{h\text{PM}}$); the electric motor part represents simply a 6/8 tooth-coil SPMSM, which is known to have not only good efficiency but also a reasonable power factor. For a small-speed drive, the MG-REM, which represents a pseudodirect drive, this advantage is paramount and may offset the additional mechanical complexity and the larger initial costs.

Not so with the outer stator configuration (Figure 13.4b), which implies a distributed winding, in general, and behaves like a Vernier PM machine with a moving flux modulator. Here, the power factor [5] is an issue, and its solution would be similar to that applied to Vernier PM machines (a dual stator). It should also be noted that the torques on the low- and high-speed rotor are not independent (Equation 13.5). So, in general, only the low-speed shaft may be available (unless the application allows for the fixed ratio between the two torques).

To obtain independent torque control of the two rotors, one more electric (stator) part is required. Radial and, respectively, axial airgap topologies for such an electric continuously variable transmission (ECVT) are illustrated in Figure 13.5a and b [5–6].

The pole pair relationships are, in principle, the same as above:

$$\text{Fig.13.5a:} \quad \begin{cases} (P_{ai})_{\substack{\text{interior}\\ \text{stator}}} = P_{i\text{PM}} - P_{o\text{PM}}; \text{ stationary flux modulator:} \\ -P_{i\text{PM}} \times \Omega_{i\text{PM}} + P_{o\text{PM}} \cdot \Omega_{o\text{PM}} + (p_{ai}) \cdot \Omega_{ai} = 0; \\ (p_{ai}) \cdot \Omega_{ai} = \omega_{ai} \\ p_{o\text{PM}} = (p_{ao})_{\text{outer stator mmf; synchronous machine}} \end{cases} \tag{13.6}$$

$$\text{Fig.13.5b:} \quad \begin{cases} (P_{a1})_{\substack{\text{primary}\\ \text{winding}}} = (P_{F-M})_{\substack{\text{primary}\\ \text{ferromagnetic}\\ \text{segmented}\\ \text{rotor}}} - (P_{\text{PM}s})_{\substack{\text{secondary PM}\\ \text{rotor}}} \\ P_{\text{PM}s}\Omega_{\text{PM}s} + P_{a1} \cdot \Omega = P_{F-M}\Omega_{F-M} \\ (P_{a2})_{\substack{\text{secondary}\\ \text{winding}}} = P_{\text{PM}s}; \text{ synchronous machine.} \end{cases} \tag{13.7}$$

It may be inferred that the dual-rotor dual-stator MG-REM includes a Vernier PM machine and a synchronous machine, while independent stators are fed through their inverters. This way, four-quadrant operation is feasible in a brushless configuration. Both machines (stators) may be motoring or generating, one motoring and the other generating, or one may be idle.

The key to four-quadrant operation is the four-quadrant circulation of electric power in each stator independently. The MG effect is manifest, in the sense that the secondary rotor speed is, in general, lower than the primary rotor speed and thus the mechanical transmission is replaced by this

(a)

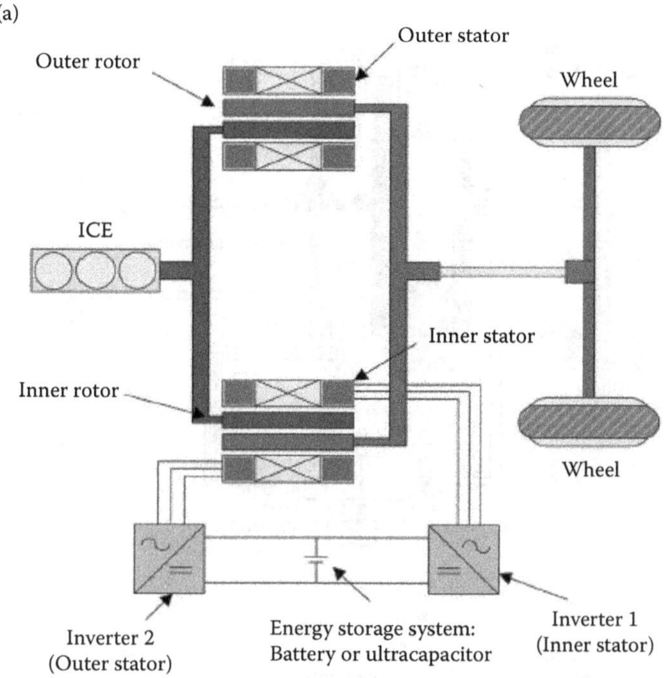

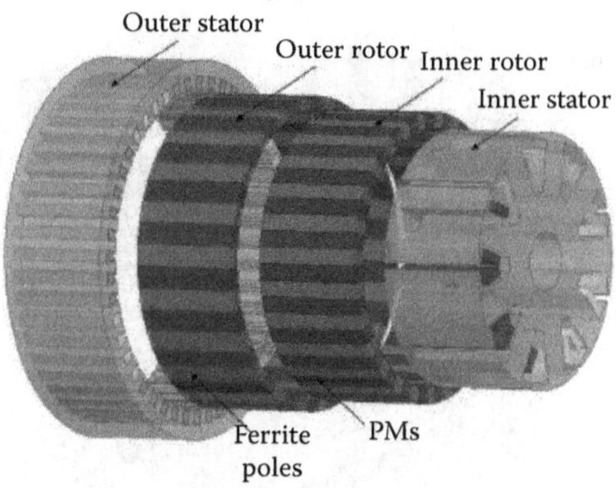

FIGURE 13.5 Dual-rotor, dual-electric part MG-REM: (a) with radial airgap. (After S. Niu et al., *IEEE Trans.*, vol. MAG–49, no. 7, 2013, pp. 3909–3912; Y. Liu et al., *IEEE Trans.*, vol. MAG–50, no. 11, 2014. [5,6]) *(Continued)*

e (electrical) continuously variable transmission (e-CVT). The jury is still out on whether e-CVT is globally (initial and ownership costs) better than a mechanically variable transmission (MVT) and a single electric motor drive. In an effort to further reduce the volume of such a four-quadrant but hybrid-electric CVT, an axial airgap two-rotor, single–stator and inverter PMSM was also introduced (Figures 13.6 and 13.7) [7].

The principle here is relying on two stator mmfs, one for $2p_1$ poles (f_1) and the other for $2p_2(f_2)$ poles, produced by a dedicated two-frequency PWM inverter that drives the two rotors independently. The $N_s = 12$ slots Gramme ring coils (for example), with $2p_1 = 10$ and $2p_2 = 14$, provide above 0.945 winding factors for both mmf space harmonics. Independent four-quadrant speed control of the two rotors simultaneously has been proven by digital simulations [7] but not yet in experiments. The drastic hardware simplification in Figures 13.6 and 13.7 comes at a price: both frequencies (and their currents) that run the two rotors produce copper losses in the entire unique stator winding,

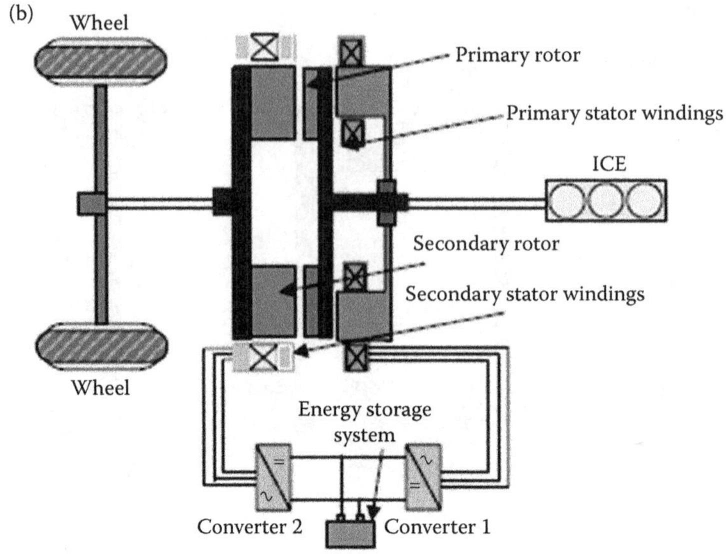

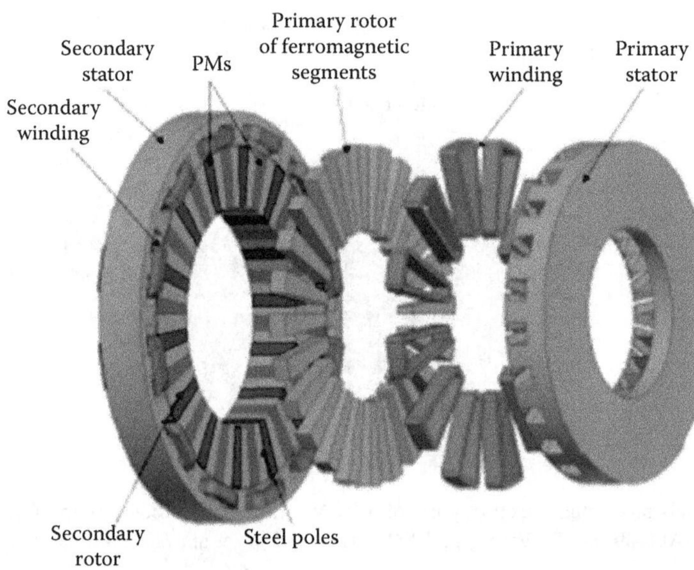

FIGURE 13.5 (Continued) Dual-rotor, dual-electric part MG-REM: (b) with axial airgap. (After S. Niu et al., IEEE Trans., vol. MAG–49, no. 7, 2013, pp. 3909–3912; Y. Liu et al., *IEEE Trans.*, vol. MAG–50, no. 11, 2014. [5,6])

though only one axial face interacts with one rotor; also, the inverter has to handle all kVA power in a dedicated, say, dual, FOC. It is feasible to build two coil rows (two windings) on the same stator core (Figure 13.7) and add a power switch to disconnect the temporarily idle winding (if both need not be active all the time).

It is also possible to apply other principles for a single-stator dual-rotor machine such as using IPM or variable reluctance rotors (with $2p_1 \neq 2p_2$ poles) in an axial airgap kind of dual-rotor BLDC-MRM (Chapter 7).

If the two rotors need to handle notably different powers, a radial airgap configuration of single toroidally (or radially) wound stator dual PM rotor may be applied [8].

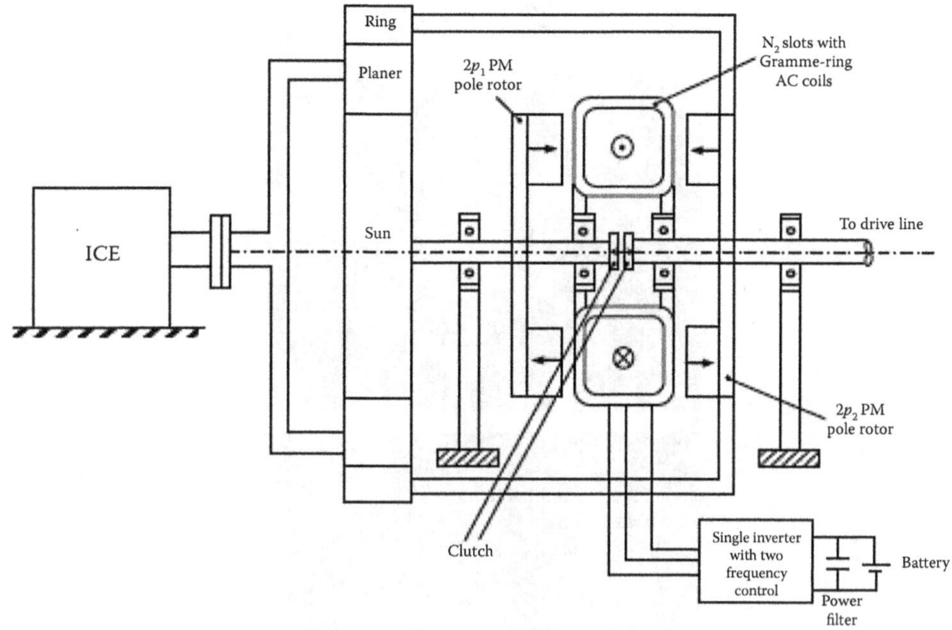

FIGURE 13.6 Single-stator single-inverter dual-rotor hybrid CVT with axial airgap, Gramme-ring coils and, say, 12 stator slots, $2p_1 = 10$ poles, and $2p_2 = 14$ poles rotors. (After I. Boldea et al., *Record of IEEE–ESARS*, 2012, Bologna, Italy. [7])

Here, also, magnetic anisotropy may be used to improve the ratio performance/costs.

Yet another, rather complex, topology of single electric part dual-rotor magnetic-geared REM was recently proposed for a pseudodirect drive for wind turbine generators [9]; see Figure 13.8.

The topology includes:

- An outer stator core with $p_a = 13$ pole pairs/section (there are 20 sections).
- Hallbach PMs on stator (four per stator pole pair).
- The low-speed flux modulator pole pieces rotor per section:

$$P_{F-M} = 15.$$

- High-speed interior PM rotor with $P_{h\text{PM}} = 2$ per section. As expected:

$$P_{h\text{PM}} = P_{F-M} - P_a \tag{13.8}$$

- The high-speed PM rotor runs free, as expected, while the low-speed (flux modulator) shaft is coupled to the wind turbine shaft.
- The magnetic gear ratio $G = P_{F-M}/P_{h\text{PM}} = 15/2 = 7.5/1$.
- The 7.5/1 magnetic gear ratio renders the pseudodirect drive here to the class of one-stage mechanical-geared (multibrid) wind generators, which are known to be competitive because they strike a good compromise between reducing the mechanical gear stages (complexity and volume) while not increasing the machine volume and cost too much. Unfortunately, for the 10-MW, 9.65-rpm (10-MNm) case, the thorough optimal design example in [9] ends up with a minimum of 45 tons of active weight at an electrical efficiency of 98.5%, but for 13.5 tons of NdFeB sintered PMs ($B_r = 1.25$ T). This is twice as much as the value calculated for the same specifications with a three-slot/pole surface direct-drive PMSG at 65 tons of active weight [10].

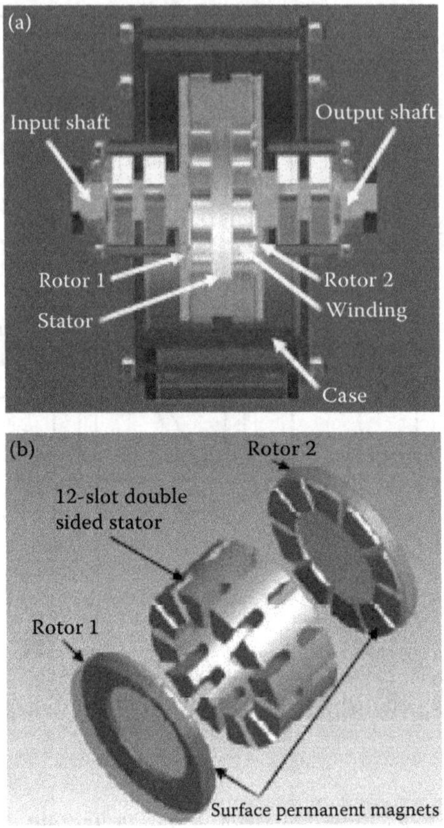

FIGURE 13.7 Single-stator dual-rotor hybrid-electric CVT (the dual stator windings with independent inverter configuration) (a) cross-section; (b) exploded view.

- There are PMs both on the stator and rotor and thus the flux reversal and Vernier principles are combined in the configuration of Figure 13.8 [9]; the excessive PM weight may prove an insurmountable demerit in the case in point.
- As in all flux-modulation machines, the power factor—which defines the PWM inverter kVA, costs, and losses—has to be checked; in all probability, it will not be large for the design in [9], but refinements in design might increase the power factor to acceptable values (0.75–0.8).
- For a 6-MW 11.3-rpm TF-PMSG, optimal design has produced an electrical efficiency of 0.967, power factor 0.67 for 16.8 tons of active weight with only 1.7 tons of NdFeB sintered magnets (Chapter 12). Extrapolating to 10 MW, the active weight of TF-PMSG, at the same power factor but, perhaps, at above 0.97 electrical efficiency, will be only 27 tons, from which 5 tons represent the NdFeB sintered PMs, in a notably less complicated (for fabrication) topology.
- This is not to say that the dual-rotor MG-REM in [9] is not to be pursued; it may also be practical in vehicular propulsion, and when the cost factor is not paramount, the configuration may be improved further.

From the above introductory remarks, we should grasp the wide spectrum of torque and speed range applications of dual-rotor MG-REMs. From all of the competing topologies, two will be investigated here in some detail:

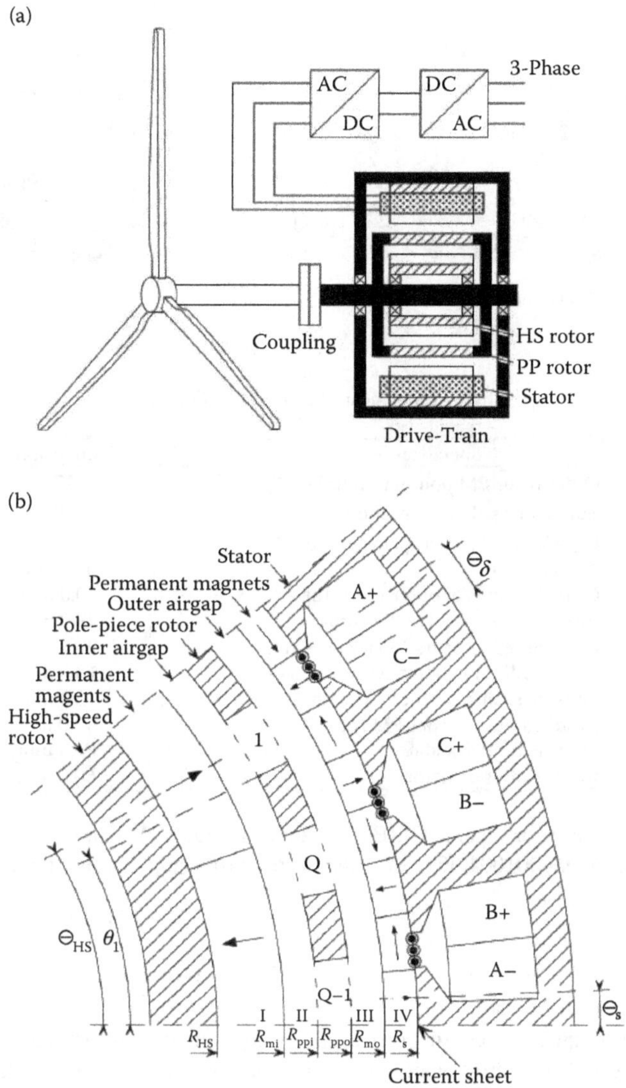

FIGURE 13.8 Dual-rotor MG-REM with additional (stator) PMs (a) integration with a wind turbine; (b) 40° periodic symmetry cross-section. (After A. Penzkofer, K. Atallah, *IEEE Trans.*, vol. MAG–51, no. 12, 2015, pp. 8700814. [9])

- Dual-rotor interior-stator MG-REMs for EVs and generators for good power factor and torque density
- Axial-airgap dual-rotor dual (electric part) MG-REMs for HEVs (for good torque density and four-quadrant independent rotor control)

13.2 DUAL-ROTOR INTERIOR STATOR MAGNETIC-GEARED RELUCTANCE ELECTRIC MACHINES

Let us present again for convenience the cross-section of a high-power-factor dual-rotor interior stator MG-REM (Figure 13.9) [3].

With high-speed PM rotor $p_{ePM} = p_a = 2$ pole pairs, flux-modulator pole count (pair) $p_{F-M} = 19$, and lower-speed PM rotor pole pair $p_{ePM} = 17$, the gear ratio $G = p_{ePM}/p_a = 19/2 = 9.5$.

(a)

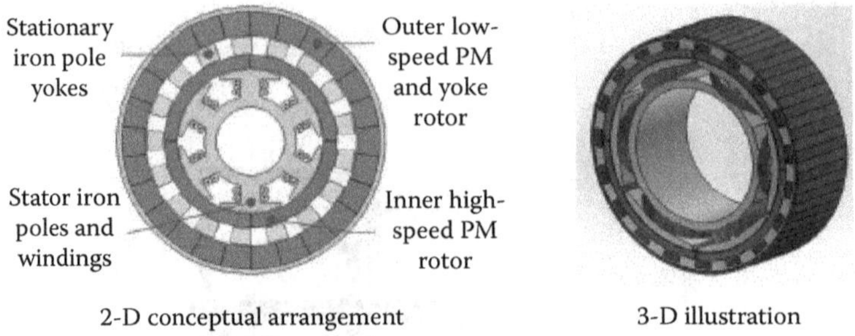

(b)

Table I
Physical specifications of MDPMG prototype

Specification	Dimension
Outer rotor PM pole-pair number, P_{ro}	17
Stationary pole yoke number, P_{vk}	19
Inner rotor PM pole-pair number, P_{ri}	2
Axial length	100.0 mm
Outside radius of outer PM rotor	150.0 mm
Inside radius of outer PM rotor	129.0 mm
Outside radius of stationary pole yokes	127.5 mm
Inside radius of stationary pole yokes	113.5 mm
Outside radius of inner PM rotor	112.0 mm
Inside radius of inner PM rotor	99.0 mm
Outside radius of stator	97.5 mm
Inside radius of stator	60.0 mm

FIGURE 13.9 Dual-rotor interior stator MG-REM. (After C-T. Liu, H.-Y. Chung, C.-C. Hwang, *IEEE Trans.*, vol. MAG–50, no. 1, 2014, p. 4001004. [3]): (a) cross-section; (b) sample prototype data.

At 1800 rpm speed of the interior four-pole SPMSM components, the low-speed (load) PM rotor runs only at $1800/9.5 \approx 189.5$ rpm.

Surface PMs have been adopted in both high-speed and the low-speed rotors to secure small machine inductance, good for a higher power factor and for low voltage regulation in generating.

Dividing the magnets in pieces, tangentially and axially, is advised to reduce PM eddy current losses.

The four PM poles on the high-speed rotor (adjacent to the inner six-slot stator) should cover only 0.7–0.75 of the pole pitch span τ_{PM}, as the cogging torque may be drastically reduced this way. There will thus be space between PM magnets on the high-speed glass-shaped rotor where a nonconducting (siluminium or stainless steel) ladderlike shape rotor frame may be adopted without excessive additional eddy currents in the latter. The PM segments have parallel—easier to inflict—magnetization, that is, lower PM costs. The stationary flux modulator and outer PM rotor construction are not an easy task, either.

Though an analytical field model may be adopted for this kind of complex topologies [9], it is felt here that direct 3D (2D)-FEM modeling is more practical, even for optimal design. Making use of 3D-FEM (Figure 13.10 [3]), the PM magnetic flux linkage versus electric cycle shows some space harmonics, but its THD is less than 5%.

The static torque (at zero speed) with DC stator currents ($I_a = -2I_b = -2I_c$) has produced torque on the high- and low-speed rotors (Figure 13.11) [3].

The pull-out torques are 170.5 Nm and $170.5/9.5 \approx 18$ Nm, respectively, for the low- and high-speed PM rotors.

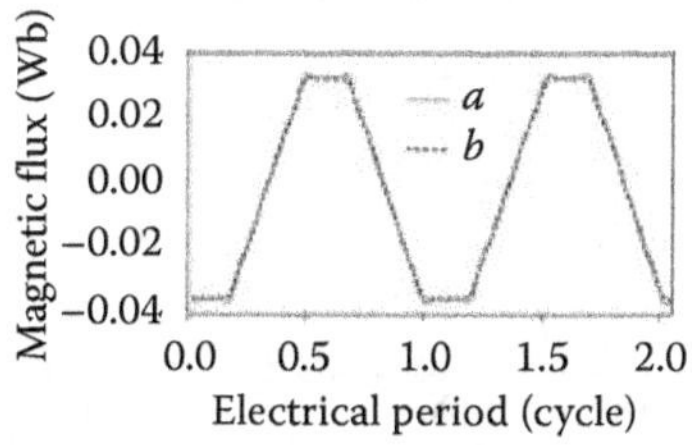

FIGURE 13.10 PM flux linkage in the four-pole inner stator phase versus electric cycle (two cycles per revolution: $p_a = 2$). (After C-T. Liu, H.-Y. Chung, C.-C. Hwang, *IEEE Trans.*, vol. MAG–50, no. 1, 2014, p. 4001004. [3])

With a high (0.65) slot filling factor, for the six coils that make the two-layer tooth-wound inner stator windings, the number of turns per phase (two coils/phase) may be calculated after setting the coil mmf that produces the desired pull-out torque (by FEM, after a preliminary standard design produces a start-up machine geometry).

The design of the 6/4 SPMSM is standard in calculating the number of turns per phase from the phasor diagram.

An assumed motor pure iq control would correspond to MTPA conditions due to small saliency $L_d = L_q = L_s$, but for an autonomous generator, the machine has to produce at a certain AC output voltage, load power factor, and active power. For pure resistive load, the phasor diagram of the machine as generator is rather straightforward, as in Figure 13.12:

$$V_{s1} - R_s I_{s1} = \sqrt{E_1^2 - (\omega_1 L_s i_{s1})^2}; \text{ rms values} \tag{13.9}$$

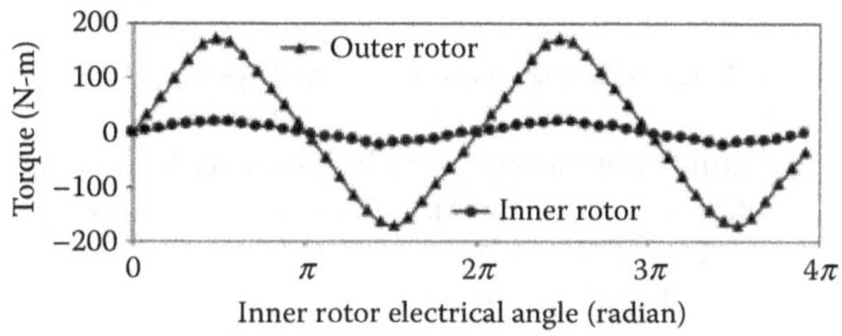

FIGURE 13.11 Torques on the outer (low-speed) and inner (high-speed) PM rotors. (After C-T. Liu, H.-Y. Chung, C.-C. Hwang, *IEEE Trans.*, vol. MAG–50, no. 1, 2014, p. 4001004. [3])

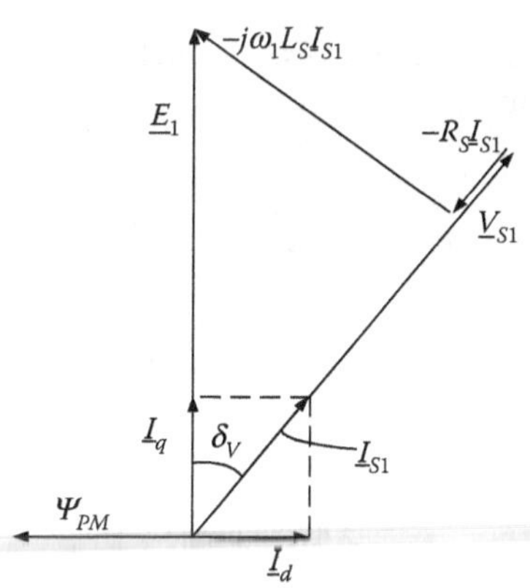

FIGURE 13.12 The phasor diagram of MG-REM as generator on resistive load.

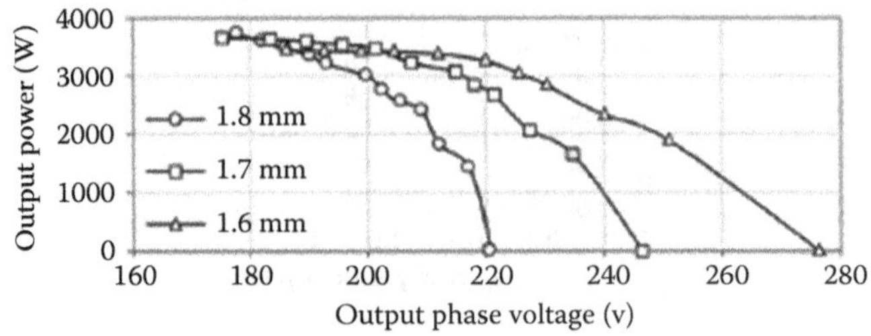

FIGURE 13.13 Generator power versus voltage for $W1 = 138, 123, 110$ turns/phase and given machine geometry. (After C-T. Liu, H.-Y. Chung, C.-C. Hwang, *IEEE Trans.*, vol. MAG–50, no. 1, 2014, p. 4001004. [3])

For already-known Φ_{PM1} (PM flux per pole pitch of the $2p_a$ pole motor):

$$E_1 = (\Phi_{PM1})_{max} W_1 k_{W_1} \omega_1 / \sqrt{2} \tag{13.10}$$

$$L_s = k_L W_1^2; \quad W_1 I_{S1} = \text{known} \tag{13.11}$$

W_1—turns per phase; k_{W1}—fundamental winding factor.

The machine inductance for $W_1 = 1$ may be calculated analytically or extracted from FEM calculations, while $W_1 I_{s1}$ is already known from analytical (or FEM) design as computed for the given full torque and power angle δ_V. Some iterations are needed here, as the presence of I_d produces only losses, but its presence is imposed by the unity power factor (resistive) load considered here.

In [3], $W_1 = 138$, 123, 110 turns/phase were tried, with power/voltage generator curves as in Figure 13.13 [3].

It is evident that a smaller number of turns ($W_1 = 110$) produces less voltage regulation:–(4–7)% at 2.5 kW. There is a compromise here, as a too-large number of turns per phase tends to produce a larger (inadmissible) voltage. Here, 220 V is the targeted phase voltage (RMS). Also, a rated power angle around $\delta_V = \pi/4$ is to be selected for the scope.

The acceptable voltage regulation (at unity power factor) (Figure 13.14) [3] is a clear consequence of a reasonably low machine inductance (in p.u.).

It goes without saying that some kind of voltage closed-loop control is required, even if the speed is constant, because the autonomous load power factor varies when a $\pm$ 5% voltage regulation is imposed across the board.

Solutions to this closed-loop control range, from switched AC capacitors to parallel power filters for AC output, may be appropriate.

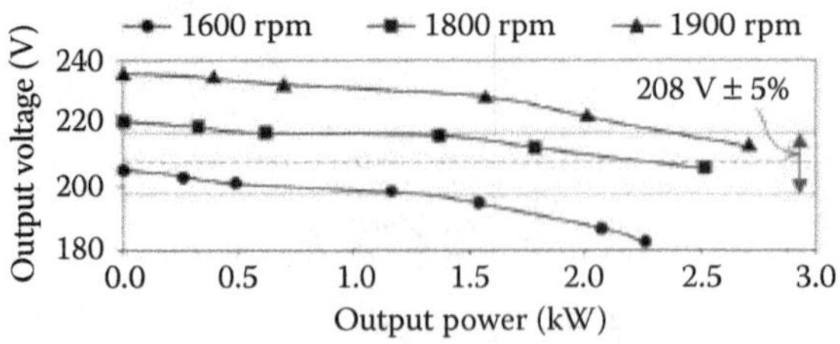

FIGURE 13.14 Voltage versus power at three different speeds. (After C-T. Liu, H.-Y. Chung, C.-C. Hwang, *IEEE Trans.*, vol. MAG–50, no. 1, 2014, p. 4001004. [3])

A PWM active rectifier is required for controlled DC output voltage. For the case in point, a loss breakdown was performed and at 1800 rpm for the high-speed rotor (and 189 rpm for the low-speed rotor), for 2522 W output power (in resistive load), the line current was 4.88 A at 220 V/phase with 71 W of iron losses, 87.25 W of copper losses, but 295 W of lower-speed rotor mechanical losses and 285 W for high-speed rotor mechanical losses. A total efficiency of around 80% was obtained. The very large mechanical losses by the two rotors should be drastically reduced to make the solution practical.

For EV propulsion, a design, refined in a few stages, with full-size experiments at 60 kW (gear ratio $G = 9/1$) (Figure 13.15) [2, 11], has been performed.

The high-speed (inner) PM rotor has eight PM poles on both sides ($p_a = p_{h\mathrm{PM}} = 4$), initially with surface magnets in titanium shells, but later as IPMs (Figure 13.16) [2]. The fixed $F - M$ (outer cylinder) has 64 salient poles ($p_{F-M} = 64$), while the outer IPM rotor has 60 pole pairs [2]: $p_{e\mathrm{PM}} = 60 = p_{F-M} - p_a$.

The fixed flux modulator with a segmented cylinder was finally made of one-piece laminations with salient poles and dummy slots for reducing the core losses and providing a more mechanically rugged lower core loss (Figures 13.16 and 13.17) [2].

Interior PMs on the outer low-speed rotor have been adopted to increase outer airgap flux density (and torque) by PM flux concentration.

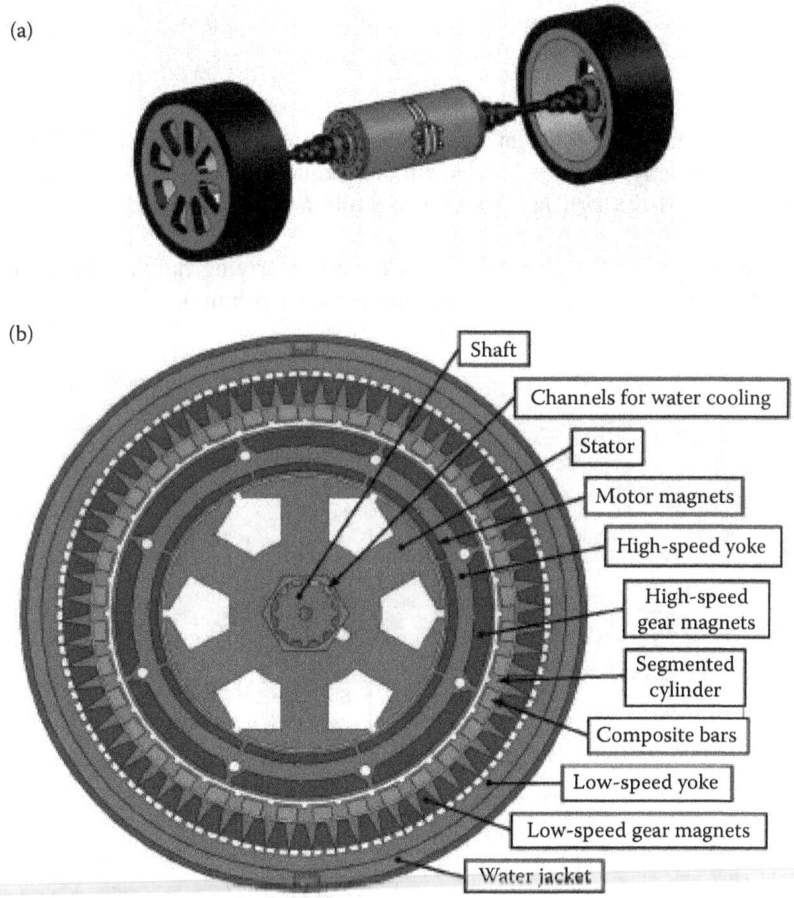

FIGURE 13.15 Dual rotor interior-stator MG-REM for EVs: (a) twin motor on EV axle; (b) cross-view. (After T. V. Frandsen, P. O. Rasmussen, K. K. Jensen, *Record of IEEE–ECCE,* 2012, pp. 3332–3339. [2])

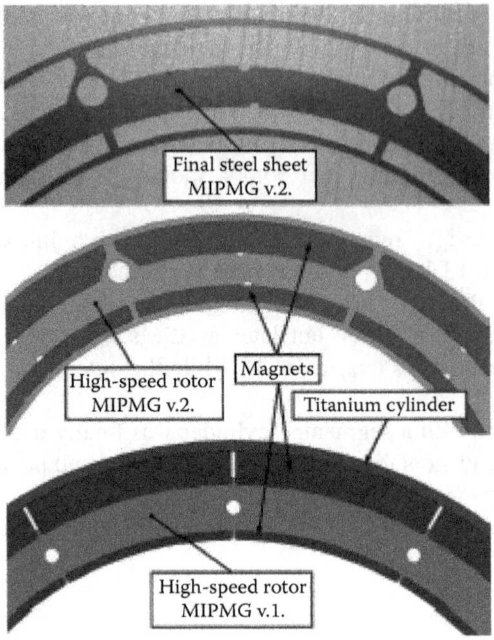

FIGURE 13.16 Old (12-pole) and new (8-pole) high-speed PM rotor. (After T. V. Frandsen, P. O. Rasmussen, K. K. Jensen, *Record of IEEE–ECCE,* 2012, pp. 3332–3339. [2])

A calculated efficiency of 95% at 248 Nm and $n_p = 8000$ rpm ($n_{low} = 8000/9$ rpm) was reported in an 8.6-liter active volume [2]. Maximum torque of 600 Nm was needed and produced to accelerate the EV to 100 km/h in 15 seconds, while a maximum stall torque of 857 Nm was measured.

The torques versus time and speed conditions for urban driving have all been met at a record (so far) 100 Nm/liter of active volume (64 Nm/liter per total volume).

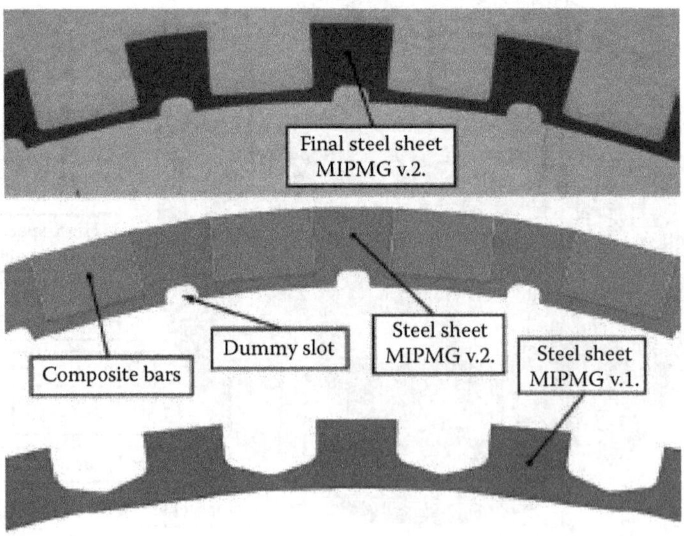

FIGURE 13.17 Old and improved segmented-cylinder (flux modulator). (After T. V. Frandsen, P. O. Rasmussen, K. K. Jensen, *Record of IEEE–ECCE,* 2012, pp. 3332–3339. [2])

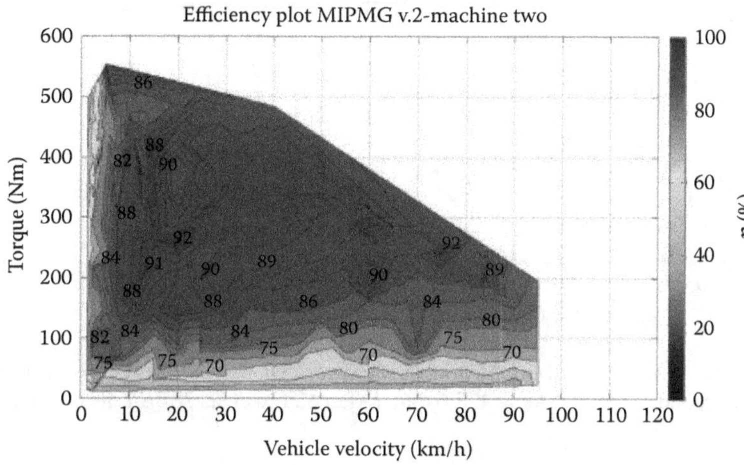

FIGURE 13.18 Measured efficiency versus torque and automobile speed. (After T. V. Frandsen et al., *IEEE Trans.*, vol. IA–51, no. 2, 2015, pp.1516–1525. [11])

Measured efficiency with torque and speed (Figure 13.18) is satisfactory but smaller than calculated values by (3–4%) [11]. This is a sign that more effort on loss control and reduction is required; also, the large PM weight, cost and PM demagnetization risks have to be addressed further.

13.3 BRUSHLESS DUAL-ROTOR DUAL-ELECTRIC PART MAGNETIC-GEARED RELUCTANCE ELECTRIC MACHINES

As already discussed briefly in "Introduction," the brushless axial-airgap dual-rotor dual-electric part MG-REM [6] (Figure 13.5b) is "copied" here for convenience (Figure 13.19).

As evident in Figure 13.18, there is only one spoke-shaped magnet rotor of $p_{\text{PMs}} = 14$ pole pairs that interacts with both the PM secondary stator winding of the same $p_{a2} = p_{\text{PMs}} = 14$ pole pairs and

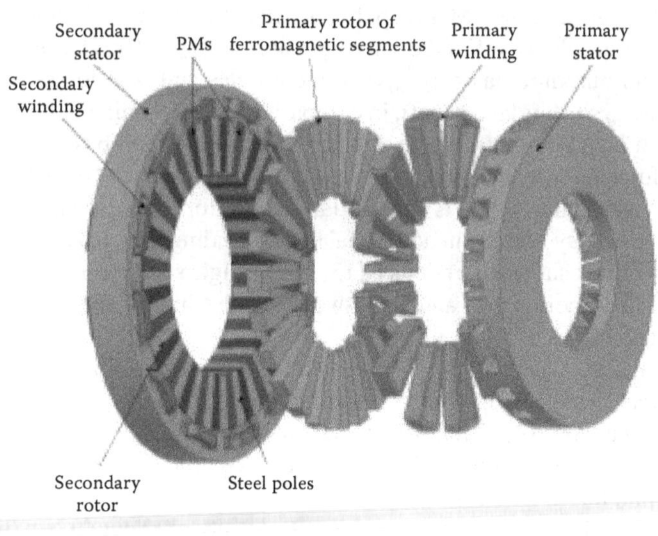

FIGURE 13.19 Dual-rotor dual-electric-part MG-REM with axial airgap. (After Y. Liu et al., *IEEE Trans.*, vol. MAG–50, no. 11, 2014. [6])

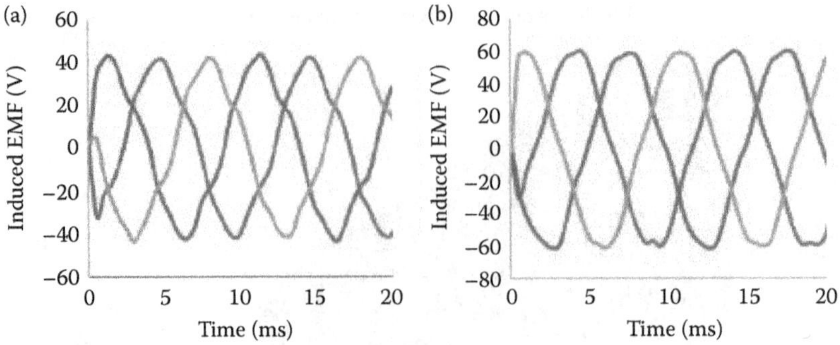

FIGURE 13.20 Emfs (under no load) in the primary (a); and secondary (b), windings, with primary $(F - M)$ rotor at stall and the secondary (PM) rotor at 428.6 rpm. (After Y. Liu et al., *IEEE Trans.*, vol. MAG–50, no. 11, 2014. [6])

the primary rotor (flux modulator), which has 24 salient poles (pole pairs: $p_{F\text{–}M} = 24$). The latter interacts with the primary winding pole pairs $p_{a1} = 10$ ($p_{a1} = p_{F\text{–}M}\text{–}p_{\text{PMs}} = 24\text{–}14 = 10$). The speed synchronism relationship is given in Equation 13.7. With 24 slots in the two stators, the two fractionary windings with 10 and 14 pole pairs may be built for $q_1 = 24/60$ and $q_2 = 24/84$ slots/pole/phase. Such $q < 1$ windings mean shorter end coils and thus lower copper losses and shorter frame (machine) radial height. The principle of this complex machine is illustrated in a test where the primary rotor (*F–M*) speed $\Omega_{F-M} = 0$ and the second rotor is rotating under no load; the emfs in the two windings should have the same frequency. Equation 13.7 yields: $f_{a1} = f_{a2} = p_{\text{PMs}} \cdot n_{\text{PMs}} = 14 \times 428.6/60 = 100$ Hz (Figure 13.20 [6]) for 428.6 rpm in the low-speed rotor and zero speed for the flux modulator.

It should be noted that the secondary (PM) rotor interacts with both stators and with the flux-modulator (primary) rotor using the same magnets. Consequently, there are PM flux lines in three directions and thus a soft magnetic composite should be sandwiched between the spoke-shaped magnets of the second rotor.

It is also recognized that up to 30% of the PM flux of the secondary rotor is leaking at the low radial end (toward the shaft), but, still, the PM airgap flux density is large enough (with more PM weight) to secure a competitive torque density. For the same case of $f_{a1} = f_{a2} = 100$ Hz, secondary rotor speed of $n_{\text{PMs}} = 428.6$ rpm, and flux modulator $n_{F-M} = 0$, the torques and core losses of the two shafts, for $j_{\text{core}} = 8\ \text{A/mm}^2$ (pure i_q sinusoidal current), in comparison with a similar machine with radial airgap and single stator with dual windings, are shown in Figure 13.21a and b [6].

It is evident that for about the same machine volume, PM volume, and copper and core losses (167 W and 29 W, respectively), the torque is substantially larger for the axial-airgap dual-stator electric-part dual-rotor MG-REM. By adding an additional stator winding with p_{PM} pole pairs to the magnetically geared dual-rotor mechanical port device [13, 14], a single stator winding and inverter with dual PM rotor (with p_1 and p_2 pole pairs) and tooth-wound coil winding can also provide independent torque control for each rotor.

Experimental results are needed to prove the practicality of such a complex machine for HEV propulsion, while optimal design with limited temperatures and demagnetization avoidance are also worth pursuing. Optimal control of dual-rotor dual electric-part MG-REM is still to be developed. The full load torque density would be 30 Nm/liter active volume at an electrical efficiency of about 90% at only 428.6 rpm, for 45 Nm (2 kW) on a secondary machine (the first machine is not activated in this example). With both stators active, a total torque density of about 45 Nm/liter (active volume) may be available with an outer secondary stator radius of 120 mm. Also, the PM volume was only 0.254 liters (about 1.80 kg of sintered NdFeB magnets with $B_r = 1.1$ T).

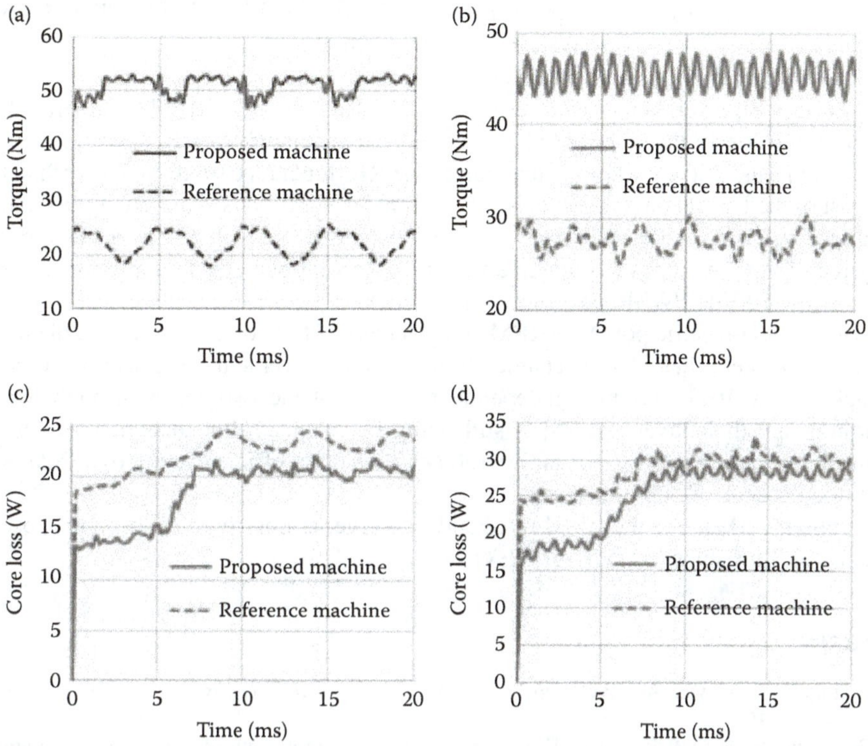

FIGURE 13.21 Full load torques of primary and secondary rotors of axial-airgap versus radial airgap (with single stator) dual-rotor dual electric part MG-REM: (a) full torque of primary rotor; (b) full torque of secondary rotor; (c) core losses for the primary stator; (d) core losses for the secondary stator, at $j_{con} = 8$ A/mm^2, $n_{F-M} = 0$, $n_{PMs} = 428.6$ rpm. (After Y. Liu et al., *IEEE Trans.*, vol. MAG–50, no. 11, 2014. [6])

13.4 SUMMARY

- Dual-rotor single (and dual) electric part magnetically geared reluctance electric machines fall into the general category of flux-modulation reluctance machines.
- Dual-rotor single-wound-stator single electric parts, in general, with single field-oriented control do not provide independent rotor control (motoring and generating). So they are suitable for applications that do not need dual mechanical ports. One rotor may be free but will experience torque and participate in magnetic gear torque magnification [2, 3–4, 12].
- Torque densities up to 90 Nm/liter for efficiency above 0.92 at 600 rpm have been reported for such machines at 30 kW [2, 11].
- Independent control of the two rotors, with a single stator winding and inverter, has been introduced in an axial airgap configuration where a 12-slot central stator with Gramme-ring windings "serves" two rotors, one of 10 poles and the other of 14 poles, while the single inverter uses dual-frequency FOC [7].
- A dual-rotor single outer electric port, with PMs both on the stator surface and the inner (free) high-speed PM rotor, with the flux modulator at the low-speed (working) rotor, has been thoroughly investigated for a pseudodirect-drive, 9.655-rpm, 10-MNm (10-MW) wind generator [9].

Calculations [9] claim an electric efficiency of 98.5% for 45 tons of active weight from which 13.5 tons are of sintered NdFeB magnets. That is, 60% more PMs than in a regular three-slot/pole SPMSG (at above 0.9 power factor). A TF-PMSG (Chapter 12) is credited with 0.967 efficiency and 0.67 power factor and 27 tons (active weight) with 5 tons of magnets. Power factor is not expected to be large in the dual-rotor single-electric-part MG-REM proposition, but it is yet to be calculated.

- As known, low power factor increases losses in the machine and its PWM converter, but also increases the cost of the inverter (which depends on machine kVA). However, in low-speed applications, the initial cost of the machine (MG-REM included) ends up larger than the initial cost of the PWM converter; this is why optimal design codes should include the inverter cost per kVA and the PWM converter losses cost when the global cost of the apparatus is considered in the composite objective function. This rationale is also true for low voltage (say, below 50 Vdc PWM converter applications).
- Dual-rotor dual-electric port MG-REMs may act as electric continuously variable transmissions and thus eliminate the mechanical transmission in vehicular applications (gear ratio, though, below 10/1). Optimal independent control of the two rotors through two stator windings and their inverters [15] is still being investigated, but theoretical investigation showed competitive torque densities (50 Nm/liter) for efficiencies above 90% starting from load speeds of 400 rpm or so.
- Also, two PMSMs, one with dual-rotor and two-inverter control, have been proposed very recently as a brushless e-CVT for HEVs [16].

REFERENCES

1. K. Atallah, D. Howe, A novel high performance gear, *IEEE Trans.*, vol. MAG–37, no. 4, 2001, pp. 2844–2846.
2. T. V. Frandsen, P. O. Rasmussen, K. K. Jensen, Improved motor integrated permanent magnet gear for traction applications, *Record of IEEE–ECCE*, 2012, pp. 3332–3339.
3. C-T. Liu, H.-Y. Chung, C.-C. Hwang, Design assessments of a magnetic-geared double-rotor permanent magnet generator, *IEEE Trans.*, vol. MAG–50, no. 1, 2014, p. 4001004.
4. J. Bai, P. Zheng, L. Cheng, S. Zhang, J. Liu, Z. Liu, A new magnetic-field-modulated brushless double-rotor machine, *IEEE Trans.*, vol. MAG–51, no. 11, 2015, p. 8112104.
5. S. Niu, S. L. Ho, W. N. Fu, A novel double-stator double-rotor brushless electrical continuously variable transmission system, *IEEE Trans.*, vol. MAG–49, no. 7, 2013, pp. 3909–3912.
6. Y. Liu, Sh. Niu, S. L. Ho, W. N. Fu, A new hybrid-excited electric continuous variable transmission system, *IEEE Trans.*, vol. MAG–50, no. 11, 2014, pp. 810404.
7. I. Boldea, L. N. Tutelea, S. I. Deaconu, F. Marignetti, Dual rotor single-stator axial airgap PMSM motor/generator drive for HEVs: A review of comprehensive modeling and performance characterization, *Record of IEEE–ESARS*, 2012, Bologna, Italy.
8. Y. H. Yeh, M. F. Hsieh, D. G. Dorrell, Different arrangements dual-rotor, dual-output radial-flux motors, *IEEE Trans.*, vol. IA–48, no. 2, 2012, pp. 612–622.
9. A. Penzkofer, K. Atallah, Analytical modeling and optimization of pseudo-direct drive PM machines for large wind turbines, *IEEE Trans.*, vol. MAG–51, no. 12, 2015, pp. 8700814.
10. H. Polinder, D. Bang, R. P. J. O. M. van Rooij, A. S. McDonald, M. A. Mueller, 10 MW wind turbine direct drive generator design with pitch or active speed stall control, *Record of IEEE—IEMDC*, 2007, pp. 1390–1395 (IEEEXplore).
11. T. V. Frandsen, L. Mathe, N. I. Berg, R. K. Holm, T. N. Matzen, P. O. Rasmussen, K. K. Jensen, Motor integrated PM gear in a battery electrical vehicle, *IEEE Trans.*, vol. IA–51, no. 2, 2015, pp.1516–1525.
12. M. Fukuoka, K. Nakamura, H. Kato, O. Ichinokura, A novel flux-modulated dual-axis motor for HEVs, *IEEE Trans.*, vol. MAG–50, no. 11, 2014, pp. 8202804.
13. D. Li, R. Qu, X. Ren, Y. Gao, Brushless dual-electrical part dual mechanical part machines based on flux-modulation principles, 2016 IEEE Energy Conversion Congress and Exposition (ECCE), pp. 1–8.
14. Y. Wang, S. Niu, W. Fu, Electrical continuously variable transmission system based on doubly fed flux-bidirectional modulation, *IEEE Trans.*, vol. IE–64, no. 4, 2017, pp. 2722–2731.
15. M. Bouheraoua, J. Wang, K. Atallah, Rotor position estimation of a pseudo direct drive PM machine using extended Kalman filter, *IEEE Trans.*, vol. IA–53, no. 2, 2017, pp. 1088–1095.
16. L. Sun, M. Cheng, H. Wen, L. Song, Motion control and performance evaluation of a magnetic—geared dual—rotor motor hybrid power train, *IEEE Trans.*, vol. IE–64, no. 3, 2017, pp. 1863–1872.

14 Direct Current + Alternating Current Stator Doubly Salient Electric Machines

Analysis, Design, and Performance

14.1 INTRODUCTION

Switched reluctance machines [1] are typically double salient and have one- or multiphase armature windings (on the stator) that are connected in sequence and triggered by the measured (or estimated) rotor position (Figure 14.1).

The torque of an SRM, magnetic saturation and phase interaction neglected, is simply:

$$T_e = \frac{1}{2}\sum_{i=1}^{m} i_i^2 \frac{dL_i}{d\theta_\gamma} \tag{14.1}$$

So the current polarity is not important in torque production. Based on this observation, multiphase DC–DC converters in parallel (one for each phase, in principle) have been applied to control the current pulses.

The simplicity (ruggedness) of the SRM with tooth-wound coils is staggering, which explains the enormous interest in it since its revival in the 1970s (S.A. Nasar apparently coined the acronym SRM) [1]. However, it was soon recognized that:

- Good torque density comes with higher peak flux density in the airgap (1.4–1.5 T) with only one or two phases active at any time; thus, peak radial forces, noise, and vibration are high.
- In an effort to reduce the radial flux density (normal force) but encourage tangential (torque) force on the rotor, a segmented glass-shaped rotor was sandwiched between two stators with diametrical coils and corresponding outer and inner phases in series in a two-airgap structure. Though with much longer end coils, notably more torque is obtained for the same geometry, copper, and current density than with tooth-wound coils. But the topology is more complicated [2] and thus not easy to fabricate.
- On top of that, the ratio of active power/kVA (RMS and peak) is notably lower [1] than with 4- to 8-pole IMs, SMs, or PMSMs; this means larger peak and RMS kVA in the converter, which, for unipolar current, is different from the off-the-shelf PWM inverters with sinusoidal or trapezoidal current control.
- While the carefully designed regular SRM was recently proven capable of 45 Nm/l, for 400 Nm, efficiency above 0.92 from 2000 rpm, the peak kVA in the converter is still large for the new standards in HEV and EV applications [3]. IMs and IPMSM [4] for the same application proved to be pretty equivalent in Wh/km in standard urban driving of HEV (EV) with the IM efficiency lower at low speed but better at high speeds (flux weakening) at peak torque densities of 45 Nm/liter and above 60 (70) Nm/liter (active volume), respectively. These solutions use DC–DC boosting converters plus off-the-shelf IGBT 500 Vdc PWM inverters. SRM R&D will continue. The incipient commercialization decades ago slowed

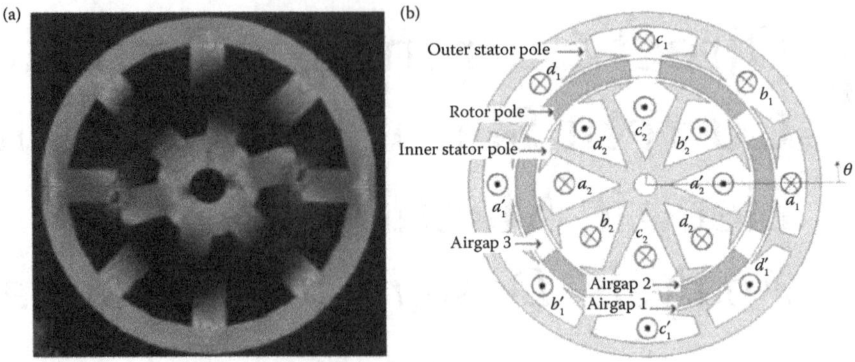

FIGURE 14.1 Switched reluctance motors: (a) with single stator and tooth wound coils; (b) with dual stator and segmented glass-shaped rotor in between. (After M. Abassian, M. Moallem, B. Fahimi, *IEEE Trans on*, vol. EC–25, No. 3, 2010, pp. 589–597. [2])

down, but it may revive in niche applications. To further reduce losses and increase the kW/kVA ratio in SRMs, the doubly salient cores of the latter have been used to:

- Add DC excitation coils on the stator.
- Now make use of AC (armature torque) coils in same slots as the DC coils; or the DC coils are placed in an inner stator with a salient-pole rotor in between.
- The DC excitation and AC armature coils may be tooth- or multitooth-wound (Figure 13.2a and b) [5–7] in the search for higher torque density, lower torque ripple, better power factor, and lower core losses.
- Such machines behave like flux-modulation machines but, alternatively, as DC-excited (cageless) rotor synchronous machines.
- It is to be borne in mind that DC coil flux in the AC coils reverses polarity in two-tooth-wound coil configurations (Figure 14.2b), but it varies (in general) from a maximum to minimum positive value for single tooth-wound coil topologies. The situation is, however, rather similar to the first case where only half the AC coil span is active, while in the second case, the single slot pitch AC coil span is active. So there is a kind of "homopolar" effect with stator DC excitation in the sense that only about half of it is useful.

  But the airgap is kept small for good magnetic coupling and flux modulation through the two saliencies (of rotor and of stator). This means, however, a large machine

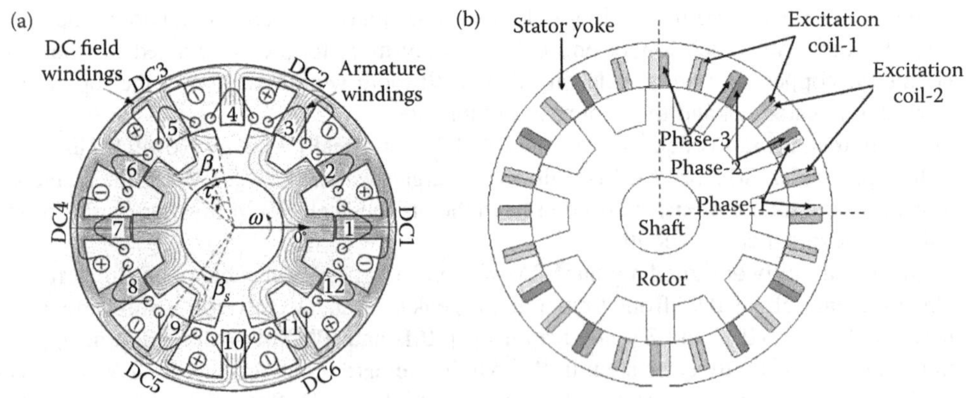

FIGURE 14.2 DC + AC stator doubly salient machines (DC + AC DSM): (a) with tooth-wound DC and AC coils; (b) with two-tooth-wound DC and stator coils. (After E. Sulaiman, T. Kosaka, N. Matsui, *Record of EPE—2011*, Birmingham, U.K. [7])

inductance, which may be useful in motoring for wide CPSR but detrimental in generating due to large voltage regulation.

- In both cases, when high torque density is targeted, the machine saturates. The power factor, considering sinusoidal emfs and ($L_s = L_d \approx L_q$), becomes low (0.6) due to the inherently large synchronous inductance L_s in p.u. A typical such configuration (DC-stator-excited, AC SRM), stemming directly from SRM (Figure 14.3) has been introduced recently as a constant DC output voltage autonomous generator with diode rectifier by using DC excitation current-only control [8,9]. 3.2 kW, 62 Vdc, at 3600 rpm, for 11.2 kg total weight (shaft and frame included), however at 79% efficiency (0.21 kW diode rectifier loss included), was reported in full-scale experiments [9].
- Despite these demerits, the machine simplicity and ruggedness appeal for quite a few applications from low-speed high-torque ones to moderate speed (even up to 6–10 krpm in HEV or EVs). Balancing the lower ownership costs of the electric machine with larger kVA inverter costs is key in a design suitable to a certain application, often with notable CPSR.

In this case, however, the phases carry bipolar currents. However, the output emf voltages are far from sinusoids unless geometry changes are made. So, its usage as a motor or AC output autonomous generator is still in question.

Note: The literature uses different names for all these "DC + AC stator doubly salient machines," each with some justification. The name adopted here is deemed to expose the principle and topology quickly and differentiate from "brushless doubly fed reluctance machines" that have DC + AC distributed windings in the stator but only one (rotor) saliency (Chapter 8). The treatment of these "flux-switching," "flux reversal," "variable-flux" doubly salient, or "flux modulation" magnetless machines in a dedicated chapter is justified here by the need to go into detail (usually where the devil is) and expose their full capabilities. This rugged (brushless) PM-less breed of electric machinery for variable speed drives in motoring and/or generating seems to hold real industrial potential, at least for niche applications [9].

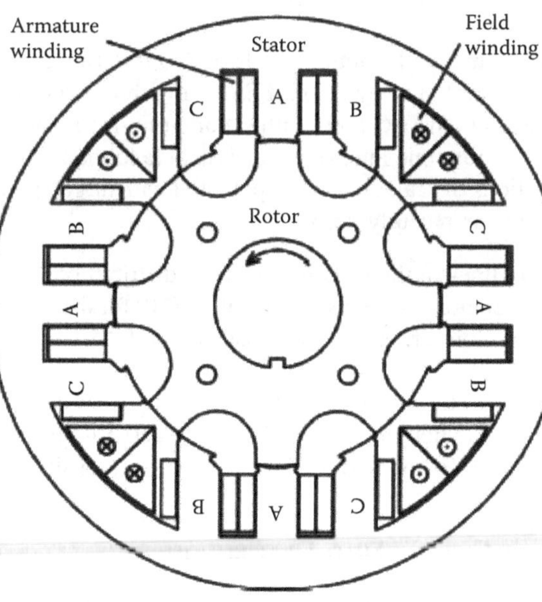

FIGURE 14.3 DC + AC stator SRM.

In what follows, we will treat the following issues in some detail:

- Two-slot-span coil stator DC + AC winding doubly salient machines.
- One tooth-wound stator DC + AC winding doubly salient machines (with single and dual stator).
- The DC + AC winding SRM with bipolar AC currents. The main common ground of all electric machines treated here is the use of sinusoidal current control (FOC) by off-the-shelf PWM inverters.

14.2 TWO-SLOT-SPAN COIL STATOR DIRECT CURRENT + ALTERNATING CURRENT WINDING DOUBLY SALIENT MACHINES

Two-slot-span stator coils, DC and AC, may be used also in FS-PMSMs with hybrid excitation and thus are considered here first (Figure 14.2b). Although there are many stator and rotor slot (pole) combinations N_s and N_r, only those that correspond to the condition:

$$N_r = N_s/2 \pm j;\ j = 1, 2 \tag{14.2}$$

are investigated here, with j the number of layers in the AC coil winding. The case of 24 stator slots (12 slots for the DC coils and 12 for the AC coils) and $N_r = 10$ rotor poles, $j = 2$ is considered here in detail.

A thorough characterization of such a DC + AC double salient machine (DC + AC DSM) requires the investigation of:

- A simplified analytical theory to produce a start-up machine geometry
- DC coil field distribution and their flux and emf in the AC coil waveforms for different values of DC current (magnetic saturation has to be allowed for)
- AC coil field distribution, flux waveforms in the coils (phases), and inductances—self and mutual—versus rotor position (magnetic saturation has to be considered)
- Torque waveform for given DC and AC coil mmfs (pure i_q control)
- Core loss under no load and under load in the stator and rotor at various torques and speeds
- Normal force on the stator rim distribution

The above targets could be met with enough precision (considering the correct airgap permeance variation and magnetic saturation) only by 2(3)D-FEM using the frozen permeability method when calculating DC-produced emfs in the AC coils with stator currents in place (under load). As known, such a method is still computation time intensive. Analytical field methods such as polar coordinates with Fourier harmonic solutions of scalar magnetic potential with airgap permeance function for infinite core permeability have been recently derived:

- With semicircular flux lines in the slots for DC field distribution
- With scalar magnetic potential Fourier series solution [10] for the stator (and then rotor) slotting regions for the AC coil (armature) field distribution [11]

Despite their neglect of magnetic saturation, the analytical methods in [10,11] pertinently describe the phenomena, at least at small enough currents (loads), to serve in pre-optimal design. In optimal design codes, either nonlinear MEC or 2(3)FEM machine modeling should be performed anyway.

14.2.1 No-Load Direct Current Field Distribution

Figure 14.4 illustrates the position of DC coils and their flux line direction reversal in the AC coil (of phase A), with their almost sinusoidal flux linkage in the AC coil.

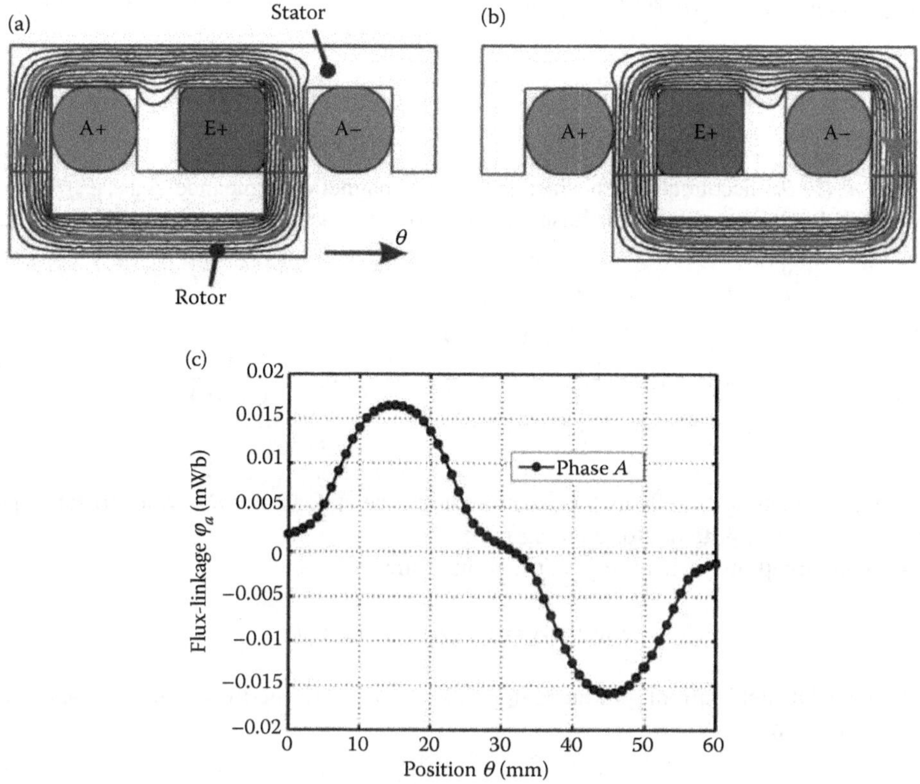

FIGURE 14.4 DC + AC DSM: DC coils, (a); their flux lines for two rotor positions, (b); their flux linkage in the phase A versus rotor position, (c). (After B. Gaussens et al., *IEEE Trans.*, vol. MAG–48, no. 9, 2012, pp. 2505–2517. [10])

The trapezoidal shape of DC mmf shows space harmonics $F_{dc}(v)$, but so does the airgap function $g(v, \theta_r)$ [10]:

$$g(v, \theta_r) = g_0 + g_s(v) + g_r(v, \theta_r) \tag{14.3}$$

With quarter-circular flux lines in the stator slots, the stator contribution to airgap permeance variation $g_r(v)$ is:

$$g_s(v) = \begin{cases} \pi R_{si} \cdot \dfrac{\sin(v/2)\sin\big(\beta_s(\theta_s/2) - (v/2)\big)}{\sin(\beta_s\theta_s/4)\cos\big((v/2) - (\beta_s\theta_s/4)\big)}; & 0 < v \le \beta_s\theta_s \\ 0 & \end{cases} \tag{14.4}$$

Using Fourier series decomposition, $g_s(v)$ becomes: $\beta_s\theta_s < v < \theta_s$

$$g_s(v) = a_0^{(\mathrm{I})} + \sum_{n=1}^{\infty} a_n^{(\mathrm{I})} \cos(nN_s v) + b_n^{(\mathrm{I})} \sin(nN_s v) \tag{14.5}$$

Similarly, for the rotor slotting [10]:

$$g_r(v, \theta_r) = a_0^{(\mathrm{II})} + \sum_{n=1}^{\infty} a_n^{(\mathrm{II})} \cos(nN_r(v - \theta_r)) + \sum_{n=1}^{\infty} b_n^{(\mathrm{II})} \sin(nN_r(v - \theta_r)) \tag{14.6}$$

Finally, the DC mmf $F_{dc}(v, \theta_r)$ becomes:

$$F_{dc}(v, \theta_r) = k_c(v, \theta_r) + \sum_{n=1,3,5}^{\infty} b_n \sin\left(n\frac{N_s}{4}v\right) \tag{14.7}$$

The term $k_c(v, \theta_r)$ accounts for the fact that some magnetic permeance $\Lambda_g(v, \theta_r) = 1/g(v, \theta_r)$ harmonics will have an order equal those of $F_{dc}(\gamma, \theta_r)$, so that an offset is introduced (based on the condition div $\overline{B} = 0$), [10]:

$$k_c(v, \theta_r) = \frac{\int_0^{2\pi} F_{dc}(v)\Lambda_g(v, \theta_r)}{\int_0^{2\pi} \Lambda_g(v, \theta_r)dv}; \; \Lambda_g(v, \theta_r) = \frac{1}{g(v, \theta_r)} \tag{14.8}$$

In the above equations, v reflects the position along the stator bore; θ_r—rotor to stator position, θ_s—stator slot pitch, β_s—stator slot opening ratio.

The no-load airgap flux density $B_{g0}(v, \theta_r)$ is thus simply:

$$B_{g0}(v, \theta_r) = F_{dc}(v, \theta_r) \cdot \Lambda_g(v, \theta) \tag{14.9}$$

To obtain the DC field current–produced flux linkage in an AC one-turn coil, φ_{sc1}, we need to simply do an integration:

$$\varphi_{sc1} = \frac{D_{is}}{2} l_{\text{stack}} \int_{\theta_{s/2}}^{5\theta_{s/2}} B_{g0}(v, \theta_r)dv \tag{14.10}$$

For single-layer windings and:

$$\varphi_{sc1} = \frac{D_{is}}{2} l_{\text{stack}} \int_{(\theta_s/2)(1+(\beta_s/2))}^{(\theta_s/2)(5-(\beta_s/2))} B_{g0}(v, \theta_r)dv \tag{14.11}$$

for dual-layer windings.

A summation of such AC coil flux linkages per each phase is then required considering their position along the stator periphery (v angle) for single- and multilayer multiphase AC windings [10].

Comparisons of such calculations with FEM are offered in Figure 14.5 for the phase flux linkage and emf in a 24-slot/10-pole machine [10].

There are notable second and sixth harmonics in the emf and thus geometry optimization of slot shaping or rotor skewing may be necessary to produce a more sinusoidal emf.

The extension of the study above in [10] for the 24/11, 24/13, 24/14 machines revealed that the 24/11, 24/13 machines exhibit uncompensated normal forces ($f_n = B_{g0}^2/2\mu_0$(in (N/m^2))), which will produce noise and vibration.

The interaction between $F_{dc}(v, \theta_r)$ and $B_{g0}(v, \theta_r)$ produces cogging torque T_{cogg}:

$$T_{\text{cogg}} = l_{\text{stack}} \int_0^{2\pi} \frac{\partial F_{dc}(v, \theta_r)}{\partial \theta_r} B_{g0}(v, \theta_r)dv \tag{14.12}$$

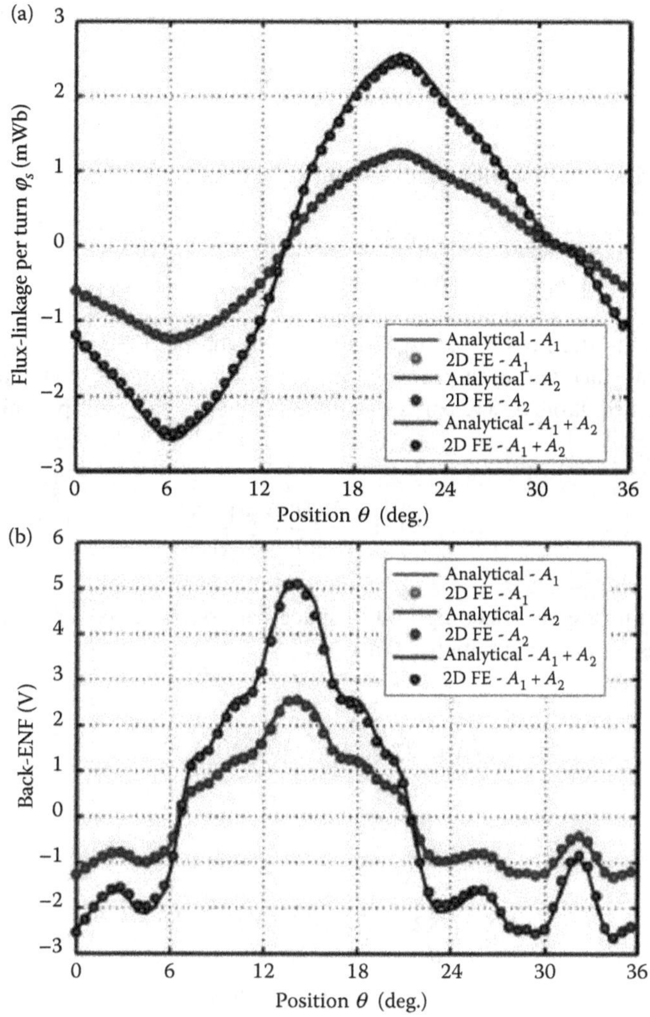

FIGURE 14.5 Flux linkage (for one-turn coils), (a); and emf, (b); in the coils A_1 and A_2 of phase A for a single-layer AC winding (24/10 slot/pole three-phase machine) for $F_{dc} = 1200$ Aturns. (After B. Gaussens et al., *IEEE Trans.*, vol. MAG–48, no. 9, 2012, pp. 2505–2517. [10])

The cogging torque turns out to be smaller for the 24/11, 24/13 machines. All in all, the 24/10 machine should in general be produced, because it has no uncompensated radial forces and its lower frequency for a given speed ($f_1 = N_r \cdot n$); that is, probably, lower core losses.

14.2.2 Armature Field and Inductances: Analytical Modeling

A similar analytical field method was developed in [11] to calculate the field distribution and self and mutual inductances of AC phases. This time, the scalar magnetic potential equations have been solved in polar coordinates in the airgap and stator slots, respectively, and the rotor slot zone where the Laplace equation holds.

$$\begin{aligned} \Delta V^{(\mathrm{I})} &= 0 \qquad \text{Region 1 (airgap)} \\ \Delta V^{(\mathrm{II})} &= 0 \qquad \text{Region 2 (slot area)} \end{aligned} \tag{14.13}$$

The scalar magnetic potential, V, is considered unity on rotor bore and zero on stator edges. Also:

$$\frac{\partial^2 V}{\partial \gamma^2} + \frac{1}{\gamma} \cdot \frac{\partial V}{\partial r} + \frac{1}{r^2} \cdot \frac{\partial^2 V}{\partial v^2} = 0; \; V(r, \gamma) = \rho(r) \cdot \theta(\gamma) \tag{14.14}$$

Using Fourier series decomposition and solving separately for stator and rotor slotting, with:

$$H_v = -\frac{1}{v}\frac{\partial V}{\partial v}; \; H_r = -\frac{\partial V}{\partial r}; \; r - \text{radius} \tag{14.15}$$

And, by normalizing the field H_r on the rotor bore to the field $H_r(D_{or}/2, \theta_s/2)$, the equivalent airgap magnetic permeance variation produced by the stator slotting is obtained, $\Lambda_{sg}(v)$. Similarly, the rotor slotting contribution $\Lambda_{rg}(v, \theta_r)$ is calculated. Finally, the total p.u. airgap permeance is: $\Lambda_g(v, \theta_r)$ is:

$$\Lambda_g(v, \theta_r) = \frac{A_{sg}(v)\Lambda_{rg}(v, \theta_r)}{\Lambda_{sg}(v) + A_{rg}(v, \theta_r)} \tag{14.16}$$

The AC phase mmfs are also decomposed in space harmonics. This depends heavily on the coil connection in phases, to obtain symmetric emfs.

But, essentially [11]:

$$F_q(v, \theta_r) = F_g(v) + k_c(v, \theta_r) \tag{14.17}$$

where

$$K_c(v, \theta_r) = \frac{\int_0^{2\pi} F_g(v) \cdot \Lambda_g(v, \theta_r) dv}{\int_0^{2\pi} \Lambda_g(v, \theta_r) dv} \tag{14.18}$$

So, the armature flux density $B_{ag}(\gamma, \theta_r)$ is:

$$B_{agq}(v, \theta_r) = \mu_0 F_g(v, \theta_r) \cdot \Lambda_g \tag{14.19}$$

The flux linkages in the various coils of the same or different phases are calculated to get the self and mutual phase inductances. Sample results on p.u. airgap permeance variation for a 24-stator slot structure $\Lambda_{sg}(v)$ are shown in Figure 14.6 for $\beta_s = 0.2, 0.5, 0.8$ (slot opening ratio) in comparison with 2D-FEM, only to show very good matching [11] for thin and wide slot openings.

The self and mutual phase inductances versus rotor position for the 24/10 machine, for single- and double-layer 2-slot-pitch-span coils, are shown in Figure 14.7 [11].

It is evident that only for double-layer winding is the variation of self and mutual inductances with rotor position symmetrically periodic, so that they may serve as a basis for rotor position estimation in encoderless variable-speed drive applications. But magnetic saturation (not considered here) may alter the situation notably.

Also, the variation of self and mutual inductance with rotor position is not large, so the machine has small functional dq saliency ($L_d/L_q \approx 1$) and may switch from $L_d/L_q > 1$ for low flux (flux weakening) operation to $L_d/L_q < 1$ for heavy loading.

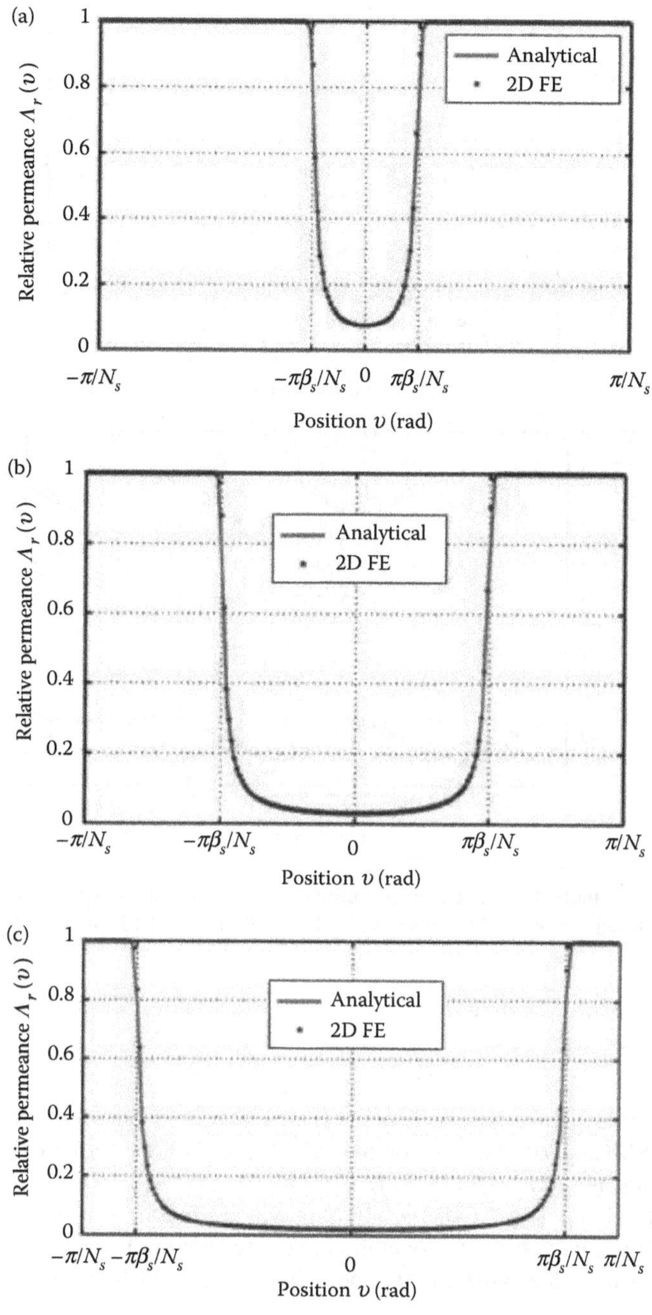

FIGURE 14.6 Relative (p.u.) stator airgap permeance function versus stator angle ν for three stator slot opening ratios β_s: (a) $\beta_s = 0.2$; (b) $\beta_s = 0.5$; (c) $\beta_s = 0.8$. (After B. Gaussens et al., *IEEE Trans.*, vol. MAG-49, no. 1, 2013, pp. 628–641. [11])

14.2.3 Practical Design of a 24/10 Direct Current + Alternating Current Double Salient Machine for Hybrid Electric Vehicles

This time, a tentative geometry for given specifications was obtained by comparison with an existing IPMSM for Lexus RX 400h (Table 14.1) and by an analytical technical theory [7].

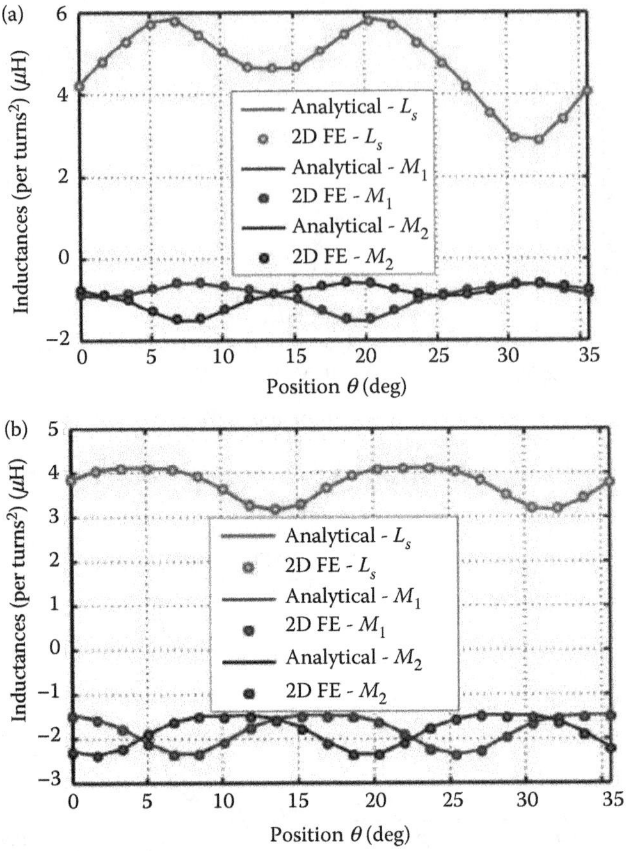

FIGURE 14.7 Self and mutual inductances of single- (a); and double-layer windings, (b), of the 24/10 machine. (After B. Gaussens et al., *IEEE Trans.*, vol. MAG–49, no. 1, 2013, pp. 628–641. [11])

It should be noticed that:

- Both motors develop the same power at the same DC voltage but at different speeds (12,400 rpm for IPMSM and 20,000 rpm for the AC + DC DSM), so a higher reduction ratio mechanical transmission is required (4/1 instead of 3.478/1).
- The torque of the DC + AC DSM is 210 Nm, while it is 333 Nm for the IPMSM.
- The two machines have the same overall active size (volume).
- The power density is the same, but the torque density is 33% less for the DC + AC DSM.

A geometrical pre-optimization study with FEM verification finally yielded a machine geometry capable of producing the required torque and power, but for a power factor of only 0.64 ($j_e = j_a = 21\ \text{A/mm}^2$).

The emf/phase and the full torque pulsations with pure I_q are shown in Figure 14.8 [7].

The torque and power factor dependence on the two current densities is given in Figure 14.9 [7].

The torque/speed envelope and the efficiency calculated in eight key operation points (1–8) are shown in Figure 14.10 [7].

The efficiency is acceptable but, again, the power factor is rather low (0.64 at full torque) and thus the converter kVA and its losses are larger. No experimental confirmation is available so far.

TABLE 14.1
FEFSSM Design Restrictions and Specifications for HEV Applications

Items	IPMSM RX400h	FEFSSM
Max. DC-bus voltage inverter (V)	650	650
Max. inverter current (A_{rms})	Conf.	250
Max. current density in armature winding. $J_a(A_{rms}/mm^2)$	Conf.	21
Max. current density in excitation winding. $J_e(A/mm^2)$	NA	21
Stator outer diameter (mm)	264	264
Motor stack length (mm)	70	70
Shaft radius (mm)	30	30
Air gap length (mm)	0.8	0.8
Permanent magnet weight (kg)	1.1	0.0
Maximum speed (r/min)	12,400	20,000
Maximum torque (Nm)	333	>210
Reduction gear ratio	2.478	4
Max. axle torque via reduction gear (Nm)	825	>840
Max. power (kW)	123	>123
Power density (kW/kg)	3.5	>3.5

Source: E. Sulaiman, T. Kosaka, N. Matsui, *Record of EPE—2011*, Birmingham, U.K. [7]

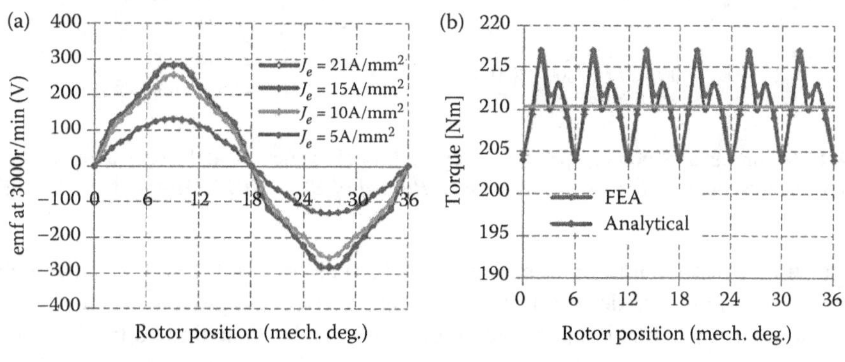

FIGURE 14.8 Emf at 3000 rpm for four field current densities (a) and torque versus rotor position (j_T = 21 A/mm^2) (b). (After E. Sulaiman, T. Kosaka, N. Matsui, *Record of EPE—2011*, Birmingham, U.K. [7])

The efficiency of this machine is better in the flux weakening zone (point 2) than for full flux full torque (point 1), but also high at low torque and low speeds. This machine is thus better than the IPMSM at low torque and high speeds. But, again, this is at 66% of torque/volume.

14.3 12/10 DIRECT CURRENT + ALTERNATING CURRENT DOUBLE SALIENT MACHINE WITH TOOTH-WOUND DIRECT CURRENT AND ALTERNATING CURRENT COILS ON STATOR

Using the 12/10 slot/pole combination has been shown to be a practical solution for high torque tooth-wound coils PMSMs when considering energy conversion, noise, and vibration.

This is one more reason to use it for DC stator-excited DSM. It has been proposed both with a single stator and with dual stators (Figure 14.11) [12].

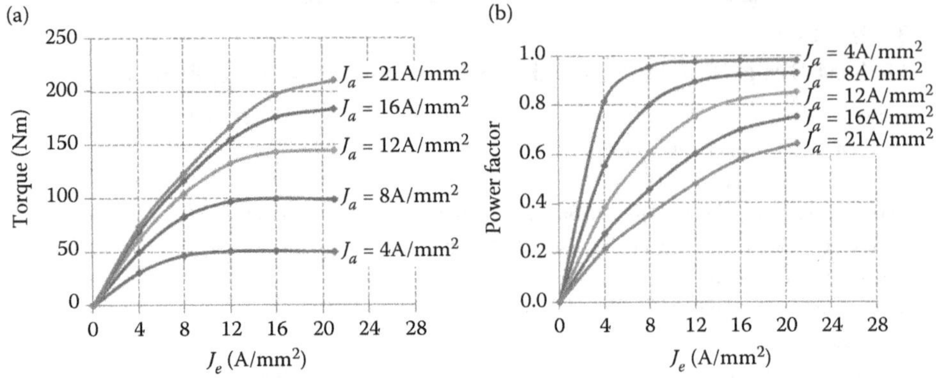

FIGURE 14.9 Torque (a); and power factor (b), versus current densities in the DC (j_e) and AC (j_c) windings for pure i_q control. (After E. Sulaiman, T. Kosaka, N. Matsui, *Record of EPE—2011*, Birmingham, U.K. [7])

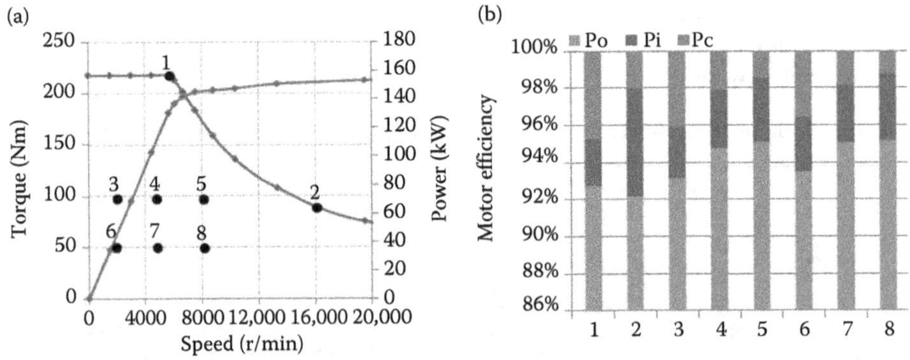

FIGURE 14.10 Torque and power/speed envelopes, (a); efficiency (and losses) in eight key operation points, (b). (After E. Sulaiman, T. Kosaka, N. Matsui, *Record of EPE—2011*, Birmingham, U.K. [7])

As the partitioned stator configuration (Figure 14.11b) did not, in the end, produce notably better performance (at least in torque density [12]) and is by far more complex, we do not treat it here but suggest that on outer rotor, inner stator, and single stator such a machine be tried, to better use the volume closer to the shaft; also, the airgap diameter may be larger, for more torque density.

The DC plus AC stator-excited DSM ultimately acts as a nonsalient pole-synchronous motor with stator DC multipolar excitation and concentrated stator AC winding. But let us show in Figure 14.12 a precursor of it that was called a "flux-bridge" or "transfer-field" machine in 1977 [14–15].

Note: In a thorough comparative study of such DC plus AC DSMs [16], it was confirmed the two-slot-pitch span coil DC + AC coil topology produces superior torque density, but the single tooth-wound DC + AC coil configuration is still more manufacturable, and this is one more reason to investigate it here.

14.3.1 12/10 Tooth-Wound Direct Current + Alternating Current Coils Double Salient Machine Practical Investigation

References 6, 12, 13, and 16 offer a plethora of studies for various combinations of stator slots/rotor poles (12/5, 2/7, 12/8, 6/4, 6/5, 6/7, etc.) and their average torque, torque pulsations, and emfs, with experimental results. Here, we review the 12/10 configuration in [5] as it refers to 55 kW

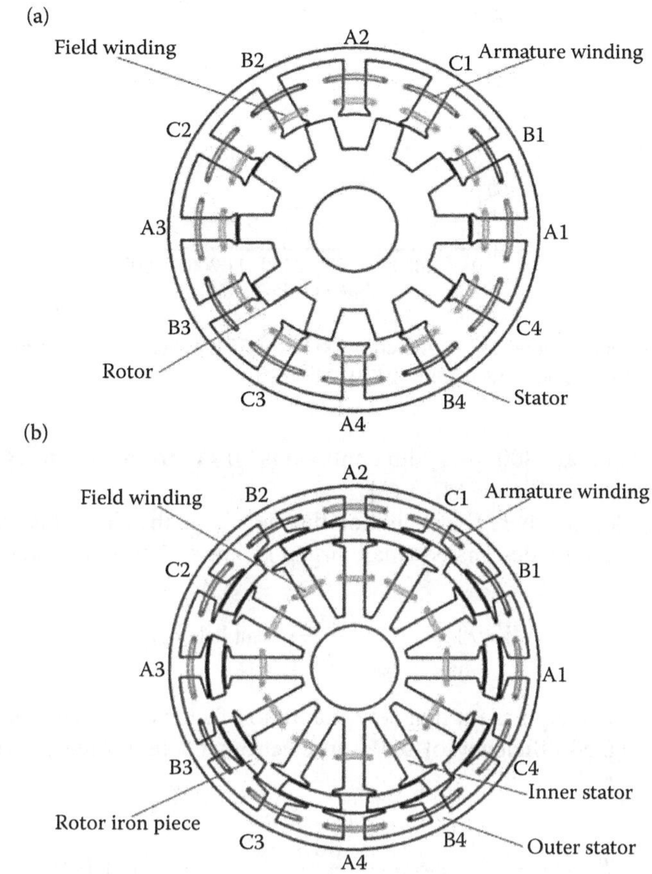

FIGURE 14.11 Tooth-wound 12/10 (DC + AC) DSM, (a) with single stator (it holds both DC and AC coils); (b) with dual (partitioned) stator: one for AC and one for DC coils (to increase torque/volume). (After Z. Q. Zhu, Z. Wu, X. Liu, *IEEE Trans.*, vol. EG–31, no. 1, 2016, pp. 78–92. [12])

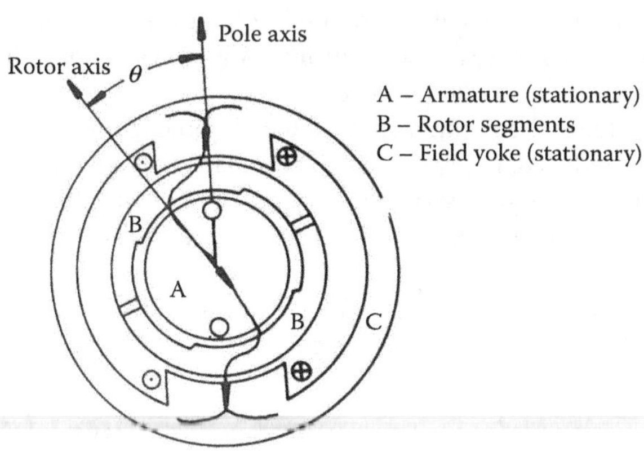

FIGURE 14.12 The flux-bridge machine. (After L. A. Agu, *an international Quarterly*, vol. 1, no. 2, 1977, pp. 185–194. [14])

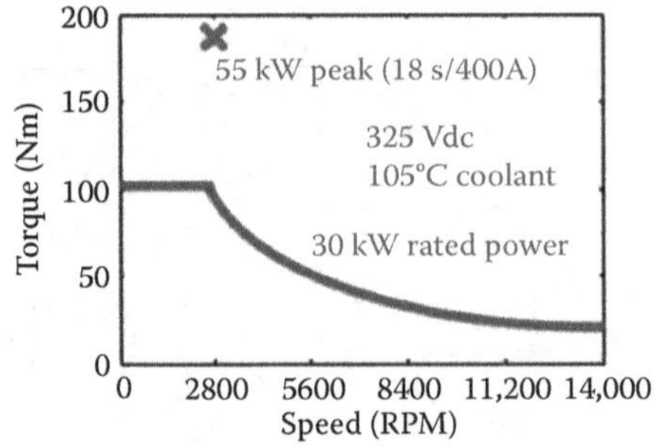

FIGURE 14.13 12/10 tooth-wound DC + AC coils DSM for 55 kW peak at 2800 rpm. (After T. Raminosoa et al., *IEEE Trans.*, vol. IA–52, no. 3, 2016, pp. 2129–2137. [5])

peak power for 18 seconds at 2800 rpm, with continuous 30 kW from 2800 to 14,000 rpm machine, destined to HEV traction; see Figure 14.13 [5].

After a preliminary design where DC excitation flux density in the airgap (stator teeth)—$B_{g0\max}$—is fixed around 1.4–1.5 T for a designated small airgap g_0, the DC coil mmf $w_F i_F$ can be computed:

$$(W_F i_F)_{\text{d.c. coil}} \approx \frac{B_{g0\max}}{\mu_0} \cdot g_0(1 + k_{sct}) \tag{14.20}$$

The coefficient k_{sat} is an equivalent magnetic saturation factor recalculated later for peak torque conditions. For an assigned efficiency of 92%, an electromagnetic torque T_{en} (for 30 kW at 2800 rpm) is:

$$T_{en} = \frac{P_{en}}{\eta \cdot 2\pi \cdot n_n} = \frac{30 \times 10^3}{0.92 \cdot 2\pi \cdot (2800/60)} = 111 \text{ Nm} \tag{14.21}$$

For continuous rated low-speed torque (111 Nm) a torque/volume of $t_{VC} \approx 13$ Nm/liter is assigned and, with a core length $l_{\text{stack}} = 90$ mm, the outer stator diameter D_{os} is calculated:

$$D_{os} = \sqrt{\frac{T_{en}}{(\pi/4) l_{\text{stack}} \cdot t_{vc}}} = \sqrt{\frac{111}{(\pi/4) \times 0.09 \times 9 \times 10^3}} = 0.348 \text{ m} \tag{14.22}$$

This torque density (calculated at stator core total volume) for continuous operation is not very large, but the DC + AC copper losses need a strong cooling system and thus, as initial cost is not large (no PMs), the design is relaxed on purpose.

As known, at no load, the AC coil turns close to the airgap experience (as in PMSMs) proximity copper losses which, to be reduced, imply a recess (open space) toward the stator slot tap of a few mm, as the maximum fundamental frequency should go up to 2.3 kHz for the max. speed of 14,000 rpm. At least 20 kHz switching frequency in the PWM inverter would be required to produce a quasi-sinusoidal current of 2.35 kHz.

To keep the machine size reasonably small, for peak power (55 kW at 2800 rpm or 200 Nm), the saturation level is high ($k_{S\max} = 1$, for a 1-mm airgap), and thus (from Equation 14.20):

$$(W_F i_F)_{\text{d.c. coil peak}} \approx \frac{1.5}{1.256 \times 10^{-6}} \times 1.0 \times 10^{-3}(1 + 1) = 2388.5 \left(\frac{\text{Aturns}}{\text{d.c. coil}}\right) \tag{14.23}$$

With $i_{F\text{peak}} = 20$ A, there will be 119 turns/DC coil.

The $i_{F\text{peak}} = 20$ A was chosen based on the available DC voltage that supplies the DC–DC converter that controls the field current, with about 2.5 kW allocated for peak-torque DC copper losses.

The $B_{g0} = 1.5$ T does not correspond to very high saturation on no load, but the armature reaction field contributes heavily to it for peak torque: with armature airgap flux density $B_{ag} = 0.7 \times B_{g0} = 1.0$ T, the peak tooth flux density would be large: $B_{\text{teeth peak}} \approx \sqrt{B_{g0}^2 + B_{ag}^2} = \sqrt{1.5^2 + 1.0^2} = 1.8$ T. Also, we have to consider the fact that the fundamental frequency will be large, even with 0.2-mm-thick laminations. Already, $f_1 = 466$ Hz at 2800 rpm and thus core losses have to be limited.

It has to be noted that the DC coil flux in the AC coils—both with same single tooth span—is homopolar and thus varies from maximum to minimum, which leads to a no-load fundamental airgap flux density (B_{g01}) AC of:

$$(B_{g01})_{\text{a.c.}} \approx \frac{B_{g0}}{2(1 + k_{\text{fringe}})}; \tag{14.24}$$

With $k_{\text{fringe}} = 0.75$, $(B_{g01})_{a.c.} = 1.5/(2 \cdot (1 + 0.75)) = 0.652$ T

This explains the not-so-large torque/volume; not to mention the very large radial forces (referred to as the maximum airgap flux density value).

A tentative peak AC coil mmf value $W_{ac}I_{ac}$ may now be calculated (neglecting the small reluctance torque contribution):

$$(T_{ek})_{\text{peak}} \approx \frac{3}{2}\left((2/\pi)B_{g01} \cdot \tau_{\text{slot}} \cdot k_{W1} \cdot l_{\text{stack}}\right)4(W_{\text{a.c.}}I_{\text{a.c.}}) \cdot N_r \tag{14.25}$$

With N_r—rotor poles, k_{W1}—fundamental (for $N_r = 10$ pole pairs) winding factor of a two-layer AC winding with three phases in 12 slots:

$$205 = \frac{3}{2} \cdot \frac{2}{\pi} \cdot 0.652 \cdot 0.0622 \times 0.933 \times 0.09 \times 4 \times 10 \times (W_{\text{a.c.}}I_{\text{a.c.}}) \tag{14.26}$$

With

$$\tau_{\text{slot}} = \frac{\pi D_{is}}{N_s} = \frac{\pi D_{os}}{N_s} \cdot k_{oi} = \frac{\pi \cdot 0.348 \times 0.684}{12} = 0.0622 \text{ m} \tag{14.27}$$

$k_{oi} = 0.684$ is the adopted split stator diameter ratio. The 62.2-mm stator slot pitch allows for enough room for the stator slot, with a balanced ratio of stator tooth/slot pitch of 0.466 (14°/30° stator pole ratio).

Finally, from Equation 14.26, the peak value of AC coil mmf ($W_{\text{a.c.}}I_{\text{a.c.}}$) for the peak torque $T_{ek} =$ 205 Nm is:

$$W_{\text{a.c.}}I_{\text{a.c.}} = 1575.2 \text{ Aturns/peak value}$$

With this very preliminary data on machine design, a thorough 2(3)D-FEM inquiry is to be performed to mitigate quite a few compromises, such as high average torque and small torque ripple by optimizing the rotor pole ratio. The results in Figure 14.14 [5]—for a machine similar to the one discussed above—are self explanatory in this respect: the best compromise between average and ripple torque is obtained with 14° stator pole and 11.5° rotor pole spans.

The 205 Nm peak torque with small torque ripple is confirmed by 3D-FEM, where a small i_d (for positive reluctance torque, corresponding to tan7.6° $= I_d/I_q$) was adopted (Figure 14.15) [5].

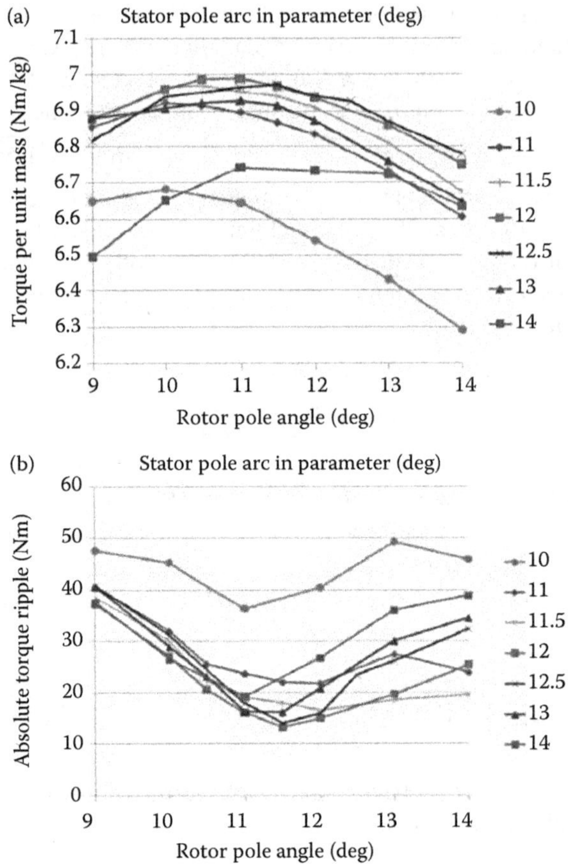

FIGURE 14.14 Average torque (a); and torque ripple variation with rotor pole angles for a few values of stator pole angles, (b). (After T. Raminosoa et al., *IEEE Trans.*, vol. IA–52, no. 3, 2016, pp. 2129–2137. [5])

Finally, by using 2D-FEM, the machine was fully characterized by adding info to the circuit model to calculate, for continuous operation, the torque-speed envelope, the voltage (phase, RMS), phase current (RMS), power, and efficiency (Figure 14.16a and b [5]).

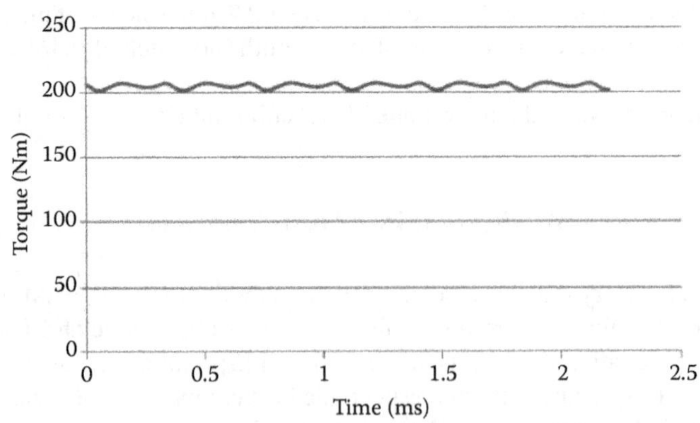

FIGURE 14.15 Torque versus time variation at 2800 rpm obtained for $(I_s)_{peak} = 252$ A (RMS) and $i_F = 20$ A. (After T. Raminosoa et al., *IEEE Trans.*, vol. IA–52, no. 3, 2016, pp. 2129–2137. [5])

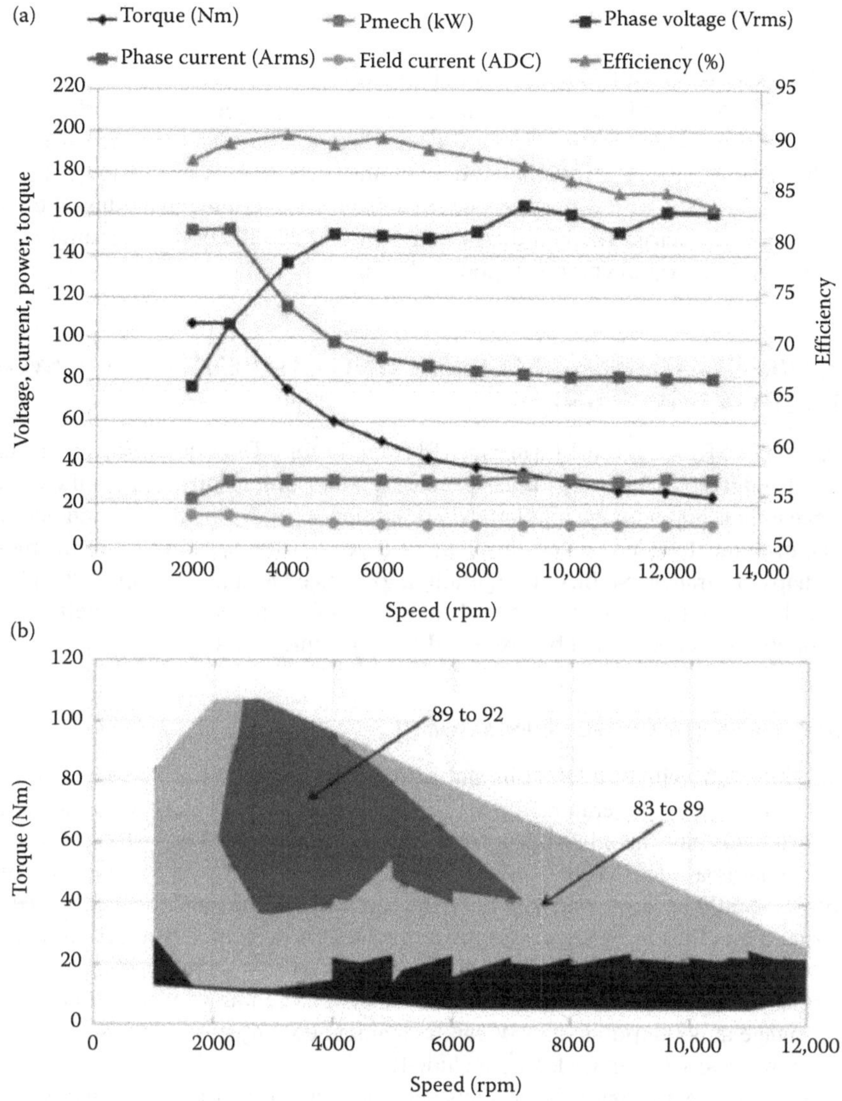

FIGURE 14.16 Machine continuous operation performance versus speed, (a); and measured efficiency map, (b). (After T. Raminosoa et al., *IEEE Trans.*, vol. IA–52, no. 3, 2016, pp. 2129–2137. [5])

The results in Figure 14.16 warrant remarks such as:

- A notable reserve in voltage for continuous operation torque is allowed for, to provide enough (full) voltage for peak torque at base speed (2800 rpm).
- The DC field current decreases mildly from 2800 to 14,000 rpm, while the AC current decreases notably with speed.
- The only 89% efficiency at 2800 rpm and 30 kW (111 Nm) is mainly due to the large copper loss in the 12 DC coils.
- Though a stator water jacket is provided to cool the stator, the DC coils (placed closer to stator yoke) are better cooled than the AC coils.
- The 83–89% efficiency in the high speed (frequency)—flux weakening—is not enough to compete with the IPMSM for traction, though even the latter is a bit worse than the copper cage IM in the flux weakening zone. The two-slot-pitch DC + AC coil DSM

(in section 14.2), in calculations, has shown better results, but experiments are needed to prove its claimed superiority.

- All in all, it seems that still further strong improvements are necessary to make the tooth-wound DC + AC DSM fully competitive at 50–60 Nm/liter, 60 kW, efficiency above 94% at 3000 rpm, with IPMSM. But, still, we have to consider the reduced initial costs due to the elimination of PMs. Perhaps adopting the partitioned stator to house the DC coils in the inner stator will be a way to reduce the DC losses and thus, by adding more copper, increase efficiency across the board. An outer-rotor single-inner-stator configuration may also provide more torque/volume at good efficiency.

14.4 THE DIRECT CURRENT + ALTERNATING CURRENT STATOR SWITCHED RELUCTANCE MACHINE

The DC + AC stator SRM (Figure 14.3) proposed recently for a DC output voltage generator, with DC field control and a diode rectifier [9], has produced interesting results, though the large synchronous inductance inflicts high voltage regulation to be compensated by the field current increasing with load. Also, the machine, by adequate design, may be used as a low-speed high-torque motor or generator, but perhaps in a transverse flux configuration, to reduce DC and AC copper weight and losses when sinusoidal current field-oriented control may be applied for an off-the-shelf PWM inverter drive. Both opportunities above will be presented here in some detail.

14.4.1 Direct Current Output Generator

A three-phase 12/8 SRM with four DC coils and 12 AC coils (Figure 14.17) may be a simple solution to DC output autonomous generators, operating at constant (or slightly variable) speed (stator PM-assisted SRMs/Gs have also been proposed, but have not reached industrialization yet).

Four or more phases are also feasible, but then the fundamental frequency for given machine volume and output power should be larger and so will be the core losses. The machine inductance now varies with rotor position and thus a reluctance torque component is present. The emfs are not sinusoidal; thus, finally, the currents contain even harmonics as well as AC phase voltages (Figure 14.17) [9].

The performance illustrated in Figure 14.18 shows that the machine voltage regulation is large, and constant DC voltage at 3600 rpm and 3 kW can be maintained only around 60 V, at an efficiency of 79% (diode rectifier and DC copper losses included).

With design refinements to make the currents and voltages symmetric and torque pulsation reduction, mitigating for an optimum airgap, this simple and rugged generator may become practical for autonomous applications. Auxiliary generators on the ground—such as small wind generators with battery DC-backed output—or on board vehicles may benefit this technology soon.

14.4.2 Transverse-Flux Direct Current + Alternating Current Switched Reluctance Motor/Generator

As the DC copper losses seem to be the main problem in DC + AC DSMs (section 14.3.1), the use of transverse flux topologies might help in reducing them drastically to allow, eventually, a larger airgap and thus a lower machine inductance (better power factor and lower voltage regulation) (Figure 14.19). It retains the homopolar character of DC (no load) flux linkage in the AC coils.

The DC field coils do not interact notably with the AC coils in the regular configuration (Figure 14.18a), but they do interact 100% in the transverse flux topology. In a multiphase modular, transverse-flux topology (Figure 14.18b), if all DC coils are connected in series, the voltage induced by the AC coils in the DC circuit is ideally zero. Thus a drastic reduction of kVA requirements of the DC circuit—to the losses level in the DC coils—is obtained.

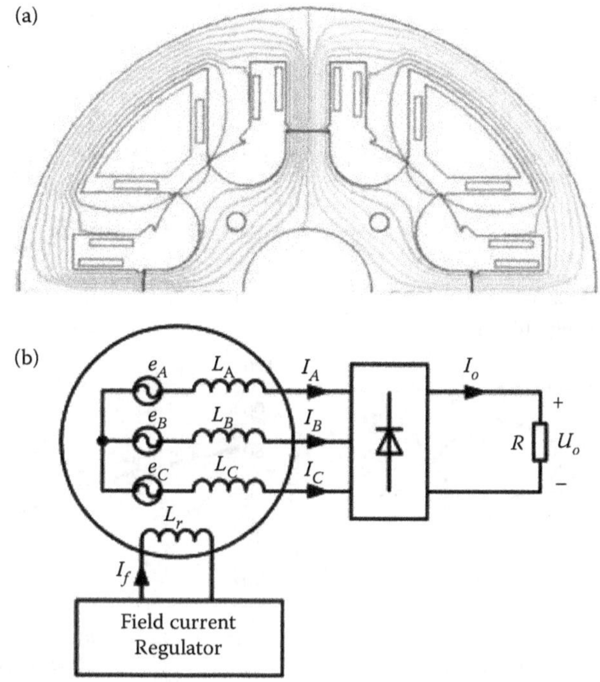

FIGURE 14.17 Three phase DC + AC SRM (G) with DC field distribution with maximum flux in phase B (in the middle), (a); and electric scheme, (b). (After L. Yu, Z. Chen, Y. Yan, *IEEE Trans.*, vol. IE–61, no. 12, 2014, pp. 6655–6663. [9])

Still, each DC coil will experience induced voltage, and it may be used to estimate rotor position in an encoderless PWM inverter drive.

The investigation of the two configurations in Figure 14.19 for a 6-MW 12-rpm direct-drive wind generator, first analytically, then by FEM, have eliminated the regular configuration (Figure 14.19a), as the total generator losses ended up at 1500 kW. For the DC + AC TF-SRM, an equivalent linear sector was investigated by 2D-FEM with results in airgap flux distribution as in Figure 14.20 [17]. An airgap of 4 mm was allowed for a stator bore diameter $D_{is} = 9$ m.

The large airgap and rather large number ($N_s = N_r = 375$) of poles (same in the stator and rotor) yield a fundamental frequency of 75 Hz at 12 rpm, which leads to a not-so-large variation of AC coil flux with position (Figure 14.20c). Still, the machine is proved capable in a six-phase configuration at $D_{is} = 9$ m, 89 tons, of producing more than the required 6.4-MNm average torque, with perhaps acceptable torque ripple (Figure 14.21) for 428 kW total generator losses (efficiency 0.93).

This super-high-torque example should be indicative of the strong potential of DC + AC TF-SRMs, mainly based on the "blessing of circular coils" and torque magnification in TF machines (Chapter 12), in addition to lower DC and AC copper losses per Nm for given machine volume.

The sinusoidal current control (FOC) in six phases here could be approached by two off-the-shelf 50% rating PWM inverters, while the AC voltage induced in any of the DC coils (in series) can serve as a basis for encoderless FOC. Finally, due to the homopolar DC current flux in the AC coils and small airgap, the power factor should not be expected to be above 0.7–0.75 for high torque densities.

14.5 SUMMARY

- DC + AC doubly salient machines may be considered DC + AC stator coil flux-switching, flux reversal, flux modulation, or synchronous PM-less electric machines: the acronym here aims to expose the principle and topology in the fewest words.

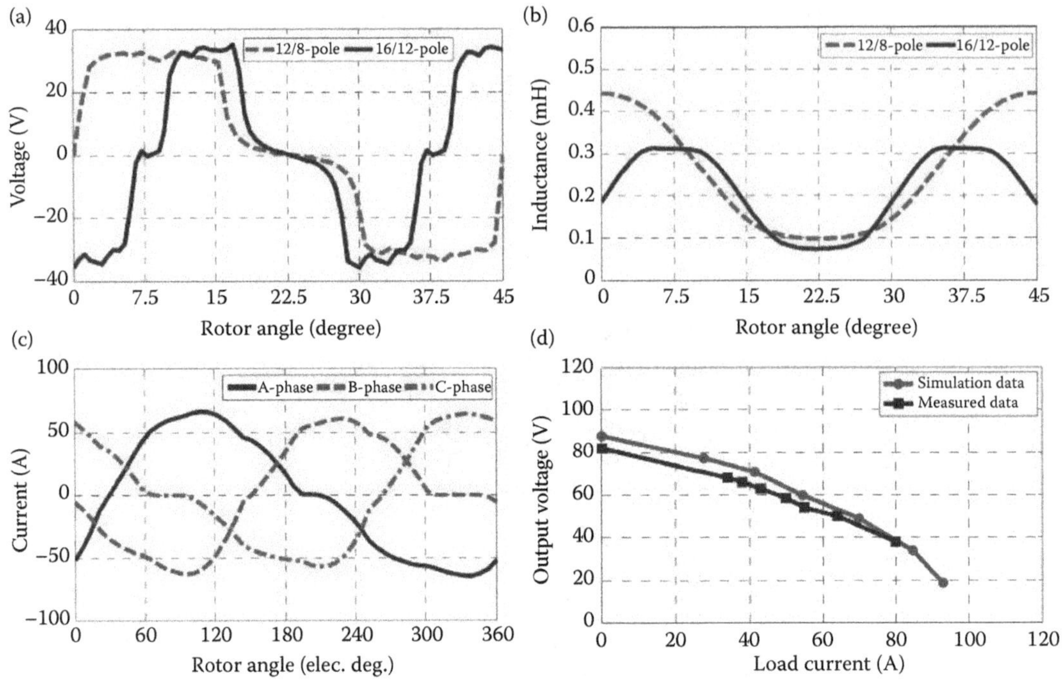

FIGURE 14.18 Three- and four-phase DC + AC FRGenerator with diode rectifier output: (a) emf of phase A; (b) phase A self-inductance; (c) three-phase currents; (d) output (DC) voltage versus current for $I_f = 17$ A. (After L. Yu, Z. Chen, Y. Yan, *IEEE Trans.*, vol. IE–61, no. 12, 2014, pp. 6655–6663. [9])

- Also, DC + AC DSMs may be considered to stem from SRMs by adding stator DC coils and running them with AC sinusoidal currents to make use of off-the-shelf PWM inverters for variable speed drives.
- Equal—one stator slot pitch or two stator slot pitch—DC and AC coils may be used, and they will be treated in detail with the aim to increase torque density for moderate efficiency.
- Different stator slot pitch span DC coils + tooth-wound AC coils in the stator on regular SRMs may also be used for scope and are treated in some detail in this chapter.
- Transverse-flux DC + AC doubly salient multiphase modular topologies have been proven by an extreme torque (6.4-MNm) 12-rpm example to drastically reduce copper losses with

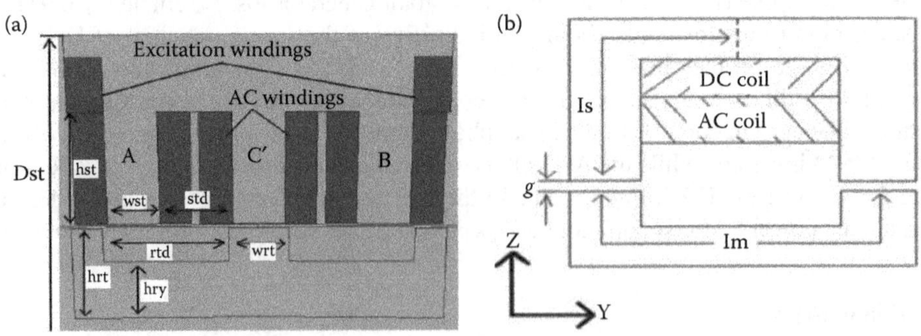

FIGURE 14.19 Three-phase DC + AC RSMs: (a) with regular configuration; (b) with transverse flux (DC coils in series). (After C. R. Bratiloveanu, D. T. C. Anghelus, I. Boldea, *Record of OPTIM*, 2012, pp. 535–543, IEEEXplore. [17])

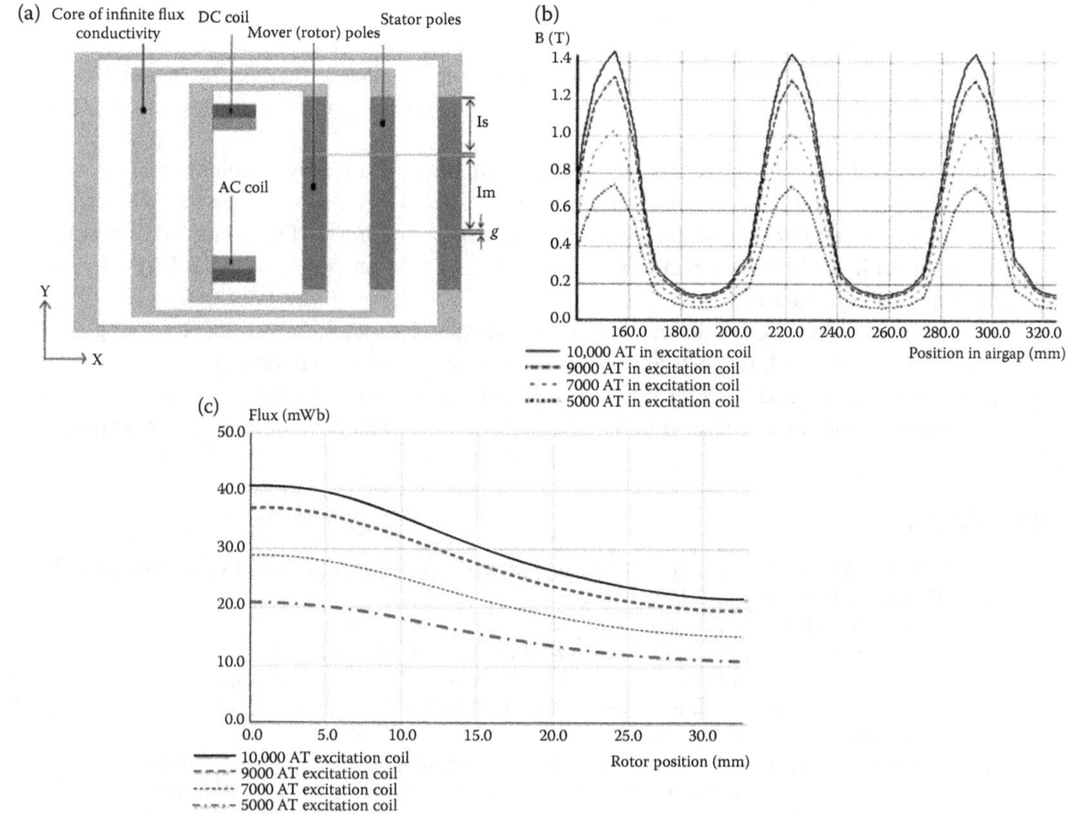

FIGURE 14.20 DC + AC TF-SRM: (a) 2D FEM linear section; (b) airgap flux density versus rotor position for four different mmfs in the DC coils; (c) airgap flux in the AC coils versus rotor position for an alleged 6-MW, 12-rpm wind generator. (After C. R. Bratiloveanu, D. T. C. Anghelus, I. Boldea, *Record of OPTIM*, 2012, pp. 535–543, IEEEXplore. [17])

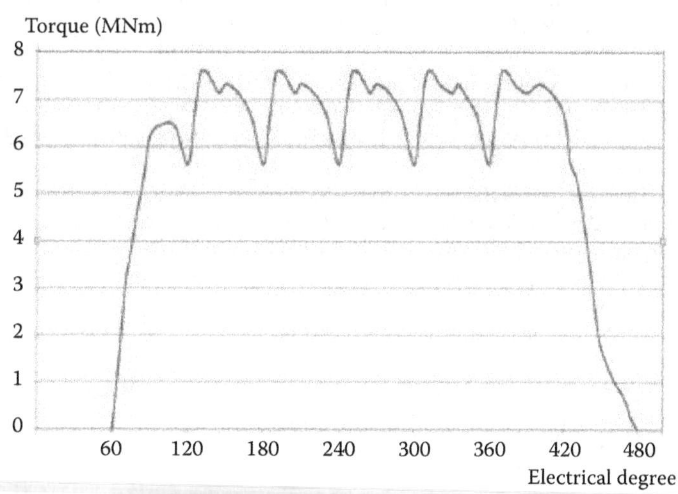

FIGURE 14.21 DC + AC TF-SRM total six-phase torque versus rotor position for sinusoidal symmetric AC currents and 6-MW, 12-rpm, with 428 kW total machine losses. (After C. R. Bratiloveanu, D. T. C. Anghelus, I. Boldea, *Record of OPTIM*, 2012, pp. 535–543, IEEEXplore. [17])

respect to the regular SRM topology. This configuration may also be associated with transverse-flux (circular coil) SRMs.

- Though the DC stator excitation helps in torque production, it is limited by the fact that, basically, only half of the DC flux that the machine is capable of is used in producing emf (torque) by the AC coils.
- But the rugged, simple, less-costly magnetless machine with passive (salient) rotor is very tempting in cost-sensitive variable speed drives.
- A particularly suitable application is represented by autonomous DC-controlled output, a generator with diode power rectifier and only DC excitation (low power) PWM DC–DC converter voltage control.
- Still, the large inductance of the machine in p.u. leads to inevitably lower than desired power factor in motoring and, respectively, large voltage regulation in generating.
- Aggressive future R&D is still needed to make this electric machine breed available in high performance, cost competitive, variable-speed drives, especially at moderate and low speeds.

REFERENCES

1. T. J. E. Miller, *Switched Reluctance Motors and Their Control*, book, Magna Physics Publishing, Hillsboro, Oregon; Clarendon Press, Oxford, 1993.
2. M. Abassian, M. Moallem, B. Fahimi, Double stator switched reluctance machines (DSSRM); fundamentals and magnetic force analysis, *IEEE Trans on*, vol. EC–25, no. 3, 2010, pp. 589–597.
3. A Chiba, M. Takeno, N. Hoshi, M. Takemoto, S. Ogasawara, Consideration of number of series turns in switch—reluctance traction motor competitive to HEV-IPMSM, *IEEE Trans.*, vol. IA–48, no. 6, 2012, pp. 2333–2340.
4. V. T. Buyukdegirmenci, A. M. Bazzi, P. T. Krein, Evaluation of induction and PMSMS using drive-cycle energy and loss minimization, *IEEE Trans.*, vol. IA–50, no. 1, 2014, pp.395–403.
5. T. Raminosoa, D. A. Torrey, A. M. El– Refaie, K. Grace, D. Pan, S. Grubic, K. Bodla, K.-K. Huh, Sinusoidal reluctance machine with d.c. winding: an attractive non PM option, *IEEE Trans.*, vol. IA–52, no. 3, 2016, pp. 2129–2137.
6. X. Liu, Z. Q. Zhu, Winding configuration and performance investigations of 12-stator pole variable flux reluctance machines, *Record of IEEE ECCE*, 2013, IEEEXplore.
7. E. Sulaiman, T. Kosaka, N. Matsui, A new structure of 12 slot-10 pole field-excitation flux switching synchronous machine for HEV, *Record of EPE—2011*, Birmingham, U.K.
8. Z. Chen, B. Wang, Z. Chen, Y. Yan, Comparison of flux regulation ability of the hybrid excitation doubly salient machines, *IEEE Trans.*, vol. IE–61, no. 7, 2014, pp. 3155–3166.
9. L. Yu, Z. Chen, Y. Yan, Analysis and verification of the doubly fed salient brushless d.c. generator for automobile auxiliary power unit application, *IEEE Trans.*, vol. IE–61, no. 12, 2014, pp. 6655–6663.
10. B. Gaussens, E. Hoang, O. de la Barriere, J. Saint-Michel, M. Lecrivain, M. Gabsi, Analytical approach for airgap modeling of field–excited flux-switched machine: no load operation, *IEEE Trans.*, vol. MAG–48, no. 9, 2012, pp. 2505–2517.
11. B. Gaussens, E. Hoang, O. de la Barriere, J. Saint-Michel, Ph. Manfe, M. Lecrivain, M. Gabsi, Analytical armature reaction field prediction in field excited flux-switching machines using an exact relative permeance function, *IEEE Trans.*, vol. MAG–49, no. 1, 2013, pp. 628–641.
12. Z. Q. Zhu, Z. Wu, X. Liu, A partitioned stator variable flux reluctance machine, *IEEE Trans*, vol. EG–31, no. 1, 2016, pp. 78–92.
13. Z. Q. Zhu, X. Liu, Novel stator electrically excited synchronous machines without rare-earth magnet, *IEEE Trans*, vol. MAG–51, no. 4, 2015, pp. 8103609.
14. L. A. Agu, "Correspondence: The flux-bridge machine", Electric machines and electro-mechanics, *An International Quarterly*, vol. 1, no. 2, 1977, pp. 185–194.
15. L. A. Agu, "The transfer field electric machine", Electric machines and electro-mechanics, *An International Quarterly*, vol. 2, no. 4, 1978, pp. 403–418.
16. Y. J. Zhou, Z. Q. Zhu, Comparison of wound-field switched-flux machines, *IEEE Trans.*, vol. IA–50, no. 5, 2014, pp. 3314–3324.
17. C. R. Bratiloveanu, D. T. C. Anghelus, I. Boldea, A comparative investigation of three PM-less MW power range wind generator topologies, *Record of OPTIM*, 2012, pp. 535–543, IEEEXplore.

Index

Chapter 1

Chapter 2

Chapter 3

Chapter 4

Chapter 5

Chapter 6

Chapter 7

Chapter 8

Chapter 9

Chapter 10

Chapter 11

Chapter 12

Chapter 13

Chapter 14